T0215648

Lecture Notes in Computer Science **12272**

More information about this series at http://www.springer.com/series/7409

Ilana Nisky · Jess Hartcher-O'Brien ·
Michaël Wiertlewski · Jeroen Smeets (Eds.)

Haptics: Science, Technology, Applications

12th International Conference, EuroHaptics 2020
Leiden, The Netherlands, September 6–9, 2020
Proceedings

Editors
Ilana Nisky
Ben-Gurion University of the Negev
Beer Sheva, Israel

Jess Hartcher-O'Brien
Delft University of Technology
Delft, The Netherlands

Michaël Wiertlewski
Delft University of Technology
Delft, The Netherlands

Jeroen Smeets
Vrije Universiteit Amsterdam
Amsterdam, The Netherlands

ISSN 0302-9743 ISSN 1611-3349 (electronic)
Lecture Notes in Computer Science
ISBN 978-3-030-58146-6 ISBN 978-3-030-58147-3 (eBook)
https://doi.org/10.1007/978-3-030-58147-3

LNCS Sublibrary: SL3 – Information Systems and Applications, incl. Internet/Web, and HCI

This Springer imprint is published by the registered company Springer Nature Switzerland AG
The registered company address is: Gewerbestrasse 11, 6330 Cham, Switzerland

Preface

In this volume, you will find the proceedings of the EuroHaptics 2020 conference, held during September 6–9, 2020, in Leiden, The Netherlands. EuroHaptics is a major international conference, organized biannually, that features the latest advances in haptics science, technology, and applications. The 2020 edition of the meeting had a special focus on accessibility, and in the spirit of the focus, we have made the proceedings open access.

The organization of the conference faced a unique challenge, as the start of the review process coincided with the beginning of the COVID-19 pandemic. Like in all aspects of our lives, we also had to reinvent our scientific life, and find a new way to work. Conferences were canceled, moved to later dates, or re-organized as online virtual conferences. Eurohaptics 2020 was rescheduled; instead of late spring (June) it was held in September. Many of the authors, reviewers, and editors faced challenges: lab shutdowns, the transition to online teaching, and caring for our families to name just a few. Considering the event, we are thankful for the effort and dedication to scientific rigor of the 220 external reviewers and 35 associate editors as they juggled additional tasks and limited resources to finish the editorial process. We also want to acknowledge the diligence of authors to improve their papers in the midst of uncertainty and, for many of them, without access to their usual resources. Their commitment though the hardship of the situation made this conference possible.

All aspects of haptics, including neuroscience, psychophysics, perception, engineering, computing, interactions, virtual reality, and the arts were covered in the initial submissions for the full-paper category. The 111 submissions came from 24 countries: 50% from Europe, 37% from Asia, and 10% from North America. The Application, Science and Technology track chairs invited experts to the Program Committee and provided initial advice on three or four papers. The Program Committee members invited three independent reviewers and provided an assessment on the basis of a minimum of two reviews. To be accepted, we expected an innovative idea as well as a diligent experimental evidence, and 60 of the submissions (55%) were accepted for publication. These papers were presented either as oral (36) or as poster presentations (24). In the proceedings, the two types of presentations are published with no distinction between them.

The conference also featured work in progress abstracts which allowed for presentation of late-breaking results as a poster, receiving feedback from the community. The abstracts of these posters are not featured in these proceedings.

Finally, we are delighted to be part of such a dynamic and innovative community, and we believe that the proceedings reflect its creativity and depth of thought. We hope that the works presented in these proceedings will serve as inspiration and spark new ideas for a better future where everyone is haptically connected.

September 2020

Ilana Nisky
Jess Hartcher-O'Brien
Michaël Wiertlewski
Jeroen Smeets

Organization

General Chairs

Irene Kuling	Eindhoven University of Technology, The Netherlands
Jan van Erp	TNO Soesterberg, The Netherlands

Program Chair

Jeroen Smeets	Vrije Universiteit Amsterdam, The Netherlands

Local Chair

Dirk Heylen	University of Twente, The Netherlands

Communication Chair

Krista Overvliet	Utrecht University, The Netherlands

Proceedings Editors and Track Chairs

Jess Hartcher-O'Brien	TU Delft, The Netherlands
Ilana Nisky	Ben-Gurion University of the Negev, Israel
Michaël Wiertlewski	TU Delft, The Netherlands

Accessory Chairs

Sponsor

Frédéric Giraud	University of Lille, France

Accessibility

Eric Velleman	Accessibility Foundation, The Netherlands

Accessibility Sponsor

Vincent Hayward	Sorbonne Université, France

Demos

William Frier	Ultrahaptics, UK

Student Volunteers

Sabrina Paneels	CEA Tech, France
Femke van Beek	Facebook Reality Labs, USA

Webmaster

Camille Sallaberry	University of Twente, The Netherlands

Program Committee

Cagatay Basdogan	Koç University, Turkey
Matteo Bianchi	University of Pisa, Italy
Luca Brayda	Acoesis, Italy
Manuel Cruz	Immersion, Canada
Massimiliano Di Luca	University of Birmingham, UK
Ildar Farkhatdinov	Queen Mary University London, UK
Francesco Ferrise	Politecnico di Milano, Italy
David Gueorguiev	Sorbonne Université, France
Matthias Harders	University of Innsbruck, Austria
Jess Hartcher-O'Brien	TU Delft, The Netherlands
Hsin-Ni Ho	NTT, Japan
Hiroyuki Kajimoto	The University of Electro-Communications, Japan
Ayse Kucukyilmaz	University of Nottingham, UK
Yoshihiro Kuroda	University of Tsukuba, Japan
Scinob Kuroki	NTT, Japan
Vincent Levesque	École de Technologie Supérieure, Canada
Monica Malvezzi	University of Siena, Italy
Sarah McIntyre	Linköping University, Sweden
Alejandro Melendez-Calderon	The University of Queensland, Australia
Anna Metzger	Justus Liebig University Giessen, Germany
Alessandro Moscatelli	University of Rome, Italy
Ilana Nisky	Ben-Gurion University of the Negev, Israel
Shogo Okamoto	Nagoya University, Japan
Claudio Pacchierotti	IRISA, Inria Rennes, France
Sabrina Paneels	CEA Tech, France
Gunhyuk Park	Gwangju Institute of Science and Technology, South Korea
Hannes Saal	The University of Sheffield, UK
Evren Samur	Boğaziçi University, Turkey
Jeroen Smeets	Vrije Universiteit Amsterdam, The Netherlands
Massimiliano Solazzi	Scuola Superiore Sant'Anna, Italy
Luigi Tame	University College London, UK
Femke van Beek	Facebook Reality Labs, USA
Michaël Wiertlewski	TU Delft, The Netherlands
Tae-Heon Yang	Korea National University of Transportation, South Korea
Mounia Ziat	Bentley University, USA

External Reviewers

Jake Abbott
Diar Abdlkarim
Arsen Abdulali
Rochelle Ackerley
Yasuhiro Akiyama
Noor Alakhawand
Felipe Almeida
Visar Arapi
Fernando Argelaguet
Chris Awai
Yusuf Aydin
Mehmet Ayyildiz
Stephanie Badde
Alfonso Balandra
Giulia Ballardini
Sandra Bardot
Angelo Basteris
Edoardo Battaglia
Philipp Beckerle
Niek Beckers
Ayoub Ben Dhiab
Corentin Bernard
Gemma Bettelani
Tapomayukh Bhattacharjee
Jeffrey Blum
Christian Bolzmacher
Luca Brayda
Jonathan Browder
Joshua Brown
Gavin Buckingham
Ozan Caldiran
Antonio Cataldo
Luigi Cattaneo
Müge Cavdan
Sonny Chan
Jacob Cheeseman
Katie Chin
Francesco Chinello
Manolis Chiou
Youngjun Cho
Inrak Choi
Roger Cholewiak
Jean-Baptiste Chossat
Simone Ciotti
Ed Colgate
Patricia Cornelio Martinez
Celine Coutrix
Manuel Cruz
Heather Culbertson
Thomas Daunizeau
Brayden De Boon
Anouk de Brouwer
Victor Adriel de Jesus Oliveira
Xavier de Tinguy
Davide Deflorio
John de Grosbois
Benoit Delhaye
Arthur Dewolf
Charles Dhong
Massimiliano Di Luca
Mariama Dione
Kouki Doi
Lionel Dominjon
Knut Drewing
Lucile Dupin
Laura R. Edmondson
Mohamad Eid
Sallnas Eva-Lotta
Grigori Evreinov
Simone Fani
Carlo Fantoni
Ildar Farkhatdinov
Ahmed Farooq
Rebecca Fenton Friesen
Manuel Ferre
Francesco Ferrise
Joshua Fleck
Euan Freeman
Ilja Frissen
Daniel Gabana
Chiara Gaudeni
Frédéric Giraud

James Goodman
Roman Grigorii
David Gueorguiev
Taku Hachisu
Reza Haghighi Osgouei
Matthias Harders
Jess Hartcher-O'Brien
Yuki Hashimoto
Zhenyi He
Seongkook Heo
Souta Hidaka
Hsin-Ni Ho
Thomas Howard
Hsien-Yung Huang
Barry Hughes
Thomas Hulin
Ali Israr
Daisuke Iwai
Bernard Javot
Seokhee Jeon
Carey Jewitt
Wafa Johal
Lynette Jones
Fabian Just
Hiroyuki Kajimoto
Christoph Kanzler
Zhanat Kappassov
Astrid Kappers
Kazuo Kiguchi
Heba Khamis
Vahid Khoshkava
Sunjun Kim
Yeongmi Kim
Ryo Kitada
Christian Koch
Ayse Kucukyilmaz
Shivesh Kumar
Yoshihiro Kuroda
Scinob Kuroki
Julien Lambert
Hojin Lee
Daisy Lei
Betty Lemaire-Semail
Fabrizio Leo
Daniele Leonardis
Vincent Levesque
Weihua Li
Juhani Linna
Tommaso Lisini Baldi
Daniel Lobo
Céphise Louison
Jani Lylykangas
Yasutoshi Makino
Monica Malvezzi
Ahmad Manasrah
Maud Marchal
Juan Martinez
Sarah McIntyre
Alejandro Melendez-Calderon
Mariacarla Memeo
Luke Miller
Roberto Montano
Taha Moriyama
Cecily Morrison
Selma Music
Abdeldjallil Naceri
Hikaru Nagano
Saad Nagi
Takuto Nakamura
Frank Nieuwenhuizen
Ilana Nisky
Andreas Noll
Cara Nunez
Seungjae Oh
Shogo Okamoto
Victor Oliveira
Bukeikhan Omarali
Ata Otaran
Lucie Pantera
Stefano Papetti
Gunhyuk Park
Roshan Peiris
Nuria Perez
Luka Peternel
Evan Pezent
Thomas Pietrzak
Myrthe Plaisier
Mithun Poozhiyil

Maria Pozzi
Roope Raisamo
Roberta Roberts
Carine Rognon
Daniele Romano
Robert Rosenkranz
Jee-Hwan Ryu
Hannes Saal
Jamal Saboune
Satoshi Saga
Justine Saint-Aubert
Mike Salvato
Harpreet Sareen
Katsunari Sato
Immo Schuetz
Hasti Seifi
Lucia Seminara
Irene Senna
Yitian Shao
Majid Sheikholeslami
Anatolii Sianov
Stephen Sinclair
Nish Singh
Jeroen Smeets
Giovanni Spagnoletti
Adam Spiers
Angelica Torres
Matteo Toscani
Ugur Tumerdem
Femke van Beek
Vonne van Polanen
Yasemin Vardar
Mickeal Verschoor
Yem Vibol
Fernando Vidal Verdú
Julie Walker
Ben Ward-Cherrier
Antoine Weill-Duflos
Michaël Wiertlewski
Michele Xiloyannis
Heng Xu
Hiroaki Yano
Yongjae Yoo
Juan Zarate
Jacopo Zenzeri
Claudio Zito
Talee Ziv
Aaron Zoeller

Sponsors

ultraleap

Contents

Haptic Applications

Haptic Science

The EmojiGrid as a Rating Tool for the Affective Appraisal of Touch

Alexander Toet[1(✉)] and Jan B. F. van Erp[1,2]

[1] TNO Human Factors, Soesterberg, The Netherlands
{lex.toet, jan.vanerp}@tno.nl
[2] University of Twente, Enschede, The Netherlands

Abstract. We evaluated the convergent validity of the new language-independent EmojiGrid rating tool for the affective appraisal of perceived touch events. The EmojiGrid is a rectangular response grid, labeled with facial icons (emoji) that express different degrees of valence and arousal. We previously showed that participants can intuitively and reliably report their affective appraisal of different sensory stimuli (e.g., images, sounds, smells) by clicking on the EmojiGrid, without additional verbal instructions. However, because touch events can be bidirectional and are a dynamic expression of action, we cannot generalize previous results to the touch domain. In this study, participants (N = 65) used the EmojiGrid to report their own emotions when looking at video clips showing different touch events. The video clips were part of a validated database that provided corresponding normative ratings (obtained with a 9-point SAM scale) for each clip. The affective ratings for inter-human touch obtained with the EmojiGrid show excellent agreement with the data provided in the literature (intraclass correlations of .99 for valence and .79 for arousal). For object touch events, these values are .81 and .18, respectively. This may indicate that the EmojiGrid is more sensitive to perspective (sender versus receiver) than classic tools. Also, the relation between valence and arousal shows the classic U-shape at the group level. Thus, the EmojiGrid appears to be a valid graphical self-report instrument for the affective appraisal of perceived touch events, especially for inter-human touch.

Keywords: Affective touch · Valence · Arousal · Emotions · EmojiGrid · SAM

1 Introduction

Next to serving us to discriminate material and object properties, our sense of touch also has hedonic and arousing qualities [1]. For instance, soft and smooth materials (e.g., fabrics) are typically perceived as pleasant and soothing, while stiff, rough, or coarse materials are experienced as unpleasant and arousing [1]. This affective component of touch plays a significant role in social communication. Interpersonal or social touch has a strong emotional valence that can either be positive (when expressing support, reassurance, affection or attraction: [2]) or negative (conveying anger,

I. Nisky et al. (Eds.): EuroHaptics 2020, LNCS 12272, pp. 3–11, 2020.
https://doi.org/10.1007/978-3-030-58147-3_1

frustration, disappointment: [3]). Affective touch can profoundly influence social interactions [4]. For example, touch can lead to more favorable evaluations of the toucher [5], can persuade [6], and can regulate our physical and emotional well-being [7]. Since it is always reciprocal, social touch not only emotionally affects the receiver [7] but also the touch giver [8]. Touch is the primary modality for conveying intimate emotions [7, 9]. Current technological advances like the embodiment of artificial entities, the development of advanced haptic and tactile display technologies also afford mediated social touch [10]. To study the emotional impact of touch and to design effective haptic social communication systems, validated and efficient affective self-report tools are needed.

In accordance with the circumplex model of affect [11], the affective responses elicited by tactile stimuli vary mainly over the two principal affective dimensions of valence and arousal [12]. Most studies on the emotional response to touch apply two individual one-dimensional Likert scales [13] or SAM (Self-assessment Mannikin: [14]) scales [15] to measure both affective dimensions separately. Although the SAM is a validated and widely used tool, it also has some practical drawbacks. People often fail to understand the emotions it depicts [16]. While the SAM's valence dimension is quite intuitive (a facial expression going from a frown to a smile), its arousal dimension (which looks like an 'explosion' in the figure's stomach) is often misunderstood [16], also in the context of affective touch [15]. Hence there is a need for new rating scales to measure the subjective quality of affective touch [17].

We developed a new intuitive and language-independent self-report instrument called the EmojiGrid (see Fig. 1): a rectangular response grid labeled with facial icons (emoji) expressing different levels of valence (e.g., angry face vs. smiling face) and arousal (e.g., sleepy face vs. excited face) [16]. We previously found that participants can intuitively and reliably report their affective response with a single click on the EmojiGrid, even without verbal instructions [16]. This suggested that the EmojiGrid might also be a general instrument to assess human affective responses.

In this study, we evaluated the convergent validity of the EmojiGrid as a self-report tool for the affective assessment of perceived touch events. We thereto used the EmojiGrid to measure perceived valence and arousal for various touch events in video clips from a validated affective image database, and we compared the results with the normative ratings that were obtained with a conventional validated affective rating tool (the SAM) and that are provided with this database. It appears that the brain activity patterns elicited by imagined, perceived and experienced (affective) touch are highly similar. To some extent, people experience the same touches as the ones they see: they have the ability to imagine how an observed touch would feel [18]. This affords the use video clips showing touch actions to study affective touch perception.

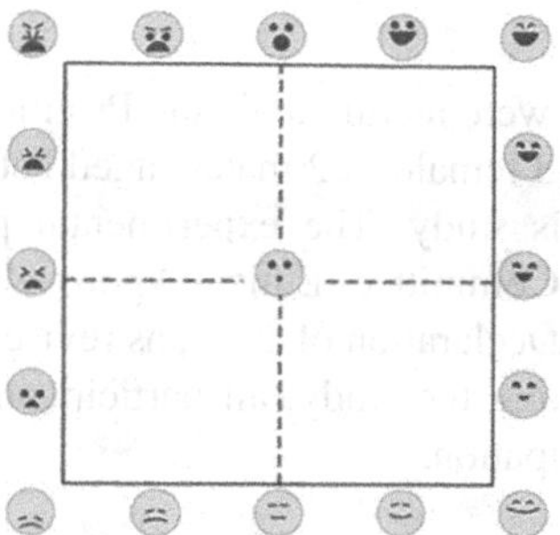

Fig. 1. The EmojiGrid. The facial expressions of the emoji along the horizontal (valence) axis gradually change from unpleasant via neutral to pleasant, while the intensity of the facial expressions gradually increases in the vertical (arousal) direction.

2 Methods and Procedure

2.1 Stimuli

The stimuli used in this study are all 75 video clips from the validated Socio-Affective Touch Expression Database (SATED: [15]). These clips represent 25 different dynamic touch events varying widely in valence and arousal. The interpersonal socio-affective touch events (N = 13) show people hugging, patting, punching, pushing, shaking hands or nudging each other's arm (e.g., Fig. 2, left). The object-based (non-social) touch events (N = 12) represent human-object interactions with motions that match those involved in the corresponding social touch events, and show people touching, grabbing, carrying, shaking or pushing objects like bottles, boxes, baskets, or doors (e.g., Fig. 2, right). Each touch movement is performed three times (i.e., by three different actors or actor pairs) and for about three seconds, resulting in a total of 75 video clips. All video clips had a resolution of 640 × 360 pixels.

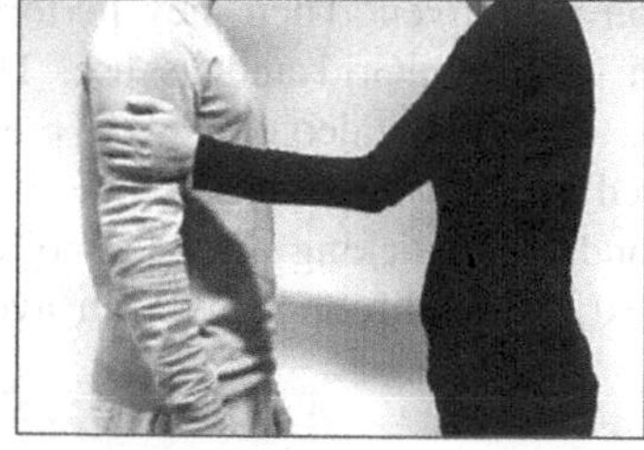

Fig. 2. Screenshots showing an interpersonal (socio-affective) touch event (left) and a corresponding object-based touch event (right).

2.2 Participants

English speaking participants were recruited via the Prolific database (https://prolific.ac). A total of 65 participants (43 females, 22 males) aged between 18 and 35 (M = 25.7; SD = 5.0) participated in this study. The experimental protocol was reviewed and approved by the TNO Ethics Committee (Ethical Approval Ref: 2017–011) and was in accordance with the Helsinki Declaration of 1975, as revised in 2013 [19]. Participation was voluntary. After completing the study, all participants received a small financial compensation for their participation.

2.3 Measures

Demographics. Participants reported their age, gender, and nationality.

Valence and Arousal. Valence and arousal were measured with the EmojiGrid (see Fig. 1; this tool was first introduced in [16]. The EmojiGrid is a square grid that is labeled with emoji showing different facial expressions. Each side of the grid is labeled with five emoji, and there is one (neutral) emoji located in its center. The facial expressions of the emoji along a horizontal (valence) axis vary from disliking (unpleasant) via neutral to liking (pleasant), and their expression gradually increases in intensity along the vertical (arousal) axis. Users can report their affective state by placing a checkmark at the appropriate location on the grid.

2.4 Procedure

The experiment was performed as an (anonymous) online survey created with the Gorilla experiment builder [20]. The participants viewed 75 brief video clips showing a different touch event and rated for each video how the touch would feel. First, the participants signed an informed consent and reported their demographic variables. Next, they were introduced to the EmojiGrid response tool and were told how they could use this tool to report their affective rating for each perceived touch event. The instructions stated: "*Click on a point inside the grid that best matches how you think the touch event feels*". No further explanation was given. Then they performed two practice trials to familiarize themselves with the EmojiGrid and its use. The actual experiment started directly after these practice trials. The video clips were presented in random order. The rating task was self-paced without imposing a time-limit. After seeing each video clip, the participants responded by clicking on the EmojiGrid (see Fig. 1). Immediately after responding the next video clip appeared. On average the experiment lasted about 10 min.

Data Analysis

IBM SPSS Statistics 25 (www.ibm.com) for Windows was used to perform all statistical analyses. Intraclass correlation coefficient (ICC) estimates and their 95% confident intervals were based on a mean-rating (k = 3), absolute agreement, 2-way mixed-effects model [21]. ICC values less than .5 are indicative of poor reliability, values between .5 and .75 indicate moderate reliability, values between .75 and .9

indicate good reliability, while values greater than .9 indicate excellent reliability [21]. For all other analyses, a probability level of $p < .05$ was considered to be statistically significant.

For each of the 25 touch scenarios, we computed the mean valence and arousal responses over all three of its representations and over all participants. We used Matlab 2019a (www.mathworks.com) to investigate the relation between the (mean) valence and arousal ratings and to plot the data. The Curve Fitting Toolbox (version 3.5.7) in Matlab was used to compute a least-squares fit of a quadratic function to the data points.

3 Results

For each touch-scenario, the mean and standard deviation response for valence and arousal was computed over each of its three representations and over all participants.

To quantify the agreement between the ratings obtained with the EmojiGrid (present study) and with the 9-point SAM scales [15] we computed Intraclass Correlation Coefficient (ICC) estimates with their 95% confidence intervals for the mean valence and arousal ratings between both studies, both for all touch events, and for social and non-social events separately (see Table 1). For all touch events, and for social touch events in particular, the valence ratings show excellent reliability and the arousal ratings show good reliability. For object-based touch events, the valence ratings also show good reliability, but the arousal ratings show poor reliability.

Figure 3 shows the correlation plots between the mean valence and arousal ratings obtained with the EmojiGrid in this study and those obtained with the SAM in [15]. This figure shows that the mean valence ratings for all touch events closely agree between both studies: the original classification [15] into positive, negative and neutral scenarios also holds in this result. A Mann-Whitney U test revealed that mean valence was indeed significantly higher for the positive scenarios (Mdn = 7.27, MAD = .31, n = 6) than for the negative ones (Mdn = 2.77, MAD = .35, n = 6), U = 0, z = −2.88, p = .004, r = .58 (large effect size). Additionally, mean valence differed significantly between positive social touch scenarios and object-based touch scenarios (Mdn = 5.13 MAD = .35, n = 12), U = 0, z = −3.37, p = 0.001) and between negative social touch scenarios and object-based touch scenarios, U = 0, z = −3.42, p = 0.001, r = .68 (large effect size).

Whereas the mean valence ratings closely agree between both studies for all touch events, the arousal ratings only agree for the social touch events, but not for the object-based touch events. The mean arousal ratings for object-based touch events are consistently higher with the EmojiGrid than with the SAM. We also compared mean arousal between social touch and object touch. The results revealed that social touch was rated as more arousing (Mdn = 5.35, MAD = 0.89) than object-based (non-social) touch (Mdn = 4.10, MAD = 0.37, U = 27.0, z = −2.77, p ≤ 0.009, r = .55 (large effect size). Social touch is more arousing than object-based (non-social) touch. This agrees with Masson & Op de Beeck's conclusion that completely un-arousing social touch does not seem to exist [15].

Figure 4 shows the relation between the mean valence and arousal ratings for all 25 different SATED scenarios, as measured with the EmojiGrid in this study and with a 9-point SAM scale in [15]. The curves represent least-squares quadratic fits to the data points. The adjusted R-squared values are respectively .75 and .90, indicating good fits. This figure shows that the relation between the mean valence and arousal ratings provided by both self-assessment methods is closely described by a quadratic (U-shaped) relation at the nomothetic (group) level.

All results are available at the OSF repository with https://doi.org/10.17605/osf.io/d8sc3.

Table 1. Intraclass correlation coefficients with their 95% confidence intervals for mean valence and arousal ratings obtained with the SAM and with the EmojiGrid (this study), for video clips from the SATED database [15].

Stimuli	N	ICC valence	ICC arousal
All touch events	25	.99 [.97 – 1.0]	.71 [.33 – .87]
Social touch events	13	1.0 [.99 – 1.0]	.79 [.32 – .94]
Object touch events	12	.81 [.32 – .95]	.18 [–.04 – .59]

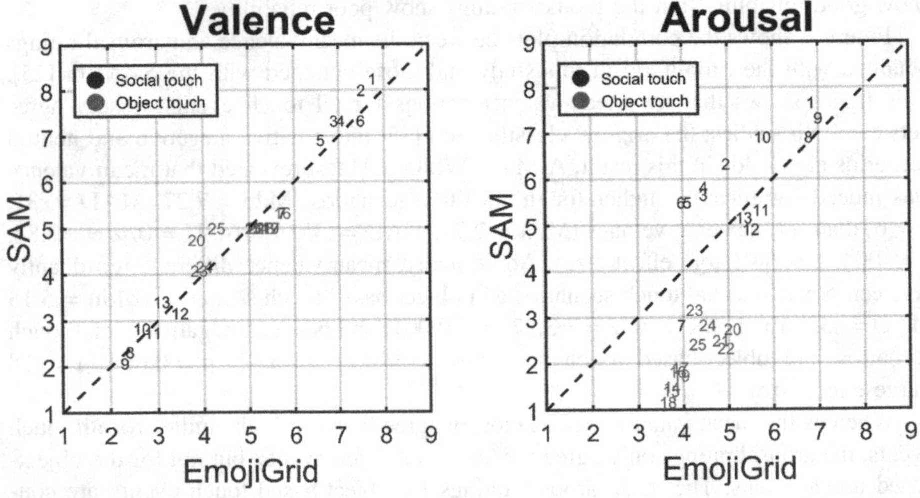

Fig. 3. Correlation plots illustrating the relationship between the valence (left) and arousal (right) ratings provided with a 9-point SAM scale in [15] and those obtained with the EmojiGrid in the present study. The numbers correspond to the original scenario identifiers in the SATED database. Scenarios 1–6 were originally classified as positive, scenarios 8–13 as negative, and scenarios 7 and 14–25 as neutral.

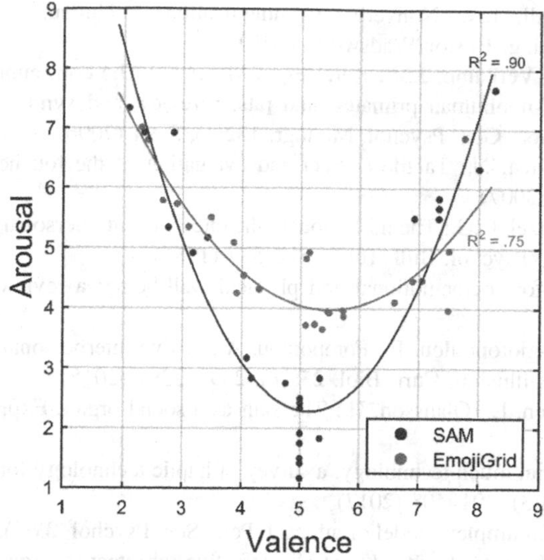

Fig. 4. Relation between the mean valence and arousal ratings for affective touch video clips from the from the SATED database, obtained with the a 9-point SAM rating scale (blue dots: [15]) and with the EmojiGrid (red dots: this study). The curves represent fitted polynomial curves of degree 2 using a least-squares regression between valence and arousal. (Color figure online)

4 Conclusion and Discussion

The affective (valence and arousal) ratings obtained with the EmojiGrid show good to excellent agreement with the data provided with the database for inter-human (social) touch events. Also, our results replicate the U-shaped (quadratic) relation between the mean valence and arousal ratings, as reported in the literature [15]. Thus, we conclude that the EmojiGrid appears to be a valid graphical self-report instrument for the affective appraisal of perceived social touch events. The agreement for object touch events is good for valence but poor for arousal. This may be related to the instruction to the participants to rate "*how the touch would feel*". Although not explicitly stated to do so, participants may have preferred the receiver perspective above the sender perspective of the interaction. This perspective has less meaning in the object touch events probably resulting in a switch to the sender perspective. Classic rating tools may be less sensitive to the different perspectives which is only relevant in touch events, hence the extremely low arousal scores with the SAM tool and the low intraclass correlations.

References

1. Essick, G.K., McGlone, F., Dancer, C., et al.: Quantitative assessment of pleasant touch. Neurosci. Biobehav. Rev. **34**(2), 192–203 (2010)
2. Jones, S.E., Yarbrough, A.E.: A naturalistic study of the meanings of touch. Commun. Monogr. **52**(1), 19–56 (1985)

3. Knapp, M.L., Hall, J.A.: Nonverbal Communication in Human Interaction, 7th edn. CENGAGE Learning, Boston/Wadsworth (2010)
4. Hertenstein, M.J., Verkamp, J.M., Kerestes, A.M., et al.: The communicative functions of touch in humans, nonhuman primates, and rats: a review and synthesis of the empirical research. Genet. Soc. Gen. Psychol. Monogr. **132**(1), 5–94 (2006)
5. Erceau, D., Guéguen, N.: Tactile contact and evaluation of the toucher. J. Soc. Psychol. **147**(4), 441–444 (2007)
6. Crusco, A.H., Wetzel, C.G.: The midas touch: the effects of interpersonal touch on restaurant tipping. Pers. Soc. Psychol. Bull. **10**(4), 512–517 (1984)
7. Field, T.: Touch for socioemotional and physical well-being: a review. Dev. Rev. **30**(4), 367–383 (2010)
8. Gentsch, A., Panagiotopoulou, E., Fotopoulou, A.: Active interpersonal touch gives rise to the social softness illusion. Curr. Biol. **25**(18), 2392–2397 (2015)
9. Morrison, I., Löken, L., Olausson, H.: The skin as a social organ. Exp. Brain Res. **204**(3), 305–314 (2010)
10. Huisman, G.: Social touch technology: a survey of haptic technology for social touch. IEEE Trans. Haptics **10**(3), 391–408 (2017)
11. Russell, J.A.: A circumplex model of affect. J. Pers. Soc. Psychol. **39**(6), 1161–1178 (1980)
12. Hasegawa, H., Okamoto, S., Ito, K., et al.: Affective vibrotactile stimuli: relation between vibrotactile parameters and affective responses. Int. J. Affect. Eng. **18**(4), 171–180 (2019)
13. Salminen, K., Surakka, V., Lylykangas, J., et al.: Emotional and behavioral responses to haptic stimulation. In: SIGCHI Conference on Human Factors in Computing Systems (CHI 2008), pp. 1555–1562. ACM, New York (2008)
14. Bradley, M.M., Lang, P.J.: Measuring emotion: the self-assessment manikin and the semantic differential. J. Behav. Ther. Exp. Psychiatry **25**(1), 49–59 (1994)
15. Lee Masson, H., Op de Beeck, H.: Socio-affective touch expression database. PLOS ONE, **13**(1), e0190921 (2018)
16. Toet, A., Kaneko, D., Ushiama, S., et al.: EmojiGrid: a 2D pictorial scale for the assessment of food elicited emotions. Front. Psychol. **9**, 2396 (2018)
17. Schneider, O.S., Seifi, H., Kashani, S., et al.: HapTurk: crowdsourcing affective ratings of vibrotactile icons. In: 2016 CHI Conference on Human Factors in Computing Systems, pp. 3248–3260. ACM, New York (2016)
18. Keysers, C., Gazzola, V.: Expanding the mirror: vicarious activity for actions, emotions, and sensations. Curr. Opin. Neurobiol. **19**(6), 666–671 (2009)
19. World Medical Association: World Medical Association declaration of Helsinki: ethical principles for medical research involving human subjects. J. Am. Med. Assoc. **310**(20), 2191–2194 (2013)
20. Anwyl-Irvine, A., Massonnié, J., Flitton, A., et al.: Gorilla in our Midst: an online behavioral experiment builder. bioRxiv, 438242 (2019)
21. Koo, T.K., Li, M.Y.: A guideline of selecting and reporting intraclass correlation coefficients for reliability research. J. Chiropractic Med. **15**(2), 155–163 (2016)

A 2-DoF Skin Stretch Display on Palm: Effect of Stimulation Shape, Speed and Intensity

Ahmad Manasrah(✉) and Shahnaz Alkhalil

Al Zaytoonah University of Jordan, Amman 11947, Jordan
{ahmad.mansrah,shahnaz.k}@zuj.edu.jo
https://sites.google.com/view/ahmadmanasrah/home

Abstract. Skin stretch has been widely utilized as a tactile display in different haptic applications. However, there has been little research focusing on skin stretch as a modality on the palm of the hand. In this study, a two dimensional tactor apparatus was designed and built to investigate the effects of stimulation speeds, shapes and intensities of skin stretch display on the palm. The tactor moved across the palm at different speeds to create stimulation shapes on the skin. Subjects reported the intensity of perceived stimuli and predicted speed rate of the tactor and stimulation shape and size. The results showed that there were statistically significant differences in the intensity of perceived tactile displays between different stimulation shapes and sizes. The results also showed the sizes and intensity of the stimulus grow larger with slower tactor speeds.

Keywords: Skin stretch · Palm · Stimulation intensity · Tactile display

1 Introduction

There have been significant advances in the field of tactile displays. The literature shows that tactile displays have the potential to improve the user's experience in gaming, teleoperation, and virtual environment simulation. For instance, tactile displays were utilized in delivering shape and material information to users [10] and, also, to deliver instructions and navigational commands [16]. Most of these haptic feedback technologies transfer information via vibrotactile actuators due to their effective perception on the skin and simple implementation [3,7]. However, the vibrotactile feedback does not necessarily provide directional information unless multiple actuators are utilized [17]. This type of tactile display may also create desensitization and discomfort to users at relatively high intensities [21,26].

Recently, skin stretch has been introduced and implemented as an alternative tactile feedback modality in multiple studies. The skin stretch approach has the

I. Nisky et al. (Eds.): EuroHaptics 2020, LNCS 12272, pp. 12–24, 2020.
https://doi.org/10.1007/978-3-030-58147-3_2

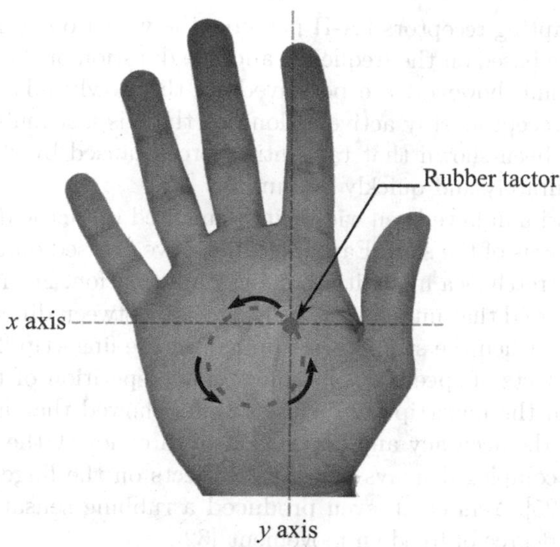

Fig. 1. Rubber tactor moves in two dimensions on the palm of the hand with different speeds to create a shape. In this case, a circle.

potential to deliver directional information especially with continuous stimuli using only one actuator [2]. The moving actuator applies a directional shear force on the surface of the skin, thus, activating the mechanoreceptors that are responsible of detecting and monitoring moving objects on the skin [15]. Shear forces, applied tangentially to the surface of the skin, result from the friction between the tip of the tactor and the skin [34]. Larger friction forces, for instance, trigger the mechanoreceptors with higher rates, thus increase the intensity of stimulation. In this study, we hypothesize that the speed of the actuator as well as the shape that is created on the palm of the hand can affect the user's perception of the intensity and area of stimulation. A two dimensional plotter mechanism was designed to move a rubber tactor across the skin as illustrated in Fig. 1. The haptic cue is conveyed when the tactor creates deformations of the skin in the direction at which it is traveling, hence, users can identify the drawn shapes and estimate their areas comparatively.

The purpose of this research is to investigate this hypothesis related to how users perceive tactile displays of different shapes and speeds. The results of this investigation may provide a new perspective on skin stretch as a method of delivering information and help the development of tactile display systems in many applications.

2 Background

Different tactile displays are perceived by different mechanoreceptors in the skin. Each type of mechanoreceptor is excited with the presence of certain stimuli. For instance, fast adapting receptors FA-II perceive the vibration stimulus with different sensitivities based on the frequency and the duration of the vibration [30]. Skin stretch stimuli, however, are perceived by the slowly adapting receptors SA-II [6]. These receptors stay active as long as there is a stimulus on the skin. Moreover, it has been shown that tangential forces caused by skin stretch can be perceived accurately and quickly by humans [4].

Skin stretch stimuli have been widely implemented in tactile display systems on different locations of the skin. Earlier studies have focused on the fingertip to investigate skin stretch as a modality to deliver information. For instance, a previous research showed that humans can differentiate between different tangential and normal forces when the stimulus is applied on the fingertip [20]. Gleeson et al, studied the effects of speed, displacement, and repetition of tangential skin stretch stimuli on the fingertip [12]. Their results showed that higher stimulus speeds improved the accuracy and perception of direction. Other studies investigated creating complex displays of virtual objects on the fingertip using skin deformation [14,25]. Yem et al. even produced a rubbing sensation on the fingertip using one degree of freedom movement [32].

There are also other studies that investigated skin stretch displays on the forearm [8] and on the lower extremities [7]. However, in comparison to other locations on the skin, few studies are found regarding skin deformation on the palm of the hand. Even though the palm is less sensitive than the fingertips, it has similar densities of SA-II receptors [29] and provides a larger display area for skin stretch stimulation [13]. Studies have shown that tactile displays which are applied on the fingers and the palm simultaneously improve users' perception of relatively large virtual objects [27]. Skin stretch stimuli on the palm have been used to deliver driving information and direction to drivers through the steering wheel [23].

Other studies have investigated several factors that may affect skin stretch and skin deformation. For example, Edin et al. applied tangential forces at different locations on the body including the palm of the hand. Their results proved that the speed and direction of the stimulus affect the perception of skin stretch displays [9]. A more recent study also showed that the intensity of skin stretch is affected by the speed and displacement of the stimulus [13]. The method presented here investigated how the intensity and size created by the stimulus were perceived based on different speeds and stimulation shapes.

3 Method

This investigation consisted of one set of experiments studying several tactor speeds and stimulation shapes. In each experiment, subjects perceived a random stimulation shape with a certain size and speed. Three stimulation shapes were used: a circle, a square, and an equilateral triangle at two different sizes and two tactor speeds.

3.1 Apparatus

A two dimensional plotter was designed and built specifically for the purpose of this study. Unlike conventional 2D plotters, the stylus was replaced with a tactor and directed upwards so that the palm can rest on it. The tactor's tip was made out of rubber and had a spherical shape with a diameter of $4mm$ which is about the size of a board pin. The plotter moved the tactor using two stepper motors attached in an H-bot connection capable of producing different speeds as shown in Fig. 2. The tactor was mounted on a 3D printer base that was connected to the plotter contact surface. All the internal components and connections of the two dimensional plotter were covered with a white box which was cut out in the center to allow the tactor to move freely in a 7 × 7 cm space which is slightly less than the size of the average palm for an adult [24].

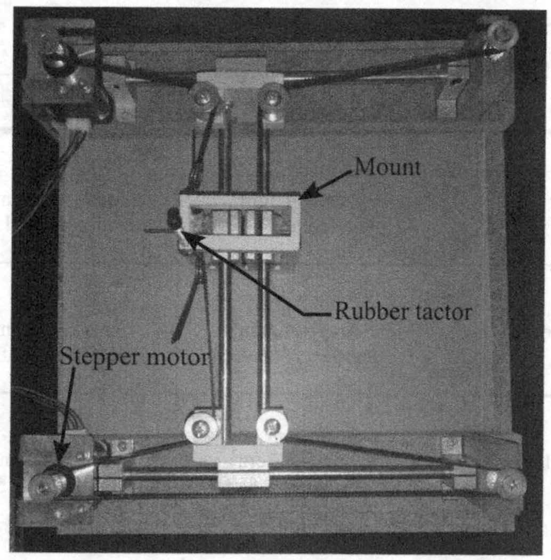

Fig. 2. Two dimensional plotter was built using two stepper motors. The rubber tactor is installed on the mount.

3.2 Experimental Setup and Procedure

At the beginning of the experiments, the nature of the study and procedures were generally explained to the subjects. After that, they were asked to sit in front of a screen and rest their right hand on the apparatus without applying extra force. Participants were also asked to wear headphones playing steady music to block noises generated from the apparatus and surroundings. Before the experiments began, a test-experiment was given to subjects where a stimulation shape is randomly applied on the participants' palms at two speeds 10 mm/s and 20 mm/s. Participants were then asked to rate the intensity of stimuli on a scale from zero to four with zero being the weakest and four the strongest. The reason of conducting the test-experiment is to ensure that subjects had a reference intensity point to compare it to the stimuli that they were about to experience at the experiments.

During the experiments, three stimulation shapes were tested, a circle, square, and an equilateral triangle. Each shape was tested with two speeds (10 mm/s and 20 mm/s) and two areas. The stimulation areas were divided into "small" and "large" areas based on the shape. However since three stimulation shapes were used, the areas were not exactly equal for all the shapes. For instance, the "small" area of the circular stimulus was about 4.5 cm^2 ($r = 1.2$ cm), while as for the square, the area was 4.4 cm^2 ($w = 2.1$ cm) and for the equilateral triangle the area was 1.7 cm^2 ($l = 2$ cm). Table 1 shows the areas and perimeters for all three shapes.

Table 1. Stimulation shapes, speeds, and sizes.

Shape	Speed	Size	
		Small	Large
Circle	10 mm/s	*r=1.2cm* Area = 4.5cm^2 Circumference = 7.5cm	*r=2.5cm* Area = 19.6cm^2 Circumference = 15.7cm
	20 mm/s		
Square	10 mm/s	*w=2.1cm* Area = 4.4cm^2 Perimeter = 8.4cm	*w=3.2cm* Area = 10.2cm^2 Perimeter = 12.8cm
	20 mm/s		
Triangle	10 mm/s	*l=2cm* Area = 1.7cm^2 Perimeter = 6cm	*l=3.5cm* Area = 5.3cm^2 Perimeter = 10.5cm
	20 mm/s		

With each experiment, a random stimulation shape was applied counterclockwise on the palm with a certain speed and size. The screen in front of the participants displayed all six stimulation shapes (two sizes for each shape). The sizes were exaggerated so that subjects can identify large and small shapes on the screen easily. The display also showed the intensity scale that was described previously. After each stimulation shape was applied, participants recorded their response using the information on the screen. A total of 12 shape, speed, and size combinations were randomly applied on each participant. The experiments took 15 min in total, including a few minutes break. Figure 3 shows the full experimental setup with the screen.

3.3 Participants

A total of ten subjects participated in the experiments, six males and four females. All of them were healthy, right handed, and between the age of 18 and 50. Each subject read and signed a consent form agreeing to participate in the experiments.

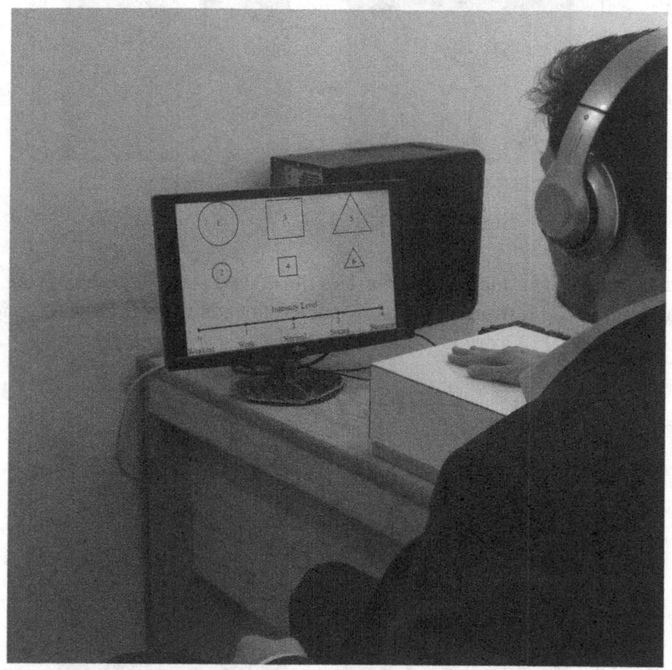

Fig. 3. Experiment setup where subject's right hand rests on the apparatus. The screen shows the stretch intensity scale and the shapes and sizes used in the experiments.

4 Results

In this study the collected data was analyzed using ANOVA with a dependent variable of stretch intensity and four independent variables of tactor speed, stimulation shape and size, and subject. Another ANOVA with a dependent variable of subject answers (shape, size) and three independent variables of stretch intensity, tactor speed, and subject was also conducted. When the results showed statistical differences, a Tukey's honest significant difference (HSD) test was performed as a post-hoc test. All statistical tests were based on alpha value of 0.05.

The results of the first analysis showed that the "slow" tactor speed (10 mm/s) had a statistically significantly smaller stretch intensity ($F(1,106) = 58.06$, $p < 0.001$) than the "fast" stimuli (20 mm/s) in all stimulation shapes.

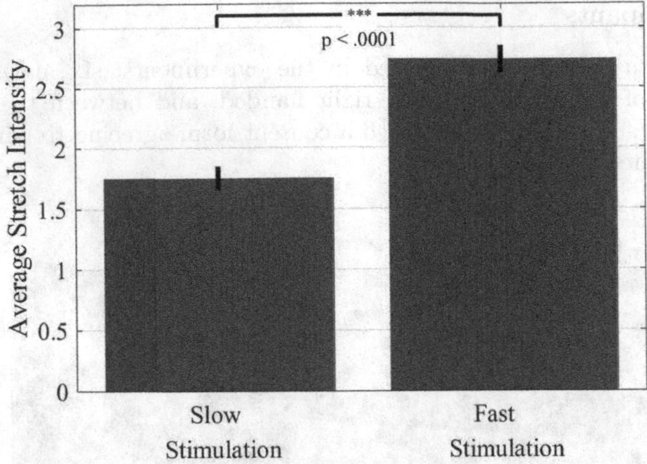

Fig. 4. The means and standard errors for the perceived stretch intensity at two tactor speeds. The average stretch intensity of slow tactor speeds was statistically significantly less than fast speeds.

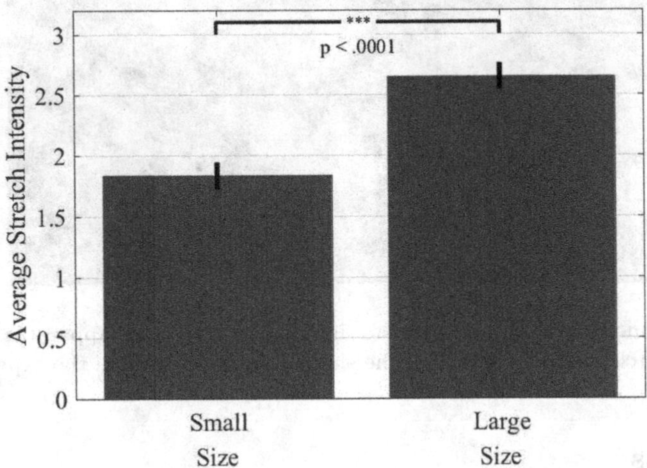

Fig. 5. The means and standard errors for the perceived stretch intensity at two stimulation sizes. The average stretch intensity of small stimulation sizes was statistically significantly less than large sizes.

Figure 4 shows the mean intensity of stimuli for the "slow" and "fast" tactor speeds. The slow tactor speed recorded a mean stretch intensity of 1.75 which was between "weak" and "normal". The mean stretch intensity of the fast tactor speed, however, was around 2.7, closer to "strong" on the stretch intensity scale. Moreover, the results showed that "small" stimuli sizes also had statistically significantly less stretch intensity ($F(1, 106) = 58.67$, $p < 0.001$) than "large" stimuli sizes within all the applied shapes as illustrated in Fig. 5. The results of stimuli shapes versus stretch intensity showed that the circular shape had statistically significantly less stretch intensity ($F(2, 106) = 26.27$, $p < 0.001$) than the other rectangular and triangular shapes regardless of the size. Figure 6 shows the stretch intensities of the three stimulation shapes. The results did not, however, show any statistical significant differences between subjects.

The second analysis focused on the perception of shapes and sizes of the stimuli. The results showed that subjects perceived the "slow" tactor speed statistically significantly "smaller" ($F(1, 106) = 15.79$, $p < 0.0001$) than the "fast" tactor speed as shown in Fig. 7. Furthermore, seven out of ten subjects miss-identified the "small" square at "fast" tactor speed as a circle of the same size.

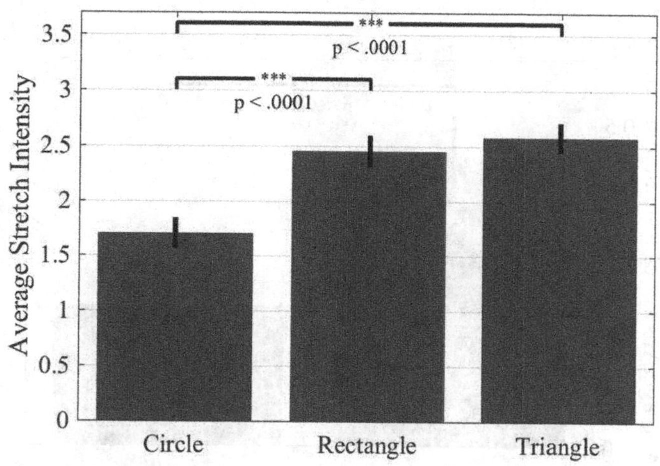

Fig. 6. The means and standard errors for the perceived stretch intensity at three stimulation shapes. The average stretch intensity of the circular shape was statistically significantly less than other stimulation shapes.

5 Discussion

In this study, the shape, speed, and intensity of skin stretch stimulation were investigated on the palm. The experimental results showed that the "slow" tactor speed was perceived with statistically significantly less intensity than the "fast" speed. These results agree with previous findings where the intensity of skin stretch perception increases as tactor speed increases [13]. They concluded that the relation between speed and intensity, however, was nonlinear. Other studies showed that an increase in the tactor speed enhances the intensity and accuracy of the perceived stimulus on the finger [12].

Further, the results of our experiments showed that "small" stimuli shapes were perceived with statistically significantly less intensity than "large" stimuli shapes. A small stimulation shape is created via short stimuli distance, thus, producing less stimulation intensity. Caswell et al. showed that shorter tactor displacement yielded a higher accuracy in predicting the direction of stimuli since their intensity increased [5]. The same results were also found in [13]. However, these findings can be related to the spatial distribution of the receptors perceiving the stimuli. Studies have shown that mechanoreceptors are not evenly distributed on the palm as their densities increase towards the fingertips [1,29]. Moreover, the ridges on the palm may have played a role on increasing the intensity of perception for the "large" stimulation shapes. Previous studies indicated that the microstructure of skin strongly affect the perception of skin stretch displays [22].

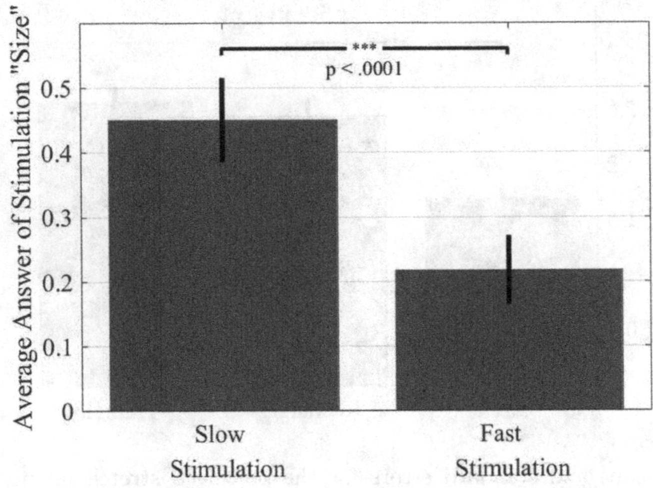

Fig. 7. The means and standard errors of average stimulus sizes at two tactor speeds. The areas created by slow tactor speeds were perceived statistically significantly larger than fast speeds.

Despite the statistical differences between stimuli sizes, the experiments showed that the "circle" stimulation shape had statistically significantly less intensity than the "square" and "triangle" as illustrated in Fig. 6. It seems that the presence of corners in the latter shapes amplified the intensity of stimulation even though the tactor did not stop at those corners during the experiments. These corners might have increased the friction coefficient between the tip of the tactor and skin, thus, increasing the intensity. It has been shown that tangential displacements with friction is perceived with higher sensitivity than frictionless motion [19,28]. Moreover, the triangular shape was perceived with the highest intensity out of the three stimulation shapes. This suggests that acute angles of a stimulus may have an impact on its intensity. Although this effect is distinctly different than the priming effect, where multiple stimuli in the same direction increase the sensitivity of perception [11], it is however worth investigating in the future.

In addition to the statistical differences in the intensity between tactor speeds, the stimulation size, and shape, the experimental results showed that "slow" tactor speeds were perceived statistically significantly larger than "fast" tactor speeds even though the 12 experiments were randomly conducted. Whitsel et al. studied the velocity of a stimulus against its perceived motion and found that distances can be perceived shorter as the stimulus velocity increased [31]. Others have concluded that a stimulus length is perceived as result of speed and duration [33]. Such a phenomenon can occur due to the spatial properties of the receptors in the skin [18]. The time duration in our experiments varied from approximately 7 seconds to 15 seconds for each experiment depending on the tactor speed and stimulation shape. It is possible that subjects may have perceived the stimulation shapes of "fast" tactor speeds as smaller. Moreover, seven out of ten subjects perceived the "small" square at "fast" tactor speed as a small "circle". This might be related to the similarities between the small square and circle in the perimeters and areas as shown in Table 1.

6 Conclusion

In this paper, the effects of stimulation speed, shape and intensity on skin stretch were investigated by conducting a series of experiments on the palm of the hand. Ten subjects participated in these experiments where two tactor speeds and three stimulation shapes were tested. The results showed that stimuli intensities of relatively slow tactor speeds were statistically significantly less than the fast ones. The experimental results also showed that relatively small stimulation sizes and circular shapes were perceived with less intensity than large stimulation sizes and other shapes respectively. Further, there were also differences in the perceived sizes of stimuli among stimulation shapes. Future work will focus on studying the direction of the stimulus and stimulation angles of different shapes on the skin.

Acknowledgment. The authors would like to acknowledge Al Zaytoonah University of Jordan for funding this research under the grant number 2017-2016/64/17.

References

1. Asamura, N., Shinohara, T., Tojo, Y., Shinoda, H.: Necessary spatial resolution for realistic tactile feeling display. In: ICRA, pp. 1851–1856. Citeseer (2001)
2. Bark, K., Wheeler, J., Lee, G., Savall, J., Cutkosky, M.: A wearable skin stretch device for haptic feedback. In: World Haptics 2009-Third Joint EuroHaptics Conference and Symposium on Haptic Interfaces for Virtual Environment and Teleoperator Systems, pp. 464–469. IEEE (2009)

3. Bark, K., Wheeler, J.W., Premakumar, S., Cutkosky, M.R.: Comparison of skin stretch and vibrotactile stimulation for feedback of proprioceptive information. In: 2008 Symposium on Haptic Interfaces for Virtual Environment and Teleoperator Systems, pp. 71–78. IEEE (2008)
4. Biggs, J., Srinivasan, M.A.: Tangential versus normal displacements of skin: relative effectiveness for producing tactile sensations. In: Proceedings 10th Symposium on Haptic Interfaces for Virtual Environment and Teleoperator Systems. HAPTICS 2002, pp. 121–128. IEEE (2002)
5. Caswell, N.A., Yardley, R.T., Montandon, M.N., Provancher, W.R.: Design of a forearm-mounted directional skin stretch device. In: 2012 IEEE Haptics Symposium (HAPTICS), pp. 365–370. IEEE (2012)
6. Chambers, M.R., Andres, K., Duering, M.V., Iggo, A.: The structure and function of the slowly adapting type ii mechanoreceptor in hairy skin. Q. J. Exp. Physiol. Cognate Med. Sci. **57**(4), 417–445 (1972)
7. Chen, D.K., Anderson, I.A., Walker, C.G., Besier, T.F.: Lower extremity lateral skin stretch perception for haptic feedback. IEEE Trans. Haptics **9**(1), 62–68 (2016)
8. Chinello, F., Pacchierotti, C., Tsagarakis, N.G., Prattichizzo, D.: Design of a wearable skin stretch cutaneous device for the upper limb. In: 2016 IEEE Haptics Symposium (HAPTICS), pp. 14–20. IEEE (2016)
9. Edin, B.B., Essick, G.K., Trulsson, M., Olsson, K.A.: Receptor encoding of moving tactile stimuli in humans. i. temporal pattern of discharge of individual low-threshold mechanoreceptors. J. Neurosci. **15**(1), 830–847 (1995)
10. Gallo, S., Son, C., Lee, H.J., Bleuler, H., Cho, I.J.: A flexible multimodal tactile display for delivering shape and material information. Sensors Actuators A: Phys. **236**, 180–189 (2015)
11. Gardner, E.P., Sklar, B.F.: Discrimination of the direction of motion on the human hand: a psychophysical study of stimulation parameters. J. Neurophysiol. **71**(6), 2414–2429 (1994)
12. Gleeson, B.T., Horschel, S.K., Provancher, W.R.: Perception of direction for applied tangential skin displacement: effects of speed, displacement, and repetition. IEEE Trans. Haptics **3**(3), 177–188 (2010)
13. Guzererler, A., Provancher, W.R., Basdogan, C.: Perception of skin stretch applied to palm: effects of speed and displacement. In: Bello, F., Kajimoto, H., Visell, Y. (eds.) EuroHaptics 2016. LNCS, vol. 9774, pp. 180–189. Springer, Cham (2016). https://doi.org/10.1007/978-3-319-42321-0_17
14. Hayward, V.: Display of haptic shape at different scales. In: Proceedings of Eurohaptics, vol. 2004, pp. 20–27. Citeseer (2004)
15. Johnson, K.O.: The roles and functions of cutaneous mechanoreceptors. Current Opin. Neurobiol. **11**(4), 455–461 (2001)
16. Jones, L.A., Ray, K.: Localization and pattern recognition with tactile displays. In: 2008 Symposium on Haptic Interfaces for Virtual Environment and Teleoperator Systems, pp. 33–39. IEEE (2008)
17. Kim, Y., Harders, M., Gassert, R.: Identification of vibrotactile patterns encoding obstacle distance information. IEEE Trans. Haptics **8**(3), 298–305 (2015)
18. Nguyen, E.H., Taylor, J.L., Brooks, J., Seizova-Cajic, T.: Velocity of motion across the skin influences perception of tactile location. J. Neurophysiol. **115**(2), 674–684 (2015)
19. Olausson, H., Norrsell, U.: Observations on human tactile directional sensibility. J. Physiol. **464**(1), 545–559 (1993)
20. Paré, M., Carnahan, H., Smith, A.M.: Magnitude estimation of tangential force applied to the fingerpad. Exp. Brain Res. **142**(3), 342–348 (2002)

21. Petermeijer, S.M., De Winter, J.C., Bengler, K.J.: Vibrotactile displays: a survey with a view on highly automated driving. IEEE Trans. Intell. Transp. Syst. **17**(4), 897–907 (2016)
22. Pham, T.Q., Hoshi, T., Tanaka, Y., Sano, A.: Effect of 3D microstructure of dermal papillae on SED concentration at a mechanoreceptor location. PloS One **12**(12), e0189293 (2017)
23. Ploch, C.J., Bae, J.H., Ju, W., Cutkosky, M.: Haptic skin stretch on a steering wheel for displaying preview information in autonomous cars. In: 2016 IEEE/RSJ International Conference on Intelligent Robots and Systems (IROS), pp. 60–65. IEEE (2016)
24. Rhodes, J., Clay, C., Phillips, M.: The surface area of the hand and the palm for estimating percentage of total body surface area: results of a meta-analysis. Br. J. Dermatol. **169**(1), 76–84 (2013)
25. Schorr, S.B., Okamura, A.M.: Fingertip tactile devices for virtual object manipulation and exploration. In: Proceedings of the 2017 CHI Conference on Human Factors in Computing Systems, pp. 3115–3119. ACM (2017)
26. Shi, S., Leineweber, M.J., Andrysek, J.: Examination of tactor configurations for the design of vibrotactile feedback systems for use in lower-limb prostheses. In: ASME 2018 International Design Engineering Technical Conferences and Computers and Information in Engineering Conference, p. V008T10A006. American Society of Mechanical Engineers (2018)
27. Son, B., Park, J.: Haptic feedback to the palm and fingers for improved tactile perception of large objects. In: The 31st Annual ACM Symposium on User Interface Software and Technology, pp. 757–763. ACM (2018)
28. Sylvester, N.D., Provancher, W.R.: Effects of longitudinal skin stretch on the perception of friction. In: Second Joint EuroHaptics Conference and Symposium on Haptic Interfaces for Virtual Environment and Teleoperator Systems (WHC 2007), pp. 373–378. IEEE (2007)
29. Vallbo, A.B., Johansson, R.S., et al.: Properties of cutaneous mechanoreceptors in the human hand related to touch sensation. Hum. Neurobiol. **3**(1), 3–14 (1984)
30. Verrillo, R.T.: Investigation of some parameters of the cutaneous threshold for vibration. J. Acoust. Soc. Am. **34**(11), 1768–1773 (1962)
31. Whitsel, B., et al.: Dependence of subjective traverse length on velocity of moving tactile stimuli. Somatosensory Res. **3**(3), 185–196 (1986)
32. Yem, V., Shibahara, M., Sato, K., Kajimoto, H.: Expression of 2DOF fingertip traction with 1DOF lateral skin stretch. In: Hasegawa, S., Konyo, M., Kyung, K.-U., Nojima, T., Kajimoto, H. (eds.) AsiaHaptics 2016. LNEE, vol. 432, pp. 21–25. Springer, Singapore (2018). https://doi.org/10.1007/978-981-10-4157-0_4
33. Yusoh, S.M.N.S., Nomura, Y., Sakamoto, R., Iwabu, K.: A study on the duration and speed sensibility via finger-pad cutaneous sensations. Procedia Eng. **41**, 1268–1276 (2012)
34. Zhang, M., Turner-Smith, A., Roberts, V.: The reaction of skin and soft tissue to shear forces applied externally to the skin surface. Proc. Inst. Mech. Eng. Part H: J. Eng. Med. **208**(4), 217–222 (1994)

BY

User-Defined Mid-Air Haptic Sensations for Interacting with an AR Menu Environment

Lawrence Van den Bogaert(✉) and David Geerts

Meaningful Interactions Lab (Mintlab), KU Leuven, Leuven, Belgium
{lawrence.vandenbogaert,david.geerts}@kuleuven.be

Abstract. Interfaces that allow users to interact with a computing system by using free-hand mid-air gestures are becoming increasingly prevalent. A typical shortcoming of such gesture-based interfaces, however, is their lack of a haptic component. One technology with the potential to address this issue, is ultrasound mid-air haptic feedback. At the moment, haptic sensations are typically designed by system engineers and experts. In the case of gestural interfaces, researchers started involving non-expert users to define suitable *gestures* for specific interactions. To our knowledge, no studies have involved – from a similar participatory design perspective – laymen to generate mid-air haptic sensations. We present the results of an end-user elicitation study yielding a user-defined set of mid-air haptic sensations to match gestures used for interacting with an Augmented Reality menu environment. In addition, we discuss the suitability of the end-user elicitation method to that end.

Keywords: Mid-air haptics · End-user elicitation study · Participatory design

1 Introduction

While advancements in gesture recognition technologies (e.g., Microsoft Kinect, Leap Motion) open up exciting new ways of human-computer interaction, one important constraint they face is their lack of a haptic component. One technology that carries the potential to address this shortcoming, without impeding the user with wearables or hand-held devices, is ultrasound mid-air haptic feedback [2,3]. By generating pressure fields in mid-air via an array of ultrasound transducers, this technology allows the user to experience a sense of touch on the palm and fingers of the unencumbered hand. The ultrasound waves that create these pressure fields can be meticulously altered, opening up a quasi-infinite set of sensations, shapes and patterns. Typically, it is experts and specialists who are tasked with their design. For gestures, however, it has already been argued that even though designer-designed gestures might be expertly crafted, end-user elicited gestures are more intuitive, guessable and preferred [4]. Accordingly, we believe this might be true for mid-air haptic sensations too. To our knowledge,

I. Nisky et al. (Eds.): EuroHaptics 2020, LNCS 12272, pp. 25–32, 2020.
https://doi.org/10.1007/978-3-030-58147-3_3

so far only one other study has involved end-users to create mid-air haptic sensations [5]. However, whereas their goal was for participants to create mid-air haptic sensations to convey and mediate emotions to others, we wanted to test whether novice users can ideate haptic sensations to match gestures used for operating a menu in Augmented Reality (AR). To this end, we turned to Wobbrock et al.'s *end-user elicitation methodology* [12]. Typically, such elicitation studies present participants with one or more *referents* for which they are asked to ideate a suitable *symbol*. A referent is the effect of an interaction with an interface (e.g., an increase in volume on a music player). The symbol is the corresponding action (input) that is required to invoke the referent (e.g., turning the volume knob clockwise). In principle, a symbol can be anything that computing systems are able to register as input. This method has been used to design symbols (input) to actuate, among others, touchscreens [13], virtual & augmented reality [6], smart glasses [9], home entertainment systems [10] and public displays [7]. Whereas up until now participants of end-user elicitation studies were tasked with the ideation of the symbol, we assigned ours with the elicitation of what we suggest to call *intermediary referents*. 'Intermediary' means it concerns not the main system output (referent), but rather a form of feedback that accompanies it. Common examples might be the vibrotactile feedback we feel when typing on our smartphone, or the beeping tones we hear when adjusting the volume.

With this work, we present a set of user-defined intermediary referents in the form of mid-air haptic sensations to match a set of gestures used for interacting with an AR interface. We discuss the eligibility of the end-user elicitation method for their design.

2 Study Setup

Twenty-four non-specialist participants were invited individually to our lab in Leuven, Belgium to elicit a set of five mid-air haptic sensations to match five gestures actuating an AR menu. Given the novelty of mid-air haptics, participants were recruited from a list of people who had previously taken part in a study involving ultrasound mid-air haptics. They were still non-experts, but had at least a basic understanding of what the technology does and how it feels. Ages ranged from 19 to 56 ($\mu = 26,2$). Ten participants were male, fourteen were female.

2.1 Apparatus and Gesture Selection

For our participants to interact with an AR menu, we chose Microsoft's HoloLens as it is by default gesture-controlled and free of hand-held controllers or wearables. Three proprietary gestures allow for its actuation: 1) 'bloom' (used to evoke the home menu at any time to easily navigate in and out of applications); 2) 'air tap' (used to select things, equivalent to a mouse click on a desktop); 3) 'tap and hold' (equivalent of keeping a mouse-button pressed, e.g., for scrolling on a page). 'Air tap' was left out because it is similar to 'tap and hold' in every aspect except for its duration ('air tap' being more brief and thus less suitable for mid-air haptic actuation). In addition to the default HoloLens commands, we turned to Piumsomboon et al.'s user-defined AR gesture set [6] to supplement the HoloLens gestures. From it, we selected three additional gestures for our study: a) 'swipe' (used to rotate a carousel); b) 'stop' (used to pause, e.g., a song or video); and c) 'spin [index finger]' (used to increase or decrease the volume). We chose these three gestures because they were directly applicable to interactions that are possible in the existing HoloLens menu environment (e.g., by using the multimedia player) and because they leave at least part of the palm and fingers unadorned and thus free to be actuated by ultrasound waves (whereas, e.g., a clenched fist would occlude each operable area of the hand).

2.2 Procedure

The goal of the study was explained to each participant as follows: while wearing a Microsoft HoloLens, they would execute five basic menu-related tasks in AR by using bare-hand gestures in mid-air. For each gesture, they would ideate a mid-air haptic sensation that suits this gesture best and feels most 'logical' for that interaction. At the start of each session, participants were presented a set of ten distinct mid-air haptic sensations to (re)acquaint themselves with the different adjustable properties (i.e., location on the hand, shape of the pattern, duration, dynamics and single-point vs. multi-point feedback).

Using the five gestures described above, participants then completed a set of tasks in AR. For each gesture, participants were encouraged to think out loud, depict and describe what mid-air haptic sensation they would want it to be accompanied by. Sessions had a duration of approximately 45 min. Because wearing an AR headset for a prolonged time becomes cumbersome and considering the time required to elicit and explain in detail each mid-air haptic sensation, we deliberately kept the set of tasks relatively small in comparison to other elicitation studies. For each participant, the entire process was video and audio recorded. In addition, the study moderator took notes; inquired for clarification, and reminded participants to go over all the adjustable properties that constitute a mid-air haptic pattern (cfr. supra).

Naturally, the HoloLens is by default unable to detect Piumsomboon's gestures and respond to them. This is why during each session, the proprietary HoloLens gestures were always visited first. After having performed the two 'functional' HoloLens gestures and having ideated a matching mid-air haptic sensation for them, participants were able to more easily perform the 'non-functional' gestures from Piumsomboon's user-defined set and imagine them having the targeted effect. No participants indicated having difficulties with this.

As to not curb creativity, we emphasized that technical limitations and feasibility did not have to be considered. Participants were told to simply imagine what they would want to feel and not worry about how the mid-air haptic sensation would be emitted on the hand. For example, we explained that the mid-air haptics could be emitted from the HoloLens itself (as in a similar setup presented by [8]) rather than from a unit standing on the table.

3 Analysis

All elicited sensations were analyzed and categorized based on the researcher transcripts and audiovisual recordings made during the elicitation phase. Because we allowed participants a high degree of freedom, often minor variations on similar ideas and patterns were proposed. Rather than discriminating sensations from each other for each identifiable objective inconsistency, we assessed and categorized them on a conceptual level. For example, to summon the 'home menu tile' in AR by making a 'bloom' gesture, 11/24 participants suggested a single short mid-air haptic sensation felt on the palm of the hand right after finishing the movement. Whether this sensation had the shape of, e.g., a square or a circle was not deemed imperative to the conceptualization and thus all 11 designs in which 'a short sensation on the palm to confirm that the gesture was well registered' was described, were placed in the same conceptual group.

As such, all 120 ideated sensations were classified based on their conceptual similarity. This resulted in 26 different conceptual groups, each representing a distinct conceptual model of mid-air haptic feedback. To understand the degree of consensus on the conceptual groups among participants, the revised agreement rate *AR* by Vatavu & Wobbrock [11] was calculated for each gesture:

$$AR(r) = \frac{|P|}{|P|-1} \sum_{P_i \subseteq P} \left(\frac{|P_i|}{|P|} \right)^2 - \frac{1}{|P|-1}$$

where P is the total amount of elicited sensations per task and Pi is the subset of sensations with a similar conceptual model for that task. To revisit 'bloom' as an example; the conceptual groups contained respectively 11, 8, 2, 2 and 1 user-elicited mid-air haptic design(s). The agreement score for this gesture is then calculated as follows:

$$AR_{bloom} = \left(\frac{|24|}{|23|}\right) \times \left(\frac{|11|}{|24|}\right)^2 + \left(\frac{|8|}{|24|}\right)^2 + \left(\frac{|2|}{|24|}\right)^2 + \left(\frac{|2|}{|24|}\right)^2 + \left(\frac{|1|}{|24|}\right)^2 - \left(\frac{|1|}{|23|}\right) = 0.308$$

4 Results

Each participant ideated one mid-air haptic sensation for each of the five gestures, resulting in a total of 120 sensations. Table 1 shows the agreement rate for each gesture/task.

Bloom. The bloom gesture, used to summon the home menu tile at any given moment in the HoloLens environment requires all 5 finger tips to be pressed together and then opened into a flat hand, palm facing up, with space between

Table 1. Consensus set of user-defined mid-air haptic sensations

Gesture	Bloom	Tap and hold	Swipe	Stop	Spin
Visual depiction (darker = stronger)	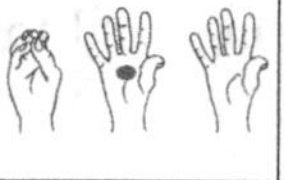	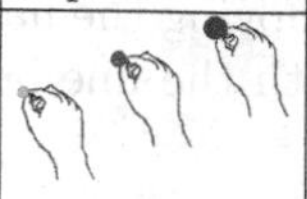	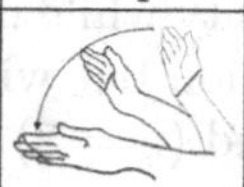	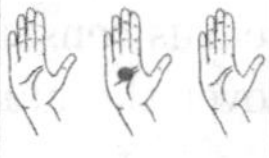	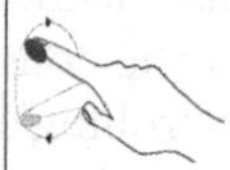
Agreement rate (AR)	0.308	0.308	0.228	0.576	0.297

each finger. Out of 24 participants, 11 suggested to receive a simple, short one-time feedback on the hand palm that immediately disappears again and confirms that the gesture has been registered well (AR = 0.308). The other conceptual groups were a) 'opening circle on the palm of the hand to mimic the menu being *spawned* from the hand (n = 8); b) 'constant sensation on the entire palm that remains present as long as the menu is opened' (n=1); c) 'horizontal line rolling from the bottom of the palm up to the fingertips' (n = 2); and d) 'mid-air haptic sensation on each fingertip that becomes sensible starting from the moment the fingertips are pressed together until the moment the gesture is completed (n = 2).

Tap and Hold. The 'tap and hold' gesture is used to scroll through pages and requires the user to pinch together their stretched-out index finger and thumb and maintain this posture in order for the scroll modus to remain activated. Users can then lift or lower the hand to move the page up and down. Scroll speed can be adjusted by moving the hand further up and down in respect to the original location. The conceptual sensation that was elicited most often here (n = 11, AR = 0.308) was 'a constant sensation on the tip of the thumb and index finger that remains sensible as long as the scroll functionality is active, with the intensity depending on the speed at which the user scrolls'. An almost similar design proposed by 8 other participants also suggested a constant sensation on the tips of the index finger and thumb. However, they made no notion of the intensity corresponding to the scrolling speed.

Swipe. To rotate through a carousel, HoloLens users normally have to point and click on elements in the far left or far right of that carousel. We, however, asked our participants to imagine that instead, they could use a full-hand 'swipe' to rotate the elements in the carousel. We clarified that this 'swipe' gesture was appropriated from Piumsomboon et al.'s user-defined gesture set for AR, and that it would not actually work with the HoloLens. Nonetheless, participants were asked to perform the gesture multiple times and imagine the carousel actually rotating. None of them indicated having trouble imagining this work and multiple participants spontaneously commented that this would indeed increase the ease of use. With an agreement rate of 0.228, the conceptual group on which there was most consensus contained the design of 10 participants who suggested to feel a 'vertical line that "rolls" over the entire length of the hand (i.e., from the bottom of the palm over to the fingertips) while making the gesture'. The two (out of five) other most popular conceptual groups were 'sensation (static) that decreased in intensity while moving the hand' (n = 4) and a concept similar to the consensus sensation but with the line being fixed to one location instead of 'rolling' over the hand (n = 5).

Stop. Using the HoloLens media player, participants watched a short movie and were asked to imagine they could pause it by briefly holding out a flat hand ('stop gesture') in front of them. This was the second non-functional gesture, yet again, none of the participants indicated having trouble to imagine it work. The consensus on elicited sensations for this gesture was 'very high' (AR = 0.576) [11]. Eighteen participants proposed 'a single short sensation to confirm that the gesture has been registered', similar to the consensus sensation for 'bloom'. The other 6 participants (n = 3, respectively) wanted to feel a) 'a sensation on the hand as long as the movie remained paused (as opposed to one short confirmation)'; and b) 'a sensation that enlarged and/or intensified while the hand was brought forward, to only be emitted at full strength when the arm was completely stretched out and the screen paused.

Spin. We asked participants to imagine that they could adjust the volume of the AR environment by spinning their stretched index finger clockwise (increase) or counter-clockwise (decrease). Seven different conceptual groups were needed to classify the heterogeneous ideas that were elicited for this gesture. The design on which there was most consensus (n = 12, AR = 0.297) was described as 'a constant sensation on the index fingertip or on the entire index finger, with intensity increasing or decreasing according to the volume'

5 Discussion

The presented sensations constitute a user-defined consensus set of mid-air haptic sensations that match gestures used to interact with an Augmented Reality menu environment. Despite the quasi-infinite range of possible mid-air haptic sensations and the idiosyncrasy of its features (timing, dynamics, location on the hand, intensity, ...), the agreement scores of our final five gestures (between 0.228 and 0.576) can be regarded as medium (0.100–0.300) to very high (>0.500) according to [11]. As the consensus set shows, the majority of participants seemed to prefer relatively simple and straightforward sensations to amplify their gestures. In addition to an inclination towards non-complex sensations, also noteworthy was how participants used mid-air haptic *intensity* in their elicited sensations. Depending on the functionality of the associated gesture, in some cases intensity was indeed a key feature. When the gesture actuated a discrete function (e.g., bloom and stop), a short monotone sensation was usually preferred, whereas more continuous menu-actions (e.g., scrolling through a page or adjusting volume) came with more continuous (i.e., changing intensity) sensations.

Administering the end-user elicitation method to generate what we propose to label as *intermediary referents* instead of *symbols*, shows promise but requires some remarks too. Our study suggests this method to be useful for the ideation of user-defined *mid-air haptic* sensations, however, for other sensory modalities it might not be as easy to have non-experts elicit novel designs and forms. Novices may not be familiar enough with the adjustable variables and properties of a sensorial modality in order to ideate variants of it. This contrasts with the end-user elicitation of, e.g., gestures, as people are apprehensive of their own physical possibilities (and limitations) and therefore inherently more capable of expressing themselves physically. Asking untrained participants to ideate, e.g., auditory intermediary referents (sound design), would assumedly require more training and/or facilitating tools. The think-aloud protocol that was used for our study, in combination with the available UltraHaptics kit on which simulations of elicitations could be depicted, was in our case proficient. However, when having end-users elicit other types of intermediary referents, we advise to assess well up front whether a modality is at all suitable for end-user elicitation and to think of the means or tools necessary to allow participants to elicit new variants of it.

Finally, one way to further validate the results of an elicitation study is what Ali et al. [1] describe as end-user *identification* studies. They are the conceptual inverse of elicitation studies: they present participants with a symbol and ask which referent would be invoked by it. In the case of intermediary referents, then, a control group would be asked to identify from a set of mid-air haptic sensations, which one suits a specific symbol (gesture) best. This would be a beneficial topic for follow-up research.

References

1. Ali, A.X., Morris, M.R., Wobbrock, J.O.: Crowdlicit: a system for conducting distributed end-user elicitation and identification studies. In: Proceedings of the 2019 CHI Conference on Human Factors in Computing Systems, CHI 2019, pp. 255:1–255:12. ACM, New York (2019). https://doi.org/10.1145/3290605.3300485, event-place: Glasgow, Scotland Uk
2. Carter, T., Seah, S.A., Long, B., Drinkwater, B., Subramanian, S.: UltraHaptics: multi-point mid-air haptic feedback for touch surfaces. In: Proceedings of the 26th Annual ACM Symposium on User Interface Software and Technology, UIST 2013, pp. 505–514. ACM, New York (2013). https://doi.org/10.1145/2501988.2502018, event-place: St. Andrews, Scotland, United Kingdom
3. Iwamoto, T., Tatezono, M., Hoshi, T., Shinoda, H.: Airborne ultrasound tactile display. In: ACM SIGGRAPH 2008 New Tech Demos. Los. ACM (2008)
4. Morris, M.R., Wobbrock, J.O., Wilson, A.D.: Understanding users' preferences for surface gestures. In: Proceedings of Graphics Interface 2010, GI 2010, pp. 261–268. Canadian Information Processing Society, Toronto, Ont., Canada (2010). http://dl.acm.org/citation.cfm?id=1839214.1839260, event-place: Ottawa, Ontario, Canada
5. Obrist, M., Subramanian, S., Gatti, E., Long, B., Carter, T.: Emotions mediated through mid-air haptics. In: Proceedings of the 33rd Annual ACM Conference on Human Factors in Computing Systems, CHI 2015, pp. 2053–2062. ACM, New York (2015). https://doi.org/10.1145/2702123.2702361, event-place: Seoul, Republic of Korea

6. Piumsomboon, T., Clark, A., Billinghurst, M., Cockburn, A.: User-defined gestures for augmented reality. In: CHI 2013 Extended Abstracts on Human Factors in Computing Systems, CHI EA 2013, pp. 955–960. ACM, New York (2013). https://doi.org/10.1145/2468356.2468527, event-place: Paris, France
7. Rodriguez, I.B., Marquardt, N.: Gesture elicitation study on how to opt-in & opt-out from interactions with public displays. In: Proceedings of the Interactive Surfaces and Spaces on ZZZ - ISS 2017, pp. 32–41. ACM Press, Brighton (2017). https://doi.org/10.1145/3132272.3134118, http://dl.acm.org/citation.cfm?doid=3132272.3134118
8. Sand, A., Rakkolainen, I., Isokoski, P., Kangas, J., Raisamo, R., Palovuori, K.: Head-mounted display with mid-air tactile feedback. In: Proceedings of the 21st ACM Symposium on Virtual Reality Software and Technology, VRST 2015, pp. 51–58. ACM, New York (2015). https://doi.org/10.1145/2821592.2821593, event-place: Beijing, China
9. Tung, Y.C., et al.: User-defined game input for smart glasses in public space. In: Proceedings of the 33rd Annual ACM Conference on Human Factors in Computing Systems, CHI 2015, pp. 3327–3336. ACM, New York (2015). https://doi.org/10.1145/2702123.2702214, event-place: Seoul, Republic of Korea
10. Vatavu, R.D.: A comparative study of user-defined handheld vs. freehand gestures for home entertainment environments. J. Ambient Intell. Smart Environ. **5**(2), 187–211 (2013). http://dl.acm.org/citation.cfm?id=2594684.2594688
11. Vatavu, R.D., Wobbrock, J.O.: Formalizing agreement analysis for elicitation studies: new measures, significance test, and toolkit. In: Proceedings of the 33rd Annual ACM Conference on Human Factors in Computing Systems, CHI 2015, pp. 1325–1334. ACM, New York (2015). https://doi.org/10.1145/2702123.2702223, event-place: Seoul, Republic of Korea
12. Wobbrock, J.O., Aung, H.H., Rothrock, B., Myers, B.A.: Maximizing the guessability of symbolic input. In: CHI 2005 Extended Abstracts on Human Factors in Computing Systems, CHI EA 2005, pp. 1869–1872. ACM, New York (2005). https://doi.org/10.1145/1056808.1057043, event-place: Portland, OR, USA
13. Wobbrock, J.O., Morris, M.R., Wilson, A.D.: User-defined gestures for surface computing. In: Proceedings of the SIGCHI Conference on Human Factors in Computing Systems, CHI 2009, pp. 1083–1092. ACM, New York (2009). https://doi.org/10.1145/1518701.1518866, event-place: Boston, MA, USA

Surface Roughness Judgment During Finger Exploration Is Changeable by Visual Oscillations

Yusuke Ota[1], Yusuke Ujitoko[1,2](✉), Yuki Ban[3], Sho Sakurai[1], and Koichi Hirota[1]

[1] The University of Electro-Communications, Tokyo, Japan
yusuke.ujitoko@gmail.com
[2] Research & Development Group, Hitachi, Ltd., Yokohama, Japan
[3] The University of Tokyo, Tokyo, Japan

Abstract. Previously, we proposed a pseudo-haptic method of changing the perceived roughness of virtual textured surfaces represented by vibrational feedback during pen-surface interaction. This study extends the method to finger-surface interactions in order to change the roughness judgment of real surface. Users watched computer-generated visual oscillations of the contact point while exploring the texture using bare fingers. The user study showed that our method could modify roughness judgment of real textured surfaces.

Keywords: Human computer interaction · Haptic interface

1 Introduction

Tactile texture rendering of virtual surfaces is important for providing the user with a highly realistic interaction experience with virtual objects. Such realistic tactile texture rendering of virtual object surfaces usually require some kind of electromechanical devices.

Recently, pseudo-haptics has attracted attention as a light-weight haptic presentation approach. The pseudo-haptic sensation [3,7] is produced by an appropriate sensory inconsistency between the physical movement of the body and the observed movement of a virtual pointer [4]. As for texture perception, some studies attempted to generate texture perception using the pseudo-haptic effects without sophisticated haptic devices [1,2,5,9,10]. Previously, we proposed a pseudo-haptic method for modulating the vibrotactile roughness of virtual surfaces during pen-surface interactions, and our user study showed the effectiveness of the proposed method [9]. In the method, users watched a visually perturbed contact point. However, since the proposed method assumed pen-surface interactions, users could not sense temperature or friction of the virtual surface.

Electronic supplementary material The online version of this chapter (https://doi.org/10.1007/978-3-030-58147-3_4) contains supplementary material, which is available to authorized users.

I. Nisky et al. (Eds.): EuroHaptics 2020, LNCS 12272, pp. 33–41, 2020.
https://doi.org/10.1007/978-3-030-58147-3_4

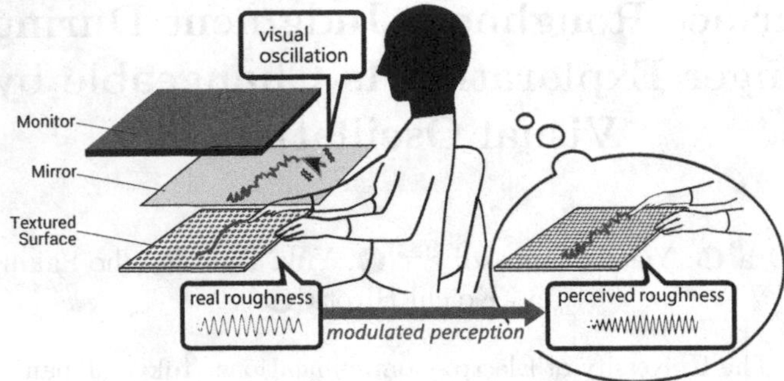

Fig. 1. We propose a method for modulating the roughness perception of real textured surfaces touched by the users using their bare fingers.

In the present study, we attempt to apply the same method to finger-surface interactions. In that environment, the users watch the visually perturbed contact point while exploring real surfaces with their bare fingers (see Fig. 1). The research question of this research is to investigate whether the method can change the roughness judgment of the real surface during finger-surface interaction.

2 Concept

2.1 Pseudo-haptic Method Using Visual Oscillations

This study adopted the same pseudo-haptic method using visual oscillation as [9]. Please see details there. The pointer translation is calculated as $X_{vis} = X_{origin} + C * \alpha * random(-1, 1) * abs(V)$ and $Y_{vis} = Y_{origin} + C * \alpha * random(-1, 1) * abs(V)$.

Here, X_{origin}, Y_{origin} are the original pointer position along the x-axis and y-axis. X_{vis}, Y_{vis} is the translated pointer position along the x-axis and y-axis, which are visualized on screen. The translation amount is obtained as a product of α, a random value sampled from the standard normal distribution, and the absolute value of the device's velocity V. C is a constant. α is the coefficient unit, which means the "size of the visual oscillation."

3 User Study 1

The first user study investigated whether visual oscillations modified the roughness judgment during finger-surface interactions. Ten participants, aged from 22 to 25 participated in the experiment. All participants were right-handed.

3.1 Experimental System

We manufactured texture plates by arranging glass particles (KENIS, Ltd) on acrylic plates following [6,8]'s approach. Glass particles of different sizes were carefully attached to a flat acrylic plate with a double-sided sticky tape. In this study, we prepared nine texture plates with uniform surfaces, shown in Fig. 2. The diameters of the glass particles were 50, 75.5, 115, 163, 210, 425, 605, 850, and 1194 μm.

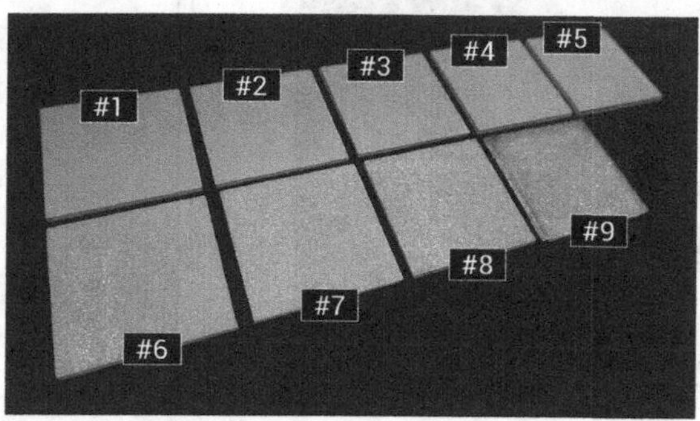

Fig. 2. Nine texture plates covered with glass particles. The diameters of the glass particles were different for different plates. The length and width of the plate were 100 mm.

Figure 3 (a) shows the experimental environment. One of the texture plates was placed about 30 cm in front of the participants. The texture plate was unseen directly from the participants, owing to a mirror. The position sensor of the tracker (Polhemus Fastrak) was attached to the participants' index fingers.

The position of the index finger was updated at 120 Hz and sent to a personal computer (PC). Unity was used for visualizing the position of a finger, according to the sensing position at 60 Hz. The contact point was visualized as a black circle, and the circle diameter was three pixels, which was equivalent to 0.93 mm on display (ASUS VG278HE, 1920 × 1080, 81.589 dpi). From the viewpoint of the participant, the pointer appeared to be positioned as the same position as their index finger. The participants touched the texture plates with their index fingers, while watching two separate areas on the screen, and judged which area felt rougher (Fig. 3 (b)). The haptic information about both areas, acquired by touching the texture plate, was the same, since the glass particles' diameters across the texture plate were the same. The width and height of the texture area displayed were adjusted to be the same as those of the real texture plate.

The pointer oscillated, and the position of the pointer was updated using the expression in section II. We conducted preliminary experiment to narrow down the α value where the pseudo-haptics were effective to some extent. According to the preliminary experiment, the user study used three different α values ($\alpha = 1, 1.5, 2$) to investigate the effect of size of α values. We refer to the three different α values as different "visual conditions". C was configured so that the translation distance was 4.82 pixels (1.5 mm) when the pointer moved at 200 mm/s and α was 1. C was constant for all three α.

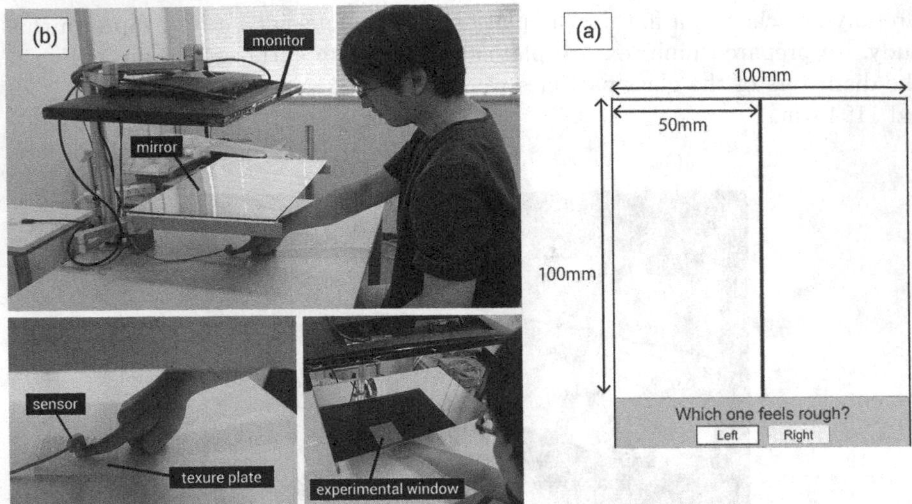

Fig. 3. (a) Environment of the user study. Top: Overview. Bottom Left: The sensor was attached to the participants' index fingers. Bottom Right: Participants watched their finger movement through pointer on mirrored display. (b) Experimental window. There were left and right side.

We refer to the nine texture plates (from #1 to #9) as different "texture conditions".

3.2 Task Design

This experiment used a within-participants design. The participants touched a texture plate under the mirror, watching the pointer in the left or right area on display. In One area, the pointer oscillated while in the other area, the pointer did not oscillated.

The assignment of the visual oscillation to the specific area (left or right) was randomly determined. After freely touching the plate, the participants were asked to report which area they felt to be rougher, by pushing the button. Since we wanted to reproduce the same situation as when touching a surface as usual, we did not define the time limit for each trial. We told participants to focus on the haptic roughness of the plates while watching the pointer on the screen and asked them to judge the roughness based on the haptic roughness instead of visual information. We allowed participants only to slide the pen on the surface and prohibited tapping or flicking gesture with pen. Also, participants were told that the texture plates were homogeneous.

There were nine texture conditions (from #1 to #9) and three visual conditions ($\alpha = 1, 1.5, 2$). For each condition, the participants repeated the trial four times. Thus, each participant conducted 108 trials ($= 9 \times 3 \times 4$). The texture and visual conditions were presented in the random order, and were counterbalanced across the participants.

3.3 Result and Discussion

First of all, to determine whether the participants' answers were dissimilar from chance (50%), a chi-square goodness-of-fit test was performed on the frequency of the participants' selections for two texture plates, against the null hypothesis that the two areas were equally selected. Figure 4 shows the number of times and a percentage a selection was made for the areas with visual oscillation for each visual and texture condition. Out of the 40 selections (10 participants multiplied by 4 times) under each condition for the question, the difference was observed to be statistically significant ($p < 0.05$) for all conditions, except only one condition ($\alpha = 1$ and texture #1, $p = 0.057$). The result of chi-square test and the biased selection to the areas with visual oscillation implied that the participants reported that the vibrotactile texture felt rougher compared with the case in which there was no visual oscillation.

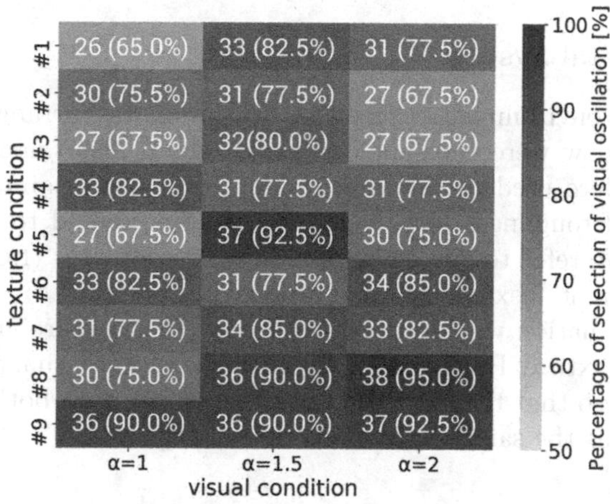

Fig. 4. Matrix along texture condition axis and visual condition axis on the frequency and percentage of selection for the areas with visual oscillation.

Next, we conducted an analysis using the generalized linear model with the binomial distribution. We compared the full model and each of the three reduced models which had variables of (1) only visual condition or (2) texture condition or (3) visual and texture condition but without interaction effect. Then, we conducted a likelihood ratio test (L.R.T) against the null hypothesis that each reduced model's deviation and the full model's deviation were same. We generated it using the chi-square test which is equivalent to L.R.T when performing logistic regression.

According to the result, the reduced model without texture condition was significantly different from the full model ($p < 0.01$) and thus, the texture condition significantly affected the response. On the other hand, the reduced model without visual condition or interaction effect was not different from the full model ($p > 0.05$). Thus, visual condition and interaction effect was not shown to affect the response.

The result of this user study suggests that visual oscillations may change the roughness judgment of real textures. However, under the experimental conditions, the participants may have answered only the presence or absence of the visual oscillations. Therefore, we conducted a user study 2 to evaluate the physical equivalent point from the viewpoint of tactile roughness, and indirectly confirmed the effect of visual oscillations.

4 User Study 2

The second user study investigated what size of visual oscillation is optimal to provide the desired, modified roughness. Six participants, aged from 22 to 25 participated in the experiment. All participants were right-handed.

4.1 Experimental System

The experimental environments except for texture plates' arrangement and the experimental window were the same as user study 1. The same texture plates from #1 to #9 were used in this user study. The closest pairs of two texture plates in terms of roughness (e.g. [#1, #2]) were presented to participants at the same time. We refer to the eight pairs of texture plates (from [#1, #2] to [#8, #9]) as different "texture pair conditions".

The visual oscillation was disabled for the rougher texture and was enabled for the smoother texture. Participants adjusted the size of visual oscillations α of smoother texture so that they felt the same roughness from both texture plate. The value of C was the same as previous user study.

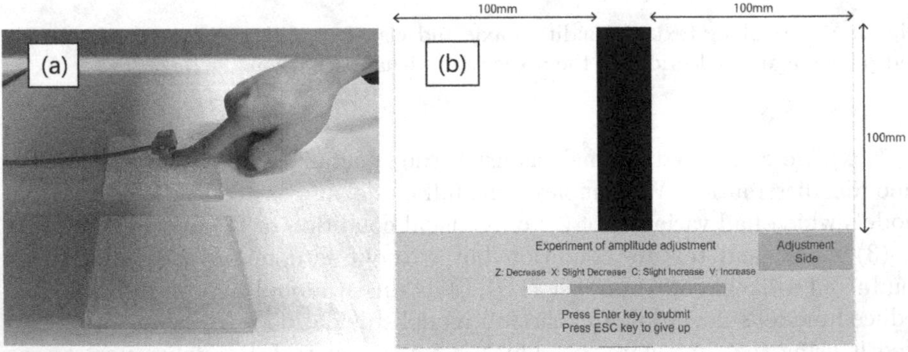

Fig. 5. (a) Participants touch two different rough texture plates. (b) Experimental window. There were left and right side, each corresponding to two plates.

4.2 Task Design

This experiment used a within-participants design. The participants touched two texture plates under the mirror. One of the two plates was rougher and another was smoother. Though the assignment of the left or right to the plates was randomly determined, which side was the target of adjustment as illustrated in the experimental window. The pointer visually oscillated when the smoother plate was touched, but the pointer did not oscillate when the rougher plate was touched. The participants were asked to adjust the size of the visual oscillations α of the smoother plate so that they felt the same roughness from both the texture plates. Participants pushed a button from "decrease (−0.04)", "slight decrease (−0.02)", "slight increase (+0.02)", or "increase (+0.04)". The initial value of α was set at 0. There was no time limit for adjustment and participants could push buttons repeatedly.

There were eight texture pair conditions. For each condition, the participants repeated the trial twice. The texture pair conditions were presented in the random order.

4.3 Result and Discussion

Figure 6 shows the results. The horizontal axis shows the smoother texture plate of texture pairs. The vertical axis shows the adjusted size of visual oscillation for the smoother texture of texture pairs.

If the visual oscillation does not affect the roughness, the adjustment value should be zero. To determine whether the adjustment value is different from zero, we conducted t-tests for each texture pair condition. We conducted the normality by the Shapiro-Wilk test for each texture pair condition in advance. The results of the t-test for all shows that the adjusted size of visual oscillation was significantly different from zero ($p < 0.05$) and adjusted values ranged from 0.8 to 1.5.

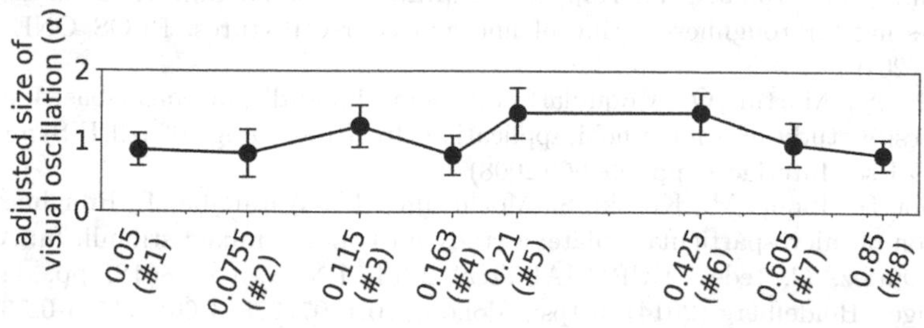

Fig. 6. The adjusted visual oscillation size for smoother texture plate for equalizing roughness.

Next, we performed a one-way ANOVA with factors of texture pair condition on the adjusted size. We conducted a Mauchly's test to check the sphericity criteria in advance of the ANOVA test. According to the ANOVA results, the texture pair condition did not significantly affect the adjusted size ($F(7, 84) = 2.12, p = 0.61$).

5 Conclusion

In this study, we applied the pseudo-haptic method using visual oscillations, to modify the subjective roughness judgment while scanning real textured plates with bare fingers. We manufactured various texture plates, with roughness ranging from several tens of micrometers to several millimeters. We conducted user study where users judged the roughness of plate under conditions with and without visual oscillations. The results showed that visual oscillations can modify the perceived roughness of real textured surfaces.

References

1. Costes, A., Argelaguet, F., Danieau, F., Guillotel, P., Lécuyer, A.: Touchy: a visual approach for simulating haptic effects on touchscreens. Front. ICT **6**, 1 (2019)
2. Hachisu, T., Cirio, G., Marchal, M., Lécuyer, A., Kajimoto, H.: Virtual chromatic percussions simulated by pseudo-haptic and vibrotactile feedback. In: Proceedings of the 8th International Conference on Advances in Computer Entertainment Technology (20), pp. 1–5 (2011)
3. Lécuyer, A.: Simulating haptic feedback using vision: a survey of research and applications of pseudo-haptic feedback. Presence **18**, 39–53 (2009)
4. Lecuyer, A., Coquillart, S., Kheddar, A., Richard, P., Coiffet, P.: Pseudo-haptic feedback: can isometric input devices simulate force feedback? In: Proceedings of IEEE Virtual Reality, pp. 83–90 (2000)
5. Lécuyer, A., Burkhardt, J.M., Etienne, L.: Feeling bumps and holes without a haptic interface: the perception of pseudo-haptic textures. In: Proceedings of the SIGCHI Conference on Human Factors in Computing Systems, vol. 6, pp. 239–246 (2004)
6. Natsume, M., Tanaka, Y., Kappers, A.M.L.: Individual differences in cognitive processing for roughness rating of fine and coarse textures. PLOS ONE **14**(1), 1–16 (2019)
7. Pusch, A., Martin, O., Coquillart, S.: Hemp-hand-displacement-based pseudo-haptics: a study of a force field application. In: Proceedings of IEEE Symposium on 3D User Interfaces, pp. 59–66 (2008)
8. Tsuboi, H., Inoue, M., Kuroki, S., Mochiyama, H., Watanabe, J.: Roughness perception of micro-particulate plate: a study on two-size-mixed stimuli. In: Auvray, M., Duriez, C. (eds.) EUROHAPTICS 2014. LNCS, vol. 8618, pp. 446–452. Springer, Heidelberg (2014). https://doi.org/10.1007/978-3-662-44193-0_56
9. Ujitoko, Y., Ban, Y., Hirota, K.: Modulating fine roughness perception of vibrotactile textured surface using pseudo-haptic effect. IEEE Trans. Vis. Comput. Graph. **25**(5), 1981–1990 (2019)
10. Ujitoko, Y., Ban, Y., Hirota, K.: Presenting static friction sensation at stick-slip transition using pseudo-haptic effect. In: Proceedings of IEEE World Haptics Conference, pp. 181–186 (2019)

Identifying Tactors Locations on the Proximal Phalanx of the Finger for Navigation

Justine Saint-Aubert(✉)

Paris, France

Abstract. Vibrotactile stimulation has been investigated to provide navigational cues through belts, vests, wrist-bands and exotic displays. A more compact solution would be the use of a ring type display. In order to test its feasibility, a user-centered experiment is conducted. The ability of participants to identify cardinal directions and inter cardinal directions by vibrotactile stimulation on the proximal phalanx is investigated. The results indicate that participants achieved 96% accuracy for cardinal directions and 69% accuracy for cardinal plus inter-cardinal directions using a static stimulation. The identification rates of dynamic stimulation are lower than that of static stimulation.

Keywords: Vibrotactile displays · Perception · Navigation · Ring device

1 Introduction

1.1 Navigation Using Tactile Stimulation

Navigation refers to the process of finding a way from one place to another. It implies asserting one's position in the environment and planning the next direction. In the following, we take an interest in the planning step only.

We usually rely on visual feedback in order to navigate but holding a map or a mobile phone while walking affect the awareness of the pedestrian [13]. Tactile has shown benefits in substituting vision since information can be conveyed to users without altering their perceptions of the surrounding environment [22]. Among tactile cues, vibrations are one of the most convenient. They can be generated using small actuators (tactors), resulting in easy-to-carry devices. Vibratory displays have shown their efficiency for in-vehicle and pedestrian navigation (e.g [3,10]), but are hardly used in everyday life.

Among existing solutions, hand-held devices have been proposed. Mobile phones that integrate one tactor are the most frequently used and handles composed of an array of tactors have also been advanced [4] but they prevent manual interactions and can induce fatigue. In a different strategy, hand-free systems have then been developed. Vibrotactile matrix have been integrated in gloves

I. Nisky et al. (Eds.): EuroHaptics 2020, LNCS 12272, pp. 42–50, 2020.
https://doi.org/10.1007/978-3-030-58147-3_5

[21], body trunk systems (e.g [6]), anklets [2], forearm systems [16] and others. All these solutions have shown to support navigational tasks but are unlikely to be used for everyday life. Furthermore, the body trunk and the head are considered as personal areas and users acceptance can be an issue. An alternative can be found in wristbands displays. Prototypes that include one or two tactors [1,14] or watch with 6 tactors [19] have been proposed[1]. The wrist is however one of the less sensitive part of the body [23] and a more compact solution can be suggested through a ring device.

A ring display would be easy to carry, discreet and versatile since it can be adapted to any finger. The finger is a socially established part so user acceptance should not be an issue. Finally, it implies in part a tactile stimulation on the glabrous skin that is higly sensitive to vibration [18]. The solution then seems promising but the feasibility of such a device has never been investigated.

1.2 Feasibility of a Ring Device

A ring display would be made of tactors placed all around the finger and the location of the activated tactors interpreted as directional information. In order to be used for navigation, users have then to correctly localize the tactors, which is not guaranteed. The stimulation would be on the proximal phalanx of the finger and at this level the glabrous skin of the palmar surface is next to the less sensitive skin of the dorsal surface, the perception of vibrations could then be altered. The tactors would also be close from each other while localization rates are better with greater inter-disances (e.g [15,23]).

These drawbacks could be compensated by the spatial repartition of the tactors on the finger that makes information map body coordinates. A dynamic strategy could also be exploited. While a single tactor is activated during static stimulation, several ones are sequentially triggered during dynamic stimulation and recognition rates can be greatly improved using this strategy [20]. To test these hypotheses, a psycho-physical experiment is conducted. Four tactors are placed on the proximal phalanx of participants in order to map body coordinates. Their ability to identify directions using static and dynamic stimulation studied. The identification of cardinal directions is investigated (1). The identification of cardinal plus inter-cardinal directions is also examined (2).

2 Psychophysical Experiment

2.1 Set up

Four pancake tactors (*GoTronic* VM834, ⌀ 8 mm, 3 V, 1200 rpm) are glued at ring level in the middle of the dorsal, volar and side surfaces of the index finger (Fig. 1 (a)). They are activated with a *Arduino Uno* card[2] controlled by a C++ program running on a Linux computer. The overall frequency is up to 3.5 kHz.

[1] The latter is not specifically dedicated to navigation but it is a potential application.
[2] https://store.arduino.cc/arduino-uno-rev3.

Signals sent to tactors are 300 ms pulse at 3 V. The resulting vibration is about 200 Hz in the range of optimal sensitivity of Pacinian corpuscles (200 Hz $\leftrightarrow$ 250 Hz) [11]. The tactors are fixed on the non-dominant hand of the participants, leaving the other hand free to provide feedback via a computer located on a table in front of them. The hand with tactors is placed palm down and the index finger raised using foam in order to prevent tactors/table and tactors/skin contact. The complete set up is shown in Fig. 1 (b).

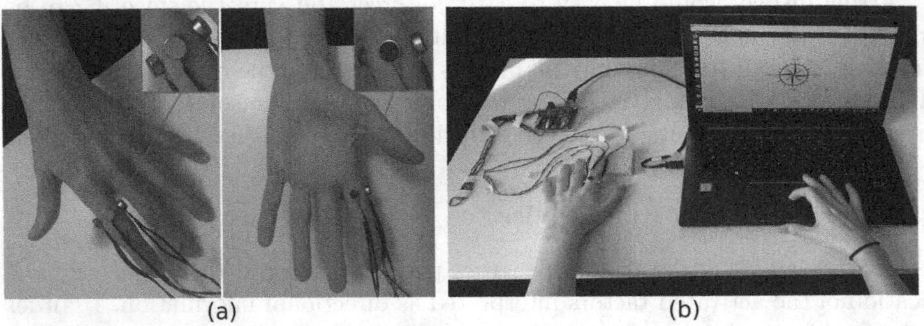

Fig. 1. (a) Pictures and close ups of tactors fixed on the proximal phalanx of the index finger. (b) Picture of the complete set up (tactors and computer) with a particpant.

2.2 Method

A compass and buttons corresponding to directions are displayed on the computer. Vibro-tactile stimulations are transmitted to the participants who have to identify the encoded directions, using a touch-pad to make responses.

During a session, the order of the stimulations is randomly predefined so it is the same between participants. Stimuli are tested in a row and no breaks are allowed. Once a participant gives an answer, no feedback is provided and the next stimulus appears after 3 s. Pink noise is played to them through headphones and they are not blindfolded but are asked to focus on the computer screen.

2.3 Participants

Eight participants (2 males), ages 27 to 43 (mean age 31 years ± 5.8), volunteer. They are selected because they do not report any visual impairment or physical issues. Seven of them are right-handed according to Coren's handedness test [5].

3 Identifications of Cardinal and Inter-cardinal Directions

3.1 Static and Dynamic Stimulations of Cardinal Directions

The ability of the participants to identify cardinal directions is examined. In a first session, static stimulation is tested. The tactors and the compass keep the same orientation (Fig. 2 (a)). The activation of the tactor on the dorsal surface of the finger corresponds to a North command, the tactor on the volar surface describes the South, a tactor on the side encodes the East and so on (Fig. 2 (b)).

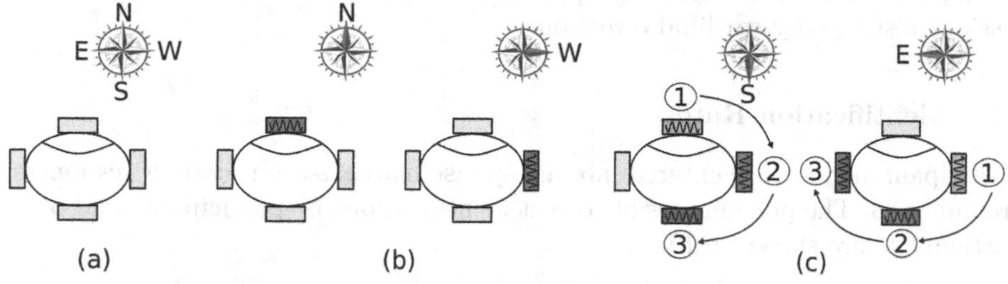

Fig. 2. (a) Relative directions between the index finger from face and the compass. Static (b) and dynamic (c) stimulation of cardinal directions. Sinusoids represent activated tactors and red color apparent locations of vibrations. Numbers and shades of red indicate a sequence of activation. (Color figure online)

In another session, dynamic stimulation is investigated. The tactors and the compass remains in the same positions, however three tactors are activated sequentially. For instance, the North is encoded by the activation of the tactor on the volar surface, the side and finally on the dorsal surface. The same scheme, always clockwise is chosen for the other directions (Fig. 2 (b)).

All the participants performed both sessions using a counterbalanced design in order to control potential effects of the order. During each session, each cardinal direction is tested 30 times for a total of 120 trials.

3.2 Static and Dynamic Stimulations with Inter-cardinal Directions

The ability of participants to identify inter-cardinal directions is investigated.

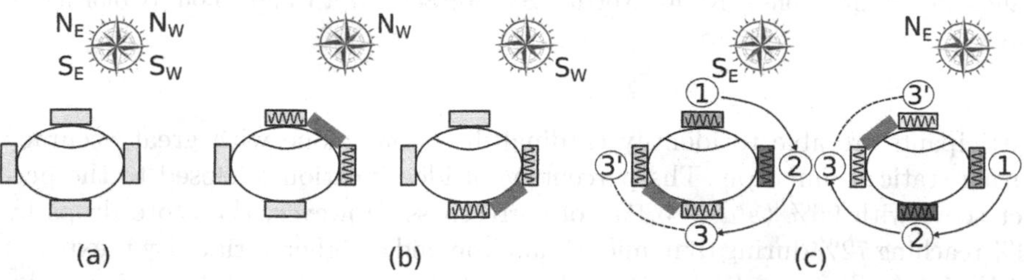

Fig. 3. Stimulations of inter-cardinal directions. See the caption of Fig. 2 for details.

While the tactors are only located at cardinal locations, inter-cardinal directions are simulated by activating two tactors simultaneously. The vibrations should be perceived in the middle [12]. Static and dynamic stimulations of inter cardinal directions using such dual tactors activations are explained in Fig. 3.

All participants tested static and dynamic stimulation. A counterbalanced design is again employed. During each session, each stimulation is tested 30 times for a total of 240 trials. This part of the experiment is conducted after the sessions testing only cardinal directions.

3.3 Identification Rates

Participant answers are entered into a response matrix so that correct responses are analysed. The percentages of correct identifications by participants and over participants are shown in Fig. 4.

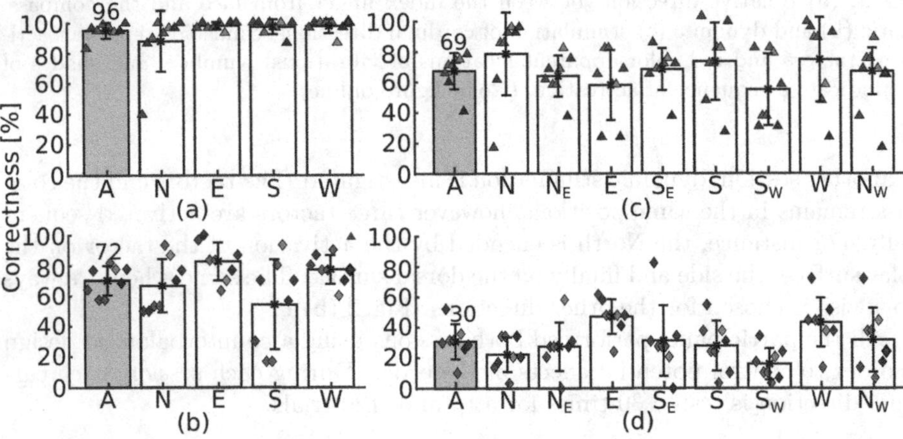

Fig. 4. Correct identifications of cardinal directions during static (a) and dynamic (b) stimumaltions, and of cardinal/inter-cardinal directions during static (c) and dynamic (d) ones. Individual results are exhibited by markers using one color by participant. Bars and errorbars represent the means and standard deviations over participants. "A" stands for All directions, "N" for North, "S_E" for South-East and so on. (Color figure online)

Participants are able to identify cardinal directions alone with great accuracy during static stimulation. The percentage of identification is closed to the perfect score with 96% (*std* = 5.1%) of correctness. However, the score drops by 24% reaching 72% during dynamic stimulation and a higher variability is exposed (*std* = 5.1%). To test whether the difference between static and dynamic simulations is significant, a non-parametric paired Wilcoxon signed-rank test (percentages don't follow a normal distribution) is used. The mean percentages correct over all directions are compared. The null hypothesis states that there is no difference between static and dynamic stimulations is rejected ($p = 0.008$). The alternative hypothesis then supports that the type of stimulations has an effect on correct identifications.

Other results are related to the tests of cardinal directions plus inter-cardinal ones. In static stimulation, the global identification is uncertain with an overall score of 69% and variability ($std = 11.5\%$). Inter-cardinal directions are not well recognized. In the dynamic case, the identification rate is even lower with only 30% ($std = 11.2\%$) of correct identifications. The statistical analysis is performed on these data and the null hypothesis is rejected ($p = 0.012$), again supporting that the type of stimulations has an effect on correct identifications.

4 Discussions

4.1 Cardinal Identifications

Following the results of the experiment, the feasibility of a ring finger display for navigation is discussed. Cardinal directions can be transmitted efficiently using a static stimulation as shown by identification rates closed to the perfect score. A ring display could then be employed to navigate in cities that apply the Hippodamian model (or grid plan) and the tactile feedback will not need to be supplied by vision. Identification rates can reach perfect score during more casual simulations as it is the case during everyday navigation or by using a repeated command strategy.

The feasibility of the system should however be examined during a navigation task since conditions are likely to differ. For instance, results have been obtained while the hand orientation remains constant. According to past studies, wrist motions have a little negative effect on direction identification [17] and while they affect a little bit the detection, factors such as intensity can be adjusted to compensate [9]. Mobility should then not be an issue. However as phalanxes have a higher mobility than the wrist, this point should be further explored.

4.2 Inter-cardinal Identifications

The use of a ring display in places that do not apply Hippodamian model is however challenging since the transmissions of inter-cardinals directions are approximate. Identification rates of cardinals directions are even lower when inter-cardinal are displayed than when they are not. Participants have then difficulty to differentiate the activation of one and two tactors. Additional tactors placed at inter-cardinal locations instead of dual tactors activation should lead to the same results, or even wors. The higher the number of tactors, the harder the identification [8], especially since they will no longer be on noticeable loci.

The use of a dynamic pattern does not help. Results for dynamic simulations are associated with poorer identification rates than static simulations in both part of the experiment. The simultaneous activation of three tactors was confusing for participants and this issue may results from the short inter-distance between tactors or a wrong choice of pattern. This pattern was chosen as a compromise between long motion and limited activation of tactors. Based on the results, different patterns should be explored. In the same way different type of dynamic stimulation have been tested on other body parts (e.g the back [7]), the same should be done on the phalanx.

An alternative solution could be to exploit temporal pattern. The identifications of cardinal directions only during static stimulation are accurate and correspond to discrete feedback. Cardinal directions displayed with short breaks could then encode inter-cardinal directions. North-East would correspond to the activation of the North, a few milliseconds break, then the activation of the East only. This solution could also answer the variability issue. Indeed dynamic patterns and direct presentations of inter-cardinal directions both increase between-subject variability. This issue can be raised because of different phalanx dimensions or type of skin that may have a detrimental effect on vibrations propagation. The static stimulation however minimizes the issue.

5 Conclusions and Perspectives

The feasibility of a ring vibro-tactile display for navigation have been investigated in a user-centered experiment. Participants were able to identify accurately cardinal directions transmitted by static stimulation of four tactors at ring level. The transmission of inter-cardinal directions was more challenging and a dynamic stimulation was not beneficial. A research will work towards the test of patterns that could be recognized accurately as inter-cardinal directions.

References

1. Alarcon, E., Ferrise, F.: Design of a wearable haptic navigation tool for cyclists. In: 2017 International Conference on Innovative Design and Manufacturing, pp. 1–6 (2017)
2. Baldi, T.L., Paolocci, G., Barcelli, D., Prattichizzo, D.: Wearable haptics for remote social walking. arXiv preprint arXiv:2001.03899 (2020)
3. Baldi, T.L., Scheggi, S., Aggravi, M., Prattichizzo, D.: Haptic guidance in dynamic environments using optimal reciprocal collision avoidance. IEEE Robot. Autom. Lett. **3**(1), 265–272 (2017)
4. Bouzit, M., Chaibi, A., De Laurentis, K., Mavroidis, C.: Tactile feedback navigation handle for the visually impaired. In: ASME 2004 International Mechanical Engineering Congress and Exposition, pp. 1171–1177. American Society of Mechanical Engineers Digital Collection (2004)
5. Coren, S.: The left-hander syndrome: the causes and consequences of left-handedness, Vintage (1993)
6. Elliott, L.R., van Erp, J., Redden, E.S., Duistermaat, M.: Field-based validation of a tactile navigation device. IEEE Trans. Haptics **3**(2), 78–87 (2010)
7. Jones, L.A., Kunkel, J., Torres, E.: Tactile vocabulary for tactile displays. In: Second Joint EuroHaptics Conference and Symposium on Haptic Interfaces for Virtual Environment and Teleoperator Systems (WHC 2007), pp. 574–575. IEEE (2007)

8. Jones, L.A., Ray, K.: Localization and pattern recognition with tactile displays. In: 2008 Symposium on Haptic Interfaces for Virtual Environment and Teleoperator Systems, pp. 33–39. IEEE (2008)
9. Karuei, I., MacLean, K.E., Foley-Fisher, Z., MacKenzie, R., Koch, S., El-Zohairy, M.: Detecting vibrations across the body in mobile contexts. In: Proceedings of the SIGCHI Conference on Human Factors in Computing Systems, pp. 3267–3276. ACM (2011)
10. Kiss, F., Boldt, R., Pfleging, B., Schneegass, S.: Navigation systems for motorcyclists: exploring wearable tactile feedback for route guidance in the real world. In: Proceedings of the 2018 CHI Conference on Human Factors in Computing Systems, p. 617. ACM (2018)
11. Konietzny, F., Hensel, H.: Response of rapidly and slowly adapting mechanoreceptors and vibratory sensitivity in human hairy skin. Pflügers Archiv **368**(1–2), 39–44 (1977)
12. Lederman, S.J., Jones, L.A.: Tactile and haptic illusions. IEEE Trans. Haptics **4**(4), 273–294 (2011)
13. Madden, M., Rainie, L.: Adults and cell phone distractions (2010)
14. Ng, G., Barralon, P., Dumont, G., Schwarz, S.K., Ansermino, J.M.: Optimizing the tactile display of physiological information: vibro-tactile vs. electro-tactile stimulation, and forearm or wrist location. In: 2007 29th Annual International Conference of the IEEE Engineering in Medicine and Biology Society, pp. 4202–4205. IEEE (2007)
15. Oakley, I., Kim, Y., Lee, J., Ryu, J.: Determining the feasibility of forearm mounted vibrotactile displays. In: 2006 14th Symposium on Haptic Interfaces for Virtual Environment and Teleoperator Systems, pp. 27–34. IEEE (2005)
16. Orso, V., et al.: Follow the vibes: a comparison between two tactile displays in a navigation task in the field. PsychNol. J. **14**(1), 61–79 (2016)
17. Panëels, S., Brunet, L., Strachan, S.: Strike a pose: directional cueing on the wrist and the effect of orientation. In: Oakley, I., Brewster, S. (eds.) HAID 2013. LNCS, vol. 7989, pp. 117–126. Springer, Heidelberg (2013). https://doi.org/10.1007/978-3-642-41068-0_13
18. Pasterkamp, E.: Mechanoreceptors in the glabrous skin of the human hand. Arch. Physiol. Biochem. **107**(4), 338–341 (1999)
19. Pezent, E., et al.: Tasbi: multisensory squeeze and vibrotactile wrist haptics for augmented and virtual reality. In: 2019 IEEE World Haptics Conference (WHC), pp. 1–6. IEEE (2019)
20. Piateski, E., Jones, L.: Vibrotactile pattern recognition on the arm and torso. In: First Joint Eurohaptics Conference and Symposium on Haptic Interfaces for Virtual Environment and Teleoperator Systems, World Haptics Conference, pp. 90–95. IEEE (2005)
21. Roberts, O., et al.: Research magazine summer 2003-focus on innovation (2003)
22. Ross, D.A., Blasch, B.B.: Wearable interfaces for orientation and wayfinding. In: Proceedings of the Fourth International ACM Conference on Assistive Technologies, pp. 193–200. ACM (2000)
23. Sofia, K.O., Jones, L.: Mechanical and psychophysical studies of surface wave propagation during vibrotactile stimulation. IEEE Trans. Haptics **6**(3), 320–329 (2013)

BY

Tactile Perception of Objects by the User's Palm for the Development of Multi-contact Wearable Tactile Displays

Miguel Altamirano Cabrera(✉), Juan Heredia, and Dzmitry Tsetserukou

Skolkovo Institute of Science and Technology (Skoltech), Bolshoy Boulevard 30, bld. 1, 121205 Moscow, Russia
{miguel.altamirano,juan.heredia,d.tsetserukou}@skoltech.ru

Abstract. The user's palm plays an important role in object detection and manipulation. The design of a robust multi-contact tactile display must consider the sensation and perception of the stimulated area, aiming to deliver the right stimuli at the correct location. To the best of our knowledge, there is no study to obtain the human palm data for this purpose. The objective of this work is to introduce a method to investigate the user's palm sensations during the interaction with objects. An array of fifteen Force Sensitive Resistors (FSRs) was located at the user's palm to get the area of interaction, and the normal force delivered to four different convex surfaces. Experimental results showed the active areas at the palm during the interaction with each of the surfaces at different forces. The obtained results were verified in an experiment for pattern recognition to discriminate the applied force. The patterns were delivered in correlation with the acquired data from the previous experiment. The overall recognition rate equals 84%, which means that user can distinguish four patterns with high confidence. The obtained results can be applied in the development of multi-contact wearable tactile and haptic displays for the palm, and in training a machine-learning algorithm to predict stimuli aiming to achieve a highly immersive experience in Virtual Reality.

Keywords: Palm haptics · Cutaneous force feedback · Tactile force feedback · Wearable display

1 Introduction

Virtual Reality (VR) experiences are used by an increasing number of people, through the introduction of devices that are more accessible to the market. Many VR applications have been launched and are becoming parts of our daily life, such as simulators and games. To deliver a highly immersive VR experience, a significant number of senses have to be stimulated simultaneously according to the activity that the users perform in the VR environment.

I. Nisky et al. (Eds.): EuroHaptics 2020, LNCS 12272, pp. 51–59, 2020.
https://doi.org/10.1007/978-3-030-58147-3_6

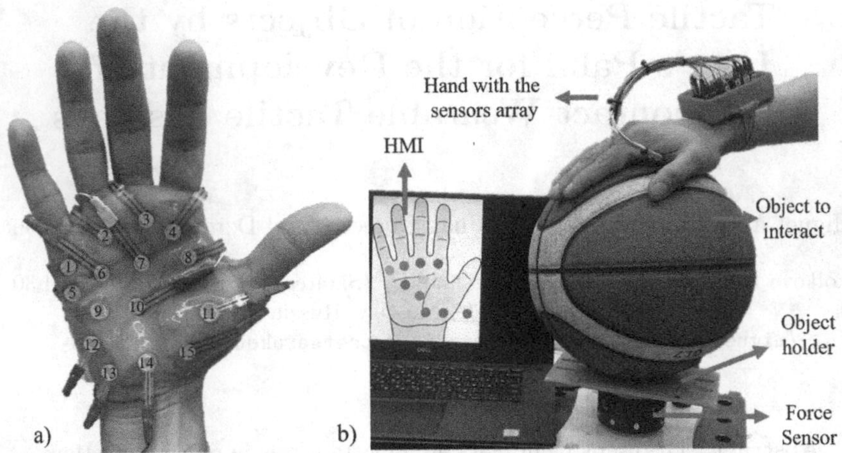

Fig. 1. a) Sensor array at the user's palm to record the tactile perception of objects. Fifteen FSRs were located according to the physiology of the hand, using the points over the joints of the bones, and the location of the pollicis and digiti minimi muscles. b) Experimental Setup. Each object was placed on the top of the FT300 force sensor to measure the applied normal force. The data from the fifteen FSRs and the force sensor are visualized in real-time in a GUI and recorded for future analysis.

The tactile information from haptic interfaces improves the user's perception of virtual objects. Haptic devices introduced in [1–5], provide haptic feedback at the fingertips and increase the immersion experience in VR.

Many operations with the hands involve more that one contact point between the user's fingers, palm, and the object, e.g., grasping, detection, and manipulation of objects. To improve the immersion experience and to keep the natural interaction, the use of multi-contact interactive points has to be implemented [6,7]. Choi et al. [8,9] introduced devices that deliver the sensation of weight and grasping of objects in VR successfully using multi-contact stimulation. Nevertheless, the proposed haptic display provides stimuli only at the fingers and not on the palm.

The palm of the users plays an essential role in the manipulation and detection of objects. The force provided by the objects to the user's palms determines the contact, weight, shape, and orientation of the object. At the same time, the displacement of the objects on the palm can be perceived by the slippage produced by the forces in different directions.

A significant amount of Rapidly Adapting (RA) tactile receptors are present on the glabrous skin of the hand, a total of 17,023 *units* on average [10]. The density of the receptors located in the fingertips is 141 units/cm^2, which is bigger than the density in the palm 25 units/cm^2. However, the overall receptor number on the palm is compensated by its large area, having the 30% of all the RA receptors located at the hand glabrous skin. To arrive at this quantity, we should cover the surface of the five fingertips. For this reason, it is imperative to take advantage of the palm and to develop devices to stimulate the most significant area with multi-contact points and multi-modal stimuli. Son et al. [11] introduced a haptic device that provides haptic feedback to the thumb, the middle finger, the index finger, and on the palm. This multi-contact device provides kinesthetic (to the fingers) and tactile stimuli (at the palm) to improve the haptic perception of a large virtual object.

The real object perception using haptic devices depends on the mechanical configuration of the devices, the contact point location on the user's hands, and the correctness of the information delivered. To deliver correct information by the system introduced by Pacchierotti et al. [12], their haptic display was calibrated using a BioTac device.

There are some studies on affective haptics engaging the palm area. In [13], the pressure distribution on the human back generated by the palms of the partner during hugging was analyzed. Toet et al. [14] studied the palm area that has to be stimulated during the physical experience of holding hands. They presented the areas of the hands that are stimulated during the hand holding in diverse situations. However, the patterns extraction method is not defined, and the results are only for parent-child hand-holding conditions.

Son et al. [15] presented a set of patterns that represent the interaction of the human hand with five objects. Nevertheless, the investigated contact point distribution is limited to the points available in their device, and the different sizes of the human hands and the force in the interaction are not considered.

In the present work, we study the tactile engagement of the user's palms when they are interacting with large surfaces. The objective of this work is to introduce a method to investigate the sensation on the user's palm during the interaction with objects, finding the area of interaction, and the normal force delivered. This information is used to the reproduction of tactile interaction, and the development of multi-contact wearable tactile displays. The relation between the applied normal force and the location of the contact area should be found, considering the deformation of the palm interacting with different surfaces.

2 Data Acquisition Experiment

A sensor array was developed to investigate the palm area that is engaged during the interaction with the different surfaces. Fifteen FSRs were located according to the physiology of the hand, using the points over the bones' joints, and the location of the pollicis and digiti minimi muscles. In Fig. 1 the distribution of the fifteen FSRs is shown.

The fifteen FSRs are held by a transparent adhesive contact paper that is attached to the user's skin. The transparent adhesive paper is flexible enough to allow users to open and close their hands freely. The different shape of the user's hands is considered every time the array is used by a different user. The method to attach the FSRs on the user's palms is the following: a square of transparent adhesive paper is attached to the skin of the users, the points, where the FSR must be placed, are indicated with a permanent marker. After that, the transparent adhesive paper is taken off from the hand to locate the fifteen FSRs at the marked points. The FSRs are connected to ESP8266 microcontroller through a multiplexer. The microcontroller receives the data from the FSRs and sends it to the computer by serial communication.

The hand deformation caused by the applied force to the objects is measured in this study. The force applied to the objects is detected by a Robotiq 6 DOF force/torque sensor FT300. This sensor was chosen because of its frequency of 100 Hz for data output and the low noise signal of 0.1 N in F_z, which allowed getting enough data for the purposes of this study. The sensor was fixed to a massive and stiff table (Siegmund Professional S4 welding table) using an acrylic base. An object holder was designed to mount the objects on top of the force sensor FT300.

2.1 Experimental Procedure

Four surfaces with different diameters were selected, three of them are balls, and one is a flat surface. The different diameters are used to focus on the position of the hand: if the diameter of the surface increases, the palm is more open. The diameters of the balls are 65 mm, 130 mm, and 240 mm.

Participants: Ten participants volunteering completed the tests, four women and six men, aged from 21 to 30 years. None of them reported any deficiencies in sensorimotor function, and all of them were right-handed. The participants signed informed consent forms.

Experimental Setup: Each object was placed on the top of the FT300 force sensor to measure the applied normal force. The data from the fifteen FSRs and the force sensor are visualized in real-time in a graphical user interface (GUI) and recorded for future analysis.

Method: We have measured the size of the participant's hands to create a sensor array according to custom hand size. The participants were asked to wear the sensors array on the palm and to interact with the objects. Subjects were asked to press objects in the normal direction of the sensor five times, increasing the force gradually up to the biggest force they can provide. After five repetitions, the object was changed, and the force sensor was re-calibrated.

2.2 Experimental Results

The force from each of the fifteen FSRs and from the force sensor FT300 was recorded at a rate of 15 Hz. To show the results, the data from the FSRs were analyzed when the normal force was 10 N, 20 N, 30 N, and 40 N. In every force and surface, the average values of each FSR were calculated. The average sensor values corresponding to the normal force for each surface are presented in Fig. 2.

From Fig. 2 we can derive that the number of contact points is proportional to the applied force. The maximum number of sensors is activated when 40 N is

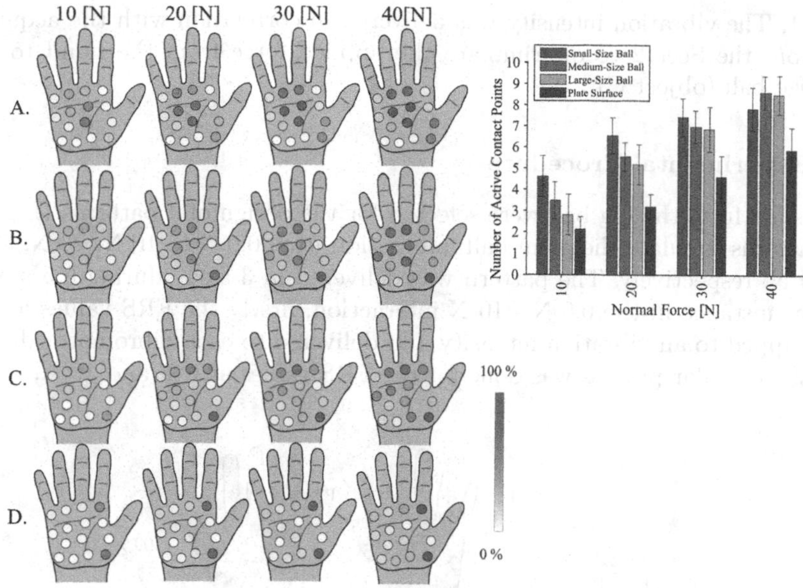

Fig. 2. Number and location of contact points engaged during the palm-object interaction. The data from the FSRs were analyzed according to the normal force in each surface. The average values of each FSR were calculated. The active points are represented by a scale from white to red. The maximum recorded value in each surface was used as normalization factor, thus, the maximum recorded value represents 100%. Rows A, B, C, D represent results for a small-size ball, a medium-size ball, a large-size ball, a flat surface, respectively. (Color figure online)

applied to the medium-sized ball, and the minimum number is in the case when 10 N is applied to large-size and flat surface.

The surface dimension is playing an important role in the activation of FSRs. It can be observed that the surfaces that join up better to the position of the hand activate more points. The shape of the palm is related to the applied normal force and to the object surface. When a normal for of 10 N is applied to a small-size ball, the number of contact points is almost twice the contact points of other surfaces, it increases only to seven at 40 N. The number of contact points with the large-size ball at 10 N is two, and increases to eight at 40 N. Instead, the flat surface does not activate the points at the center of the palm.

3 User's Perception Experiment

To verify if the data obtained in the last experiment can be used to render haptic feedback in other applications, an array of fifteen vibromotors was developed. Each of the vibromotors was located in the same positions as the FRS, as shown in Fig. 1. The vibration intensity was delivered in correlation with the acquired data from the Sect. 2 to discriminate the applied force from the hand to the large-size ball (object C).

3.1 Experimental Procedure

The results from the big ball were selected for the design of 4 patterns (Fig. 3). The patterns simulate the palm-ball interaction from 0.0 N to 10 N, 20 N, 30 N and 40 N, respectively. The pattern was delivered in 3 steps during 0.5 s each one. For instance in the 0.0 N - 10 N interaction, firstly the FRS values at 3 N were mapped to an vibration intensity and delivered to each vibromotor during 0.5 s, then similar process was done for the FRS values at 6 N and 10 N.

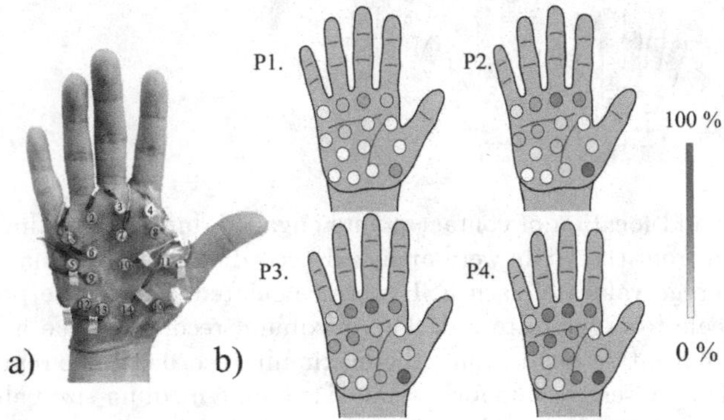

Fig. 3. Tactile presentation of the patterns with the array of vibration motors a) the arrangement of fifteen vibromotors is shown. b) P1, P2, P3, and P4 represent the actuated vibromotors on the palm and the intencity of vibration of each actuator (by red gradient color). (Color figure online)

Participants: Seven participants volunteering completed the tests, two women and five men, aged from 23 to 30 years. None of them reported any deficiencies in sensorimotor function, and all of them were right-handed. The participants signed informed consent forms.

Experimental Setup: The user was asked to sit in front of a desk and to wear the array of fifteen vibromotors on the right palm. One application was developed in Python where the four patterns were delivered and the answers of the users were recorded for the future analysis.

Method: We have measured the size of the participant's hands to create a vibromotors array according to the custom hand size. Before the experiment, a training session was conducted where each of the patterns was delivered three times. Each pattern was delivered on their palm five times in random order. After the delivery of each pattern, subject was asked to specify the number that corresponds to the delivered pattern. A table with the patterns and the corresponding numbers was provided for the experiment.

3.2 Experimental Results

The results of the patterns experiment are summarized in a confusion matrix (see Table 1).

Table 1. Confusion matrix for the pattern recognition.

%		*Answers (predicted class)*			
		1	2	3	4
Patterns	1	**97**	3	0	0
	2	0	**71**	26	3
	3	0	0	**83**	17
	4	0	3	11	**86**

The perception of the patterns was analyzed using one-factor ANOVA without replication with a chosen significance level of $\alpha < 0.05$. The $p - value$ obtained in the ANOVA is equal to 0.0433, in addition, the $Fcritic$ value is equal to 3.0088, and the F value is of 3.1538. With these results, we can confirm that statistic significance difference exists between the recognized patterns. The paired t-tests showed statistically significant differences between the pattern 1 and pattern 2 ($p = 0.0488 < 0.05$), and between pattern 1 and pattern 3 ($p = 0.0082 < 0.05$). The overall recognition rate is 84%, which means that the user can distinguish four patterns with high confidence.

4 Conclusions and Future Work

The sensations of the users were analyzed to design robust multi-contact tactile displays. An array of fifteen Interlink Electronics FSRTM 400 Force Sensitive Resistors (FSRs) was developed for the user's palm, and a force sensor FT300 was used to detect the normal force applied by the users to the surfaces. Using the designed FSRs array, the interaction between four convex surfaces with different diameters and the hand was analyzed in a discrete range of normal forces. It was observed that the hand undergoes a deformation by the normal force. The experimental results revealed the active areas at the palm during the interaction with each of the surfaces at different forces. This information leads to the optimal contact point location in the design of a multi-contact wearable tactile and haptic display to achieve a highly immersive experience in VR. The experiment on the tactile pattern detection reveled a high recognition rate of 84.29%.

In the future, we are planing to run a new human subject study to validate the results of this work by a psychophysics experiment. Moreover, we will increase the collected data sensing new objects. With the new dataset and our device, we can design an algorithm capable to predict the contact points of an unknown object. The obtained results can be applied in the development of multi-contact wearable tactile and haptic displays for the palm.

The result of the present work can be implemented to telexistence technology. The array of FSR senses objects and the interactive points are rendered by a haptic display to a second user. The same approach can be used for affective haptics.

References

1. Chinello, F., Malvezzi, M., Pacchierotti, C., Prattichizzo, D.: Design and development of a 3RRS wearable fingertip cutaneous device. In: IEEE/ASME International Conference on Advanced Intelligent Mechatronics, pp. 293–298 (2015)
2. Kuchenbecker, K.J., Ferguson, D., Kutzer, M., Moses, M., Okamura, A.M.: The touch thimble: providing fingertip contact feedback during point-force haptic interaction. In: HAPTICS 2008: Proceedings of the 2008 Symposium on Haptic Interfaces for Virtual Environment and Teleoperator Systems (March), pp. 239–246 (2008)
3. Gabardi, M., Solazzi, M., Leonardis, D., Frisoli, A.: A new wearable fingertip haptic interface for the rendering of virtual shapes and surface features. In: Haptics Symposium (HAPTICS), pp. 140–146. IEEE (2016)
4. Prattichizzo, D., Chinello, F., Pacchierotti, C., Malvezzi, M.: Towards wearability in fingertip haptics: a 3-DoF wearable device for cutaneous force feedback. IEEE Trans. Haptics **6**(4), 506–516 (2013)
5. Minamizawa, K., Prattichizzo, D., Tachi, S.: Simplified design of haptic display by extending one-point kinesthetic feedback to multipoint tactile feedback. In: 2010 IEEE Haptics Symposium, pp. 257–260, March 2010
6. Frisoli, A., Bergamasco, M., Wu, S.L., Ruffaldi, E.: Evaluation of multipoint contact interfaces in haptic perception of shapes. In: Barbagli, F., Prattichizzo, D., Salisbury, K. (eds.) Multi-point Interaction with Real and Virtual Objects. Springer Tracts in Advanced Robotics, vol. 18, pp. 177–188. Springer, Heidelberg (2005). https://doi.org/10.1007/11429555_11
7. Pacchierotti, C., Chinello, F., Malvezzi, M., Meli, L., Prattichizzo, D.: Two finger grasping simulation with cutaneous and kinesthetic force feedback. In: Isokoski, P., Springare, J. (eds.) Haptics: Perception, Devices, Mobility, and Communication, pp. 373–382. Springer, Heidelberg (2012). https://doi.org/10.1007/978-3-642-31401-8_34
8. Choi, I., Culbertson, H., Miller, M.R., Olwal, A., Follmer, S.: Grabity: a wear-able haptic interface for simulating weight and grasping in virtual reality. In: Proceedings of the 30th Annual ACM Symposium on User Interface Software and Technology - UIST 2017, pp. 119–130 (2017)
9. Choi, I., Hawkes, E.W., Christensen, D.L., Ploch, C.J., Follmer, S.: Wolverine: a wearable haptic interface for grasping in virtual reality. In: IEEE International Conference on Intelligent Robots and Systems, pp. 986–993 (2016)
10. Johansson, B.Y.R.S., Vallbo, A.B.: Tactile sensibility in the human hand: relative and absolute sensities of four types of mechanoreceptive units in glabrous skin. 283–300 (1979)

11. Son, B., Park, J.: Haptic feedback to the palm and fingers for improved tactile perception of large objects. In: Proceedings of the 31st Annual ACM Symposium on User Interface Software and Technology, UIST 2018, pp. 757–763. Association for Computing Machinery, New York (2018)
12. Pacchierotti, C., Prattichizzo, D., Kuchenbecker, K.J.: Cutaneous feedback of fingertip deformation and vibration for palpation in robotic surgery. IEEE Trans. Biomed. Eng. **63**(2), 278–287 (2016). https://doi.org/10.1109/TBME.2015.2455932
13. Tsetserukou, D., Sato, K., Tachi, S.: ExoInterfaces: novel exosceleton haptic interfaces for virtual reality, augmented sport and rehabilitation. In: Proceedings of the 1st Augmented Human International Conference, pp. 1–6. USA (2010)
14. Toet, A., et al.: Reach out and touch somebody's virtual hand: affectively connected through mediated touch (2013)
15. Son, B., Park, J.: Tactile sensitivity to distributed patterns in a palm. In: Proceedings of the 20th ACM International Conference on Multimodal Interaction, ICMI 2018, pp. 486–491. ACM, New York (2018)

From Hate to Love: How Learning Can Change Affective Responses to Touched Materials

Müge Cavdan[1(✉)], Alexander Freund[1], Anna-Klara Trieschmann[1], Katja Doerschner[1,2], and Knut Drewing[1]

[1] Justus Liebig University, 3539 Giessen, Germany
Muege.Cavdan@psychol.uni-giessen.de
[2] Bilkent University, 06800 Ankara, Turkey

Abstract. People display systematic affective reactions to specific properties of touched materials. For example, granular materials such as fine sand feel pleasant, while rough materials feel unpleasant. We wondered how far such relationships between sensory material properties and affective responses can be changed by learning. Manipulations in the present experiment aimed at unlearning the previously observed negative relationship between roughness and valence and the positive one between granularity and valence. In the learning phase, participants haptically explored materials that are either very rough or very fine-grained while they simultaneously watched positive or negative stimuli, respectively, from the International Affective Picture System (IAPS). A control group did not interact with granular or rough materials during the learning phase. In the experimental phase, participants rated a representative diverse set of 28 materials according to twelve affective adjectives. We found a significantly weaker relationship between granularity and valence in the experimental group compared to the control group, whereas roughness-valence correlations did not differ between groups. That is, the valence of granular materials was unlearned (i.e., to modify the existing valence of granular materials) but not that of rough materials. These points to differences in the strength of perceptuo-affective relations, which we discuss in terms of hard-wired versus learned connections.

Keywords: Haptics · Valence · Affect · Roughness · Granularity · Learning

1 Introduction

We constantly interact with various materials like plastic, fabric, or metal. Haptic perceptual properties of materials have been summarized by five different dimensions [1] that are softness (but cf. [2]), warmness, micro- and macro roughness, friction, and stickiness. In addition to the sensory properties that we experience while haptically exploring a material, we often also have an initial affective reaction to it. Moreover, the

Research was supported by the EU Marie Curie Initial Training Network "DyVito" (H2020-ITN, Grant Agreement: 765121) and Deutsche Forschungsgemeinschaft (DFG, German Research Foundation) – project number 222641018 – SFB/TRR 135, A5.

I. Nisky et al. (Eds.): EuroHaptics 2020, LNCS 12272, pp. 60–68, 2020.
https://doi.org/10.1007/978-3-030-58147-3_7

affective reactions that a material elicits might also influence the duration of our haptic interactions. For example, soft and smooth materials, which cause more pleasant feelings than rough and sticky materials [3–6], might be explored longer.

Previous research on the semantic structure of affective experiences postulates three basic affective dimensions [8]: valence, *arousal*, and *dominance,* where each dimension has two opposite poles [9]: arousal ranges from a very calm state and sleepiness (low) to vigilance, which is accompanied by excitement (high). Valence is a continuum from negative to positive and dominance ranges from dominant to submissive. Most of the research in haptic perception has focused on the connection between pleasantness (which can be equated with valence) and the perception of sensory dimensions. One key finding has been that smooth and soft materials are related to more pleasant feelings than rough materials [10], and that the rougher a material is rated, the more unpleasant it feels [10]. In a more recent study [11], all three basic affective dimensions and their relationship with materials' sensory characteristics have been systematically investigated: Drewing et al. [11] used a free exploration paradigm to study the sensory and affective spaces in haptics and tested the generalizability of their results to different participant groups. They found that arousal was related to the amount of perceived fluidity, that higher dominance as associated with increases in perceived heaviness and decreases in deformability, and that greater positive valence was associated with increased granularity and decreased roughness.

It is currently unknown to what extent such perceptuo-affective connections are due to learning experiences and to what extent they are hard-wired, innate mechanisms. Here we investigated directly whether existing relationships between sensory material properties and affective responses can be unlearned, and whether the extent of unlearning depends on the specific perceptuo-sensory relation, for two haptic perceptual dimensions: granularity and roughness. We speculate that hard-wired connections should be more resistant to unlearning than learned ones.

We ran a classical conditioning study that constisted of two phases: learning and experimental phases with two groups each (experimental and control). In the learning phase, participants haptically explored selected materials while watching affective images: In the experimental group rough materials were combined with positive images, granular materials with negative images and distractor materials with neutral images. In the control group participants learned instead associations of fibrous and fluid materials with arousal, which were however not subject of this paper and will not be further discussed. In the experimental phase, participants rated a representative set of 28 materials for 12 affective adjectives. We calculated perceptuo-affective correlations for valence-roughness and valence-granularity relationships per group and compared these correlations between groups. Lower correlations for the participants in the experimental group would indicate an unlearning of the relationship between valence and the respective perceptual dimension.

2 Methods

2.1 Participants

Sixty-six students (9 males; age 18–34 years, mean: 23.5 years) from Giessen University participated in our study. Four were excluded from analysis due to misunderstanding the task, technical error, or an increased threshold in the two-point touch discrimination (>3 mm at index finger). All participants were naïve to the aim of the experiment, spoke German at native-speaker level, and none reported relevant sensory, or motor impairments. All procedures were in accordance with the Declaration of Helsinki (2008), and participants provided written informed consent prior to the study.

2.2 Setup, Material, and Adjectives

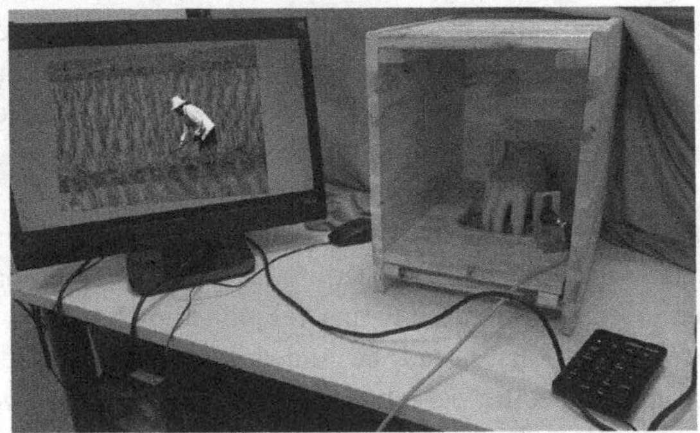

Fig. 1. Experimental setup from the experimenter's viewpoint.

Participants sat at a table in front of a big wooden box with a hand opening. Materials were presented in the box (see Fig. 1). Participants reached the materials through the hand opening, which was covered with linen to hinder participants to look inside the box. On a monitor (viewing distance about 60 cm) we presented images (visual angle 14.2°) and adjectives to the participant. Earplugs and active noise cancelling headphones (Beyerdynamic DT770 PRO, 30 O) blocked the noises that can occur from exploring the materials. All materials were presented in 16 × 16 cm plastic containers embedded in the bottom of the box. A light sensor in the box signaled when the participant's hand was on the front edge of the container, allowing to start picture presentation simultaneously to haptic exploration. Participants gave responses using a keyboard. The experimenter sat on the other side of the table in order to exchange materials guided by information presented on another monitor.

For the learning part of the experimental group, we selected materials from [11] which had a high factor value on one of the target sensory dimensions (either granularity or roughness) but did not show high factor values in any of the other dimensions (fluidity, fibrousness, heaviness, deformability). Bark and sandpaper were selected for roughness, and salt and lentils for granularity. In the control group other materials were used (jute, wadding, water, shaving foam). Additionally, for both groups (experimental and control) we added four distractor materials, which did not have high factor values on the manipulated dimensions: cork, chalk, paper, and polystyrene. We also the sensory and affective adjectives were obtained from [11]; one to two representative adjectives per sensory dimension (rough, granular, moist, fluffy, heavy and light, deformable and hard). In the learning phase, sensory ratings served to draw the participant's attention to the materials.

For establishing affective-sensory associations we used images from the International Affective Picture Systems (IAPS). The IAPS database includes 1196 colorful images of various semantic contents, that have been rated according to valence, and arousal [12, 13]. For the experimental group, we selected sixteen images with high negative valence (<2.5) and sixteen images with high positive valence (>7.5), and as diverse content as possible (excluding drastic injury images). For the control group, we used images with high or low arousal instead. We also selected 32 distractor images, which have average valence and arousal values (between 4.5 and 5.5).

In the experimental phase, participants rated 28 materials (plastic, wrapping foil, aluminum foil, fur, pebbles, playdough, silicon, paper, styrofoam, paper, sandpaper, velvet, jute, silicon, stone, bark, flour, metal, cork, polish stone, oil, shaving foam, soil, hay, chalk, salt, lentil). We assessed affective responses via adjectives and selected four high loading adjectives per affective dimension: *valence* (pleasant, relaxing, enjoyable, and pleasurable), *arousal (*exciting, boring, arousing, and attractive) and *dominance* (dominant, powerful, weak, and enormous/tremendous).

2.3 Design and Procedure

Participants were randomly assigned to either the experimental or the control group. In the learning phase of the experimental group, we coupled the exploration of the two very rough materials with positive images and the granular materials with negative images, in order to manipulate valence. In the control group, different materials were explored and coupled with different images.

The learning phase consisted of 64 trials: in the experimental group, each of the two granular materials was presented eight times coupled with one of the 16 negative images, and each of the two rough materials was coupled 8 times with one of the positive images. Also, each of the four distractor materials (cork, chalk, polystyrene, and paper) was presented eight times with a distractor image. Both, the assignment of corresponding images to materials and the order of presentation, were random.

In each trial of the learning phase (Fig. 2), an initial beep sound signaled the participant to insert the hand in the box and to start exploring the material. When participant's hand started the exploration, an image was displayed on the screen. Participants explored the materials while looking at the images for five seconds. Another beep sound signaled participants to end the exploration, and a randomly chosen sensory adjective appeared on the screen. Participants rated how much the adjective applied to the material (1*: not at all*, 4: *maybe*, 7: *very*) using a keyboard. Finally, a multiple-choice question about the main content of the image was posed. The experimenter exchanged the stimuli between trials. In total, the learning phase took about 30 min.

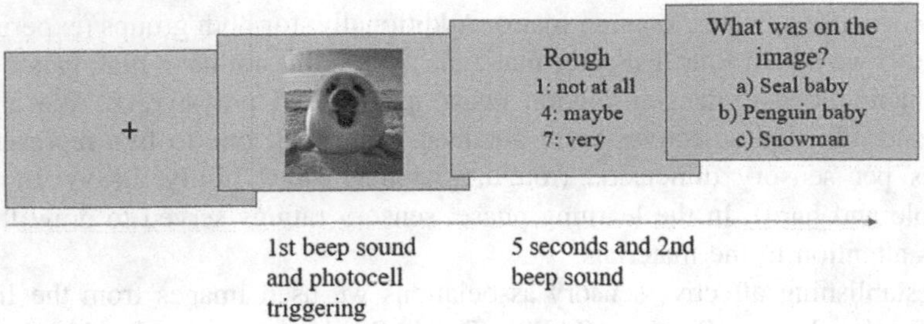

Fig. 2. Time course of one trial of the learning phase.

The experimental phase consisted of 336 trials (28 materials × 12 adjectives). Each trial started with a fixation cross on the screen (Fig. 2). Then participants reached in the box with their dominant hands and explored the material. During exploration each of the 12 affective adjectives appeared on the screen (in random order), and participants had to rate how much each adjective applied to the material (1–7). The hand was retracted, and the material exchanged after the all twelve adjectives were evaluated. The total duration of the experiment including learning phase, instructions, preparation, pauses for cleaning hands and debriefing was about 2–2.5 h.

2.4 Data Analysis

We first assessed the number of correct responses from the multiple-choice questions of the learning phase. With an average of 96.2% correct (individual minimum: 84.4%), we could verify that all participants had attended to the images as they should. Next, we used the affective ratings from the experimental phase in a covariance-based principal component analysis (PCA) with Varimax-rotation (for all adjective ratings across all materials and participants) in order to extract underlying affective dimensions. Before doing so, we assessed whether the PCA was suitable by a) checking the consistency across participants by calculating Cronbach's alpha for each affective adjective (separately for experimental and control group), b) computing the Kaiser-Meyer-Olkin (KMO) criterion, c) using Bartlett's test of sphericity [11].

Lastly, in order to test a potential unlearning of the perceptuo-affective relationships valence-roughness and valence-granularity, we determined material-specific individual factor values of the valence dimension. We calculated individual correlations of these values with previously observed average granularity and roughness values across materials (taken from Exp. 2 in [11]), and used two independent samples t-tests in order to compare the two perceptuo-affective Fisher-z transformed correlations of experimental and control group.

3 Result

3.1 PCA on Affective Dimensions

Cronbach's alpha was higher than .80 per adjective and participant group, indicating good consistency between participants. Bartlett's test of sphericity was statistically significant, □2(66, $N = 28$) = 12868.897, $p < .001$, and the KMO value, which has a range from 0–1, was 0.86 [14]. Given these results we proceeded with the PCA.

The PCA extracted three components according to the Kaiser criterion, explaining 73.1% of the variance in total. After the varimax-rotation, the first component explained 31.8% variance with the highest component loads obtained from the adjectives pleasant (score: 1.8), relaxing (1.8), enjoyable (1.5), and pleasurable (1.8). Thus, we identified this component as *valence*. The second component explained 22.4% variance with high loads from adjectives dominant (1.6), powerful (1.6), weak (−1.1), and enormous/tremendous (1.6); consequently, we called this component dominance. The last component explained 18.9% variance with high loads from exciting (1.4), boring (1.5), arousing (0.7), and attention-attracting (1.5), and therefore we labeled it arousal. All other component loads of any adjective had an absolute value below 0.7 and where thus not considered in the interpretation.

3.2 Learning Effects on Materials

For the control group ($N = 30$), correlations between roughness and valence, $r = -.37$, $p < .001$ and granularity and valence, $r = .25, p < .001$ were statistically significant after Bonferroni correction, confirming the basic perceptuo-affective relations previously observed in [11]. In order to test the effect of unlearning perceptuo-affective relationship, we compared the Fisher-z-transformed correlations of the two groups (Fig. 3).

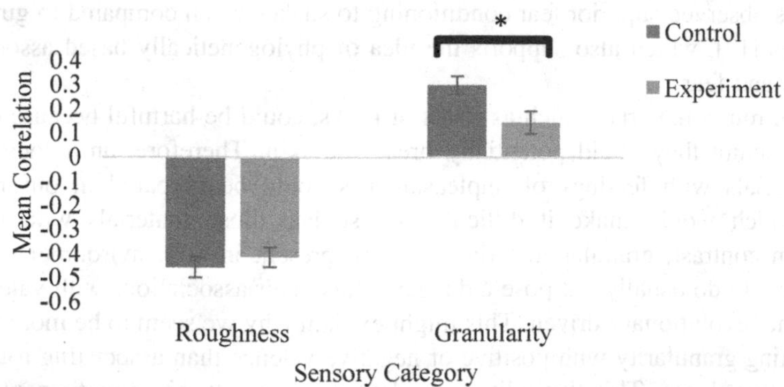

Fig. 3. Relationship between sensory category and valence for experimental (orange) and control group (blue). It shows mean correlations (inverse of average Fisher z-transforms) as a function of sensory category (roughness and granularity). Error bars show 1 ± standard errors (*Significant $p < .05$ level). (Color figure online)

There was not a statistically significant difference between experimental ($M = -.41$, $SD = .21$) and control ($M = -.46$, $SD = .23$) groups, t (60) = .763, $p = .449$ for the valence-roughness correlation. However, there was a statistically significant difference between control ($M = .30$, $SD = .23$) and experiment ($M = .14$, $SD = .26$) groups, t (60) = −2.461, $p = .017$ for the valence-granular correlation.

4 Discussion

People experience rougher materials as more unpleasant, and more granular materials as more pleasant [11]. Here we investigated whether brief learning experiences can influence the affective assessments of these two material properties. Our aim was to modify previously found affective responses towards granular and rough materials, and we found, that the perceptuo-affective correlation between granularity and valence was lowered through learning in the experimental group compared to control group. However, the valence-roughness relation was not significantly different in experimental and control group, suggesting that this connection could not be unlearned. We suggest that these results demonstrate different strengths in the perceptuo-affective connections, which relate to the degree to which connections are learned during lifetime vs being evolutionary prepared to serve a biological function.

Studies on fear conditioning suggest that some classes of stimuli are phylogenetically prepared to be associated with fear responses, while others can be hardly learned. For example, it has been shown that lab-reared monkeys easily acquire fear of snakes by observing videos of the fear that other monkeys had shown - even if they had never seen snakes before in their lives [15]. When these videos were reproduced to create similar fear against toy snakes, crocodiles, flowers, and rabbits, lab-reared monkeys showed fear against snakes and crocodiles, but not flowers and rabbits [16]. Because these monkeys had never been exposed to the stimuli before, this can be taken as evidence for a phylogenetic basis of selective learning. Furthermore, in humans, researchers observed superior fear conditioning to snakes when compared to guns with loud noises [17], which also supports the idea of phylogenetically based associations for snakes and fear.

Natural rough materials, such as rocks or barks, could be harmful because of their surface structure they could potentially break the skin. Therefore, an association of those materials with feelings of unpleasantness could be prepared in our nervous system, which would make it difficult to associate those materials with positive valence. In contrast, granular materials that are present in our environment such as sand, generally do usually not pose a danger. Thus, their associations with valence are probably not evolutionary driven. This might explain why we seem to be more flexible in associating granularity with positive or negative valence than associating roughness with positive valence. This flexibility is evident in our results since participants in the experimental group learned to associate granular materials with negative valence. We conclude that even brief learning experiences can change perceptuo-affective connections depending on the source and strength of the relationship. In the current case, the valence of granular materials was unlearned but not that of rough materials. This might mean that perceptuo-affective connections for granular materials are learned, yet for rough materials they might be hard-wired or at least prepared.

References

1. Okamoto, S., Nagano, H., Yamada, Y.: Psychophysical dimensions of tactile perception of textures. IEEE Trans. Haptics **6**, 81–93 (2013)
2. Cavdan, M., Doerschner, K., Drewing, K.: The many dimensions underlying perceived softness: how exploratory procedures are influenced by material and the perceptual task*. In: 2019 IEEE World Haptics Conference (WHC), pp. 437–442. IEEE Press (2019)
3. Essick, G., et al.: Quantitative assessment of pleasant touch. Neurosci. Biobehav. Rev. **34**, 192–203 (2010)
4. Ripin, R., Lazarsfeld, P.: The tactile-kinaesthetic perception of fabrics with emphasis on their relative pleasantness. J. Appl. Psychol. **21**, 198–224 (1937)
5. Klöcker, A., Wiertlewski, M., Théate, V., Hayward, V., Thonnard, J.: Physical factors influencing pleasant touch during tactile exploration. PLoS ONE **8**, e79085 (2013)
6. Klöcker, A., Arnould, C., Penta, M., Thonnard, J.: Rasch-built measure of pleasant touch through active fingertip exploration. Front. Neurorobot. **6**, 5 (2012)
7. Ramachandran, V., Brang, D.: Tactile-emotion synesthesia. Neurocase **14**, 390–399 (2008)
8. Russell, J., Mehrabian, A.: Evidence for a three-factor theory of emotions. J. Res. Pers. **11**(3), 273–294 (1977)
9. Osgood, C.: The nature and measurement of meaning. Psychol. Bull. **49**(3), 197–237 (1952)
10. Guest, S., et al.: The development and validation of sensory and emotional scales of touch perception. Atten. Percept. Psychophys. **73**(2), 531–550 (2010)
11. Drewing, K., Weyel, C., Celebi, H., Kaya, D.: Systematic relations between affective and sensory material dimensions in touch. IEEE Trans. Haptics **11**, 611–622 (2018)
12. Bradley, M.M., Lang, P.J.: The International Affective Picture System (IAPS) in the study of emotion and attention. In: Coan, J.A., Allen, J.J.B. (eds.) Handbook of Emotion Elicitation and Assessment, pp. 29–46. Oxford University Press (2007)
13. Lang, P.J., Bradley, M.M., Cuthbert, B.N.: International affective picture System (IAPS): affective ratings of pictures and instruction manual. Technical Report A-8. University of Florida, Gainesville, FL (2008)
14. Cerny, B., Kaiser, H.: A study of a measure of sampling adequacy for factor-analytic correlation matrices. Multivar. Behav. Res. **12**(1), 43–47 (1977)
15. Cook, M., Mineka, S.: Selective associations in the observational conditioning of fear in rhesus monkeys. J. Exp. Psychol. Anim. Behav. Process. **16**(4), 372–389 (1990)
16. Cook, M., Mineka, S.: Selective associations in the origins of phobic fears and their implications for behavior therapy. In: Martin, P. (ed.) Handbook of Behavior Therapy and Psychological Science: An Integrative Approach, pp. 413–434. Pergamon Press, Oxford (1991)
17. Cook, E., et al.: Preparedness and phobia: effects of stimulus content on human visceral conditioning. J. Abnorm. Psychol. **95**(3), 195–207 (1986)

Switching Between Objects Improves Precision in Haptic Perception of Softness

Anna Metzger(✉) and Knut Drewing

Justus-Liebig University of Giessen, Giessen, Germany
anna.metzger@psychol.uni-giessen.de

Abstract. Haptic perception involves active exploration usually consisting of repeated stereotypical movements. The choice of such exploratory movements and their parameters are tuned to achieve high perceptual precision. Information obtained from repeated exploratory movements (e.g. repeated indentations of an object to perceive its softness) is integrated but improvement of discrimination performance is limited by memory if the two objects are explored one after the other in order to compare them. In natural haptic exploration humans tend to switch between the objects multiple times when comparing them. Using the example of softness perception here we test the hypothesis that given the same amount of information, discrimination improves if memory demands are lower. In our experiment participants explored two softness stimuli by indenting each of the stimuli four times. They were allowed to switch between the stimuli after every single indentation (7 switches), after every second indentation (3 switches) or only once after four indentations (1 switch). We found better discrimination performance with seven switches as compared to one switch, indicating that humans naturally apply an exploratory strategy which might reduce memory demands and thus leads to improved performance.

Keywords: Softness · Haptic · Perception · Psychophysics

1 Introduction

We usually have to actively move our sensory organs to obtain relevant information about the world around us. Such exploratory movements are often tuned to maximize the gain of information [1–4]. In active touch perception tuning of exploratory movements is very prominent. Humans use different highly stereotypical movements to judge different object properties [2]. For instance, they move the hand laterally over the object's surface to judge its roughness, in contrast, the hand is held statically on the object to judge its temperature. For each haptic property precision is best with the habitually used *Exploratory Procedure* as opposed to others [2]. Also motor parameters of *Exploratory Procedures*, such

This work was supported by Deutsche Forschungsgemeinschaf (DFG, German Research Foundation) – project number 222641018 – SFB/TRR 135, A5.

I. Nisky et al. (Eds.): EuroHaptics 2020, LNCS 12272, pp. 69–77, 2020.
https://doi.org/10.1007/978-3-030-58147-3_8

as finger force in perception of softness and shape [5,6], are tuned to optimize performance. This fine tuning is based on available predictive signals as well as on progressively gathered sensory information [3,7]. However, it could also reflect a compensation for limitations of the perceptual system. For instance it is believed that most eye movements are not designated to gain novel information for building up an internal representation of the scene, but to obtain momentary necessary information using the world as an external memory given limited capacity of short-term working memory [8,9]. Here we study whether natural haptic exploration is tuned to compensate for memory limitations.

Integration of accumulated sensory information was in many cases shown to be consistent with Bayesian inference [10]. In this framework available sensory information and assumptions based on prior knowledge are integrated by weighted averaging, with weights being proportional to the reliability of single estimates. This integration is considered statistically optimal because overall reliability is maximized. This framework was successfully applied to describe the integration of sensory and prior information (e.g. [11]), simultaneously available information (e.g. [12]) as well as information gathered over time (e.g. [13]). For the later purpose usually a Kalman filter [14] is used: A recursive Bayesian optimal combination of new sensory information with previously obtained information, which can also account for changes of the world over time.

In haptic perception it was shown that prior information is integrated into the percept of an object's softness [15]. It was also shown that sequentially gathered information from every indentation contributes to the overall perception of an object's softness [16] and information from every stroke over the object's surface contributes to the perception of its roughness [17]. However, the contribution of these single exploratory movements to the overall percept seems to be not equal, as would be predicted if there was no loss of information and integration was statistically optimal. When two objects are explored one after the other in a two-interval forced-choice task, information from later exploratory movements on the second object contributes less to the comparison of the two objects [16,17]. These results are consistent with the idea that perceptual weights decay with progressing exploration of the second object due to a fading memory representation of the first object. Indeed it could be shown that when a stronger representation of the first object's softness is built up by longer exploration, information from later indentations contributes more than with a weaker representation [18]. Also, consistent with memory characteristics [19] mere temporal delay of 5s after the exploration of the first object does not affect the memory representation of the first object [20]. A model for serial integration of information using a Kalman filter could explain the decrease of perceptual weights in the exploration of the second object by including memory decay of the first object's representation [17]. In the modelled experimental conditions people had to discriminate between two objects that are explored strictly one after the other, i.e. in their exploration people switched only once between the two stimuli. In this case memory decay of the first object's representation should have had particularly pronounced negative effects on discrimination. However, in free explorations participants usually switch more often between the objects in order to compare them (in softness discrimination on average 4 times, given 6–14 indentations in total and an achieved performance of 85–90% correct [7,21]). This might be a strategy to cope with memory decay and improve discrimination performance.

Using the example of softness perception here we test the hypothesis that discrimination improves when participants switch more than once between the two objects given the same number of indentations of each of them, i.e. the same sensory input. In our experiment participants were instructed to indent each test object four times. There were three switch conditions: Participants switched between the two objects after every single indentation (7 switches), after every second indentation (3 switches) or only once after four indentations (1 switch).

2 Methods

2.1 Participants

Eleven volunteers (6 female, right-handed) participated in the experiment. Written informed consent was obtained from each participant and they were reimbursed with 8€/h for their time. The study was approved by the local ethics committee at Justus-Liebig University Giessen LEK FB06 (SFB-TPA5, 22/08/13) and was in line with the declaration of Helsinki from 2008.

2.2 Apparatus

The experiment was conducted at a visuo-haptic workbench (Fig. 1A) consisting of a force-sensor (bending beam load cell LCB 130 and a measuring amplifier GSV-2AS, resolution .05 N, temporal resolution 682 Hz, ME-Messsysteme GmbH), a PHANToM 1.5A haptic force feedback device, a 22"-computer screen (120 Hz, 1280 $\times$ 1024 pixel), stereo glasses and a mirror. Participants sat at a table with the head resting in a chin rest. Two softness stimuli were placed side-by-side (distance in between 2 cm) in front of them on the force sensor. To prevent direct sight of the stimuli and of the exploring hand but in the same time indicating their position, a schematic 3D representation of the stimuli and the finger spatially aligned with the real ones was shown via monitor and mirror. The finger was represented as a 8 mm diameter sphere only when not in contact with the stimuli (force $<$ 1N). Participants touched the stimuli with the right index finger. The finger's position was detected with the PHANToM. For this purpose it was attached to the PHANToM arm with a custom-made adapter (Fig. 1B) consisting of magnetically interconnected metallic pin with a round end and a plastic fingernail. We used adhesive deformable glue pads to affix the plastic fingernail to the fingernail of the participant. Connected to the PHANToM the finger pad was left uncovered and the finger could be moved with all six degrees of freedom within a workspace of $38 \times 27 \times 20$ cm. Custom-made software (C++) controlled the experiment, collected responses, and recorded finger positions and reaction forces every 3 ms. Signal sounds were presented via headphones.

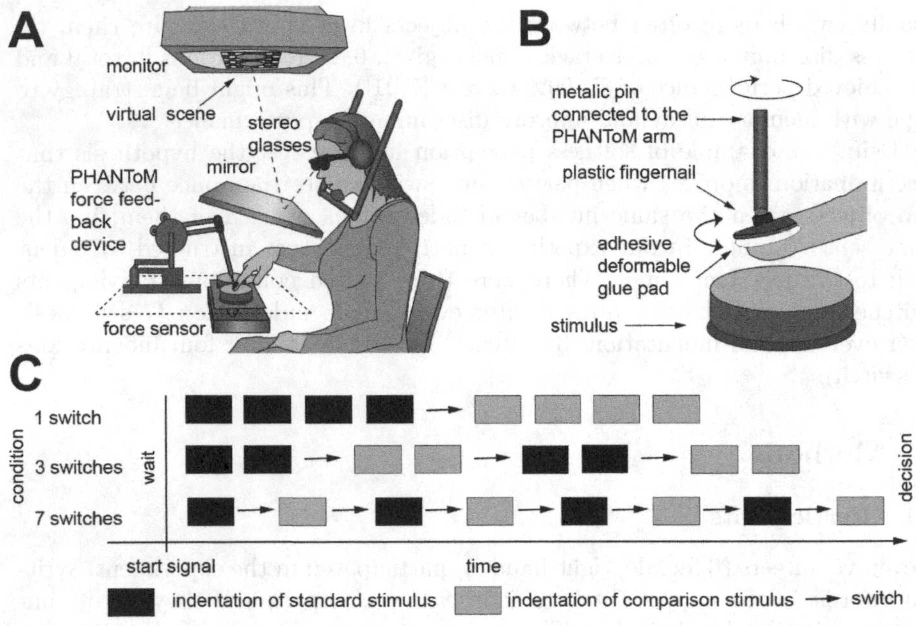

Fig. 1. A. Visuo-haptic workbench. **B.** Custom made finger PHANToM to finger connection. **C.** Experimental procedure. Timing of single trial phases was not restricted. Either the comparison or the standard stimulus could have been explored first.

2.3 Softness Stimuli

We produced softness stimuli with different elasticity by mixing a two-component silicon rubber solution (AlpaSil EH 10:1) with different amounts of silicon oil (polydimethylsiloxane, viscosity 50 mPa/s) and pouring it into cylindrical plastic dishes (75 mm diameter × 38 mm height). We produced in total 10 stimuli: 1 standard (Young's modulus of 59.16 kPa) and 9 comparison stimuli (Young's moduli of 31.23, 42.84, 49.37, 55.14, 57.04, 69.62, 72.15, 73.29 and 88.18 kPa). To characterize the elasticity of the stimuli we adopted the standard methodology proposed by [22]. From the solution mixed for every stimulus a portion was poured into a small cylinder (10 mm thick, 10 mm diameter) to obtain standardized substrates of the same material for the measurement. We used the experimental apparatus to measure elasticity. They were placed onto the force sensor and were indented by the PHANToM force feedback device. For this purpose we attached an aluminium plate of 24 mm diameter instead of the fingertip adapter. The force was increased by 0.005 N every 3 ms until a minimum force of 1 N and a minimum displacement of 1 mm were detected. Measures in which the force increased quicker than it should were considered as artefacts and removed before analysis. From the cleaned force and displacement data we calculated stress and strain. A linear Young's modulus was fitted to each stress-strain curve (only for 0–0.1 strain) in MATLAB R2017.

2.4 Design

There were three *Nr of switches* conditions: 1, 3 and 7 switches. The dependent variable was the just noticeable difference (JND), which we measured using a two-alternative-force-choice task combined with a method of constant stimuli. For each condition the standard stimulus was presented 12 times with each of the 9 comparison stimuli (overall 324 trials). The experiment was organized in 3 blocks of 108 trials, comprising 4 repetitions of each standard-comparison pairing, to balance for fatigue effects. Trials of different conditions were presented in random order. The position of the standard (right or left) and its presentation as first or second stimulus was balanced for every standard-comparison pairing. Differences between the conditions were analysed with paired two-sample *t*-tests.

2.5 Procedure

In every trial, participants sequentially explored the standard and a comparison and decided which one felt softer. A schematic of the procedure is outlined in Fig. 1C. Before a trial, participants rested the finger in the left corner of the workspace and waited until the experimenter changed the stimuli. The beginning of a trial was signaled by a tone and by the appearance of the schematic representation of the first stimulus. The number of indentations allowed to explore the stimulus (1, 2 or 4) was indicated above it. To detect and count stimulus' indentations we used the algorithm from [16]. After the necessary number of indentations the visual representation of the first stimulus disappeared and the representation of the second stimulus appeared. Participants switched to the other stimulus, and again the number of sequential indentations was indicated. Participants continued to switch between the stimuli until every stimulus was indented 4 times in total. Following this, participants indicated their decision on softness by pressing one of two virtual buttons using the PHANToM. A trial in which the stimuli were indented more then 8 times was repeated later in the block. Before the experiment participants were acquainted to the task and the setup in a training session of 12 trials. They did not receive any feedback on their performance to avoid explicit learning. Each participant completed the experiment on the same day within on average 2.5 h.

2.6 Analysis

We calculated for each participant, each condition, and each comparison stimulus the percentage of trials in which it was perceived to be softer than the standard. Combined for all comparisons these values composed individual psychometric data, to which we fitted cumulative Gaussian functions using the psignifit 4 toolbox [23]. Only the means and the standard deviations of the functions were fitted, lapse rates were set to 0. From the fitted psychometric functions, we estimated the JNDs as the 84% discrimination thresholds.

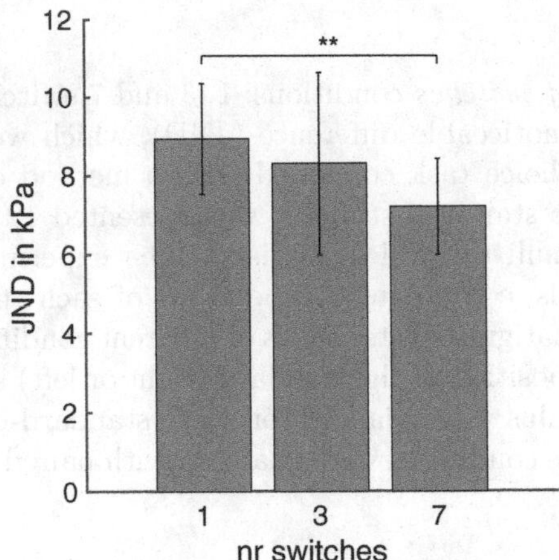

Fig. 2. Average discrimination thresholds for softness of two silicon stimuli each indented 4 times with the bare finger. Participants switched between the stimuli once after 4 indentations, after every 2. indentation (3 switches) or after every indentation (7 switches). Error bars represent within-subject standard error [24]. $**p < 0.005$

3 Results

Figure 2 depicts the average JNDs in the three different *Nr of switches* conditions (1, 3 and 7). The average Weber fractions were 15.08%, 14.01% and 12.14% for 1, 3, and 7 switches respectively. The JNDs decreased with increasing number of switches between the stimuli. Pairwise comparisons revealed that performance was significantly better, when participants switched after every indentation of the stimulus as compared to switching only once after 4 indentations, $t(10) = 4.64$, $p = 0.001$. The performance in the intermediate condition (switching after every second indentation) was numerically in between the other ones. However, we did not find significant differences in the comparisons with this condition (1 vs. 3 switches: $t(10) = 0.57$, $p = 0.578$; 3 vs. 7 switches: $t(10) = 1.07$, $p = 0.311$). We observed higher variance in this condition, possibly due to a more complicated task than in the other two conditions (Fig. 2).

4 Discussion

In the present study, we investigated whether more frequent switching between the stimuli improves discrimination performance in active touch perception of objects' softness. Participants explored two softness stimuli with the same number of indentations (4 per stimulus) but switched between the stimuli either only once after four indentations, after every second indentation or after every indentation. We showed that if participants could switch after every indentation discrimination performance was significantly better than if they could switch only once. We argue that this is due to greater memory demands, because the first stimulus' has to be remembered longer to compare it to every indentation of the second stimulus, than when switching after every indentation. It could also be argued that integration across more indentations of the first stimulus demands greater attention, leading to worse performance, especially because here we did not manipulate factors directly affecting memory i.e. inter-stimulus delay or masking. However, we previously observed that in the exploration of the first stimulus the weights of single indentations were equal, as would be predicted for optimal integration [12] while weights of single indentations of the second stimulus decreased with time, indicating that not all information about the second stimulus could be used for the comparison. We could also show that observed decay in weights was consistent with memory decay in tactile perception [17,25], suggesting that a difference in memory demands is a more likely explanation for the observed difference in performance in different switch conditions.

The Kalman model of serial integration proposed by [17], assumes that when switching only once between the stimuli, differences between each sequentially gathered estimate of the second stimulus and the overall estimate of the first stimulus are integrated in a statistically optimal way given memory decay, modelled as increasing variance of the first stimulus' estimate over time. Thus, given the same amount of information, this model predicts that precision should be higher the shorter the first stimulus needs to be remembered, which is consistent with our results. However, to apply this model to our data, we would need to extend it by the variable containing the result of one comparison. To test the predictions of this extended model we would need additional data (e.g. precision based on one indentation per stimulus).

Overall our results suggest that in active touch perception where integration of information is limited by memory [16–18,20], the naturally applied strategy of switching between the stimuli [21] can be more beneficial for performance than prolonged accumulation of information about each stimulus.

References

1. Najemnik, J., Geisler, W.S.: Optimal eye movement strategies in visual search. Nature **434**(7031), 387–391 (2005)
2. Lederman, S.J., Klatzky, R.L.: Hand movement: a window into haptic object recognition. Cogn. Psychol. **19**, 342–368 (1987)
3. Saig, A., Gordon, G., Assa, E., Arieli, A., Ahissar, E.: Motor-sensory confluence in tactile perception. J. Neurosci. **32**(40), 14022–14032 (2012)
4. Toscani, M., Valsecchi, M., Gegenfurtner, K.R.: Optimal sampling of visual information for lightness judgments. Proc. Natl. Acad. Sci. **110**(27), 11163–11168 (2013)
5. Drewing, K.: After experience with the task humans actively optimize shape discrimination in touch by utilizing effects of exploratory movement direction. Acta Psychologica **141**(3), 295–303 (2012)
6. Kaim, L., Drewing, K.: Exploratory strategies in haptic softness discrimination are tuned to achieve high levels of task performance. IEEE Trans. Haptics **4**(4), 242–252 (2011)

7. Lezkan, A., Metzger, A., Drewing, K.: Active haptic exploration of softness: indentation force is systematically related to prediction, sensation and motivation. Front. Integrative Neurosci. **12**, 59 (2018)
8. O'Regan, J.K.: Solving the "real" mysteries of visual perception: the world as an outside memory. Can. J. Psychol. **46**, 461–488 (1992)
9. Ballard, D.H., Hayhoe, M.M., Pelz, J.B.: Memory representations in natural tasks. J. Cogn. Neurosci. **7**(1), 66–80 (1995)
10. Kersten, D., Mamassian, P., Yuille, A.: Object perception as Bayesian inference. Ann. Rev. Psychol. **55**, 271–304 (2004)
11. Körding, K.P., Wolpert, D.M.: Bayesian integration in sensorimotor learning. Nature **427**(6971), 244–247 (2004)
12. Ernst, M.O., Banks, M.S.: Humans integrate visual and haptic information in a statistically optimal fashion. Nature **415**(6870), 429–433 (2002)
13. Kwon, O.S., Tadin, D., Knill, D.C.: Unifying account of visual motion and position perception. Proc. Natl. Acad. Sci. **112**(26), 8142–8147 (2015)
14. Kalman, R.E.: A new approach to linear filtering and prediction problems. J. Basic Eng. **82**(1), 35–45 (1960)
15. Metzger, A., Drewing, K.: Memory influences haptic perception of softness. Sci. Rep. **9**, 14383 (2019)
16. Metzger, A., Lezkan, A., Drewing, K.: Integration of serial sensory information in haptic perception of softness. J. Exp. Psychol.: Hum. Perception Perform. **44**(4), 551–565 (2018)
17. Lezkan, A., Drewing, K.: Processing of haptic texture information over sequential exploration movements. Attention Percept. Psychophys. **80**(1), 177–192 (2017)
18. Metzger, A., Drewing, K.: The longer the first stimulus is explored in softness discrimination the longer it can be compared to the second one. In: 2017 IEEE World Haptics Conference, WHC 2017 (2017)
19. Lewandowsky, S., Oberauer, K.: No evidence for temporal decay in working memory. J. Exp. Psychol.: Learn. Memory Cogn. **35**(6), 1545–1551 (2009)
20. Metzger, A., Drewing, K.: Effects of stimulus exploration length and time on the integration of information in haptic softness discrimination. IEEE Trans. Haptics **12**(4), 451–460 (2019)
21. Zoeller, A.C., Lezkan, A., Paulun, V.C., Fleming, R.W., Drewing, K.: Integration of prior knowledge during haptic exploration depends on information type. J. Vis. **19**(4), 1–15 (2019)
22. Gerling, G.J., Hauser, S.C., Soltis, B.R., Bowen, A.K., Fanta, K.D., Wang, Y.: A standard methodology to characterize the intrinsic material properties of compliant test stimuli. IEEE Trans. Haptics **11**(4), 498–508 (2018)
23. Schuett, H.H., Harmeling, S., Macke, J.H., Wichmann, F.A.: Painfree and accurate Bayesian estimation of psychometric functions for (potentially) overdispersed data. Vis. Res. **122**, 105–123 (2016)
24. Cousineau, D.: Confidence intervals in within-subject designs: a simpler solution to Loftus and Masson's method. Tutorial Quant. Methods Psychol. **1**(1), 42–45 (2005)
25. Murray, D.J., Ward, R., Hockley, W.E.: Tactile short-term memory in relation to the two-point threshold. Q. J. Exp. Psychol. **27**(2), 303–312 (1975)

BY

Discriminating Between Intensities and Velocities of Mid-Air Haptic Patterns

Isa Rutten[1], William Frier[2(✉)], and David Geerts[1]

[1] Mintlab, KU Leuven, Leuven, Belgium
{isa.rutten,david.geerts}@kuleuven.be
[2] Ultraleap, Bristol, UK
william.frier@ultraleap.com

Abstract. This study investigates people's ability in discriminating between different intensities and velocities of mid-air haptic (MAH) sensations. Apart from estimating the just noticeable differences (JND), we also investigated the impact of age on discrimination performance, and the relationship between someone's confidence in his/her performance and the actual discrimination performance. In an experimental set-up, involving 50 participants, we obtained a JND of 12.12% for intensities and 0.51 rev/s for velocities. Surprisingly, the impact of age on discrimination performance was only small and almost negligible. Furthermore, participants' subjective perception of their discrimination performance aligned well with their actual discrimination performance. These results are encouraging for the use of intensities and velocities as dimensions in the design of MAH patterns to convey information to the user.

Keywords: Mid-air haptic feedback · Just-noticeable difference · Intensity · Velocity

1 Introduction

Mid-air haptic (MAH) feedback aims at stimulating the sense of touch in mid-air and is predominantly rendered using ultrasound [3]. Compared to vibrotactile feedback, MAH feedback is rendered with less fidelity, but its advantage of not requiring an actuator attached to the body makes it an attractive way of providing system feedback in gesture-based interfaces. While MAH's usefulness for in-car infotainment [9] or digital kiosks [13] has already been studied, an important question remains: How can information be encoded within MAH patterns?

One way is to rely on MAH shapes, with studies showing shape identification rates ranging between 44% [18] to 60% [12] and 80% [14]. However, the shape or pattern type constitutes only one channel of information. As MAH sensations are often applied to gesture-based devices as a type of system feedback, the

Supported by the SHAKE project, realized in collaboration with imec. Project partners are Verhaert, Nokia Bell, NXP, imec and Mintlab (KU Leuven), with project support from VLAIO (Flanders Innovation and Entrepreneurship).

I. Nisky et al. (Eds.): EuroHaptics 2020, LNCS 12272, pp. 78–86, 2020.
https://doi.org/10.1007/978-3-030-58147-3_9

possibility to encode information in different channels of the MAH sensation would be valuable as it enables richer and more diverse feedback messages. For instance, in a car, one MAH channel could convey the music loudness while another channel could convey the AC fan speed. To enable such applications, we investigated two potentials information channels regarding MAH sensations, namely intensity and velocity.

To this end, we estimated people's abilities to discriminate between differences in the intensity and velocity of MAH patterns. We assessed the Just-Noticeable Difference (JND) of both the intensity and velocity of MAH sensations. In order to compare JNDs across stimulus types (MAH vs vibrotactile) and dimensions (intensity vs velocity), Weber fractions were calculated [21]. Earlier work has observed Weber fractions for the intensity of vibrotactile stimuli to range between 13% and 16% [5,8]. When considering the velocity of tactile brushing stimuli, Weber fractions between 20% and 25% have been found [7]. However, it is unknown how these values would extend to the case of MAH stimuli.

Additionally, we investigated to what extent age would influence the discrimination performance. We expected a decrease in performance with increasing age based on earlier work, showing a sharp drop in identification accuracy of MAH shapes with increasing age [18]. Finally, we assessed participants' confidence in their own discrimination performance, in order to explore whether their subjective experience would align well with their actual discrimination performance.

2 Methods

2.1 Participants

A total of 50 people, 25 men and 25 women, participated in the study, with a mean age of 44.58 years (SD = 15.93, range = 19–77). Only 7 participants were left-handed, all others were right-handed. Exclusion criteria were: Previous experience with mid-air haptic feedback and touch deficits in the upper limbs. This study was approved by the local social and societal ethics committee (G- 2018 10 1361), and participants received a voucher of 20 euro as incentive.

2.2 Device and Stimuli

The MAH stimuli were created using the sensation editor of the touch development kit, developed by Ultrahaptics©. The device produced a dial-like MAH pattern: one focal point, with amplitude modulated at 125 Hz [1], which moved around a circular path with a diameter of 5 cm (see Fig. 1-right). In the intensity discrimination task, the target stimulus levels were intensities of 100%, 95%,

[1] 125 Hz is the default modulation frequency of the Ultrahaptics' sensation editor.

90%, 85%, 80%, 75%[2], and all stimuli had a speed of 2 rev/s. In the velocity discrimination task, the target stimulus levels were velocities of 2 rev/s, 1.8 rev/s, 1.6 rev/s, 1.4 rev/s, 1.2 rev/s, and 1 rev/s, and all stimuli had an intensity of 100%. In both task, the reference stimuli were characterized by an intensity level of 100% and a speed of 2 rev/s.

Fig. 1. Left: Experimental set-up. Right: Visual representation of the MAH sensation

2.3 Tasks

Both experimental tasks were identical except for the target stimuli being used (see Device and Stimuli). The tasks were created using PsychoPy 3 [17] and were administered on a laptop. We applied the Method of Constant Stimuli in a 2-interval forced choice (IFC) task. In each trial, participants indicated which of two stimuli they perceived as the target stimulus, with the stimuli being presented successively in randomized order. In the intensity task, participants were instructed to indicate which stimulus had the lowest intensity, and in the velocity task participants had to indicate which stimulus had the slowest speed. Thus, the target stimulus was always the stimulus with the lowest intensity or slowest speed. Participants had to indicate the temporal position of the target stimulus by pressing "1" or "2" on their keyboard, with pressing 1 referring to the first stimulus they experienced in that trial, and pressing 2 to the second stimulus. No corrective feedback was provided. In line with the typical amount of stimulus levels involved [19], we chose six different target stimulus levels to compare against the reference level (see Device and Stimuli). Each level was presented 12 times, resulting in a total of 72 trials in each task. Each stimulus had a duration of one second, with an inter-stimulus interval of 0.5 s, and unlimited response time. After each response, the trial ended when participants indicated on a scale from 1–7 how confident they felt about their answer.

[2] Ultrahaptics' maximum output is limited to 155 dB SPL for health and safety reasons. The intensity percentages can be translated to the following Pascal values: 100% = 155 dB, SPL = 1124 Pa, 95% = 1067.8 Pa, 90% = 1011.6 Pa, 85% = 955.4 Pa, 80% = 899.2 Pa, 75% = 843 Pa.

2.4 Procedure

After signing the informed consent, participants filled out a demographic survey. They were asked to position their non-dominant hand above the MAH device, with their arm resting on the armrest to avoid excessive fatigue (see Fig. 1-left). The armrest ensured a stable distance of about 20 cm between the hand and the MAH device. Participants were urged to keep their hand horizontally and fixed above the device. To familiarize with the feeling of MAHs, they were presented with a single focal point on the palm during 10 s before starting with the tasks. Next, participants were told that all stimuli in both tasks would have the same pattern. To give them an idea of what to expect, a visual representation (short movie) of the dynamical stimulus pattern was presented on their screen during 5 s (see Fig. 1-right). Next, participants were assigned to both conditions in a counterbalanced order (within-participant design). Between the two discrimination tasks, a short break was provided.

3 Results

3.1 Discrimination of Intensities

Two participants were excluded from these analyses because of a technical error and a misunderstanding of the instructions, resulting in a sample of N = 48. We first checked whether the subjective confidence in one's responses would correlate with one's accuracy (=proportion of correct responses)[3], and this was indeed the case. A (repeated measures) correlation coefficient of $r_{r_m}(239) = .63, p < .001, 95\%\ CI = [.54, .70]$ was observed (R package rmcorr [1]). The correlation coefficient slightly increased in strength after removing nine influential data points based on Cook's distance: $r_{r_m}(230) = .66, p < .001, 95\%\ CI = [.58, .73]$. Secondly, we tested for an association between discrimination accuracy and age. A negative association between age and discrimination accuracy was observed, with $r = -.35, t(46) = -2.53, p = .02, 95\%\ CI = [-.58, -.07]$. However, when we removed three data points that were identified as influential outliers, the correlation coefficient diminished in strength and could only barely reach statistical significance (see Fig. 2), $r = -.30, t(43) = -2.03, p = .049, 95\%\ CI = [-.54, -.002]$.

We modeled the count data (how many times the target stimulus was picked as having the lowest intensity) using a generalized linear mixed model (GLMM) with a probit link function, and with the intensity of the target stimulus as predictor[4] (R package lme4 [2]). The model included person-based random intercepts, as they significantly improved the model's fit, showing a $SD = 0.14$, $X^2(1) = 16.43$, $p < .0001$, with $AIC = 1158$, $BIC = 1169$, $LL = -576$ for the model with random intercepts, and $AIC = 1172.4$, $BIC = 1179.8$, $LL = -584.21$ for the model without random intercepts. The final model did

[3] Within all participants, the mean confidence ratings were correlated with the proportion of correct responses over stimulus levels.

[4] The predictor "intensity" was rescaled by dividing it by 100.

not include random slopes, as this resulted in singular fit, which often indicates overfit. No influential outliers were observed, using R package influence.ME [16].

Next, the JND was estimated from the GLMM model using the Delta method, without any lapse rate or bias parameters included in the model (R package MixedPsy [15]). The JND was defined as the inverse function of the slope (see also Dallmann, Ernst, & Moscatelli, 2015 [6]). It measured the perceptual noise and was computed as half of the difference between the stimulus values at response probabilities of 75% and 25%. We obtained a JND = 12.12%(SE = 0.62%), 95%CI = [10.90%, 13.33%]. Next, a preliminary Weber fraction was calculated as follows: $\Delta I/I = c$, 12.12/100 = 0.1212, resulting in a fraction of 12.12% [11,21].

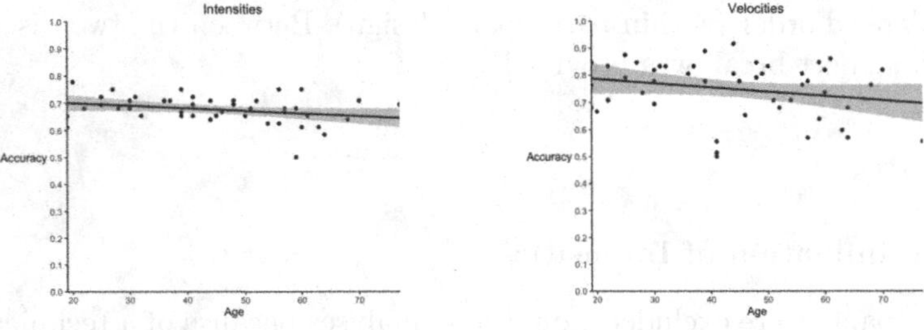

Fig. 2. Scatter plots with regression line (grey zone indicates the 95% CI) showing the relationship between accuracy and age in the discrimination task involving intensities (left) and velocities (right).

3.2 Discrimination of Velocities

No participants were excluded from this task, resulting in a sample of N = 50. A strong correlation was found between self-reported confidence in one's ratings and accuracy, $r_{r_m}(249) = .66, p < .001, 95\%\ CI = [.58, .72]$. The correlation coefficient increased in strength after removing 14 influential data points: $r_{r_m}(235) = .72, p < .001, 95\%\ CI = [.65, .78]$. Similar to intensity discrimination, a negative relationship was observed between accuracy and age, $r = -.31, t(48) = -2.29, p = .03, 95\%\ CI = [-.54, -.04]$, but disappeared after removing an outlier (based on Cook's distance), $r = -.23, t(47) = -1.65, p = .11, 95\%\ CI = [-.48, .05]$ (see Fig. 2).

Next, we again modeled the count data (how many times the target stimulus was picked as having the slowest speed) using a GLMM with a probit link function, and with the target's number of rev/sec as predictor. Using model comparison, we tested whether we should include random intercepts and slopes in the model. There was evidence for significantly improved model fit when adding person-based random intercepts, $SD = 0.35, X^2(1) = 125.60, p < .0001$, with $AIC = 1175.1, BIC = 1186.2, LL = -584.55$ for the model with random intercepts, and $AIC = 1298.7$, $BIC = 1306.1$, $LL = -647.35$ for the model without random intercepts. Given that the addition of random slopes resulted in a correlation of −1 between the random effects (often a sign of overfit), the final model only included random intercepts. No outliers were detected based on Cook's distance. When estimating the discrimination threshold between different velocities based on the GLMM model and using the Delta method [15] (no lapse rate or bias parameters were included in the model), we obtained the following values: JND = 0.51 rev/s ($SD = 0.03$ rev/s), 95% CI = [0.45 rev/s, 0.56 rev/s]. A preliminary Weber fraction of 25.5% was obtained: $\Delta V/V = c, 0.51/2 = 0.255$.

4 Discussion

This study shows JNDs of 12.12% for the intensity and 0.51 rev/s for the velocity of MAH sensations. In other words, to use MAH intensity and velocity as information channels, a minimum difference in intensity of at least 12.12%, and a minimum velocity difference of at least 0.51 rev/s needs to occur from the reference output of 100% intensity and 2 rev/s.

The estimated Weber fraction of 25.5% for MAH velocity is comparable, although slightly higher, to the fractions observed for tactile brushing stimuli, which were found to vary between 20% and 25% [7].

Our estimated Weber fraction for MAH intensity is based on acoustic pressure, while the Weber fractions for vibrotactile intensity are based on skin displacement, and therefore cannot be directly compared. Assuming the skin is mainly elastic [22], displacement is proportional to stress. Additionally, stress at the MAH pattern is proportional to the square of the acoustic pressure [4]. Thus, displacement is proportional to the square of the acoustic pressure and one can estimate the Weber fraction of MAH displacement to be 25.7% : $\Delta I^2/I^2 = 0.257$. This estimated Weber fraction is higher than the Weber fractions observed for vibrotactile stimuli (ranging between 13% [8] and 16% [5]). We hypothesise that this difference lies in the nature of the stimuli. Vibrotactile stimuli are easier to discriminate when subtle changes in intensity or velocity occur, as they are rendered with higher fidelity.

Wilson et al. [23] reported good motion perception of MAHs in the range 0.1–2 rev/s. Using our estimated Weber fraction, one could decrease the reference MAH velocity by 10 levels and still remain in this range. Similarly, previous studies reported a perceptual threshold for MAH intensities around 30% of the maximum output pressure [10, 20]. Using our estimated Weber fraction, one could decrease the reference intensity by 9 levels and still remain above this threshold. Based on these observations, we can assume that one can discriminate between 11 and 10 levels of MAH velocities and intensities, respectively (i.e. 3.5 and 3.3 bits of information, respectively). Moreover, bits of information from two dimensions can be summed up [11]. Hence we predict a maximum of 6.8 bits of information using the MAH pattern investigated in our study. However, we highlight that this value is only an upper bound for two reasons. Firstly, we relied on perceptual thresholds from the literature to estimate the bits of information, instead of data from an information transfer study. Secondly, these perceptual thresholds from the literature were observed for different types of MAH patterns, compared to the pattern we used in our study Therefore, while this estimate is encouraging for using MAH intensity and velocity as information channels, further research is needed to determine the actual value.

Interestingly, age was not found to be strongly related to discrimination accuracy of intensities or velocities. These findings are remarkable as earlier work found a strongly negative correlation between age and identification accuracy of MAH shapes, $r = -.62$ [18]. However, as we did not assess physiological characteristics of participants' skin, it is hard to explain why the correlation between age and discrimination accuracy was weak in our study. Finally, based on the confidence ratings, it became clear that people's subjective perception of their discrimination abilities aligned nicely with their actual performance[5].

With this study, we performed a first step in mapping intensity and velocity JNDs of MAH patterns. However, the current study only looked at dial-like patterns. Future work should verify the generalizability of the obtained results regarding other pattern types. Moreover, we estimated the intensity and velocity Weber fractions based on only one reference stimulus, future work should improve this estimation by including multiple reference stimuli and target stimuli with values both lower/slower and greater/faster than the reference stimulus. Additionally, future studies should investigate to what extent the start and end positions of the stimuli in the velocity discrimination task could influence participants' performance. To avoid a potential confound of start position, one could randomize the start position between trials. Despite these limitations, we believe the current study is a valuable initial step towards investigating intensity and velocity differences of MAH sensations as potential information channels.

5 Conclusion

This study is the first to report intensity and velocity JNDs for MAH sensations, which is essential when implementing MAHs as a means of system feedback. Based on our results, we recommend to only implement intensity differences greater than 12.12% and velocity differences greater than 0.51 rev/s. Intensity and velocity appear to be attractive information channels when considering MAH sensations as a means of system feedback. Future research could further elaborate on these findings, with a focus on investigating their boundary conditions.

Acknowledgements. We thank the participants for their valuable time and effort. This work was supported by the SHAKE project, realized in collaboration with imec. Project partners are Verhaert, Nokia Bell, NXP, imec and Mintlab (KU Leuven), with project support from VLAIO (Flanders Innovation and Entrepreneurship). This project has also received funding from the EU Horizon 2020 research and innovation programme under grant agreement No 801413.

[5] A final person-based characteristic we investigated was hand size (= sum of the hand width and length) but this revealed no significant correlation with the discrimination accuracy of intensities or velocities.

References

1. Bakdash, J.Z., Marusich, L.R.: rmcorr: Repeated Measures Correlation. R package version 0.3.0 (2018). https://CRAN.R-project.org/package=rmcorr/
2. Bates, D., Mächler, M., Bolker, B., Walker, S.: Fitting linear mixed-effects models using lme4. arXiv preprint arXiv:1406.5823 (2014)
3. Carter, T., Seah, S.A., Long, B., Drinkwater, B., Subramanian, S.: Ultrahaptics: multi-point mid-air haptic feedback for touch surfaces. In: Proceedings of the 26th Annual ACM Symposium on User Interface Software and Technology (2013)
4. Chilles, J., Frier, W., Abdouni, A., Giordano, M., Georgiou, O.: Laser doppler vibrometry and fem simulations of ultrasonic mid-air haptics. In: Proceeding of IEEE World Haptics Conference (2019). https://doi.org/10.1109/WHC.2019.8816097
5. Craig, J.C.: Difference threshold for intensity of tactile stimuli. Percept. Psychophys. **11**, 150–152 (1972). https://doi.org/10.3758/BF03210362
6. Dallmann, C.J., Ernst, M.O., Moscatelli, A.: The role of vibration in tactile speed perception. J. Neurophysiol. **114**(6), 3131–3139 (2015)
7. Essick, G., Franzen, O., Whitsel, B.: Discrimination and scaling of velocity of stimulus motion across the skin. Somatosens. Mot. Res. **6**(1), 21–40 (1988)
8. Francisco, E., Tannan, V., Zhang, Z., Holden, J., Tommerdahl, M.: Vibrotactile amplitude discrimination capacity parallels magnitude changes in somatosensory cortex and follows Weber's law. Exp. Brain Res. **191**(1), 49 (2008). https://doi.org/10.1007/s00221-008-1494-6
9. Harrington, K., Large, D.R., Burnett, G., Georgiou, O.: Exploring the use of mid-air ultrasonic feedback to enhance automotive user interfaces. In: Proceedings of the 10th International Conference on Automotive User Interfaces and Interactive Vehicular Applications (2018)
10. Howard, T., Gallagher, G., Lécuyer, A., Pacchierotti, C., Marchal, M.: Investigating the recognition of local shapes using mid-air ultrasound haptics. In: 2019 IEEE World Haptics Conference (WHC), pp. 503–508. IEEE (2019)
11. Jones, L.A., Tan, H.Z.: Application of psychophysical techniques to haptic research. IEEE Trans. Haptics (2013). https://doi.org/10.1109/TOH.2012.74
12. Korres, G., Eid, M.: Haptogram: ultrasonic point-cloud tactile stimulation. IEEE Access **4**, 7758–7769 (2016)
13. Limerick, H., Hayden, R., Beattie, D., Georgiou, O., Müller, J.: User engagement for mid-air haptic interactions with digital signage. In: Proceedings of the 8th ACM International Symposium on Pervasive Displays. ACM (2019)
14. Long, B., Seah, S.A., Carter, T., Subramanian, S.: Rendering volumetric haptic shapes in mid-air using ultrasound. ACM Trans. Graph. **33**(6), 1–10 (2014)
15. Moscatelli, A., Mezzetti, M., Lacquaniti, F.: Modeling psychophysical data at the population-level: the generalized linear mixed model. J. Vision **12**(11), 26 (2012)
16. Nieuwenhuis, R., Te Grotenhuis, M., Pelzer, B.: influence.Me: tools for detecting influential data in mixed effects models. R J. **4**(2), 38–47 (2012)
17. Peirce, J., MacAskill, M.: Building Experiments in PsychoPy. Sage, New York City (2018)
18. Rutten, I., Frier, W., Van den Bogaert, L., Geerts, D.: Invisible touch: how identifiable are mid-air haptic shapes? In: Extended Abstracts of the 2019 CHI Conference on Human Factors in Computing Systems (2019). https://doi.org/10.1145/3290607.3313004

19. Simpson, W.A.: The method of constant stimuli is efficient. Percept. Psychophys. **44**, 433–436 (1988). https://doi.org/10.3758/BF03210427
20. Takahashi, R., Hasegawa, K., Shinoda, H.: Tactile stimulation by repetitive lateral movement of midair ultrasound focus. IEEE Trans. Haptics **13**(2), 334–342 (2019)
21. Weber, E.H., Ross, H.E.: The sense of Touch. Academic Press for [the] Experimental Psychology Society (1978)
22. Wiertlewski, M., Hayward, V.: Mechanical behavior of the fingertip in the range of frequencies and displacements relevant to touch. J. Biomech. (2012). https://doi.org/10.1016/j.jbiomech.2012.05.045
23. Wilson, G., Carter, T., Subramanian, S., Brewster, S.A.: Perception of ultrasonic haptic feedback on the hand: localisation and apparent motion. In: Proceedings of the 32nd Annual ACM Conference on Human Factors in Computing Systems - CHI 2014 (2014). https://doi.org/10.1145/2556288.2557033

Density Estimation is Influenced More by Mass When Objects are Denser

Lara Merken[1] and Vonne van Polanen[2(✉)]

[1] Department of Neurosciences, KU Leuven, Herestraat 49, 3000 Leuven, Belgium
lara.merken@kuleuven.be

[2] Department of Movement Sciences and Leuven Brain Institute, KU Leuven, Tervuursevest 101, 3001 Leuven, Belgium
vonne.vanpolanen@kuleuven.be

Abstract. When judging the heaviness of objects, the perceptual estimate can be influenced by the object's density next to its mass. In the present study, we investigated whether density estimates might be similarly affected by object mass. Participants lifted objects of different sizes and masses in a virtual reality environment and estimated the density. We found that density perception was influenced both by density and mass, but not for the lowest density value, which could be perceived correctly. A modelling procedure on fitted slopes through the different objects revealed that density contributed 56% to the density estimate. However, if low- and high-density values were modelled separately, contributions of 100% and 41% were found for the low and high densities, respectively. These results indicate that perception of heaviness and density are closely related but can be better distinguished with objects of lower density and mass.

Keywords: Perception · Density · Mass · Heaviness

1 Introduction

When we manipulate objects, we can perceive object properties such as size, weight and material. During object lifting, we receive haptic feedback that allows us to make a heaviness estimation of the object. The brain defines the most optimal estimate (i.e. with minimal variance) based on different sensory sources, such as visual and haptic information [5]. In other words, a combination of cues can be used to make a perceptual estimate. Regarding heaviness perception, one of these cues might be object density, which is the relation between the size and weight of the object and depends on the objects' material.

Previous research has shown that density might contribute to heaviness perception. Five decades ago, Ross and Di Lollo [12] suggested that the impact

Research supported by a Fonds Wetenschappelijk Onderzoek grant to VVP (FWO post-doctoral fellowship, Belgium, 12X7118N). The authors would like to thank Ellen Vervoort for her help in data collection.

I. Nisky et al. (Eds.): EuroHaptics 2020, LNCS 12272, pp. 87–95, 2020.
https://doi.org/10.1007/978-3-030-58147-3_10

of mass and density in heaviness estimation shifts with the objects' density. The authors claimed that heaviness estimations of low-density objects were predominantly based on mass, while for high-density objects the density had more impact. In another study was suggested that the combination of density and mass controls heaviness discrimination. When both density and mass simultaneously increase/decrease, heaviness perception is more accurate. In contrast, accuracy declines when the factors change in opposite directions [9]. More recently, it was proposed that heaviness estimation is formed from a weighted combination of mass and density information, depending on the properties' reliability [3,13].

The inclusion of other object properties, such as density, into heaviness estimation might provide an explanation for some weight-related illusions. The size-weight illusion [1] shows that a smaller object is perceived as being heavier than a larger object with an identical mass, suggesting that the unexpected denser object feels heavier. Similarly, the material-weight illusion shows that visual features of the material bias the perception of heaviness. When two objects of seemingly different materials but equal size and mass are presented, the denser-looking material (i.e. metal vs wood) is perceived to be lighter [4].

These studies show that density influences heaviness perception. It has also been observed that density estimations can be influenced by object mass [7], suggesting that neural processing of these concepts is related. However, in that study, mass was not varied. Therefore, it has not been systematically examined how mass could affect density perception. Object density cannot be directly perceived but has to be inferred from both the size and mass of an object, using visual and/or haptic information. The perception of density is important to distinguish between different object materials, such as wood or metal, or, e.g. determine the content of a closed box. For this purpose, further research on density perception might give new insights into object perception. In the present study, we investigated how participants perceived the density of objects of different sizes and weights. Participants lifted objects in a virtual reality set-up and judged the density. We found that the mass of objects affected the density estimate but more so when objects were denser and heavier.

2 Methods

2.1 Set-Up and Procedure

12 participants took part in the study (age range 18–26 years, 6 females, 11 self-reported right-handed). They had normal or corrected-to-normal vision and had no known neurological impairments. Before the experiment they all signed an informed consent form. The study was approved by the local ethical committee of KU Leuven.

The experiment was performed in a virtual reality (VR) setup that simulated virtual cuboids on a checkerboard background. The visual virtual environment was projected with a 3D screen (Zalman) on a mirror, under which participants moved their hands so they were unable to see their hands. The tips of the thumb and index finger were inserted into two haptic force-feedback devices (Phantom

Premium 1.5, Sensable) to be able to haptically simulate the objects. The fingertips were visually indicated by two red spheres on the screen. More details of the VR setup can be found in [11]. Forces were measured in 3 directions and sampled at a 500 Hz frequency. 12 cuboids were used in the experiment with different size-mass combinations (Table 1). The size was varied by changing object height, while keeping the width and depth at 5 cm. Each density value (0.75, 1.25, 1.75, 2.25 g/cm^3) was presented in 3 different size-mass combinations.

Participants were seated in front of the VR setup and familiarised with the environment. In each trial, a virtual object appeared and participants had to grasp it after hearing a beep. They lifted the object with thumb and index finger up to a target height, indicated with a yellow sphere. After they had replaced the object, they were asked to estimate the density of the object on a self-chosen scale. Before the experiment, they received an explanation of the concept of density and were instructed not to report the heaviness, but only the density of the object. Participants were unaware of the number of objects or densities used in the experiment. Each object was presented 10 times, giving a total of 120 trials, which were presented in a randomised order. Two objects of the same weight or size were never presented in a row, so participants could not compare the density of two sequential objects by only varying one parameter. The experiment was divided into two sessions with a short break in between.

Table 1. Mass (g), size (cm) and density (g/cm^3) of the objects used in the experiment.

Size/Mass	112.5	157.5	202.5	262.5	337.5	472.5
3.6	1.25	1.75	2.25			
6	0.75			1.75	2.25	
8.4		0.75		1.25		2.25
10.8			0.75		1.25	1.75

2.2 Analysis

The individual density estimates of participants were normalised to z-scores. The normalised estimates were averaged over the trials for each object. Missing samples (0.03%) of the force data were interpolated. Next, force data was filtered with a 2nd order low-pass Butterworth filter with a cut-off frequency of 15 Hz. The load forces (LF) were the sum of the vertical forces and were used to calculate the loading phase duration (LPD). We used the LPD as a measure of lifting performance, since this is indicative of force planning and different for different object masses [8]. The LPD was calculated as the time between LF onset (LF > 0.1 N) and lift-off (LF > object weight). 31 (2%) trials were removed from the force analysis due to incorrect lifts.

A linear mixed model was used to statistically determine whether density and mass affected the perceived density and LPD. The z-scored estimates or the

LPD was the dependent variable, density and mass were included as fixed factors and participant number as random factor. The density × mass interaction and the intercepts were also included in the model. We used maximum likelihood for the estimation and a 1st order autoregressive model. Post-hoc comparisons were performed with a Bonferroni correction. The alpha was set to 0.05.

The linear mixed model provides estimates for each density and mass, because they were included as fixed factors, but no relative contribution of those factors could be computed. Therefore, we performed a different modelling procedure where we assumed that density estimates only relied on mass and density. The weights given to density and mass were w_d and w_m, respectively, and $w_d + w_m = 1$. In general, a relation such as $D = w_d ad + w_m bm + c$ can be expected, where D is the estimated density, d and m the density and mass values and a, b and c are constants. Since we only had one data set, we could not fit all the unknown constants and the weights. Therefore, we chose to fit the relation between D and d or m by calculating the slopes: s_d and s_m, respectively. We assumed that the slopes could be seen as a relative value of a constant value, which we called the maximum slope:

$$\begin{aligned} s_m &= w_d s_{max,m} \\ s_d &= (1 - w_d) s_{max,d} \end{aligned} \tag{1}$$

More specifically, to assess the contribution of density to the density estimates, we plotted the density estimates against the mass of the objects (Fig. 1). Through similar densities, we fitted linear regressions to obtain 4 slopes ('density slopes'). When solely relying on density, slopes will be equal to 0 (Fig. 1A). However, if density would be ignored and participants rely on mass, the slopes would be on the line between the lowest and highest mass (Fig. 1B), i.e. the maximum slope ($s_{max,d}$). The slopes will have values in between these cases when participants have a balanced weighting of mass and density. Similarly, to evaluate the contribution of mass, 6 slopes could be fitted to objects with equal mass ('mass slopes') when density estimates are plotted against object density. If participants rely solely on density, mass slopes will be maximal (Fig. 1A). However, if they ignore density and only rely on mass, mass slopes are 0 (Fig. 1B).

We set the maximum slopes as the differences between the estimate for the object with the lowest mass-density combination and the highest mass-density combination: $s_{max,d} = 6.13$ and $s_{max,m} = 1.47$. Next, the only free parameter, w_d, was obtained by minimizing the sum of the squared errors between the modelled slopes from Eq. 1 and the measured slopes from the data. Note that only slopes were fitted to estimate w_d and not density estimates. Therefore, the modelled slopes and weights cannot be used to calculate expected density estimates from the mass and density of objects. The plotted modelled slopes in the Fig. 1 and 2 only serve to compare the slope value, as the intercepts were not fitted, but estimated by dividing the maximum slope into equal steps for the number of slopes and depended on w_d.

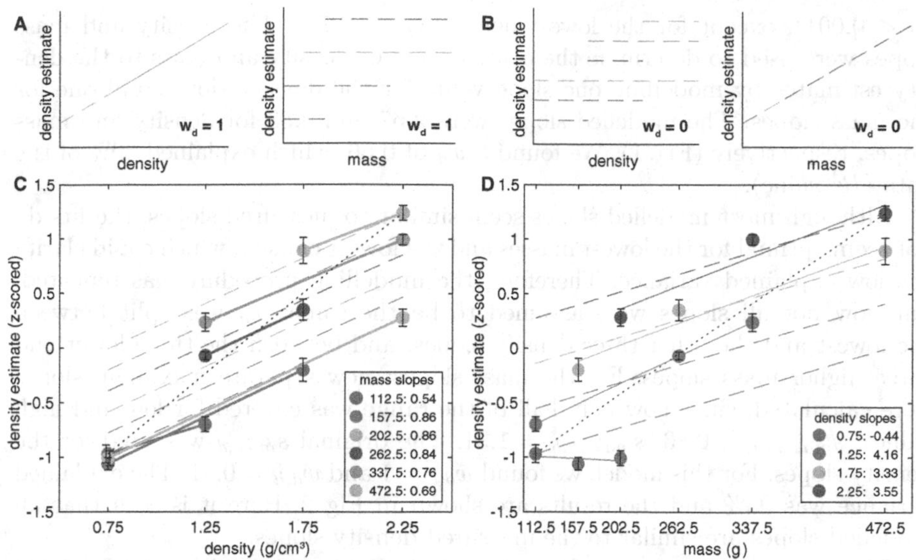

Fig. 1. Mass and density slopes. **A–B**: predicted slopes if the weight given to density (w_d) is equal to 1 or 0. **C–D**: mean density estimates (solid circles) and standard errors (error bars). Solid lines represent measured slopes (C: mass, D: density) fitted to the data, which values are indicated in the legend. Dashed and dotted lines indicate modelled slopes and maximum slopes from the model.

3 Results

Figure 1 shows the average density estimates for each object. It can be seen that the estimates increase both with density and mass. The linear mixed model on the density estimates revealed significant effects of mass ($F(3, 134) = 45, p < 0.001$) and density ($F(5, 107) = 97, p < 0.001$), but not their interaction ($p = 0.533$). Post-hoc effects showed that all density values differed significantly from each other (all $p < 0.001$). The masses all differed significantly from each other (all $p < 0.027$), except the two lowest masses ($p = 0.208$). Thus, for most objects, both mass and density affected the perceptual estimate for density.

For LPD, there was only a significant effect of mass ($F(5, 114) = 55, p < 0.001$). Except for the 157.5 g and 202.5 g objects ($p = 0.72$), and the 202.5 g and 262.5 g objects ($p = 0.75$), all other masses differed from each other (all $p < 0.003$). The effect of density or the interaction of density × mass were not significant, indicating that lifting performance was only affected by object mass, not density. Therefore, the density estimates for different densities could not be explained by alterations in lifting performance.

The measured slopes through objects with similar mass or density are shown in Fig. 1 and values are displayed in the legend. One sample t-tests of participants' slopes indicated that all mass slopes were significantly different from zero (all $p < 0.004$). The density slopes were also significantly different from zero

($p < 0.001$), except for the lowest density ($p = 0.75$). The density and mass slopes were used to determine the contribution of density and mass to the density estimates by modelling one slope value for the density slopes and one for the mass slopes. The modelled slopes were 2.67 and 0.83 for density and mass slopes, respectively (Fig. 1). We found a w_d of 0.56, which explained 39% of the data (R^2 value).

Although most modelled slopes seem similar to measured slopes, the fits do not seem optimal for the lowest masses and the lowest density, which could clarify the low explained variance. Therefore, the modelling procedure was repeated, but now not all slopes were assumed to be the same: w_d was split between the lowest and the other three density slopes, and between the three lower and three higher mass slopes. For the mass slopes, new separate maximum slopes were calculated, since now only half of the range was covered for low and high values: $s_{max,m,l} = 0.83$, $s_{max,m,h} = 1.31$. The original $s_{max,d}$ was used for the density slopes. For this model, we found $w_{d,l} = 1$ and $w_{d,h} = 0.41$. The explained variance was 96% and the results are shown in Fig. 2. Here it is seen that all modelled slopes are similar to the measured density slopes.

The procedure was also performed on the individual data. A range of weights was found with w_d ranging between 0.12–1.0 (mean = 0.55), with the explained variance ranging from 0–97%. However, if we also split the fits for the lower and higher densities, we obtained weights of $w_{d,l}$ between $0.06 - 1.0$ and $w_{d,h}$ between 0.04–1.0, with means of 0.78 and 0.43, respectively. For the low density, 8/12 participants had a $w_{d,l} > 0.98$, and for the high densities, 11/12 participants had a $w_{d,h} < 0.54$. Here, the explained variance was between 11–97%, with >50% and >90% in 10/12 and 6/12 participants, respectively. All in all, it seems that the weight that was assigned to density for density judgements varied among participants but that most participants gave higher weights to density for low densities, whereas for higher densities, the weight assigned to mass was higher.

4 Discussion

In this study, we investigated how humans perceive density when presented with objects of different size and mass. We found that density estimates depended both on density and mass. This is similar to previous studies that found that heaviness estimates depended on these two object properties as well [3,13] and the finding that participants had more difficulties reporting a difference in heaviness with objects of different compared to equal densities [9]. Hence, density and heaviness perception might be closely related.

We found a contribution of 56% of density relative to mass for density estimates. This seems slightly different from the density weight of 29% for heaviness estimation found in [13], suggesting that the weighing of density and mass can be adapted to the required percept. However, it must be noted that the contribution varied for different densities and individuals.

Interestingly, we found that the weight of mass on density perception increased with density, indicating that cue weighing can also depend on intensity. Whereas density estimates were not affected by mass for the lowest density,

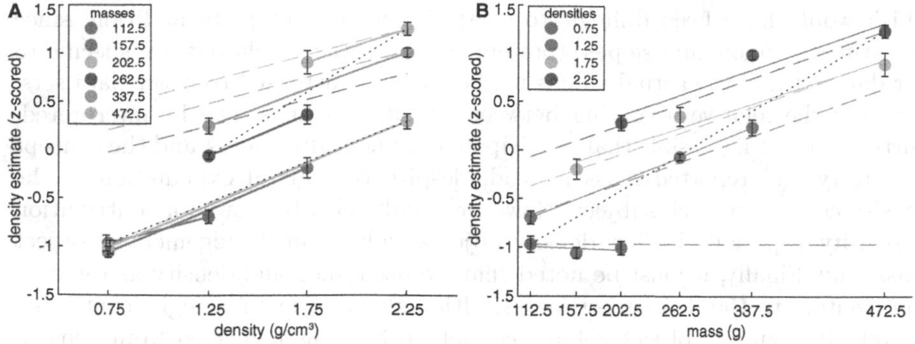

Fig. 2. Results for the mass (A) and density slopes (B) for low and high values separately. Solid circles and error bars represent density estimates mean and standard errors, respectively. Solid and dashed lines represent fitted and modelled slopes, respectively. The black dashed lines are the maximum slopes used in the model.

the influence of mass became more pronounced for objects with higher densities. In our experiment, the low-density objects all had the lowest masses as well. Therefore, it is also possible that mass influences density estimates more when objects are heavier. At first hand, a bias of mass seems logical, since in daily life we might be more familiar with judging object mass than density. However, this does not explain why low-density objects can be perceived accurately. Possibly, we are more familiar with handling objects of low densities, such as wood or plastic, and have more experience in distinguishing such objects. Furthermore, in daily life, we experience a non-linear distribution of size and weight. These learned statistics of everyday objects can be used to estimate those of novel objects. In general, small objects are seen as denser in comparison with objects larger in size, since in real life small objects tend to be denser [10]. Similarly, daily life experience might also suggest a relation between heavy and dense objects. Finally, for heaviness estimations, lower contributions of density were found for objects with lower densities and mass [12], further suggesting that for low densities and light objects, the object properties density and mass can be better disentangled.

The mutual influence of density and mass in object perception suggest that brain processes for these properties are related. A functional magnetic resonance imaging study revealed that object density is processed in a higher-order area of the brain (left ventral premotor area) and the primary motor cortex encodes object weight [2]. Density and mass perception might mutually influence each other since these areas are closely connected [6]. Therefore, it could be difficult for humans to estimate these properties independently.

The present study also has some limitations. First, we used a limited set of object sizes and weights for each density. Therefore, it is difficult to disentangle whether low densities can be accurately perceived due to the low density or due to low masses. A larger object set would include very small or large objects which would have been difficult to grasp in our VR setup. In addition, since we estimated maximum slopes from our data set, the weightings are specific to our data set and the actual values might not be generalised to other data sets, but only the relative weighting between density and mass can be interpreted. Furthermore, it is possible that participants did not fully understand the concept of density and reported mass instead, despite the careful explanation of the density concept to each subject. However, results clearly showed a contribution of density, especially for low-density objects, ruling out a judgement of object mass only. Finally, it must be noted that we used mass and density as the only contributors to the density estimates, although other factors might contribute as well. Particularly object volume cannot be fully distinguished from density, and when volume and mass were used in the model fit, similar results were seen.

In conclusion, both mass and density influence density perception, but mass affects density perception more when objects are denser and heavier. Our study supports the idea that density perception is closely related to heaviness perception, but also indicates that the weighing of these properties can vary based on their intensity. The perception of density is important for distinguishing objects of different materials or objects with different contents. This could be relevant when designing virtual environments or controlling robotic arms to interact with different objects of different materials. The interaction of density and mass in object perception should be taken into account.

References

1. Charpentier, A.: Analyse experimentale de quelques elements de la sensation de poids. Archive de Physiologie normale et pathologiques **3**, 122–135 (1891)
2. Chouinard, P.A., Large, M.E., Chang, E.C., Goodale, M.A.: Dissociable neural mechanisms for determining the perceived heaviness of objects and the predicted weight of objects during lifting: an fMRI investigation of the size-weight illusion. NeuroImage **44**(1), 200–12 (2009)
3. Drewing, K., Bergmann Tiest, W.M.: Mass and density estimates contribute to perceived heaviness with weights that depend on the densities' reliability. In: 2013 World Haptics Conference, WHC 2013, vol. 2, pp. 593–598. IEEE (2013)
4. Ellis, R.R., Lederman, S.J.: The material-weight illusion revisited. Percept. Psychophys. **61**(8), 1564–1576 (1999)
5. Ernst, M.O., Banks, M.S.: Humans integrate visual and haptic information in a statistically optimal fashion. Nature **415**(6870), 429–433 (2002)
6. Grafton, S.T.: The cognitive neuroscience of prehension: recent developments. Exp. Brain Res. **204**(4), 475–491 (2010)
7. Huang, I.: The size-weight illusion and the weight-density illusion. J. General Psychol. **33**, 65–84 (1945)
8. Johansson, R.S., Westling, G.: Coordinated isometric muscle commands adequately and erroneously programmed for the weight during lifting task with precision grip. Exp. Brain Res. **71**, 59–71 (1988)
9. Kawai, S.: Heaviness perception. II. Contributions of object weight, haptic size, and density to the accurate perception of heaviness or lightness. Exp. Brain Res. **147**(1), 23–8 (2002)

10. Peters, M.A.K., Balzer, J., Shams, L.: Smaller = denser, and the brain knows it: natural statistics of object density shape weight expectations. PloS One **10**(3), e0119794 (2015)
11. van Polanen, V., Tibold, R., Nuruki, A., Davare, M.: Visual delay affects force scaling and weight perception when lifting objects in virtual reality. J. Neurophysiol. **121**, 1398–1409 (2019)
12. Ross, J., Di Lollo, V.: Differences in heaviness in relation to density and weight. Percept. Psychophys. **7**(3), 161–162 (1970)
13. Wolf, C., Bergmann Tiest, W.M., Drewing, K.: A mass-density model can account for the size-weight illusion. PloS One **13**(2), e0190624 (2018)

Haptic Feedback in a Teleoperated Box & Blocks Task

Irene A. Kuling[1], Kaj Gijsbertse[2(✉)], Bouke N. Krom[3], Kees J. van Teeffelen[3], and Jan B. F. van Erp[1,4]

[1] Perceptual and Cognitive Systems, TNO, Soesterberg, The Netherlands
[2] Training and Performance Innovations, TNO, Soesterberg, The Netherlands
Kaj.gijsbertse@tno.nl
[3] Intelligent Autonomous Systems, TNO, The Hague, The Netherlands
[4] Human Media Interaction, University of Twente, Enschede, The Netherlands

Abstract. Haptic feedback is a desired feature in teleoperation as it can improve dexterous manipulation. Direct force feedback to the operator's hand and fingers requires complex hardware and therefore substituting force by for instance vibration is a relevant topic. In this experiment, we tested performance on a Box & Blocks task in a teleoperation set-up with no feedback, direct force feedback and substituted vibration feedback. Objective performance was the same in all conditions as was the learning effect over three sessions, but participants had a clear preference for haptic feedback over no haptic feedback. The preferred type of feedback (force or vibration or both) varied over participants. In general, this study showed that haptic feedback is preferred in teleoperation, the Box & Blocks task seems not sensitive enough for our (and most) current teleoperation set-up(s), and vibration feedback as substitute for direct force feedback works well and can be used intuitively.

Keywords: Haptic feedback · Teleoperation · Dexterous manipulation

1 Introduction

Robots are frequently deployed to perform tasks that are dull, dirty or dangerous. Especially in repetitive tasks, technical advances allow the application of autonomous systems. Despite advances in autonomy, autonomous robots will not be a viable option in all applications for the foreseeable future [1]. If the tasks of a robotic system are diverse, the environments unpredictable and the stakes of successfully performing the task are high, robots will not be able to carry out all necessary tasks with sufficient reliability without human involvement. The common solution in these use cases is teleoperation (e.g. [2, 3]). In teleoperation the task is performed by an operator controlling the robot remotely, typically in a master-slave setup.

Currently the control of teleoperated systems is often not very intuitive and the cognitive load during control of the system is high for the controller. One of the improvements is to give the operator good dexterity and haptic feedback (e.g. [4]. The most intuitive haptic feedback is to present the forces on the robot's hands and fingers as forces on the operator's hands and fingers. However, this requires complex actuators

I. Nisky et al. (Eds.): EuroHaptics 2020, LNCS 12272, pp. 96–104, 2020.
https://doi.org/10.1007/978-3-030-58147-3_11

and limits wearability. An alternative method is to provide feedback through other sensory cues like vibratory or visual cues (sensory substitution). In this study, we compared the performance in a teleoperated Box & Block test without force feedback, with direct force feedback and with substituted force feedback in the form of vibration on the fingers. We choose the Box & Block test (e.g. [5]) following Catoire et al. [6] who described a teleoperation test battery with standard tests that can be used to benchmark system dexterity, and allow to advance the design, quantify possible improvements, and increase the effectiveness of the teleoperated system. The Box & Block test is one of these tests and focuses on dexterity.

2 Teleoperation Set-Up

In this study, a teleoperation setup consisting of a telemanipulator, a haptic control interface and a visual telepresence system was used (see Fig. 1). The telemanipulator consists of a four-digit, 13 degrees of freedom humanoid robotic hand (Shadow Hand Lite, Shadow Robot Company, London, UK), equipped with 3D force sensors (Optoforce OMD 10) on its fingertips, mounted on the flange of a KUKA IIWA 7 serial link robot with 7 degrees of freedom (KUKA Robotics, Augsburg, Germany).

The haptic control interface is realized by a haptic glove (Senseglove DK1, SenseGlove, Delft, the Netherlands) that tracks finger movements in 11 degrees of freedom and can provide passive force feedback on each of the fingers. Underneath the Senseglove a custom vibrotactile glove (Elitac, Utrecht, the Netherlands) was worn, which contains 16 strategically placed pancake motors. The movements of the hand in space are recorded by an HTC Vive tracker (HTC, Xindian, Taiwan) mounted on the Senseglove. The visual system is an in-house build, closed system that transfers the images from a stereocamera mounted on a pan-tilt-roll unit (PTRU) to a custom made Head-Mounted Display (HMD), which in turn controls the movement of the PTRU [7].

The software interfacing between the telemanipulator and haptic control interface was realized using two PC's running the Robot Operating System (ROS), version Kinetic Kame on Ubuntu 16.04. The PC's, one controlling the telemanipulator hardware and one controlling the haptic control interface, were connected through a local, dedicated gigabit ethernet network, with a typical latency of a few milliseconds.

2.1 Controls

The control of the robotic system was the same throughout the experiment. The KUKA arm is controlled via the fast research interface (FRI), in impedance mode. With a 1 kHz update frequency, the robot tracks the Cartesian end-effector setpoint with a translational stiffness of 150 N/m and a rotational stiffness of 100 Nm/rad. The null space motion is uncontrolled and therefore compliant. At the start of each run, the robot was returned to the same initial position.

The setpoint of the robot is determined by the position and orientation of the Vive tracker on the back of the hand of the operator, updated with 60 Hz. The operator switches control on and off using a foot pedal, using a common 'clutching' logic: as long as the pedal is pressed, the robot follows the movements of the user. When the pedal is released, the robot motions are decoupled and it remains at the current position. When the operator decides to continue control by pressing the pedal, the current position of the user's hand is used as the reference position for new movements.

The robot hand is continually controlled by mapping the Senseglove measurements of finger positions to joint positions on the Shadow Hand at 50 Hz. This mapping was calibrated before the experiment by using two predefined hand postures (i.e. flat hand and strong fist) to obtain parameters that define the operator's hand model, which was used in finger position estimation. A linear mapping between these finger positions and the joint positions of the robot fingers was empirically determined to reach a high degree of movement mimicking.

2.2 Haptics

In the experiment, the visual feedback of teleoperation could be accompanied by haptic feedback from the robot's sensors to the operator. Two types of haptic feedback were used; braking force at the fingertips (direct force feedback; DFF) and vibrations at the fingertips (substituted force feedback; SFF). A third type of feedback involving vibrations at the back of the hand functioning as a binary signal indicating collisions was also measured in the experiment, but not analyzed for this paper.

For the DFF a fingertip was blocked when the Euclidean norm of the force vector measured at the robot finger exceeded 0.1 N. This mimics making contact with a non-compliant object. The SFF provided gradual feedback of the measured force at the robot finger in the form of vibratory feedback at the corresponding fingertip using the vibrotactile glove. A force with a norm of more than 0.05 N is linearly scaled to a maximum vibration at 2 N. The Elitac glove provides a logarithmic stimulation intensity in 16 steps. A pilot study has shown that the vibration steps are identifiable and judged linearly on a magnitude scale. Since DFF provides proprioceptive-information and SFF provides information on the applied amount of force, DFF and SFF were combined to explore potential additive effects of the feedback types.

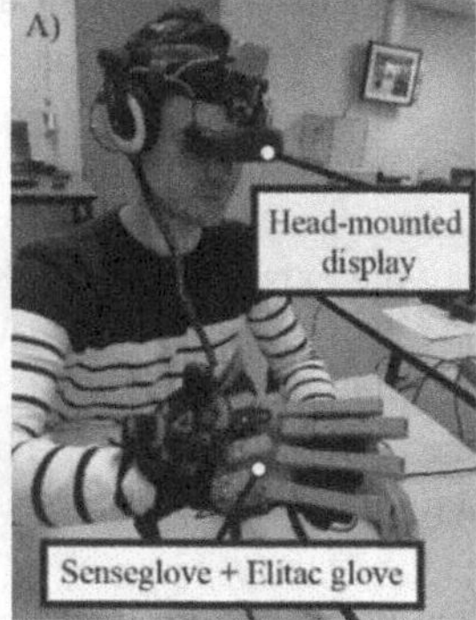

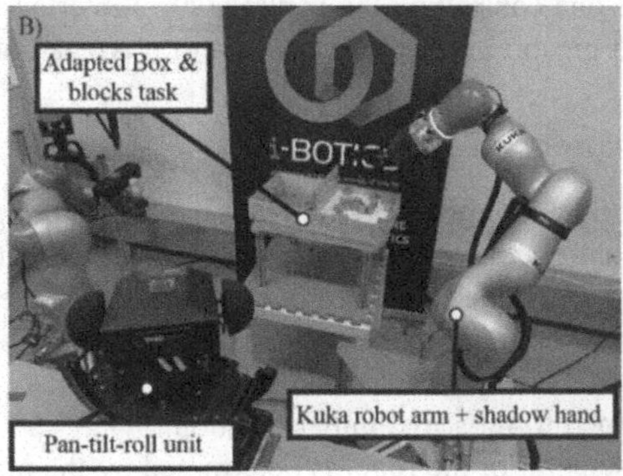

Fig. 1. The teleoperation setup. A) the master side including a subject wearing the HMD, Senseglove and Elitac glove, and B) the slave side with the teleoperated robot arm and PTRU.

3 Methods

3.1 Participants

Twenty-two participants (13 female, mean age 29.5 ± 10.0 yrs) participated in the three pretests of the experiment. Fourteen of them passed the third pre-test and were selected for the main experiment (10 female, mean age 25.6 ± 8.0 yrs). None of the participants had previous experience with teleoperation and all participants were naïve about the details of the experiment and were compensated for their time. All participants gave their written informed consent prior to the experiment. The setup and experiments were approved by the TNO Internal Review Board (registered under number 2019-026).

3.2 Experiment Design and Procedure

Adapted Box & Blocks Evaluation

For this study the classic Box & Block test was adapted to fit the capabilities and limitations of the visual part of the teleoperation setup. The depth of the start compartment was decreased to be able to more easily grasp the blocks (3.5 cm instead of the traditional 7.5 cm depth [5]). We also reduced the number of blocks in the test to 30 to give the participants a bit more space in the box, which made placing the (robot) fingers around the blocks easier (Fig. 2A).

As in the original task, the aim is to move as many blocks, i.e. cubes sized 2.5 cm square, as possible from one side of the box to the other side while using one hand and it is not allowed to throw the block over the partition. The participant is seated while performing the test and the box is placed directly in front of the participant on a table. In our reference task, we asked participants to move 30 blocks as fast as possible with their own hand directly. We did this both from the higher side to the lower side and vice versa in a randomized fashion. This gave us an idea about the effects on performance due to the adjusted depth of the box.

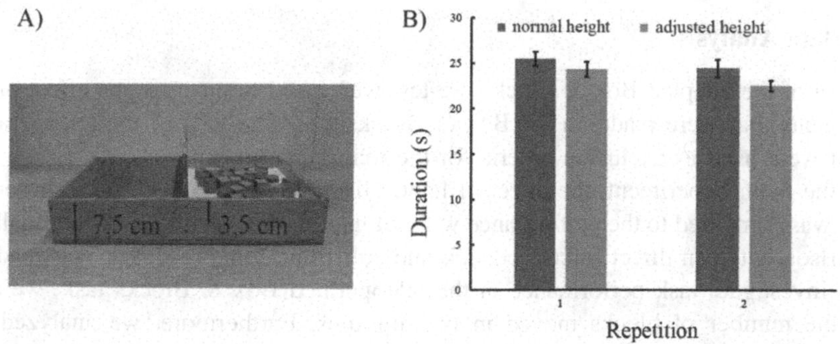

Fig. 2. A) Adapted box & blocks task, B) Difference in completion time for the standard and adjusted box & blocks test.

Teleoperation Pre-test

The teleoperation pre-test consisted of a simple displacement task with blocks from the Box & Block test while using the teleoperation system without feedback. Four blocks were located in the middle of an A3 sheet of paper with four equal squares (9 cm) at the corners. The participants had to move the blocks to the predetermined squares, one to each corner. To keep control of the duration of the pre-test, the maximum time to do this was 4 min. Participants had three attempts, and had to move (at least in one try) all four blocks to the corners within 90 s to be included in the second subtest. This criterion was not known to the participants. Regardless of the result of the teleoperation pre-test, the participant was asked to do the adapted Box & Blocks again.

Main Experiment

The main experiment started with written instructions and a practice session performing a simple teleoperation task, i.e. moving several blocks without haptic feedback. Next, the different types of haptic feedback were explained and presented. Four types of haptic feedback were analyzed in this paper: No feedback (NF), Direct finger force feedback (DFF), Substitute finger force feedback (SFF), Both DFF and SFF. The participant performed the teleoperated (modified) Box & Blocks task in all conditions, each presented once in a semi-balanced (incomplete Latin Square) order. Between the conditions there was a short (ca. 2 min) break, in which the NASA-TLX questionnaire was answered. A 10 min break was scheduled after all conditions. This was repeated a second and third time, resulting in three repetitions per condition.

Teleoperated Adapted Box & Block Task

In the teleoperated Box & Block test, participants always moved the blocks from the higher side to the lower side. They were asked to move as many blocks as possible to the other side of the box within two minutes.

Subjective Measures

The NASA Task Load Index (NASA-TLX) was used to measure workload [8]. Furthermore, the participants were asked to rank the experimental conditions in order of preference.

3.3 Data Analysis

The data of the adapted Box & Blocks pre-test were used to quantify the effect of the adjustments that were made on the Box & Blocks test. The data of the teleoperation pre-test were used as exclusion criteria for the main experiment.

In the main experiment, the effect of haptic finger force feedback on task performance was compared to the performance without haptic force feedback. Additionally, a comparison between direct force feedback and substitute force feedback was made.

To investigate task performance in the teleoperated Box & Blocks task, we analyzed the number of blocks moved in two minutes. Furthermore, we analyzed the subjective scores on the NASA-TLX test, i.e. Mental load, Physical load, Time pressure, Effort, Performance and Frustration. Participants who moved 2 or fewer blocks in one of the repetitions (regardless of whether the haptic feedback type in which this occurred was evaluate in this paper or not) were excluded from the analysis. This occurred for two participants. Next, the data were analyzed with 2 × 2 × 3 RM ANOVAs (DFF × SFF × repetition) on all metrics. Greenhouse-Geisser corrections were applied when sphericity was violated. In these analyses, main effects would show the influence of either DFF, SFF and repetition on the performance, while interaction effects would be able to reveal a potential additive effect of combining DFF and SFF and potential differences in learning for the feedback types.

Table 1. Statistical design of the main experiment

		DFF	
		Off	On
SFF	Off	"NF"	"DFF"
	On	"SFF"	"DFF+SFF"

4 Results

Adapted Box & Blocks Evaluation

Participants needed on average 24.2 s (range 18.5–39.3 s) to move all 30 blocks from one side of the box to the other. A 2 × 2 RM ANOVA (height of box (low/high) × timing (before/after)) showed significant main effects of both the height of the box, $F_{1.0,21.0} = 10.73$, $p = .004$, and the timing $F_{1.0,21.0} = 8.04$, $p = .010$ (Fig. 2C). These effects show that participants were on average about 1.5 s (ca. 6%) faster starting at the adjusted (=higher) side of the box, and that participants were about 1.3 s (ca. 5%) faster in their second repetition.

Teleoperation Pre-test

Fourteen out of the 22 participants in the pre-test fulfilled the criteria to participate in the main experiment. From the eight that were not selected, two participants were not able to perform the task at all. The other six non-selected participants exceeded the 90 s time requirement (range 100–406 s).

Main Experiment

On average the participants moved 11.2 (range 3–21) blocks in the teleoperated Box & Blocks task. In the 2 × 2 × 3 RM ANOVA (DFF × SFF × Repetition) we found that neither forms of haptic feedback on the fingers (DFF, SFF and DFF + SFF) improved the performance on the number of blocks since there were no significant main effects and no significant interaction between DFF and SFF (statistical details can be found in Table 1). It is interesting that all except one participant had a preference for one of the feedback types over NF. For DFF, 5 out of 12 participants ranked DFF over NF, however the NASA-TLX results showed that DFF scored worse than NF on several aspects (time pressure, performance and frustration). SFF was ranked higher than NF by 8 out of 12 participants, and both objective and subjective measures were similar to no feedback. These results suggest that people preferred feedback even when it did not improve performance in our task. When comparing the ranking of the different feedback types, DFF, SFF and DFF+SFF, were ranked higher than the other feedback types by respectively 4, 5, and 3 out of 12 participants. This shows that the preferences for the feedback types are highly individual, and that more feedback (DFF + SFF) is not systematically preferred.

Furthermore, for number of blocks and effort main effects of repetitions were found (respectively p = .004 and p = .045), indicating a better performance (i.e. more blocks and less effort) in the later repetitions. This shows a learning effect in the teleoperation task of ca. 8% per repetition (Table 2).

Table 2. Results of the 2 × 2 × 3 RM ANOVA's (DFF × SFF × Repetition)

	DFF	SFF	Repetition	Interaction effects
Number of blocks	$F_{1.0,11.0} = 4.22$ p = .065	$F_{1.0,11.0} = .10$ p = .756	**$F_{2,22} = 7.13$** **p = .004** ***r1 < r3*** ***p = .007***	All p's > .180
Mental load	$F_{1.0,11.0} = .037$ p = .852	$F_{1.0,11.0} = .179$ p = .681	$F_{2,22} = .624$ p = .545	All p's > .258
Physical load	$F_{1.0,11.0} = .051$ p = .825	$F_{1.0,11.0} = .306$ p = .591	$F_{1.2,13.6} = 3.43$ p = .079	All p's > .143
Time pressure	**$F_{1.0,11.0} = 4.954$** **p = .048** DFF > NF	$F_{1.0,11.0} = 2.811$ p = .122	$F_{2,22} = .145$ p = .866	All p's > .502
Effort	$F_{1.0,11.0} = .001$ p = .972	$F_{1.0,11.0} = .059$ p = .813	**$F_{1.1,12.3} = 4.78$** **p = .045** ***r1 > r2*** ***p = .030***	All p's > .471
Performance	$F_{1.0,11.0} = 4.780$ p = .051	$F_{1.0,11.0} = .001$ p = .975	$F_{2,22} = .870$ p = .433	All p's > .320
Frustration	**$F_{1.0,11.0} = 7.215$** **p = .021** DFF > NF	$F_{1.0,11.0} = .015$ p = .904	$F_{2,22} = .574$ p = .572	All p's > .210

5 Discussion and Conclusion

In this study we compared the effect of different haptic feedback types on performance in a teleoperated Box & Blocks task. No difference was found in objective performance in the conditions with or without haptic feedback, but all except one participant had a preference for one of the feedback conditions compared to no feedback. Comparing direct force feedback and substituted vibration feedback also no difference was found in objective performance, but there were differences in preference among the participants. The lack of results on objective performance might be caused by the level of difficulty and lack of sensitivity of the teleoperated Box & Blocks task; to get a significantly higher score one additional block should be moved, which corresponds to an increase of about 8%.

In all conditions participants were able to perform from the first trial, but still performance increased over repetitions. This shows that our set-up is intuitive, but that there is learning involved. The improvement over repetitions in the teleoperated Box & Blocks task was similar in size as the improvement over repetitions in the Adapted Box & Blocks pre-test (ca. 8% and 5% respectively), which shows that the learning had to do mainly with the task and dexterous manipulations itself and not that much with the teleoperation set-up or the haptic feedback. Moreover, there is no interaction between the repetitions and the type of feedback. This indicates that first shot performance and learning with SFF is as good as with DFF. Whether performance would change when learning is finished or whether the maximum performance with the different feedback types is on the same level is impossible to conclude from this study, but would be very interesting for future research.

At first sight it seems surprising that there were higher scores on frustration and time pressure for the direct force feedback compared to no feedback. However, this might be caused by the design of the direct force feedback; to be able to feel hard surfaces the feedback blocked the fingers almost immediately when force was applied at the fingertips of the robot hand. This resulted in some unfortunate situations in which the blocks were not firmly hold yet, but the fingers could not close further. Future research is planned on the effect of haptic feedback when the visual information is less reliable or on more subtle dexterous tasks and feedback.

To conclude, haptic feedback is preferred in teleoperation, the Box & Blocks task is a too coarse test for our current teleoperation set-up, and vibration feedback as substitute for direct force feedback works well and can be used intuitively.

References

1. SPARC: Robotics 2020 multi-annual roadmap for robotics in Europe (2017)
2. Van Erp, J.B.F., Duistermaat, M., Jansen, C., Groen, E., Hoedemaeker, M.: Tele presence: bringing the operator back in the loop. In: NATO RTO Workshop on Virtual Media for Military Applications (2006)
3. Pacchierotti, C., et al.: Cutaneous feedback of fingertip deformation and vibration for palpation in robotic surgery. IEEE Trans. Biomed. Eng. **63**(2), 278–287 (2016)
4. Okamura, A.: Methods for haptic feedback in teleoperated robot-assisted surgery. Ind. Robot **31**(6), 499–508 (2004)
5. Mathiowetz, V., Volland, G., Kashman, N., Weber, K.: Adult norms for the box and block test of manual dexterity. Am. J. Occup. Ther. **39**(6), 387–391 (1985)
6. Catoire, M., Krom, Bouke N., van Erp, J.B.F.: Towards a test battery to benchmark dexterous performance in teleoperated systems. In: Prattichizzo, D., Shinoda, H., Tan, H.Z., Ruffaldi, E., Frisoli, A. (eds.) EuroHaptics 2018. LNCS, vol. 10894, pp. 440–451. Springer, Cham (2018). https://doi.org/10.1007/978-3-319-93399-3_38
7. Jansen, C., Winckers, E.: TNO telepresence robot control (2015). https://www.elrob.org/files/elrob2016/TeamInformation_TNO-NLD_EODD.pdf
8. Hart, S.G., Staveland, L.E.: Development of NASA-TLX (Task Load Index): Results of empirical and theoretical research. In: Human Mental Workload, Amsterdam, The Netherlands, pp. 239–250 (1988)

Systematic Adaptation of Exploration Force to Exploration Duration in Softness Discrimination

Aaron C. Zoeller(✉) and Knut Drewing

Giessen University, 35394 Gießen, Germany
aaron.zoeller@psychol.uni-giessen.de

Abstract. When interacting haptically with objects, humans enhance their perception by using prior information to adapt their behavior. When discriminating the softness of objects, humans use higher initial peak forces when expecting harder objects or a smaller difference between the two objects, which increases differential sensitivity. Here we investigated if prior information about constraints in exploration duration yields behavioral adaptation as well. When exploring freely, humans use successive indentations to gather sufficient sensory information about softness. When constraining the number of indentations, also sensory input is limited. We hypothesize that humans compensate limited input in short explorations by using higher initial peak forces. In two experiments, participants performed a 2 Interval Forced Choice task discriminating the softness of two rubber stimuli out of one compliance category (hard, soft). Trials of different compliance categories were presented in blocks containing only trials of one category or in randomly mixed blocks (category expected vs. not expected). Exploration was limited to one vs. five indentations per stimulus (Exp. 1), or to one vs. a freely chosen number of indentations (Exp. 2). Initial peak forces were higher when indenting stimuli only once. We did not find a difference in initial peak forces when expecting hard vs. soft stimuli. We conclude that humans trade off different ways to gather sufficient sensory information for perceptual tasks, integrating prior information to enhance performance.

Keywords: Perception · Softness · Exploration duration · Prior information

1 Introduction

Humans use their hands to explore the softness of objects every day. For example when squeezing a stress ball during work, when testing if bread is still good to eat or when palpating the leg to see if it is swollen. We use our hands rather than other senses, because softness is best explored by haptic interaction [1]. To explore softness, humans indent the surface of an object repeatedly to collect sensory information over time

This work was supported by Deutsche Forschungsgemeinschaft (DFG, German Research Foundation) – project number 222641018 – SFB/TRR 135, A5.

I. Nisky et al. (Eds.): EuroHaptics 2020, LNCS 12272, pp. 105–112, 2020.
https://doi.org/10.1007/978-3-030-58147-3_12

[1, 2]. In everyday life the time during that objects can be explored may vary from free exploration to a highly constrained one (e.g., only one indentation). Here we ask how constraints in exploration duration influence softness exploration.

Perceived softness is not only, but clearly related to compliance [3]. The compliance of an object is defined as the amount of an object's deformation under a given force [mm/N]. In previous studies investigating softness perception using a 2 Alternative Forced Choice (2AFC) discrimination task paradigm, participants were reported to perform 3 to 7 indentations per stimulus, when allowed to explore freely [2, 4]. In free exploration, humans seem to stop exploring when they have sufficient information. With constant exploratory force, discrimination performance increases with the number of indentations up to a certain level. When exploration movements are restricted in duration, humans perform worse [2, 5].

However, perception relies on multiple information sources [6], and it is known that humans can use prior knowledge about object properties to adapt their exploratory behavior in haptic perception to the present task conditions [7, 8]. For example, for softness perception results from our lab showed that humans use more force in the initial indentations when they expect to discriminate two harder objects as compared to two softer objects [2, 4, 9]. Humans also use higher forces when expecting less difference between the compliance of the to-be discriminated objects. These adaptations seem to be well chosen, because more deformation of an object's surface and the exploring finger (e.g. by applying more force) improves sensory information on softness [9–11]. We wonder whether humans would also use higher force, when they know that they are constrained in their exploration duration, but free to adapt other movement parameters. High force might be a means to counterbalance lack of sensory information in shorter exploration. We hypothesize that participants explore with more force when planning a shorter exploration. In line with previous results, we also expect that participants apply more initial force when they expect harder as compared to softer stimuli [2]. Finally, we hypothesize that adaptations to compliance category are more relevant and hence also more pronounced for short explorations, where initial lack of information cannot be compensated for later.

In two experiments we used a 2 Interval Forced Choice (2IFC) softness discrimination task to investigate how initial peak forces vary in dependence of the exploration duration and prior information about object compliance. In a discrimination task we can control the success rate to be at an intermediate level. We expect that an intermediate level of success increases the effort that participants spend on the task (cf. Exp.2 in [9]). Further, stimuli were presented sequentially in a 2IFC task rather than in a 2AFC task so that participants did not perform uncontrolled switches between stimuli. Switches would have confounded an interpretation of the initial peak force by exploration duration. In each trial participants explored two objects from either a hard or soft category. Trials with harder and softer objects were presented in a blocked manner (only trials of soft or hard stimuli; prior information about compliance induced through recurring presentation) or randomly mixed (no prior information). In Experiment 1 participants had to indent in different blocks each stimulus once vs. five times. In Experiment 2 we compared one-indentation explorations to free explorations.

2 Methods

2.1 Participants

Sixteen healthy students ($N = 8$ per experiment) from Giessen University participated (6 males, average age in years: 24, range: 18–30). All participants were right-handed, reported no motor and cutaneous impairments, had normal or corrected-to normal vision, and were naive to the study's purpose. They were paid 8€/h. Methods were approved by the local ethics committee LEK FB06 and in accordance with the 2013 Declaration of Helsinki. Participants gave written informed consent.

2.2 Stimuli and Setup

Participants sat on a custom-made visuo-haptic workbench containing a 24" 3D screen (120 Hz, 1600 × 900 pixel), a force sensor (resolution 0.05 N, temporal resolution 682 Hz), to collect data of executed force, and a PHANToM 1.5A haptic force feedback device (spatial resolution: 0.03 mm, temporal resolution: 1000 Hz), to collect finger position data in a 38 × 27 × 20 cm^3 workspace (Fig. 1). The right index finger of participants was connected to the PHANToM via a spherical magnetic adapter, allowing maximum freedom in finger movements and bare-finger exploration. We used stereo glasses (Nvidia 3D Vision 2) to present a 3D virtual scene. Finger position was displayed via a green sphere (3 mm) in the scene. During contact with the stimuli the sphere disappeared to give no feedback about finger position or stimulus deformation. Participants looked at the screen through a front surface mirror (viewing distance 40 cm) aligned with the haptic scene to ensure a natural connection between haptic action and visual feedback. A chinrest was used to stabilize the heads of participants. Audio signals were given via headphones (Senheiser HD 280 Pro). All devices were connected to a PC; custom-made software controlled experiment and data collection.

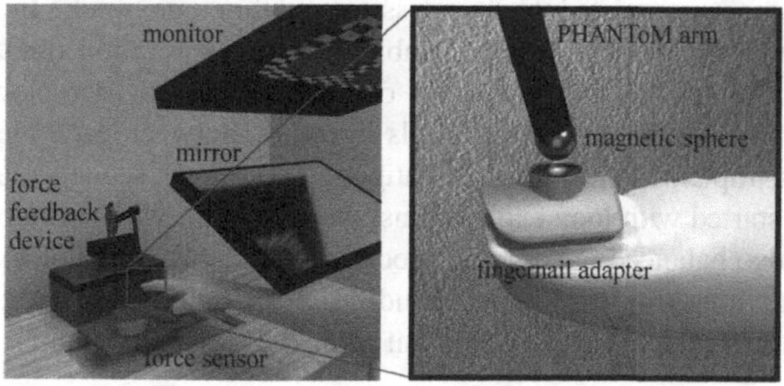

Fig. 1. Schema of the visuo-haptic workbench and the magnetic fingernail adapter.

In both experiments we used six cylindrical shaped silicone rubber stimuli (height: 38 mm, diameter: 75 mm; [9] for details of production). Stimuli were divided in a hard and a soft category. We used one stimulus per category as standard (hard: 0.16 mm/N; soft: 0.84 mm/N) and two stimuli per category as comparison stimuli (hard:

0.15 mm/N and 0.17 mm/N; soft: 0.79 mm/N and 0.88 mm/N). Stimuli were selected in a pilot study (N = 8) based on their distinctness. Comparison stimuli were equally distinguishable (about 80% in free exploration) from the corresponding standard stimulus.

2.3 Procedure and Design

In both experiments we used a 2IFC softness discrimination task. We investigated the influence of three within-participant variables. *Compliance (C)*: We presented stimuli from the soft or the hard compliance category. *PriorCompliance (PC)*: In blocked conditions prior information about stimulus compliance was induced by presenting trials with stimuli out of one compliance category only. In randomized conditions, trials with stimuli out of the hard or soft category were presented in a randomly mixed order (no prior information about compliance). *ExplorationDuration (ED)*: In Experiment 1, participants were instructed to indent each stimulus one or five times per trial (short vs. long exploration). In Experiment 2, participants indented each stimulus once or as often as they liked (short vs. free exploration). As can be seen in Fig. 1, the setup allowed participants to use their entire arm for force production. We measured initial peak forces, and perceptual performance as percentage of correct judgments.

In each trial, participants had to discriminate the softness of two stimuli (one standard and one comparison stimulus) and judge which one is softer. In the beginning one of the two stimuli was presented visually, to indicate where participants should start haptic exploration. Both starting positions were chosen equally often in randomized order. An acoustic signal indicated the start of the exploration. After participants indented the first stimulus as often as instructed (short exploration, long exploration and free exploration) it vanished from the visual scene and the second stimulus appeared on the screen. After indenting the second stimulus as often as instructed it vanished. By pressing one of two virtual buttons, participants indicated which of the two stimuli they had felt to be softer. No immediate feedback was given.

Both experiments consisted of two sessions within one week. In each session, participants conducted blocks of each combination of exploration duration and prior information conditions. Each combination overall comprised four blocks (two each day) consisting of 64 trials each (1024 trials in total). Between each block a break of one minute was implemented to counter fatigue. To counter sequence effects, half of the participants started with long explorations, the other half with short explorations on both days. We also balanced the order of blocks of randomized and blocked conditions within short and long/free exploration conditions on both days across participants, using 4 × 4 Latin squares. Each experiment took 6 h per participant.

2.4 Data Analysis

In this study we focused on analyzing initial peak forces in each trial, but also calculated the percentage of correct responses. Because initial peak forces are barely influenced by later sensory feedback, they are a good indicator of the influence of prior information on behavior [9]. In order to calculate peak forces, we initially subtracted stimulus mass from force measurements and smoothed the resulting force values over time, using a moving-averaging window with a kernel of 45 ms. To capture peaks, we identified turning points in which the derivate of force over time changed from positive to negative. The applied force at this maximum additionally had to be higher than 5 N. The time interval between two turning points was restricted to be at least 180 ms. With these restrictions we ensured the exclusion of small finger shaking movements, local maxima and movement rests while releasing the finger from the object after valid peaks. Trials with more or less indentations than the allowed number (one, five or free exploration) were excluded from the analysis. The first captured maximum in each trial was assigned to be the initial peak force.

We compared initial peak forces (the first time indenting a stimulus in each trial) for all conditions in a repeated measurement ANOVA, using the three within-participant variables. In a further ANOVA with the same three variables we compared the percentage correct. Arcsine square root transformed percentages entered analysis, because these transforms approximate a normal distribution [12].

3 Results

For initial peak forces in Experiment 1, we found a significant main effect for *ExplorationDuration (ED)*, $F(1,7) = 27.44$, $p = .001$ indicating that participants used more force in shorter as compared to longer explorations (Fig. 2A). No other main effect or interaction was significant, *Compliance (C)*, $F(1,7) = 3.94$, $p = .087$; *PriorCompliance (PC):* $F(1,7) = 0.02$, $p = .890$; *ED* × *PC*: $F(1,7) = 0.42$, $p = .539$; *ED* × *C*: $F(1,7) = 0.34$, $p = .580$; *PC* × *C*: $F(1,7) = 0.03$, $p = .859$; *ED* × *PC* × *C*: $F(1,7) < 0.01$, $p = .961$. We additionally investigated peak forces of the different indentations in the long exploration condition. For reasons of space, we cannot report each detail. An extra ANOVA with the variables *Indentation* (1st to 5th), *PriorCompliance*, and *Compliance*, revealed no significant main effect and no interaction ($\alpha = 5\%$). Five *t*-tests confirmed that peak force in each indentation of the long exploration condition (per indentation in N, 1st: 17.3, 2nd: 17.6, 3rd: 17.6, 4th: 17.4, 5th: 18.8) was significantly lower than in the short exploration condition (each $p < .001$).

In Experiment 1, on average 67% of all answers were correct (*SEM* = 0.021, range across individuals 63%–75%). In the ANOVA we found a significant effect for the variable *PC*, $F(1,7) = 10.09$, $p = .016$, indicating that participants performed better when prior information available (70%) as compared to the randomized conditions (65%). No other effects were significant ($\alpha = 5\%$).

In Experiment 2, participants indented each stimulus on average 2.9 times in the free exploration condition. For initial peak forces (Fig. 2B), we again found a significant main effect for the variable *ED*, $F(1,7) = 12.85$, $p = .009$. Other effects were not significant, *C*: $F(1,7) = 0.02$, $p = .896$; *PC*: $F(1,7) = 5.55$, $p = .051$; *ED* × *PC*: $F(1,7) = 0.46$, $p = .518$; *ED* × *C*: $F(1,7) = 1.13$, $p = .322$; *PC* × *C*: $F(1,7) = 0.02$, $p = .896$; *ED* × *PC* × *C*: $F(1,7) = 2.64$, $p = .148$. Extra analyses of peak forces of later indentations in the free exploration conditions showed some variation (range of averages about 16 to 19 N), but for each indentation in each free exploration condition peak force was lower than in the corresponding short exploration condition (16 tests, each $p < .05$).

On average, participants responded correctly in 70% of all trials ($SEM = 0.017$, range 67% to 73%). In the ANOVA comparing performance, we again found a significant effect for the variable *PC*, $F(1,7) = 21.78$, $p = .002$, indicating that participants performed better in blocked conditions (73%) as compared to randomized conditions (67%). We found no other significant effect ($\alpha = 5\%$).

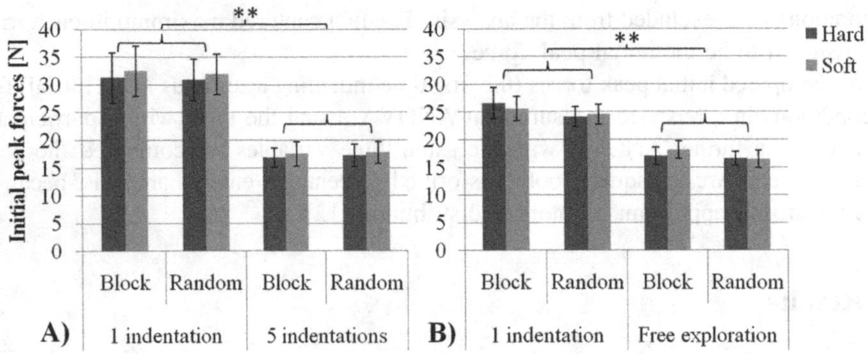

Fig. 2. Average initial peak forces (N) and standard error of mean (*SEM*) for each of the variable combinations of Experiment 1 (**2A**) and 2 (**2B**). Significant effects are marked $p < .01$ = **.

4 Discussion

In two experiments, we compared initial peak forces for different exploration durations with and without prior information about stimulus compliance. In contrast to previous studies [2, 4, 9], we did not find adaptation of initial peak forces to predictable stimulus compliances. However, in both experiments we found a strong effect for the exploration duration indicating that participants used much more initial peak force when indenting stimuli only once as compared to five times or to free exploration (3 indentations, on average). Also, peak forces of following indentations were lower as compared to the initial peak force of the short exploration condition, indicating no compensation for low initial forces in later indentations. In line with [9], it seems that humans adapt their exploration forces when expecting shorter explorations, which do not allow for repetitive gathering of sensory information. It has been speculated that with more deformation of the finger and an object's surface more sensory information on softness can be gathered [9–11], and it has been shown that higher force can improve differential sensitivity for softness up to a certain maximum (Exp. 3 in [9]). Hence, we speculate that high force has served in the present study as a means to counterbalance lack of sensory information in shorter exploration. Further, we did not find performance differences between longer/free and shorter explorations (which had been reported, when force was held constant [2, 5]). We conclude that our participants traded off force and exploration duration to gather sufficient sensory information.

Unexpectedly, prior information about an object's compliance did not have the expected effect on initial peak force. Still, participants performed better when prior information about the stimulus compliance was available. This may indicate that although no motor adaptation based on the prior information about stimulus compliance was found, the information was integrated in perception, enhancing performance.

But why did participants not use prior information on compliance for motor adaptation? One possible reason is that force adaptation based on exploration duration was dominant and rendered further force adaptation based on an object's compliance unnecessary: When sensory information input was limited due to short exploration, humans might already have used very high force to gather as much sensory information as possible. This adaptation might overshadow any useful force adaption based on stimulus compliance in the form of a ceiling effect. In contrast, when forced to indent the stimuli five times each (long exploration), participants might have expected to gather sufficient sensory information anyway and used low forces to save energy.

Notice that the presently found peak forces were higher as compared to some previous studies [e.g., 11, 13, 14]. However, contrary to the previous studies, in which participants produced forces using their finger muscles only, in the present setup participants used their entire arm. Present peak forces are comparable to those found in previous studies with the same setup [e.g. 2, 4, 9]. In [10], where participants were allowed to use their body to produce force, even higher peak forces between 80 and 400 N were found. Thus, the present peak forces do not seem exceptionally high.

However, the lack of adaptation during free exploration in Experiment 2 cannot be explained by this speculation. We speculate that the design of our study might have also interfered with natural adaptation behavior, as it is more constricted and less natural as compared to previously used designs, where participants were allowed to change back and forth between stimuli in order to compare them [2, 4, 9]. Here, movements were explicitly constrained to sequentially explore stimuli in each trial. As shown by Zoeller et al. [4], the appearance of peak force adaptation based on stimulus compliance might vanish with too explicit control of motor behavior.

Overall, we conclude that humans adapt their exploration behavior in softness discrimination to the expected exploration duration. When expecting a short exploration, humans use much more force to indent objects as compared to longer explorations. This tradeoff seems to be a sufficient strategy to enhance sensory perception. We suggest that humans trade off different ways to gather sufficient sensory information for perceptual tasks, integrating prior information to enhance performance. This, of course, has to be further tested in other perceptual domains.

References

1. Lederman, S.J., Klatzky, R.L.: Hand movements: a window into haptic object recognition. Cogn. Psychol. **19**(3), 342–368 (1987)
2. Lezkan, A., Metzger, A., Drewing, K.: Active haptic exploration of softness: indentation force is systematically related to prediction, sensation and motivation. Front. Integr. Neurosci. **12**, 59 (2018)
3. Bergmann Tiest, W.M., Kappers, A.M.L.: Physical aspects of softness perception. In: Di Luca, M. (ed.) Multisensory Softness. SSTHS, pp. 3–15. Springer, London (2014). https://doi.org/10.1007/978-1-4471-6533-0_1

4. Zoeller, A.C., Lezkan, A., Paulun, V., Fleming, R., Drewing, K.: Integration of prior knowledge during haptic exploration depends on information type. J. Vis. **19**, 20 (2019)
5. Lezkan, A., Drewing, K.: Processing of haptic texture information over sequential exploration movements. Atten. Percept. Psychophys. **80**(1), 177–192 (2017). https://doi.org/10.3758/s13414-017-1426-2
6. Ernst, M.O., Bülthoff, H.H.: Merging the senses into a robust percept. Trends Cogn. Sci. **8** (4), 162–169 (2004)
7. Lederman, S.J., Klatzky, R.L.: Haptic classification of common objects: knowledge-driven exploration. Cogn. Psychol. **22**(4), 421–459 (1990)
8. Tanaka, Y., Bergmann Tiest, W.M., Kappers, A.M., Sano, A.: Contact force and scanning velocity during active roughness perception. PloS ONE **9**(3), e93363 (2014)
9. Kaim, L., Drewing, K.: Exploratory strategies in haptic softness discrimination are tuned to achieve high levels of task performance. IEEE Trans. Haptics **4**(4), 242–252 (2011)
10. Nicholson, L.L., Maher, C.G., Adams, R.: Hand contact area, force applied and early non-linear stiffness (toe) in a manual stiffness discrimination task. Man. Ther. **3**(4), 212–219 (1998)
11. Srinivasan, M.A., LaMotte, R.H.: Tactual discrimination of softness. J. Neurophysiol. **73**(1), 88–101 (1995)
12. Claringbold, P.J., Biggers, J.D., Emmens, C.W.: The angular transformation in quantal analysis. Biometrics **9**(4), 467–484 (1953)
13. Xu, C., Wang, Y., Hauser, S.C., Gerling, G.J.: In the tactile discrimination of compliance, perceptual cues in addition to contact area are required. In: Proceedings of Human Factors Ergonomics Society, vol 42, pp. 1535–1539. SAGE Publications, Los Angeles (2018)
14. Friedman, R.M., Hester, K.D., Green, B.G., LaMotte, R.H.: Magnitude estimation of softness. Exp. Brain Res. **191**(2), 133–142 (2008). https://doi.org/10.1007/s00221-008-1507-5

Perception of Vibratory Direction on the Back

Astrid M. L. Kappers[1,2,3](✉), Jill Bay[3], and Myrthe A. Plaisier[1]

[1] Dynamics and Control, Department of Mechanical Engineering, Eindhoven University of Technology, Eindhoven, The Netherlands
{a.m.l.kappers,m.a.plaisier}@tue.nl

[2] Control Systems Technology, Department of Mechanical Engineering, Eindhoven University of Technology, Eindhoven, The Netherlands

[3] Human Technology Interaction, Department of Industrial Engineering & Innovation Sciences, Eindhoven University of Technology, Eindhoven, The Netherlands

Abstract. In this study, we investigated the accuracy and precision by which vibrotactile directions on the back can be perceived. All direction stimuli consisted of two successive vibrations, the first one always on a centre point on the spine, the second in one of 12 directions equally distributed over a circle. Twelve participants were presented with 144 vibrotactile directions. They were required to match the perceived direction with an arrow they could see and feel on a frontoparallel plane. The results show a clear oblique effect: performance in terms of both precision and accuracy was better with the cardinal directions than with the oblique ones. The results partly reproduce an anisotropy in perceived vertical and horizontal distances observed in other studies.

Keywords: Vibrotactile stimulation · Direction perception · Haptic matching.

1 Introduction

Vibrotactile displays provide ways to convey information in circumstances where vision or audition are occupied with different tasks or are not available at all. For persons with deafness, blindness or even deafblindness such devices might be helpful in daily tasks such as navigation and communication. In many situations a hands- and head-free device is preferred, and then the back is an obvious choice. Although there certainly has been done some research on the perception of vibrotactile stimulation on the back, the fundamental knowledge at this stage is far from sufficient to design an optimal device. Therefore, the current paper focuses on vibrotactile stimulation on the back, and more in particular, on the perception of direction.

This work was supported by the European Union's Horizon 2020 research and innovation programme under Grant 780814, Project SUITCEYES.

I. Nisky et al. (Eds.): EuroHaptics 2020, LNCS 12272, pp. 113–121, 2020.
https://doi.org/10.1007/978-3-030-58147-3_13

There are a few concepts that are of relevance here, and one of these is anisotropy. Weber [9] already found that vertical two-point pressure thresholds on the back are larger than the horizontal thresholds. Although pressure and vibration do not stimulate the same receptors, and thus Weber's observation on pressure thresholds does not necessarily apply to vibratory stimulation, the study by Hoffmann and colleagues [3] points in the same direction: they found that the accuracy of determining the direction of two subsequent vibration stimuli was higher for horizontal than for vertical directions.

Another relevant concept is the "oblique effect". Although this term has been used in many different experimental settings (both visual and haptic), the basic idea is that performance with oblique stimuli is worse than with stimuli oriented in cardinal (i.e. horizontal and vertical) directions. Performance can apply to both accuracy and precision. A task is performed accurately if the setting of the participant is close to the intended physical setting, so this is related to bias or systematic directional error. A task is performed precisely if subsequent measurements consistently lead to the same setting that is not necessarily the correct physical setting. So precision is related to variability or spread. Appelle and Gravetter [1], for example, found that rotating a bar to a specified orientation led to larger *variable* but not systematic *directional* errors for oblique orientations in both visual and haptic conditions. Lechelt and Verenka [6] asked participants to match the orientation of a test bar with that of a reference bar in the frontoparallel plane, again in both visual and haptic conditions. They also found much larger *variable* errors for the oblique orientations, but no *directional* bias. On the other hand, Kappers [5] reported systematic *directional* errors when the orientation of a bar had to be matched haptically to a bar at a different location in the horizontal plane. It remains to be seen how representative all these findings are for vibrotactile stimuli on the back.

Finally, it is important to take the difference between simultaneous and successive presentation into account. For pressure stimuli, Weber [9] already observed that the thresholds were smaller if stimuli were presented one after another. Similarly, for vibrotactile stimulation, both Eskilden and colleagues [2] and Novich and Eagleman [7] found better performance with sequential stimulation. Therefore, in the current study, we will only make use of successive stimulation.

In this study, we will investigate the perception of vibrotactile directions on the back. More in particular, we will investigate whether there are biases in the perception of direction, and whether there are differences in spread between the settings for cardinal and oblique directions.

2 Methods

2.1 Participants

Twelve students (7 female, 5 male) of Eindhoven University of Technology participated in this experiment. Their ages ranged between 18 and 23 years. Ten participants were right-handed, two were left-handed (self-report). They were unfamiliar with the research questions and the set-up. Before the experiment they gave written informed consent. They received a small financial compensation for participation. The experiments were approved by the Ethical Committee of the Human Technology Interaction group of Eindhoven University of Technology, The Netherlands.

2.2 Set-Up, Stimuli and Procedure

Twelve tactors (coin-style ERM vibration motors from Opencircuit, 8 mm diameter) were placed in velcro pockets at every 30° on a circle with a radius of 110 mm on the back of an office chair; an identical pocket with tactor was placed in the centre of the circle (see Fig. 1). A distance of 110 mm is well above the vibrotactile two-point discrimination threshold of 13–60 mm reported in several papers (e.g. [4,7,8]), and is about the maximum radius that could be presented on the back. Each trial consisted of a 1-s vibration of the centre tactor, followed by a 1-s break and a 1-s vibration of one of the other 12 tactors. This timing guaranteed that all vibration motors were always easy to distinguish. The vibrations were strong enough to be easily perceived for all locations on the back, but they were not necessarily perceived to be equally strong. During the practice session it was ensured that participants could indeed perceive all motors. Random blocks of the 12 different stimuli were presented 12 times, so the total number of trials per participant was 144.

The task of the participant was to indicate the direction in which the stimulus was felt by means of rotating an arrow located on a frontoparallel plane at about eye height and within easy reach (see Fig. 1d). They were explicitly instructed to touch the arrowhead with one of their finger tips. The participants put on blurred glasses that still allowed them to see the arrow, but prevented them from reading off the degrees on the protractor. Participants were not informed about the actual directions, nor the number of different directions. The experimenter was able to read off the adjusted orientation with a precision of 1°. Noise-cancelling headphones with white noise and earplugs were used to mask the sound of the vibrators.

Participants were asked to wear thin clothing to guarantee they could feel the vibratory stimulation. At the start of the experiment, the tactors on the chair were covered with a cloth so that the participant remained unware of the actual locations. The participants had to sit down on the chair with their back pressed against the back of the chair. With the help of a line marked on the chair (Fig. 1b), the experimenter made sure that the spine of the participant was aligned with a vertical line through the centre of the circle. The back of the chair was not adjusted in height for the individual participants, but as a difference in body height would result in at most a few cm difference on the back, the stimulated back areas were still quite similar. The participant was instructed explicitly that s/he should not move to ensure both contact and alignment were kept constant; the experimenter made sure they indeed remained with the back centered on the back rest throughout the experiment.

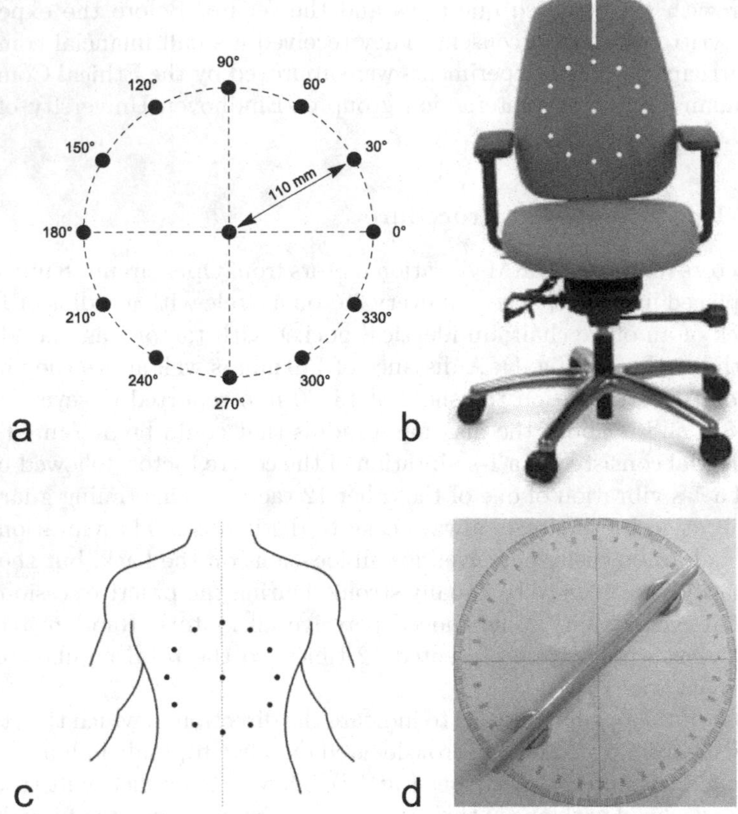

Fig. 1. Set-up. a) Circle with the 12 directions; b) Chair with the positions of the vibration motors (white dots) and the reference line for the spine (white line) indicated; c) Location of the circle of tactors on the back of the participant when seated on the chair; d) Arrow and protractor used to indicate the perceived direction.

The experiment started with a block of 12 different practice trials, after which the participant could ask remaining questions. Neither during the practice trials nor during the actual experiment feedback was given.

2.3 Data Analysis

In total there were 1728 (12 participants × 144) trials. 14 trials were discarded due to technical problems with the tactors. In 17 occasions the matched directions were about 180° off. This could either be due to a misperception of the participant or ignorance of the arrowhead. As the latter explanation seems much more likely than the former (some participants indeed confessed that they sometimes forgot to attend to the arrowhead), we decided to correct these cases. Finally, there were 16 clear outliers (leaving out such points led to a reduction in the standard deviation by at least a factor 2, but often much more). As these would have enormous effects on the standard deviations without being representative, it was decided to discard these 16 trials (less dan 1%).

For all analyses, we first computed mean and standard deviations per participant and per direction. Subsequently, we computed means and standard deviations over participants but per direction. We also compared results for cardinal (0°, 90°, 180° and 270°) and oblique (all other) directions.

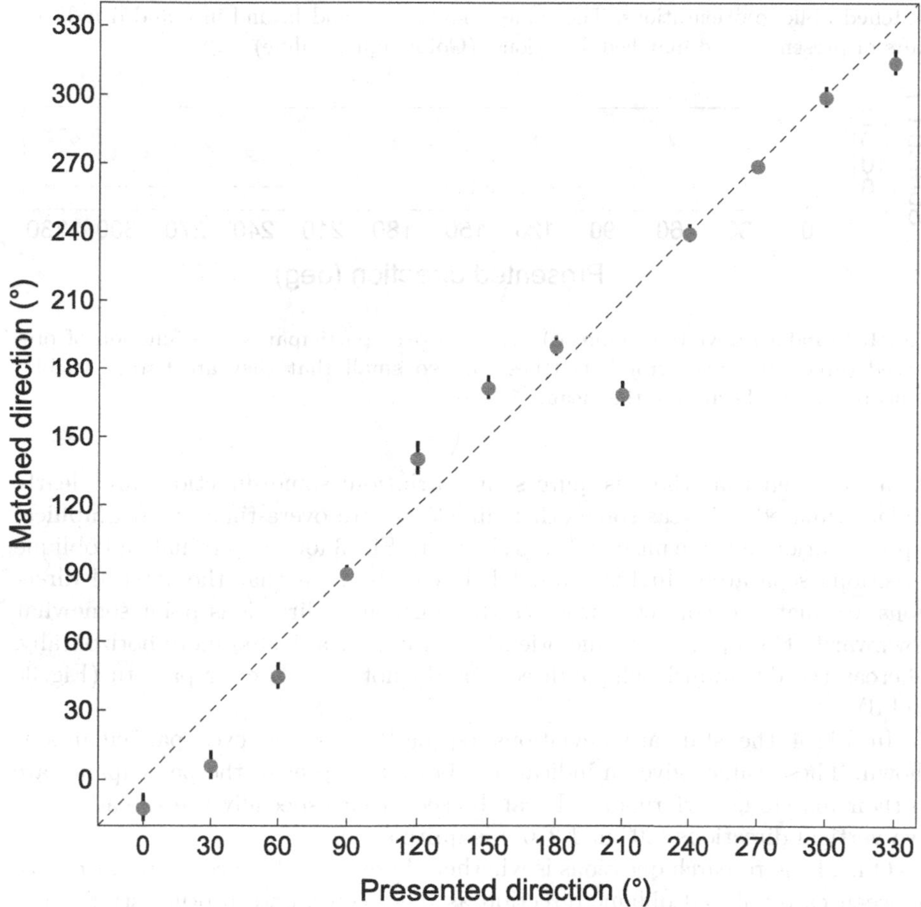

Fig. 2. Matched directions as a function of presented directions averaged over all participants. The error bars indicate standard errors of the mean and the dashed line the unity line.

3 Results

In Fig. 2 the matched directions are shown as a function of the presented directions. The error bars indicate standard errors over the averages of participants.

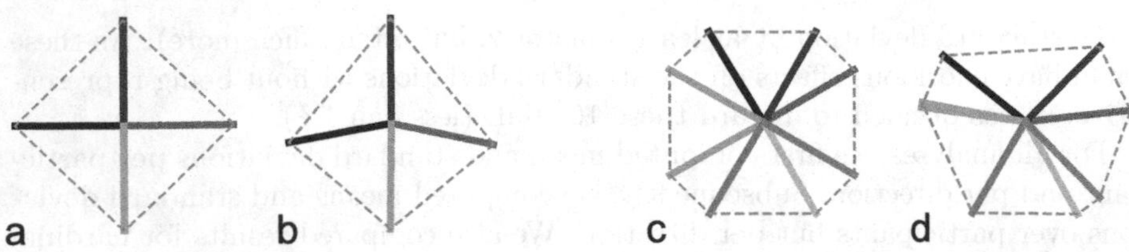

Fig. 3. Graphical representation of the deviations shown in Fig. 2. a) Presented cardinal directions; b) Matched cardinal directions; c) Presented oblique orientations; d) Matched oblique orientations. The same colours in a and b, and in c and d indicate pairs of presented and matched directions. (Color figure online)

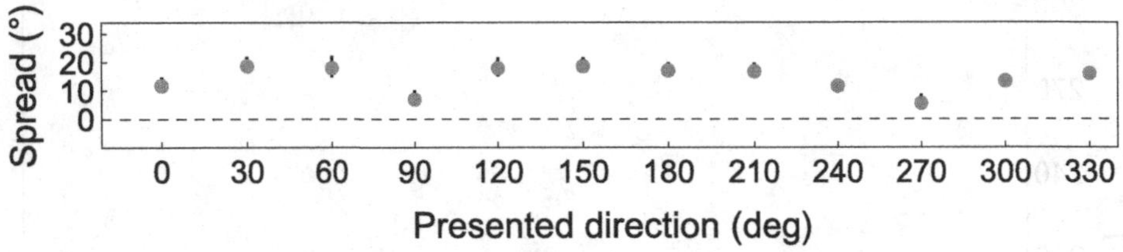

Fig. 4. Standard deviations (spread) averaged over participants as a function of presented directions. The error bars (these are so small that they are hardly visible) indicate standard errors of the mean.

It can be seen that there is quite some variation: some directions are clearly underestimated, whereas some other directions are overestimated. A graphical representation of these mismatches is shown in Fig. 3 for the cardinal and oblique directions separately. In Fig. 3a and b it can be seen that the vertical directions are matched correctly, whereas the horizontal directions point somewhat downward. The upward oblique orientations are all adjusted more horizontally, whereas the downward oblique directions do not show a clear pattern (Fig. 3c and d).

In Fig. 4 the standard deviations (spread) averaged over participants is shown. These values give an indication about how precise the participants are in their matching performance. It can be seen that especially the spread of the two vertical directions (90° and 270°) is quite small.

One of the research questions is whether there are differences in performance between cardinal and oblique direction as has been found in other studies not using vibrotactile stimuli and not presented on the back (e.g. [1,5,6]). To investigate this, we need to look at the *absolute* values of the mean deviations per participant, because *signed* values might average out over the various directions (see Fig. 2). In Fig. 5a we show the absolute mean deviations averaged over participants for both the cardinal (M = 9.4, SD = 6.2) and oblique (M = 20.2, SD = 5.5) directions. It can clearly be seen that the values of the oblique directions are higher than those of the cardinal directions. A paired t-test shows that this difference is highly significant: $t(11)=5.5$, $p < 0.0002$. In Fig. 5b we compare the

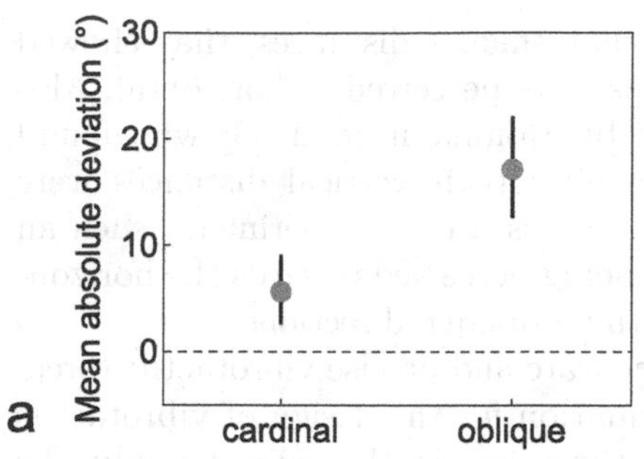

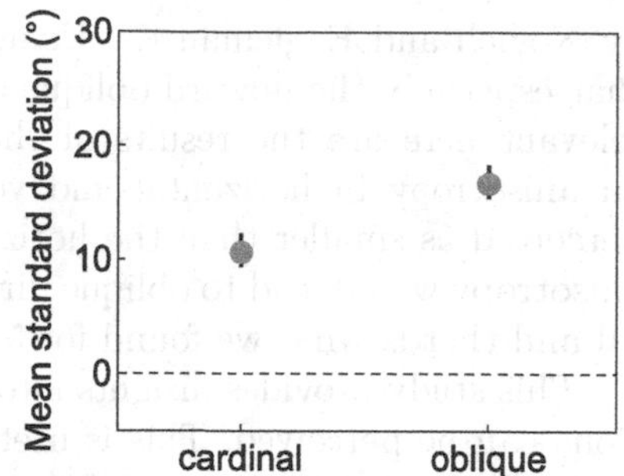

Fig. 5. Comparison of performance on cardinal and oblique directions. a) Absolute mean deviations averaged over participants for both the cardinal and oblique directions; b) Standard deviations in the cardinal and oblique directions averaged over participants. The error bars indicate standard errors of the mean.

spread of the deviations in the cardinal (M = 10.8, SD = 4.4) and oblique (M = 16.8, SD = 4.5) directions. Also this difference is significant: $t(11) = 4.2$, $p <$ 0.002.

4 Discussion and Conclusions

The aim of this study was to investigate the perception of vibrotactile directions presented on the back. The results show that the participants were well able to do this task, although they made some systematic directional errors. In Figs. 2 and 3, it can be seen that vertical directions (90° and 270°) were perceived veridical and in Fig. 4 it can be seen that also the variable errors for these directions were small. As the tactors used to generate these directions were all located on the spine of the participants, it is likely that perception was helped by the spine serving as anchor point. Other studies on vibrotactile perception also mention improved performance on or near the spine (e.g. [3,8]).

For the horizontal directions (especially 0°) the directional and variable errors are also relatively small, although perception is not veridical. Both horizontal directions are perceived as somewhat downward. Interestingly, Weber [9] already observed a somewhat oblique orientation for a two-point pressure threshold measurement on the back, albeit that this seems a rather informal observation without a mention of the actual direction. Novich and Eagleman [7] did not find confusions of their horizontal vibrotactile stimuli with oblique stimuli. However, in their 8-alternatives forced-choice experiment participants had the choice of 4 cardinal directions and 4 diagonal directions. Confusing horizontal with oblique would imply a misperception of 45° which is probably a too large difference.

The oblique directions caused both larger directional errors (biases) and larger variability of the errors than the cardinal directions. The type of deviations can best be appreciated in Fig. 3d. Especially the upward oblique directions appear to be perceived as closer to horizontal. A similar finding was reported by Novich and Eagleman [7]. Using somewhat smaller distances, they showed that especially the upward oblique directions were perceived as horizontal. Also relevant here are the results of the study by Hoffmann et al. [3] who found an anisotropy in horizontal and vertical acuity: their vertical distances were perceived as smaller than the horizontal distances. In our experiment, such an anisotropy would lead to oblique directions being perceived towards the horizontal and that is what we found for 5 out of the 8 oblique directions.

This study provides insights into how accurate and precise vibrotactile directions can be perceived. This is useful information for the design of vibrotactile devices intended to convey information to the users. In the current study, the first active tactor was always located on the spine. As the spine may have served as an anchor point, it remains a question whether the results are representative for a similar off-centre presentation of directions.

References

1. Appelle, S., Gravetter, F.: Effect of modality-specific experience on visual and haptic judgment of orientation. Perception **14**(6), 763–773 (1985). https://doi.org/10.1068/p140763
2. Eskildsen, P., Morris, A., Collins, C.C., Bach-y-Rita, P.: Simultaneous and successive cutaneous two-point thresholds for vibration. Psychonomic Sci. **14**(4), 146–147 (1969). https://doi.org/10.3758/BF03332755
3. Hoffmann, R., Valgeirsdóttir, V.V., Jóhannesson, Ó.I., Unnthorsson, R., Kristjánsson, Á.: Measuring relative vibrotactile spatial acuity: effects of tactor type, anchor points and tactile anisotropy. Exp. Brain Res. **236**(12), 3405–3416 (2018). https://doi.org/10.1007/s00221-018-5387-z
4. Jóhannesson, Ó.I., Hoffmann, R., Valgeirsdóttir, V.V., Unnþórsson, R., Moldoveanu, A., Kristjánsson, Á.: Relative vibrotactile spatial acuity of the torso. Exp. Brain Res. **235**(11), 3505–3515 (2017). https://doi.org/10.1007/s00221-017-5073-6
5. Kappers, A.M.L.: Large systematic deviations in a bimanual parallelity task: further analysis of contributing factors. Acta Psychologica **114**(2), 131–145 (2003). https://doi.org/10.1016/S0001-6918(03)00063-5
6. Lechelt, E.C., Verenka, A.: Spatial anisotropy in intramodal and cross-modal judgments of stimulus orientation: the stability of the oblique effect. Perception **9**(5), 581–589 (1980). https://doi.org/10.1068/p090581
7. Novich, S.D., Eagleman, D.M.: Using space and time to encode vibrotactile information: toward an estimate of the skin's achievable throughput. Exp. Brain Res. **233**(10), 2777–2788 (2015). https://doi.org/10.1007/s00221-015-4346-1
8. Van Erp, J.B.F.: Vibrotactile spatial acuity on the torso: effects of location and timing parameters. In: First Joint Eurohaptics Conference and Symposium on Haptic Interfaces for Virtual Environment and Teleoperator Systems. World Haptics Conference. pp. 80–85 (2005). https://doi.org/10.1109/WHC.2005.144
9. Weber, E.H.: E.H. Weber on the Tactile Senses. Ross, H.E., Murray, D.J. (eds.) Erlbaum (UK), Taylor & Francis (1834/1986)

Comparing Lateral Modulation and Amplitude Modulation in Phantom Sensation

Tao Morisaki(✉), Masahiro Fujiwara, Yasutoshi Makino, and Hiroyuki Shinoda

The University of Tokyo, 5-1-5 Kashiwanoha, Kashiwa-shi, Chiba-ken 277-8561, Japan
morisaki@hapis.k.u-tokyo.ac.jp, Masahiro_Fujiwara@ipc.i.u-tokyo.ac.jp, {yasutoshi_makino,hiroyuki_shinoda}@k.u-tokyo.ac.jp

Abstract. Phantom Sensation (PhS) is a tactile illusion in which a single sensation is elicited by stimulating two distant points. That sensation moves continuously between the two stimuli by changing the amplitude ratio. In this paper, we compared PhS for two types of tactile stimuli: lateral modulation (LM) of 20 Hz and amplitude modulation (AM) of 200 Hz. In LM, a stimulus point is moved periodically and laterally by several millimeters. In AM, the pressure is changed periodically at a fixed stimulation point. LM and AM are produced by ultrasound radiation pressure, where the force intensity of LM and AM were changed with the same temporal pattern. The results showed that the continuity and the localization of PhS elicited by LM at 20 Hz were significantly smaller than those elicited by AM at 200 Hz in 18 out of 24 conditions. However, PhS remained in all conditions that we used, regardless of LM or AM, even for an extremely long duration of 7.5 s and a short duration of 0.5 s.

Keywords: Phantom sensation · Tactile receptor · Ultrasound.

1 Introduction

When stimulating two distant points on the human skin, a single sensation is felt between them. This phenomenon is called Phantom Sensation (PhS). In PhS, a continuous tactile motion is presented by changing the amplitude ratio between the distant stimulation points [7]. The illusion is useful to minimize the number of stimulators that simplify the tactile display device [1,9]; however, the mechanism and properties are not well clarified.

In this study, we examined the conditions to elicit PhS using ultrasound radiation pressure that can control the stimulus quantitatively with high reproducibility and controllability. In particular, we compared PhS elicited by lateral

This work was supported in part by JSPS Grant-in-Aid for Scientific Research (S) 16H06303 and JST CREST JPMJCR18A2.

I. Nisky et al. (Eds.): EuroHaptics 2020, LNCS 12272, pp. 122–130, 2020.
https://doi.org/10.1007/978-3-030-58147-3_14

modulation (LM) [6] and amplitude modulation (AM) [4] presented by ultrasound. LM is the stimulus in which an ultrasound focus is moved periodically and laterally by several millimeters. AM is the stimulus in which the pressure amplitude of an ultrasound focus is modulated periodically at a fixed point. The details of LM and AM are described in Sect. 2.2. LM is clearly perceived even below 50 Hz, and AM is strongly perceived above 100 Hz. As the authors' subjective view, LM is perceived to be small and localized near the stimulus point, and AM is perceived as more widespread.

In the experimental design, we considered that low-frequency LM and high-frequency AM are typical stimuli that selectively stimulate type-I and FAII mechanoreceptors, respectively, where type-I includes fast-adapting (FAI) and slowly adapting (SAI) mechanoreceptors and FAII indicates fast-adapting type-II mechanoreceptors. The selectivity has not been evaluated quantitatively. However, we tentatively assumed it because it is plausible from the temporal characteristics of the threshold and the perceived spatial area. In the following experiments, we compared the difference in the perceived tactile motion continuity and localization of PhS elicited by LM and AM, changing the intensity of LM and AM equally with the same temporal pattern.

Some studies have been conducted on PhS using non-vibratory stimuli as well as various vibrotactile stimuli at 100 Hz and 30 Hz [1,2]. However, there is no comparison between the different types of stimuli under the same conditions. In this paper, we carefully measured the pressure pattern on the skin, and equalized the temporal profile of the stimulus between LM and AM.

2 Principle

2.1 Phantom Sensation

In our experiment, we simultaneously stimulated two points on the palm to elicit PhS by airborne ultrasound phased array (AUPA) [5]. AUPA creates an ultrasound focus (stimulus point) of approximately 1 cm^2. The maximum force of AUPA we used is approximately 4.8 g, as shown in Fig. 5 (right) and higher than the perception threshold on the hand [8]. AUPA is described in *Appendix*.

Figure 1 (C) shows a schematic illustration of the PhS evaluated in this paper. The presented pressure was changed linearly and logarithmically by 5 and 6 described in *Appendix*. Linear and logarithmic changes have been widely used in PhS experiments [1,9,10]. In Fig. 1, L_p is the length between two stimulated points, and $\boldsymbol{u}_p$ is the moving direction vector of PhS ($\|\boldsymbol{u}_p\| = 1$). P_1^{Linear} and P_2^{Linear} are the pressures presented at $\boldsymbol{r}_{\text{f}1}$ and $\boldsymbol{r}_{\text{f}2}$, respectively, when changing the pressure linearly. When changing the pressure logarithmically, the corresponding pressure is defined as $P_1^{\log}$ and $P_2^{\log}$. P_k^{Linear} and $P_k^{\log}$ ($k = 1, 2$), are shown in Fig. 1 (A and B) and given by

$$P_k^{\text{Linear}}(t) = P_k^{\max}(2 - k + \frac{t}{d_p}), \tag{1}$$

$$P_k^{\log}(t) = P_k^{\max} \log_2[3 - k + (-1)^k \frac{t}{d_p}], \tag{2}$$

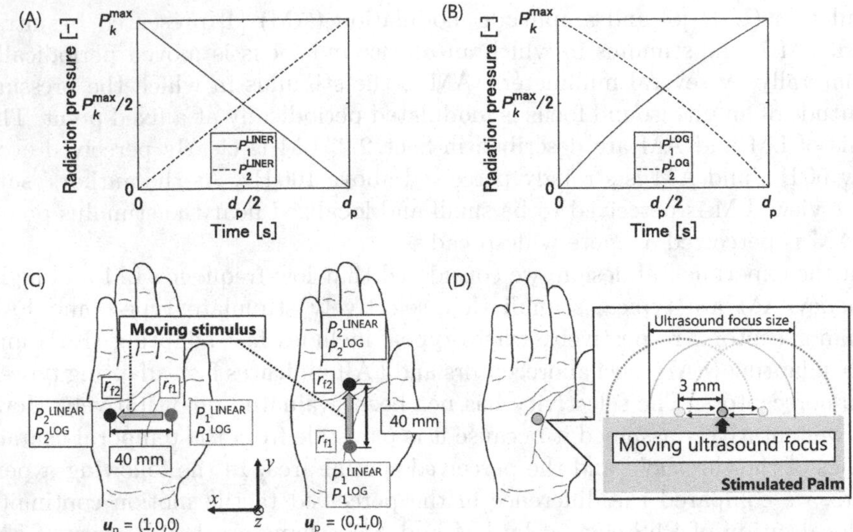

Fig. 1. A) & B) Changes in pressure linearly and logarithmically (P_k^{Linear} and P_k^{log}). **C)** Schematic illustration of Phantom Sensation evaluated in this paper. **D)** Schematic illustration of lateral modulation.

where d_p is the duration of PhS and $P_k^{\max}$ is the maximum pressure that AUPAs can present. The range of t is $0 \leq t \leq d_p$.

2.2 Lateral Modulation and Amplitude Modulation

Amplitude modulation: AM is the method to present a vibrotactile stimulus by periodically modulating the pressure amplitude of an ultrasound focus [4]. Previous research showed that the sensitivity of RAII is highest at approximately 200 Hz and higher than that of RAI and SAI at the same frequency [3]. Therefore, AM at 200 Hz is considered to stimulate RAII more strongly than RAI and SAI.

Lateral modulation: LM is the ultrasound stimulus method that moves an ultrasound focus periodically and laterally several millimeters while maintaining the intensity constant [6] (Fig. 1-D). LM at 20 Hz is more strongly perceived than AM at the same frequency [6]. Moreover, at 20 Hz, the sensitivity of RAII is lower than that of RAI and SAI [3]. Therefore, we used LM at 20 Hz to stimulates RAI and SAI more strongly than RAII.

3 Experimental Methods

In this experiment, we compared PhS elicited by LM and AM in terms of continuity and localization. Continuity and localization are typical elements of PhS that have been evaluated in many previous studies [1,2,9,10].

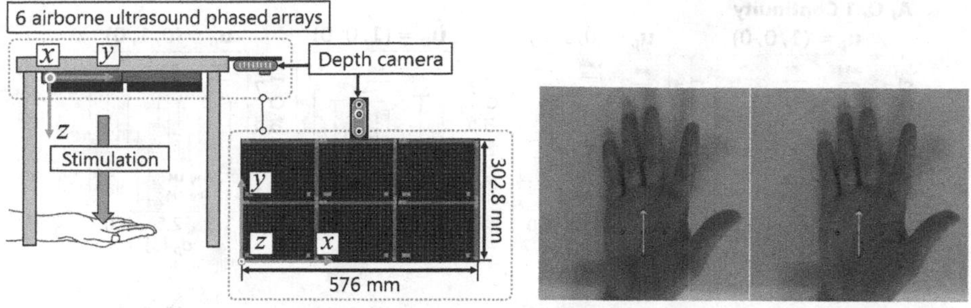

Fig. 2. Left: Experimental setup presenting Phantom Sensation. **Right**: Video of participants' hand captured by a depth camera with markers.

Stimulus: The experimental setup is shown in Fig. 2 (left). The setup consists of 6 AUPAs and a depth camera (Intel Real Sense D435). The coordinate system is a right-handed system with an origin at the lower left of AUPAs. We presented two ultrasound foci to $\boldsymbol{r}_{\mathrm{f}1} = (308, 151.4, z_{h1})$ mm and $\boldsymbol{r}_{\mathrm{f}2} = (268, 151.4, z_{h2})$ mm, then $\boldsymbol{u}_p = (1, 0, 0)$. We also presented the foci to $\boldsymbol{r}_{\mathrm{f}1} = (288, 131.4, z_{h1})$ mm and $\boldsymbol{r}_{\mathrm{f}2} = (288, 171.4, z_{h2})$ mm, then $\boldsymbol{u}_p = (0, 1, 0)$. z_{h1} and z_{h2} are the heights of the participants' hand measured by the depth camera. In PhS, we used $L_p = 40$ mm, $d_p = 1.5, 2.5, 4.0$ s, and a linear and logarithmic function shown in Fig. 1 (A and B) as the amplitude change. The stimulus type was LM and AM stimuli. The pressure of AM at a 200 Hz sinusoidal wave, where the perception threshold of RAII is less than -10 dB [3]. In LM, the ultrasound focus was moved laterally at a constant speed and periodically at 20 Hz. The movement width was 6 mm. The motion direction of LM was perpendicular to $\boldsymbol{u}_p$. AM and LM had the same maximum pressure. As a whole, we varied the stimulus type (LM or AM), the type of amplitude change (linear or logarithm), d_p, and $\boldsymbol{u}_p$. Therefore, there were $2 \times 2 \times 3 \times 2 = 24$ trials for each set. Each participant performed three sets. All trials were performed in the same order, which was randomized once.

Procedure: 10 male participants (ages 23 – 27) were included in the experiment. Participants put their hands toward the radiation surface of AUPAs. Participants saw the video of the hand captured by the depth camera. The video is shown in Fig. 2 (right). Participants adjusted the position of their hand so that the center of the circle-shaped markers added to the video was aligned with the center of their hand. The direction of PhS was also indicated by a green arrow in the video. After adjustment, PhS was presented, and participants answered the following two questions using a 7-grade Likert scale (1–7): **Q. 1) The presented stimulus is continuously moving stimulation, Q. 2) The presented stimulus consisted of only one stimulus point.** We intended that Q. 1 and Q. 2 evaluated the continuity and the localization of PhS, respectively. After answering the questions, the participants readjusted the position of their hands and proceeded to the next trial.

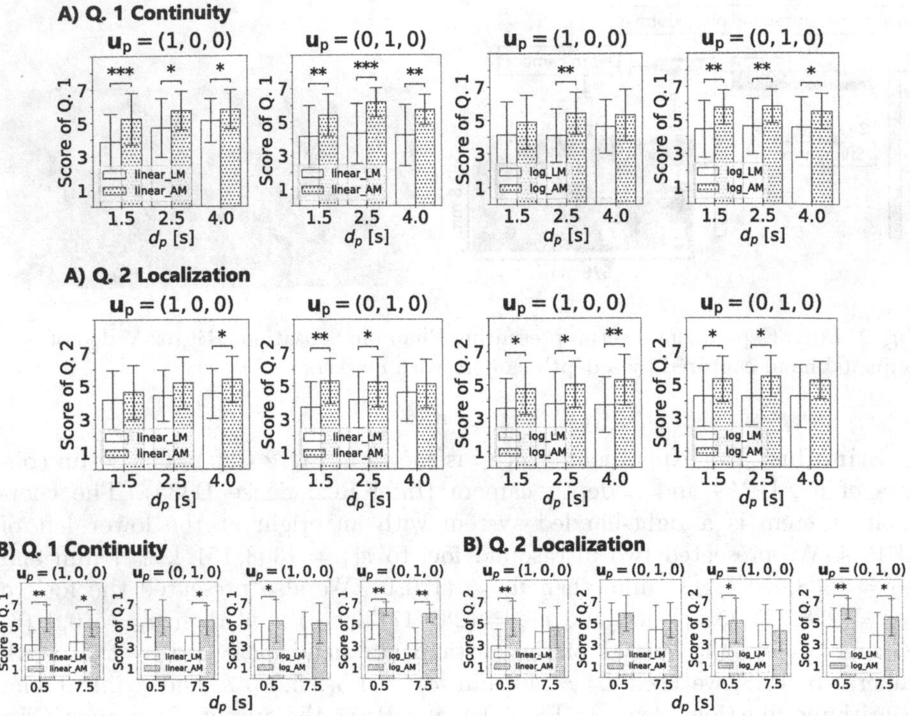

Fig. 3. Average scores of Q. 1 (continuity) and Q. 2 (localization). Error bars mean standard deviation. **A)** Stimulus duration $d_p = 1.5, 2.5, 4.0\,\mathrm{s}$. **B)** $d_p = 0.5, 7.5\,\mathrm{s}$.

3.1 Result

The average scores of Q. 1 (continuity) and Q. 2 (localization) are shown in Fig. 3 (A). The AM score was higher than the LM score in all conditions. Wilcoxon signed-rank test showed that the differences between the AM scores and the LM scores were significant in the Q. 1-linear case ($p < 0.05$). In the Q. 2-logarithm case, the differences were significant when $\{\boldsymbol{u}_p = (1, 0, 0)$ and $d_p = 2.5$ s$\}$, and $\{\boldsymbol{u}_p = (0, 1, 0)$ and $d_p = 1.5, 2.5, 4.0$ s$\}$. In the Q. 2-linear case, the differences were significant when $\{\boldsymbol{u}_p = (1, 0, 0)$ and $d_p = 4.0$ s$\}$, and $\{\boldsymbol{u}_p = (0, 1, 0)$ and $d_p = 1.5, 2.5$ s$\}$. The significant differences with $p < 0.05$, $p < 0.005$, and $p < 0.0005$ were indicated by *, **, and ***, respectively, in Fig. 3.

We applied ANOVA to the results with the fore factors: the type of stimulus, the type of amplitude change, $\boldsymbol{u}_p$, and d_p. We observed that the type of stimulus ($F(1, 714) = 173.3640$, $p = 6.6093 \times 10^{-17} < 0.0001$) and d_p ($F(2, 714) = 4.9319$, $p = 7.4604 \times 10^{-3} < 0.05$) had significant effects on the scores of Q. 1. We also observed that only the type of stimulus ($F(1, 714) = 106.2125$, $p = 2.5904 \times 10^{-23} < 0.0001$) had significant effects on the scores of Q. 2.

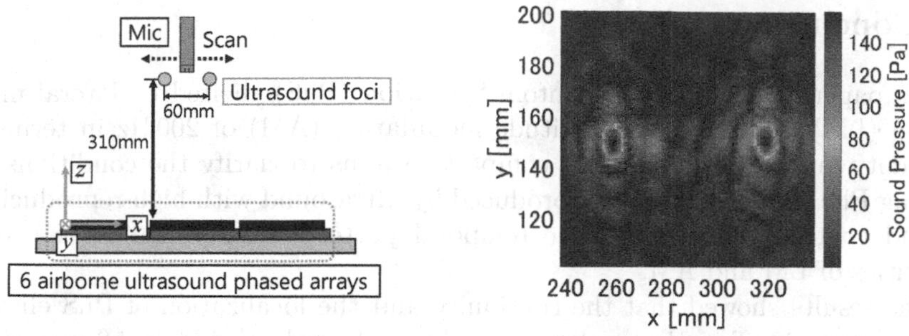

Fig. 4. Left: Experimental setup. **Right:** Measured sound distribution of the two ultrasound foci.

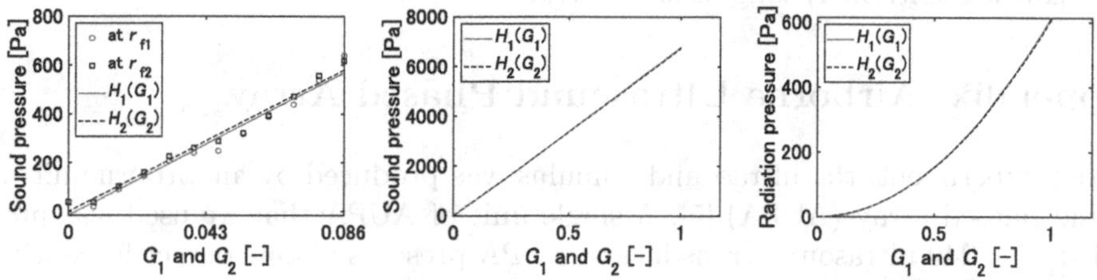

Fig. 5. Left: Measured change in sound pressure, according to G_1 and G_2. The results of linear regression H_1 and H_2 were also added. **Center & Right:** Calculated sound pressure and radiation pressure in $0 \leq G_1 \leq 1, 0 \leq G_2 \leq 1$.

3.2 Discussion

Originally, we expected that LM cannot elicit PhS because LM is perceived to be localized near the stimulus point. The results showed that AM elicited PhS more easily than LM, but PhS remained in the LM conditions. Therefore, we compared LM and AM again with extremely long or short durations of PhS ($d_p = 0.5, 7.5$ s), where eliciting PhS may be difficult. Our results and previous studies showed that the stimulus duration d_p has significant effects on PhS [10].

However, contrary to the authors' expectations, LM and AM still elicited PhS under extreme conditions. Moreover, the localization of LM was greater than that of AM in $\boldsymbol{u}_p = (1, 0, 0)$, $d_p = 7.5$ s. The results are shown in Fig 3 (B). 5 males (ages 24 – 26) participated in the second experiment. The experimental procedure is the same as that described in Sect. 3.

These results indicated that AM tends to elicit PhS more easily than LM, but both AM and LM can elicit PhS even under the extreme conditions. Note that there are differences in the stimulus area and perception intensity between LM and AM. The stimulus area of LM was larger than that of AM because of the movement of the stimulus points. The perceived intensities of LM and AM differ for the same driving power [6]. In this experiment, we did not equalize the perceived intensities but maximized it in both cases to make the participants clearly perceive the stimuli. These differences should be explored in future work.

4 Conclusion

In this paper, we compared Phantom Sensation (PhS) elicited by lateral modulation (LM) of 20 Hz and amplitude modulation (AM) of 200 Hz in terms of continuity and localization. This comparison aims to clarify the conditions for eliciting PhS. LM and AM were produced by ultrasound with high reproducibility and controllability. The same temporal pattern was applied to the force intensities of LM and AM.

The results showed that the continuity and the localization of PhS elicited by LM were significantly smaller than those elicited by AM in 18 out of 24 conditions. However, Phs remained in all conditions we used, regardless of LM or AM, even when the stimulus duration of PhS having significant effects on PhS [10] was extremely long or short (7.5 s or 0.5, s).

Appendix: Airborne Ultrasound Phased Array

In our experiment, the ultrasound stimulus was produced by an airborne ultrasound phased array (AUPA) [5]. A single unit of AUPA that we used was provided with 249 ultrasound transducers. AUPA presents a non-contact force with about 1 cm^2 circle area. The pressure is called acoustic radiation pressure. In this section, we measured the sound pressure by AUPA to change the intensity of LM and AM with the same temporal pattern (Eq. 1, 2).

Presenting PhS requires two-point stimuli [7]. When AUPA creates a focus at $\boldsymbol{r}_{\mathrm{f}1}$ and $\boldsymbol{r}_{\mathrm{f}2}$ simultaneously, the sound pressure p_{f} at $\boldsymbol{r}$ and t is as follows:

$$p_{\mathrm{f}}(t, \boldsymbol{r}, \boldsymbol{r}_{\mathrm{f}1}, \boldsymbol{r}_{\mathrm{f}2}) = \sum_{i=1}^{N_{\mathrm{trans}}} p_i(t, \boldsymbol{r})(G_1 \mathrm{e}^{-\mathrm{j}\|\boldsymbol{r}_{\mathrm{f}1}-\boldsymbol{r}_i\|} + G_2 \mathrm{e}^{-\mathrm{j}\|\boldsymbol{r}_{\mathrm{f}2}-\boldsymbol{r}_i\|}), \tag{3}$$

$$p_i(t, \boldsymbol{r}) = \frac{AD(\theta_i)}{\|\boldsymbol{r} - \boldsymbol{r}_i\|} \mathrm{e}^{-\beta\|\boldsymbol{r}-\boldsymbol{r}_i\|} \mathrm{e}^{\mathrm{j}(\nu\|\boldsymbol{r}-\boldsymbol{r}_i\| - \omega t)}, \tag{4}$$

where $\boldsymbol{r}_i$ is the position of each transducer, N_{trans} is the total number of transducers, A is the maximum amplitude of a transducer, D is the directivity of a transducer, ω is the frequency of ultrasound, ν is the wavenumber of the ultrasound we used, β is the attenuation coefficient of air, j is the imaginary unit, and θ_i is the angle from the transducer centerline to a focus. G_1 and G_2 are coefficients to determine the amplitude of two foci, respectively; thus, $0 \leq G_1, 0 \leq G_2, G_1 + G_2 \leq 1$.

We measured the sound distribution of two foci created by 6 AUPAs at $\boldsymbol{r}_{\mathrm{f}1} = (258, 151.4, 310)$ mm and $\boldsymbol{r}_{\mathrm{f}2} = (318, 151.4, 310)$ mm using Eq. 3. The experimental setup is shown in Fig. 4 (left). The sound distribution was measured by a microphone (Brüel & Kjær Type 2670) moved by a three-axis stage in 100 mm × 100 mm ($238 \leq x \leq 338, 101.4 \leq y \leq 201.4, z = 310$ mm) by 1.5 mm. The AUPAs output was 6.2% of the maximum value ($G_1 = 0.031$ and $G_2 = 0.031$) because the maximum sound pressure which the microphone can measure is about 316 Pa. The measured sound distribution is shown in Fig. 4 (right).

To change the radiation pressure accurately and obtain $P_k^{\max}$ of 6 AUPAs, we measured the change in sound pressure according to $\boldsymbol{G} = (G_1, G_2)$. The experimental setup is the same as that shown in Fig. 4 (left). 6 AUPAs created two foci at $\boldsymbol{r}_{f1} = (258, 151.4, 310)$ mm and $\boldsymbol{r}_{f2} = (318, 151.4, 310)$ mm. We changed $\boldsymbol{G}$ from $(0.086, 0)$ to $(0, 0.086)$ by $(-0.0078, 0.0078)$ and measured the sound pressure at $\boldsymbol{r}_{f1}$ and $\boldsymbol{r}_{f2}$ in each case. The measured sound pressure and the result of linear regression are shown in Fig. 5 (left). H_1 and H_2, which are sound pressure according to $\boldsymbol{G}$ at $\boldsymbol{r}_{f1}$ and $\boldsymbol{r}_{f2}$, respectively, are as follows:

$$H_1(G_1) = 1.7323 \times 10^6 G_1 - 17.0363, \tag{5}$$

$$H_2(G_2) = 1.7182 \times 10^6 G_2 - 1.3803. \tag{6}$$

The sound pressure and calculated radiation pressure [5] in $0 \le G_1 \le 1, 0 \le G_2 \le 1$ are shown in Fig. 5 (center and right).

References

1. Alles, D.S.: Information transmission by phantom sensations. IEEE Trans. Man Mach. Syst. **11**(1), 85–91 (1970)
2. Culbertson, H., Nunez, C.M., Israr, A., Lau, F., Abnousi, F., Okamura, A.M.: A social haptic device to create continuous lateral motion using sequential normal indentation. In: 2018 IEEE Haptics Symposium (HAPTICS), pp. 32–39. IEEE (2018)
3. Gescheider, A., Bolanowski, S.J., Hardick, K.R.: The frequency selectivity of information-processing channels in the tactile sensory system. Somatosens. Mot. Res. **18**(3), 191–201 (2001)
4. Hasegawa, K., Shinoda, H.: Aerial vibrotactile display based on multiunit ultrasound phased array. IEEE Trans. Haptics **11**(3), 367–377 (2018)
5. Hoshi, T., Takahashi, M., Iwamoto, T., Shinoda, H.: Noncontact tactile display based on radiation pressure of airborne ultrasound. IEEE Trans. Haptics **3**(3), 155–165 (2010)
6. Takahashi, R., Hasegawa, K., Shinoda, H.: Tactile stimulation by repetitive lateral movement of midair ultrasound focus. IEEE Transactions on Haptics **13**, 334–342 (2019)
7. Von Békésy, G.: Neural funneling along the skin and between the inner and outer hair cells of the cochlea. J. Acoust. Soc. Am. **31**(9), 1236–1249 (1959)
8. Weinstein, S.: Intensive and extensive aspects of tactile sensitivity as a function of body part, sex and laterality. The skin senses (1968)
9. Yatani, K., Truong, K.N.: Semfeel: a user interface with semantic tactile feedback for mobile touch-screen devices. In: Proceedings of the 22nd Annual ACM Symposium on User Interface Software and Technology, pp. 111–120. ACM (2009)
10. Yun, G., Oh, S., Choi, S.: Seamless phantom sensation moving across a wide range of body. In: 2019 IEEE World Haptics Conference (WHC), pp. 616–621. IEEE (2019)

Context Matters: The Effect of Textual Tone on the Evaluation of Mediated Social Touch

Sima Ipakchian Askari(✉), Antal Haans, Pieter Bos, Maureen Eggink, Emily Mengfei Lu, Fenella Kwong, and Wijnand IJsselsteijn

Eindhoven University of Technology, Eindhoven, The Netherlands
s.ipakchian.askar@tue.nl

Abstract. Mediated Social Touch (MST) promises interpersonal touch over a distance through haptic or tactile displays. Tests of the efficacy of MST often involve attempts to demonstrate that effects of social touch (e.g., on affective responses or helping behavior) can be replicated with MST. Results, however, have been mixed. One possible explanation is that contextual factors have not sufficiently been taken into account in these experiments. A touch act is accompanied by other verbal and non-verbal expressions, and whom we touch, when, and in what manner is regulated through social and personal norms. Previous research demonstrated, amongst others, effects of gender and the facial expression of the toucher on the recipients' touch experience. People can use expressions of the toucher's emotions as a cue to anticipate the meaning of the ensuing social touch. This current study examines whether emotions expressed in text (i.e., textual tone) affects the meaning and experience of MST. As expected we found textual tone to affect both the comfortableness of the touch as well as its perceived meaning. Limitations and implications are discussed.

Keywords: Mediated social touch · Textual tone · Affective haptic devices · Computer mediated communication · Haptic feedback

1 Introduction

Social touch plays an important role in human development, interpersonal communication, and well-being [1]. However, circumstances exist when it is not possible to have skin-to-skin contact, e.g. due to geographical separation. Mediated Social Touch (MST) devices address this problem by facilitating touch over distance through the use of tactile or haptic displays [2]. Research and design efforts that demonstrate what is possible technology-wise are numerous, and span a wide range of applications: from giving your child a hug, perform arm-wrestling, to giving a squeeze to another person [2]. Despite this work, research on the extent to which the effects of natural social touch can be replicated with MST has shown mixed results. Whereas Haans et al. [3] showed that the Midas touch—increased helping behaviour and willingness to comply to request [4]—may be replicated with vibrotactile MSTs, no effects of MST on, e.g., reducing stress have been established [5, 6], unlike prior work in the field of naturalistic social touch [1, 7]. One possible explanation is that current day tactile and haptic displays cannot mimic a real social touch with sufficient fidelity. At the same time,

I. Nisky et al. (Eds.): EuroHaptics 2020, LNCS 12272, pp. 131–139, 2020.
https://doi.org/10.1007/978-3-030-58147-3_15

social touch is more than tactile stimulation. Indeed, a touch act is always combined with other verbal and non-verbal cues alone (e.g., physical closeness [3]) which together shape its meaning and convey the perceived intention of the toucher.

Moreover, research aimed at demonstrating response similarities between real and MST has not always taken into account that where, when and whom we touch, is regulated by social norms. As a result, the potential stress-reducing effects of MST have been tested in contextual settings that appear to be rather unnatural; e.g. having male strangers hold hands through MST after having watched an emotionally charged movie [5, 6]. Existing research suggests that such contextual factors as the gender of the toucher can influence touch experience. For example, Gazzola et al. [9] showed that the primary somatosensory cortex differs in response depended on whether participants believed they were stroked by a male as compared to a female actor. Similarly Harjunen et al. [8] found the facial expression of a virtual agent to affect touch perception, with participants reporting more pleasant evaluations when being touched by a happy agent in comparison to an angry one. The perceived intensity of the touch also was found to depend on the agent's expression. According to Harjunen et al. [8], people use facial expressions of emotions as a cue to anticipate the meaning of an upcoming social touch. Such emotions and the resulting anticipation do not solely rely on facial expression but can also be derived from written text [10].

Therefore, in the present paper, we test whether the tone of a textual message affects the experience of an ensuing MST in terms of social comfortability with the touch, the perceived meaning of the touch, and the physical sensation of the touch.

2 Method

2.1 Participants

Ninety-three participants were recruited through the participant database of TU/e. Eight participants were excluded due to technical errors or for neglecting to fill in the questionnaire after each received touch. Of the remaining 85 participants, 46 were male and 39 female, with a mean age of $M_{age} = 27$ years ($SD = 10$; range: 19 to 64). Participants' self-reported ethnicity include 60% Dutch, 18.8% Indian, and 21.2% others. Participants received 5 euros as compensation (7 euros for externals).

2.2 Design

We conducted a two-condition (receiving a MST in a friendly vs. dominant textual tone context) within-subject design. Dependent variables were perceived comfortableness, smoothness, and hardness of the touch, as well as its perceived meaning and match with the tone of the message. Participants received two MSTs from a female confederate during an online question and answer (Q&A) conversation. The conversation was fully scripted around the topic of childhood (i.e., all questions and corresponding answers were fixed; see Table 1). The participant asked the questions, which the confederate answered. The confederate was one out of three randomly assigned females, due to limited availability of confederates. The textual tone of one answer was designed to be friendly and the tone of another to be dominant (see Table 1). The remaining four answers were written in a neutral tone. The same MST was given to the participant after the friendly and the dominant answers. The order of these two touches —and thus the order of the questions with the dominant and friendly answers—were counterbalanced across participants. For this purpose, two Q&A scripts were designed. In one set the 2nd answer was friendly, and the 4th dominant, in the other vice versa.

Table 1. Q&A script.

Question	Answer
Q1: Do you think your childhood had a major impact on who you are today?	Yes my childhood did influence me. How I was raised, which people I considered my friends, things that have happened, it all shaped me
Q2: What was your favorite toy as a child? and why?	Lego. I played with that a lot. I really enjoyed building complete cities with my brothers. Each year we all would get a new set for birthday. it was something we really looked forward to doing together as a family
Q3: Are you still in touch with the people you were close with during your childhood?	Yeah, I tried to keep in touch with a bunch of them. I still hang out or chat with a few of them, but most of them do not live close by so it is hard to see each other regularly
Q4: What was your favorite subject at elementary school? and why?	Maths, not just because it was easy, but the teacher was useless! He couldn't even answer the more complex questions in the back of the book. He told me to wait and ask the teacher in the following year. That is when I learnt that I am better than the rest
Q5: What was your dream job when you were young? Has that dream job changed since? and why?	I wanted to be a physician. Yes it has changed. You learn more about the world, and get to know what you really like and are good at. Now I want to develop tech that helps people with aging
Q6: Do you think you had a happy childhood?	Yeah. Compared to other yes. Parents are still together, and I did not have to move to many different cities to change school. I'm grateful for that

Note: The answer to Q2 was the friendly and that to Q4 the dominant answer. The order of these two questions within the Q&A set was counterbalanced across participants.

2.3 Apparatus and Stimuli

The experimental setting consisted of two desks facings each other, separated by a full size table divider. On each desk a laptop running Skype Online, a keyboard, and mouse were placed. The confederate's laptop also contained software for sending the MSTs. The tactile stimulus consisted of a caress applied by a finger-tip-sized soft polyurethane foam to the non-glabrous skin of the participant's left forearm, at a speed of 3.1 cm/s for a duration of 3.9 s [11], and in the distal direction. The mechanism of the MST device consisted of two sprockets, a tooth belt, and a Nema 17 stepper motor (12 V), and was concealed from view by means of a cardboard box. An Arduino Uno microcontroller was used alongside a TMC2208 driverboard to control the caress. To avoid anticipation of the MST, a sound recording of the MST device was played during the experiment. In addition, participants wore earmuffs.

2.4 Procedure

Before entering the room, the participant was notified that the other participant (the confederate) was already waiting in the lab. Upon entering, the participant was introduced to the confederate, who was sitting at the participant's seat, as though she had just tested the MST device. By doing so, we aimed to make it more plausible to the participant that the confederate too was a participant.

Next, the participant was given a cover story explaining that the aim of the experiment was to investigate when MST would be used during a Skype conversation —on the topic of childhood—and how such MSTs are perceived. After assigning, seemingly randomly, the role of interviewer to the confederate and that of interviewee to the participant, the participant was instructed to ask a pre-defined set of questions, and pay attention to the confederate's answers. The full list of questions was placed on the participant's table. The confederate was asked to answer the questions and deliver touches to the participant whenever she found appropriate. In reality, the timing of the MSTs was scripted. Both were instructed to complete a short survey each time a MST was used. After signing the informed consent, the MST device was placed over the participant's forearm, and the confederate was explained how to initiate a MST. The participant then received two MSTs to familiarize with the sensation. Next, both were asked to put on their earmuffs, after which the Q&A began. After each of the two MSTs, the participant completed the touch experience questionnaire. After the last question, the participant notified the experimenter. The background sound was paused, and a general questionnaire was handed out for both to fill in. Finally, the participant was payed, and informed that the debriefing would be sent by email later.

2.5 Measures

The touch experience questionnaire consisted of several open- and closed-ended questions. The 12 closed-ended items were 7-pt semantic differentials: 6 on the comfortableness (e.g. uncomfortable vs. comfortable; unacceptable vs. acceptable), 3 on the smoothness (relaxed vs. tense; smooth vs. rough; elastic vs. rigid), and 3 on the hardness of the touch (light vs. heavy; soft vs. hard; short vs. long). Open-ended questions concerned the perceived meaning of the touch, and whether its physical characteristics matched that perceived meaning. The latter was asked alongside a 5-pt response scale (not matched - matched).

Two separate factor analyses were performed on the polychoric correlation matrix of the comfortableness, hardness and smoothness items: one on the responses after the first, the other on the responses after the second MST. We used principal (axis) factoring as extraction, and oblique oblimin as rotation method. Prior to the analysis, items were inspected for missing values, low inter-item correlations, and low KMO values. Based on parallel analysis [12], three factors were extracted in both sets of responses. Except for one item (short vs. long; $\lambda \leq .50$), all items loaded on the expected factor with $\lambda \geq .67$. Therefore the former item was excluded from the analysis. We used the summated scale method to calculate factor scores. Cronbach's alpha values were $\alpha \geq .87$ for comfortableness, $\alpha \geq .83$ for hardness and $\alpha \geq .80$ for smoothness. For one person, no smoothness score could be calculated due to missing values. All three variables were found to be normally distributed.

The general questionnaire consisted of demographical questions (i.e. age, ethnicity and gender), 2 items measuring likability of the confederate on a 7-pt scale (e.g. unfavourable - favourable), 6 items measuring touch avoidance on a 5-pt scale [13], and 2 open-ended question concerning participants' thoughts on the interview setting and the MST device. Factor analysis—using the same method as described above—demonstrated that the six touch avoidance items all loaded on a single factor. Factor scores were calculated with the summated scale method, and the reliability was $\alpha = .82$. For one person, no touch avoidance score could be calculated due to missing values. Since touch aversion was not normally distributed, we used a 1/sqrt transformation. Likability of the confederate was also calculated using the summated scale method, and the reliability was $\alpha = .87$. For one person, no likability score could be calculated due to missing values. Likability was not normally distributed and no satisfactory transformation could be found.

3 Results

3.1 Comfortableness, Smoothness, and Hardness

To test the effect of textual tone on perceived comfortableness, smoothness, and hardness of the touch, we used paired sample *t* tests. Since three tests were conducted, we set the confidence level at $\alpha = .016$. We found textual tone to affect comfortableness to a statistically significant extent, with $t(84) = 3.4$, $p = .001$. The MST was perceived to be more comfortable when combined with a friendly tone (see Table 2). No statistically significant effect was found on smoothness and hardness, with $t(83) = 1.8$, $p = .075$ and $t(84) = -2.0$, $p = .046$ respectively. Exclusion of outliers (with $|Z| > 3$) did not affect the interpretation of the results.

Table 2. Mean comfortableness, smoothness and hardness and their standard deviations (SD) for the friendly and dominant textual tone condition

	Mean (SD)	
	Friendly	Dominant
Comfortableness	4.57 (1.16)	4.23 (.98)
Smoothness	4.02 (1.15)	3.79 (1.11)
Hardness	3.45 (1.21)	3.74 (1.21)

We found self-reported likability of the confederate to be positively correlated with comfortableness (*rho* $\geq$ .31; *p* $\leq$ .010), but not with smoothness and hardness (|*rho*| $\leq$ .13; *p* $\geq$ .241). No correlations were found between these dependent variables and touch avoidance (|*r*| $\leq$ .13; *p* $\geq$ .26). To explore how likability of the confederate may affect the observed effect of textual tone on perceived comfortableness, we conducted a repeated-measures ANOVA with mean centred likability as covariate. The main effect of textual tone remained statistically significant with F (1,82) = 11.3, and *p* = .001. While likability was significantly related to comfortableness, with F(1,82) = 13.8, and *p* < . 001, there was no significant textual tone by likability interaction, with F(1,82) = 0.8, and *p* = .382.

3.2 Perceived Meaning of the Touch

We used content analysis to investigate differences in perceived meaning of the touch between the two conditions. Responses were coded into one of the following categories: positive (phrases including, e.g., happiness, excitement, nostalgia, and fun), negative (phrases including, e.g., frustration, anger, better, arrogance), and other (i.e., not fitting the other two categories). Responses were coded independently by the first and second author. Cohen's kappa showed a moderate level of agreement: 0.74 [14]. Any disagreement between the authors was resolved (see Table 3 for observed category counts). A Chi-square test showed a statistically significant difference in category counts between the friendly and dominant tone condition, with $\chi^2(2) = 25.5$, $p < 0.001$. To confirm that this difference was indeed due to a relative change in positively and negatively charged phrases, we repeated the Chi-square test, but this time with the other category removed from the analysis. This confirmed that participants used comparatively more positive than negative words in the friendly condition than in the dominant condition, and vice versa, with $\chi^2(1) = 18.8$, $p < 0.001$.

Table 3. Distribution of negative, other, and positive coded meaning of MST for the friendly and dominant textual tone condition

	Negative	Other	Positive
Friendly textual tone	1	34	50
Dominant textual tone	12	51	20

From the proportion of responses coded as "other" (see Table 1), it becomes clear that there was more confusion regarding the meaning of the touch in the dominant condition than in the friendly condition. Although some participants assigned a negative meaning to the touch (e.g., "A bit of anger and frustration about the lack of competence of the math teacher" or "To support the statement that the participant is better than the rest"), others used more positive terms (e.g., "That she felt sort of proud of being the best in maths when she was in elementary school. Despite the teacher being useless"). Many, however, were ambivalent about the meaning (e.g., "I did not really understand. I am not sure this would make sense in a similar context in real life."

and "I actually really don't know. The touch did not make sense in this case."). Responses in the friendly condition were less ambivalent, and many described the touch as positive (e.g., "A feeling of joyfulness when thought of the Lego." or "To show her warm memory with her family, the warm feeling about the family activities". Although less than in the dominant condition, the meaning of the friendly touch also remained unclear to many (e.g., "I am again not sure." and "I don't know").

With more ambivalence as to the meaning of the touch in the dominant textual tone condition, one would have expected that the physical characteristics of the MST would match less well with the dominant answer as compared to the friendly answer. However, responses to the 5-pt item tapping into the extent of such a match were rather similar between the two conditions, with M = 3.02 (SD = 1.14) and M = 2.92 (SD 1.19), respectively. Similarly, the responses to the open-format question on the match between meaning and physical characteristics yield similar explanations for why the touch matched or mismatched for both the dominant as well as the friendly conditions. The reasons participants mentioned were often attributed to tactile stimulations of the touch or emotional content of the message. Additionally, participants expressed doubt or confusion regarding the meaning of the MST.

4 Discussion

Qualitative and quantitative results showed that textual tone can influence touch experience. Consistent with existing research [8, 9], we found a significant difference in the perceived comfortableness of touch between the friendly and dominant textual tone. In contrast to previous studies [8], however, the effects of textual tone on perceived smoothness and hardness of the tactile stimulation, although in the expected direction, were not found to be statistically significant.

The content analysis revealed that, as expected, textual tone affected the perceived meaning of the MST: Participants used comparatively more positive than negative words in the friendly condition than in the dominant condition, and vice versa. However, we participants to be rather ambivalent as to the meaning of the touch in the dominant textual tone condition, where most of the responses were coded as "positive" or "other". One possible explanation is that the tactile stimulus did not match well with the dominant textual tone answer. A caress is typically used in affective settings and found to communicate emotions such as love and sympathy [15], and may thus match better with the friendly than the dominant textual tone. Apparently, both the physical characteristics of the touch as well as its context are taken into account in the processing of a touch's meaning.

There were several limitations to the present work. First, due to time constraints of the project multiple confederates were used. Second, several participants were skeptical of the cover story, believing the interview was scripted and/or the fellow participant being a confederate.

Despite these limitations, our findings demonstrate that the textual tone of a chat message can change how people experience MST and what meaning they assign to it. As such, our findings are in line with previous studies, demonstrating that contextual factors affect how a tactile stimulus provided by a MST device is experienced, and thus its effect on the receiver's behavior as well. Consequently research aimed at testing the efficacy of MST (e.g., by demonstrating response similarities with naturalistic touch) should design carefully not only the tactile stimulus but also the context in which the touch act is delivered. Nonrepresentative and unanticipated context, such as when having two male strangers holding hands after having watched an emotionally charged movie, may not elucidate the possible beneficial effects of MST.

Acknowledgments. We thank Martin Boschman, Nasir Abed, Aart van der Spank and Twan Aarts for their continued assistance during development of the MST device.

References

1. Cascio, C.J., Moore, D., McGlone, F.: Social touch and human development. Dev. Cogn. Neurosci. **35**, 5–11 (2019)
2. Huisman, G.: Social touch technology: a survey of haptic technology for social touch. IEEE Trans. Haptics **10**(3), 391–408 (2017)
3. Haans, A., de Bruijn, R., IJsselsteijn, W.A.: A virtual midas touch? Touch, compliance, and confederate bias in mediated communication. J. Nonverbal Behav. **38**(3), 301–311 (2014). https://doi.org/10.1007/s10919-014-0184-2
4. Crusco, A.H., Wetzel, C.G.: The Midas touch: the effects of interpersonal touch on restaurant tipping. Pers. Soc. Psychol. Bull. **10**(4), 512–517 (1984)
5. Erk, S.M., Toet, A., Van Erp, J.B.: Effects of mediated social touch on affective experiences and trust. PeerJ **3**, e1297 (2015)
6. Cabibihan, J.-J., Zheng, L., Cher, C.K.T.: Affective tele-touch. In: Ge, S.S., Khatib, O., Cabibihan, J.-J., Simmons, R., Williams, M.-A. (eds.) ICSR 2012. LNCS (LNAI), vol. 7621, pp. 348–356. Springer, Heidelberg (2012). https://doi.org/10.1007/978-3-642-34103-8_35
7. Ditzen, B., et al.: Effects of different kinds of couple interaction on cortisol and heart rate responses to stress in women. Psychoneuroendocrinology **32**(5), 565–574 (2007)
8. Harjunen, V.J., Spapé, M., Ahmed, I., Jacucci, G., Ravaja, N.: Individual differences in affective touch: behavioral inhibition and gender define how an interpersonal touch is perceived. Pers. Individ. Differ. **107**, 88–95 (2017)
9. Gazzola, V., Spezio, M.L., Etzel, J.A., Castelli, F., Adolphs, R., Keysers, C.: Primary somatosensory cortex discriminates affective significance in social touch. Proc. Natl. Acad. Sci. **109**(25), E1657–E1666 (2012)
10. Gill, A.J., Gergle, D., French, R.M., Oberlander, J.: Emotion rating from short blog texts. In: Proceedings of the SIGCHI Conference on Human Factors in Computing Systems, pp. 1121–1124 (2008)
11. Löken, L.S., Wessberg, J., McGlone, F., Olausson, H.: Coding of pleasant touch by unmyelinated afferents in humans. Nat. Neurosci. **12**(5), 547 (2009)
12. Dinno, A.: Implementing Horn's parallel analysis for principal component analysis and factor analysis. Stata J. **9**(2), 291–298 (2009)
13. Wilhelm, F.H., Kochar, A.S., Roth, W.T., Gross, J.J.: Social anxiety and response to touch: incongruence between self-evaluative and physiological reactions. Biol. Psychol. **58**(3), 181–202 (2001)

14. McHugh, M.L.: Interrater reliability: the kappa statistic. Biochem. Med. **22**(3), 276–282 (2012)
15. Huisman, G., Darriba Frederiks, A.: Towards tactile expressions of emotion through mediated touch. In: CHI 2013 Extended Abstracts on Human Factors in Computing Systems, pp. 1575–1580. ACM, New York City (2013)

Influence of Roughness on Contact Force Estimation During Active Touch

Kaho Shirakawa(✉) and Yoshihiro Tanaka(✉)

Nagoya Institute of Technology, Gokiso-cho, Showa-ku, Nagoya 466-8555, Japan
k.shirakawa.680@nitech.jp, tanaka.yoshihiro@nitech.ac.jp

Abstract. Haptic sensations consist of cutaneous information elicited on the skin and kinesthetic information collected by the musculoskeletal system; there is a bidirectional relationship between haptic sensations and exploratory movements. Previous researches have investigated exploratory movement strategies for active haptic perception and the influence of exploratory movements on haptic sensations. This paper investigates the influence of roughness on the estimation of contact force during the active touch of samples with different textures. Two stimuli with different roughness were prepared and the contact force was measured when the participants rubbed pairs of samples with identical and different stimuli under the instruction of keeping the contact force constant. Eleven healthy adults participated in the experiment. The results showed that the accuracy of controlling the contact force for identical samples was not significantly different between coarse and smooth textures, whereas the contact forces between the coarse and the smooth sample when rubbing pairs of them were significantly different for six of eleven participants. These participants overestimated the contact force exerted for the coarse stimulus in comparison with the smooth stimulus. Thus, the results imply that textures during rubbing can yield perceptual bias for the contact force exerted; however, there are individual differences for this effect. There might be a complex perception mechanism for kinesthetic information involving cutaneous information.

Keywords: Perceptual bias · Roughness · Contact force

1 Introduction

Haptic sensations consist of cutaneous information elicited on the skin and kinesthetic information collected by the musculoskeletal system [1]. Some types of cutaneous information are pressure, vibration, and temperature. Exploratory movements, which are used to collect kinesthetic information, affect the mechanical phenomenon related to the collection of cutaneous information. For example, Lederman and Talyor [2] demonstrated that roughness perception enhances with

This work was supported by JSPS KAKENHI Grant Number JP19K22871.

I. Nisky et al. (Eds.): EuroHaptics 2020, LNCS 12272, pp. 140–148, 2020.
https://doi.org/10.1007/978-3-030-58147-3_16

the rise of the contact force. Natume et al. [3] reported that temporal vibrotactile information, spatial pressure distribution, and friction information affect subjective roughness ratings, indicating that the cognitive loads to each parameter differ among individuals. Thus, haptic perception is complex, and relevant factors are related to each other, including individual differences in cognitive processing. Furthermore, at high-level cognition, other modalities also influence haptic perception as observed in the size-weight illusion [4].

In active touch, there is sensory-motor control; this implies a bidirectional relationship between haptic sensations and exploratory movements. We consciously and unconsciously modify exploratory movements (e.g. contact force and scanning velocity) according to haptic sensations such as hardness, texture, and temperature. Previous studies have investigated the relationship between exploratory movements and haptic perception: Lezkan et al. [5] showed that sensory signals led to lower forces for more compliant objects; Tanaka et al. [6] showed that the variance of the contact force was larger for smooth stimuli than for coarse stimuli in a discrimination task; and Smith et al. [7] demonstrated that participants used a larger contact force for detecting a small concave object than for a small convex object. Thus, many previous works have been focused on exploratory movement strategies for active haptic perception, and as can be seen in [2], the influence of exploratory movements on haptic sensations has been also investigated. However, the influence of haptic sensations on the kinesthetic sensation involved in exploratory movements has been hardly investigated. Exploratory movements like contact force have been often controlled or measured in haptics researches. However, the kinesthetic sensations involved in exploratory movements may be affected by haptic sensations. In other aspect, previous researches on motor control have showed that perception of force and weight is affected by muscle activities [8].

This paper investigates the influence of roughness on the estimation of the contact force during active touch for different textures. Our hypothesis is that roughness promotes perceptual bias for the contact force exerted when rubbing samples of different textures with the fingertip. Two stimuli with different roughness were prepared and the contact force was measured when pairs of the same stimuli and different stimuli were rubbed under the instruction of keeping the same contact force. Differences in the contact force between the pairs were calculated and investigated.

2 Materials and Methods

2.1 Participants

Eleven healthy adults (five males and six females, aged 19–32) participated in the experiment. According to Coren's dominant hand discrimination test [9], the participants were strongly right-handed. The participants were naive about the purposes of the experiment and gave their written informed consent before participating in the experiment; they were instructed to use their dominant index finger during the tests. The experiment was approved by the Ethics Committee of Nagoya Institute of Technology.

2.2 Stimuli and Experimental Setup

Two types of stimuli, a coarse sample (C) and a smooth sample (S), were used in this experiment as shown in Fig. 1. Glass beads (Toshin Riko; glass beads no. 005 and no. 1) were attached to a flat 100×60 mm acrylic plate with double-sided tape. The particle diameter of the glass bead was approximately 1.0 mm (0.991–1.397 mm) and 0.05 mm (0.037–0.063 mm) for sample C and sample S, respectively. The participants could distinguish between the roughness of these stimuli according to the results of a previous study [3]. The experiment was performed using pairs consisting of C and C, S and S, and S and C. Thus, two identical samples for each stimulus type were prepared for a total of 4 samples and were used in random order for the experiment.

The experimental setup is shown in Fig. 2. Two samples were placed on a six-axis force sensor (ATI, Gamma) that was used to measure the contact force exerted during rubbing the samples. A laptop computer and a data acquisition (DAQ) module (National Instruments, NI USB-6218) were used to collect the contact force data. The sampling frequency was 1 kHz, and a low-pass filter with a cutoff frequency of 10 Hz were used for smoothing the data collected.

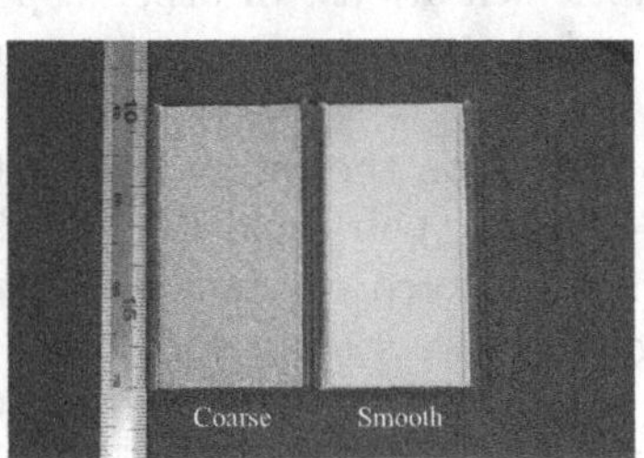

Fig. 1. Samples used in the experiment.

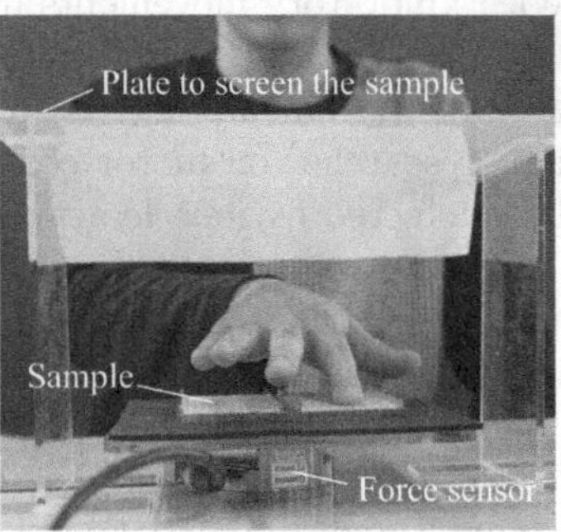

Fig. 2. Experimental setup.

2.3 Procedure

As shown in Fig. 2, two samples were arranged side by side in front of the participant. The participants were instructed to rub each sample from the back edge to the front edge. They were also instructed to rub each sample alternately, for a total of three times per sample (six strokes in total for one trial). On half of the trials, the participants firstly rub the sample placed on the left side, and on the other half of the trials, firstly rub the sample placed on the right side. The sample order was randomized within all trials for each participant. Regarding contact force, the participants were instructed to rub the two samples, exerting a constant contact force during each trial. In a preliminary test, when the exerted contact force was extremely small, it seemed that the participants relaxed and only used the weight of their own fingers and arms to exert the force. This case resembles feedforward control and implies that the participants might have not used finger motor control. Thus, the participants were allowed to use their preferred contact force, but the experimenter instructed them to slightly increase it when the exerted force was less than 0.3 N. The rubbing distance was approximately 100 mm, and the rubbing velocity was kept constant at approximately 100 mm/s by instructing the participants to listen to a metronome sound of 60 beats per minute by means of a headphone. As shown in Fig. 2, the participants could not see the samples or hear any sound produced by rubbing the sample. For these purposes, a blind plate mounted on the experimental setup and headphones were employed, respectively.

Three pairs of stimuli (C-C, S-S, and S-C) were used in the experiment. The experiment consisted of three sessions; in each session, three pairs were presented twelve times (four times each pair) in random order, considering the position (left or right) for starting the assessment and two samples for each stimulus. In total, 36 trials were conducted for each participant.

2.4 Data Analysis

Figure 3 shows an example of contact force data collected for one trial. During each trial, the two samples were rubbed alternatively three times (1st, 3rd, and 5th stroke for the firstly-rubbed sample, and 2nd, 4th, and 6th stoke for the secondly-rubbed sample). The force sensor output signals were extracted from the collected data. The rubbing period corresponded to 0.5 s within one stroke, from 0.25 s before to 0.25 s after the center point between the start and end points of the rubbing process. Approximately 50% of the whole rubbing period was used for the analysis. The start and end points of the rubbing process were detected using a 0.2 N contact force threshold for each stroke. Only for 3 trials out of the 396 trials, a threshold of 0.5 N was used because the participants kept touching the base plate of the force sensor where the two samples were placed to find the starting position.

The mean contact force was calculated from the extracted data of 3 strokes for each sample and then the difference between the mean contact force of the two samples presented (defined as α) was calculated for each trial. For the S-C pairs, the difference was calculated by subtracting the contact force of sample C from the contact force of sample S. Twelve α values were obtained for each S-C, C-C, and S-S pairs presented to each participant. Additionally, the contact force and the sensitivity were compared by using the data for S-S and C-C. The mean contact force F_z within one trial (6 strokes) and the quotient $|\alpha|/F_z$ by using the absolute difference $|\alpha|$ and F_z were calculated for the pairs consisting of identical samples (S-S and C-C). For the pairs S-S and C-C, the samples presented the same stimulus; therefore, $|\alpha|/F_z$ represents the sensitivity to keep the contact force constant.

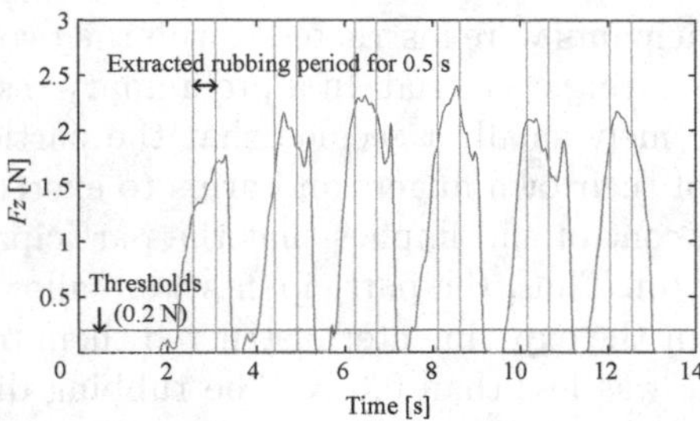

Fig. 3. Example of contact force for one trial. Two samples were rubbed alternatively for a total of 3 stroke for each. The data for the analysis was extracted for each stroke by using a threshold.

Considering individual differences, this study conducted statistical analysis for each participant. A Wilcoxon rank sum test with Bonferroni correction was performed for each participant in order to compare the difference in contact force α to zero, for each S-C, C-C, and S-S. The contact force F_z and the sensitivity $|\alpha|/F_z$ were compared between two conditions of identical samples (C-C and S-S) with a Wilcoxon rank sum test with Bonferroni correction for each participant. The significance level was set to 0.05.

3 Results

3.1 Differences in Contact Force

Figure 4 shows the difference in contact force α of the three conditions (S-C, C-C, and S-S) for each participant. It can be seen that the tendency of α was different among participants. The Wilcoxon rank sum test with Bonferroni correction was performed for each participant in order to compare α to zero. Regarding S-C, the statistical analysis showed that six participants presented significant positive values of α ($p < 0.05$ for 3 participants and $p < 0.01$ for 3 other participants), whereas, for the other participants, there were no significant differences. As for C-C and S-S, no significant differences were observed.

3.2 Contact Force and Sensitivity for Pairs of Identical Samples

Figure 5 shows the mean contact force F_z for the pairs consisting of identical samples (C-C and S-S) for each participant. The mean contact force and standard deviation for all participants were 1.24 ± 0.57 N for C-C and 1.42 ± 0.56 N for S-S. The Wilcoxon rank sum test with Bonferroni correction for each participant showed that there was no significant difference in sensitivity between the two conditions (C-C and S-S). The contact force for all trials and all participants was within 0.31–3.7 N. This range is within that reported in previous studies on texture rubbing [10]. The contact force for S-C was within the range of that for C-C and S-S.

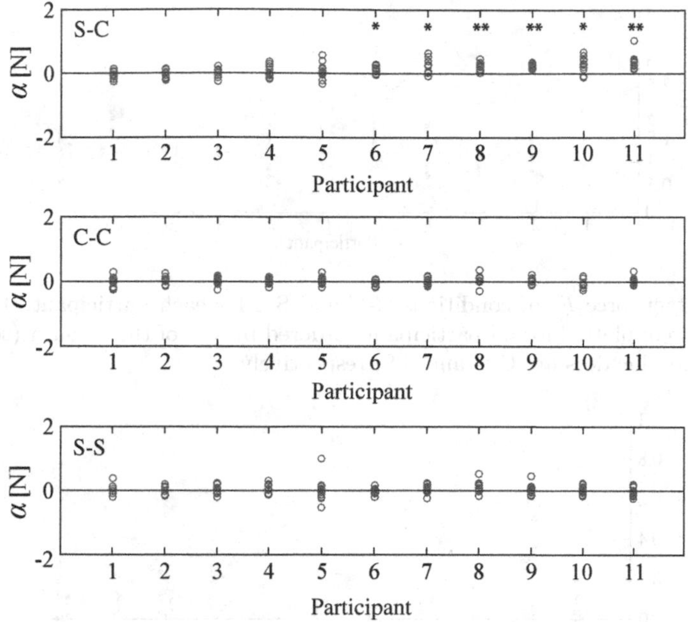

Fig. 4. Contact force differences of conditions S-C, C-C, and S-S for each participant. The results of all trials were plotted for all participants ordered by size of the mean α; * and ** denote $p < 0.05$ and $p < 0.01$, respectively.

Figure 6 shows the mean sensitivity $|\alpha|/F_z$ for the pair of coarse samples (C-C) and smooth samples (S-S) for each participant. The mean sensitivity and standard deviation were 0.08±0.02 for C-C and 0.09±0.03 for S-S. The Wilcoxon rank sum test with Bonferroni correction for each participant showed that there was no significant difference in sensitivity between the two conditions (C-C and S-S).

4 Discussion

The mean contact force results shown in Fig. 5 indicate that the contact force exerted was not significantly different between pairs of coarse samples and pairs of smooth samples, despite of being allowed to use a preferred contact force during each trial. A previous study with sandpapers demonstrated that contact force was smaller for rough samples than for smooth samples and discussed a possible reason of the discomfort [6]. In our experiment, samples were not discomfort due to using glass beads. Thus, no significant difference in the contact force was observed. For some participants, it appears that data point in Fig. 5 is distributed in two groups because they trended to use different force among each session.

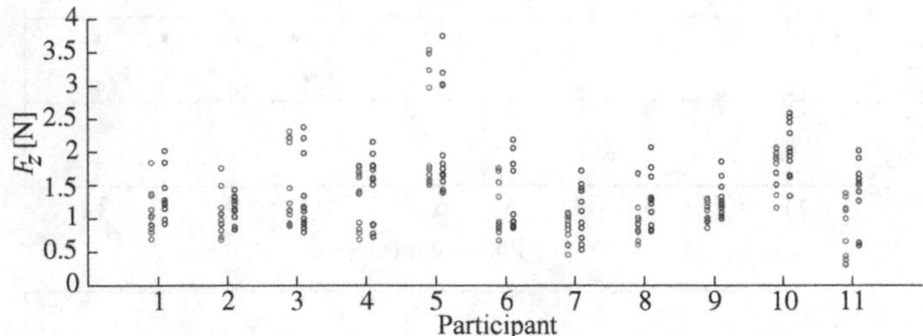

Fig. 5. Contact force F_z of conditions C-C and S-S for each participant. The results of all trials were plotted for all participants ordered by size of the mean α (see Fig. 4); Red dots and blue dots are C-C and S-S, respectively.

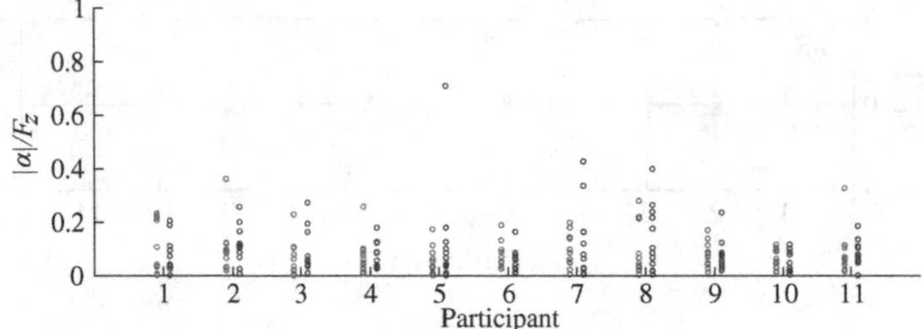

Fig. 6. Sensitivity of conditions C-C and S-S for each participant. The results of all trials were plotted for all participants ordered by size of the mean α (see Fig. 4); Red dots and blue dots are C-C and S-S, respectively. (Color figure online)

The sensitivity results shown in Fig. 6 indicate that the accuracy of the contact force exerted on the two samples presented was not significantly different between pairs of coarse samples and pairs of smooth samples. The result indicated that the participants could keep the contact force constant within an error of approximately 10% when rubbing identical samples for both coarse and smooth stimuli. Vibrotactile stimulation can promote a masking effect on haptic/tactile perception [11]; however, the experimental results indicated that the accuracy of the contact force did not become significantly low even for coarse samples, which elicited larger vibrotactile stimulation as compared with the smooth samples.

The results on contact force difference shown in Fig. 4 demonstrated that 6 out of 11 participants exerted a significantly larger contact force on the smooth sample than on the coarse sample under S-C condition, whereas the others did not exert a significantly different contact force. Moreover, none of the participants exerted a significantly different contact force under S-S and C-C conditions. The results shown in Fig. 5 indicated that the comfort did not significantly influence the contact force exerted for any participant in this experiment, and the results shown in Fig. 6 indicated that the sensitivity did not become significantly low for the coarse stimulus. Therefore, the significant differences observed in Fig. 4 imply the presence of perceptual bias for contact force; these participants overestimated the contact force exerted for the coarse stimulus. This indicates that, for some individuals, the contact force sensation based on kinesthetic perception can be affected by the roughness sensation based on cutaneous perception; there might be an integration mechanism for kinesthetic and cutaneous information regarding contact force perception. Additionally, it seems that the cognitive loads related to them differ among individuals; the results showed no significant differences for 5 out of 11 participants, whereas there were individual differences among the participants that presented significant differences, as shown in Fig. 4.

Here, as this experiment used only two different roughness stimuli, the discussion has limitations. A more diverse set of textures should be investigated to determine the physical factor that derives the perceptual bias of contact force. Furthermore, perceptual bias on exploratory movements involving the scanning velocity will be investigated in future work.

5 Conclusion

The present paper investigated contact force exerted during rubbing different samples with coarse and smooth textures. For pairs of different textures, six out of eleven participants significantly overestimated the contact force to the coarse textures whereas the others did not use significantly difference force. For identical samples of coarse or smooth textures, none of the participants used significantly different contact force. Previous studies have showed that cutaneous perceptions, like those involved in roughness rating, are affected by kinesthetic information [2]. Herein, our results implied that kinesthetic perceptions like contact force can be also affected by cutaneous information. There might be a complex perception mechanism for kinesthetic information involving cutaneous information. Our findings might be useful for the evaluation of haptic sensations and the development of haptic devices. In future work, a more diverse set of textures should be investigated to determine the driving factor of the perceptual bias of the contact force. We will also investigate the influence of other cutaneous information like friction and temperature on contact force perception, involving the scanning velocity, and the possible causes of the differences in cognitive loading between participants.

References

1. Loomis, J.M., Lederman, S.J.: Tactual perception. In: Boff, K.R., Kaufman, L., Thomas, J.P. (eds.) Handbook of Perception and Human Performance, Cognitive Processes and Performance. Wiley-Interscience, Hoboken (1986)
2. Lederman, S.J., Taylor, M.M.: Fingertip force, surface geometry, and the perception of roughness by active touch. Percept. Psychophys. **12**(5), 401–408 (1972). https://doi.org/10.3758/BF03205850

3. Natume, M., Tanaka, Y., Kappers, A.M.L.: Individual differences in cognitive proceeding for roughness rating of fine and coarse textures. PLoS One **14**(1), e0211407 (2019)
4. Flanagan, J.R., Beltzner, M.A.: Independence of perceptual and sensorimotor predictions in the size-weight illusion. Nat. Neurosci. **3**(7), 737–741 (2000)
5. Lezkan, A., Drewing., K.: Predictive and sensory signals systematically lower peak forces in the exploration of softer objects. In: Proceedings of 2015 IEEE World Haptics Conference, pp. 69–74 (2015)
6. Tanaka, Y., Bergmann Tiest, W.M., Kappers, A.M., Sano, A.: Contact force and scanning velocity during active roughness perception. PLoS One **9**(3), e93363 (2014)
7. Smith, A.M., Gosselin, G., Houde, B.: Deployment of fingertip forces in tactile exploration. Exp. Brain Res. **147**(2), 209–218 (2002). https://doi.org/10.1007/s00221-002-1240-4
8. Jones, L.A.: Perception of force and weight: theory and research. Psychol. Bull. J. **100**(1), 29–42 (1986)
9. Coren, S.: The Left-Hander Syndrome. Vintage Books, New York (1993)
10. Natsume, M., Tanaka, Y., Bergmann Tiest, W.M., Kappers, A.M.L.: Skin vibration and contact force in active perception for roughness ratings. In: 2017 Proceedings of IEEE RO-MAN, pp. 1479–1484 (2017)
11. Gescheider, G.A., Verrillo, R.T., Van Doren, C.L.: Prediction of vibrotactile masking functions. J. Acoust. Soc. Am. **72**, 1421–1426 (1982)

Green Fingers: Plant Thigmo Responses as an Unexplored Area for Haptics Research

Gijs Huisman(✉)

Amsterdam University of Applied Sciences, Digital Society School, Wibautstraat 2-4, 1091GM Amsterdam, The Netherlands
g.huisman@hva.nl

Abstract. Haptics research has been firmly rooted in human perceptual sciences. However, plants, too, possess capabilities for detecting mechanical stimuli. Here, I provide a brief overview of plant thigmo (touch) perception research with the aim of informing haptics researchers and challenging them to consider applying their knowledge to the domain of plants. The aim of this paper is to provide haptics researchers with conceptual tools, including relevant terminology, plant response mechanisms, and potential technology applications to kickstart research into plant haptics.

Keywords: Plants · Thigmo · Thigmomorphogenesis · Mechanosensing · Haptics · Indoor farming · Vertical farming

1 Introduction

The haptics research community has been primarily focused on studying haptic technology with human subjects [26], and not without reason. Haptics research has resulted in invaluable insights into human haptic perception capabilities and innumerable technological innovations [18,26].

Nevertheless, we share our daily lives with more than just humans. Haptic interactions are abundant in human-animal interaction, for example. In yet a different domain we might use our sense of touch to engage with plants, whether it be manipulating a pair of scissors to carefully sculpt a bonsai tree, or using one's hands to take down vines. Still, these are both examples from a human haptic perception perspective. Research in the biological sciences has shown that plants, too, are capable of perceiving 'touch' [36]. Think, for example, of how a bonsai tree changes its growing pattern when parts of its trunk are bound to create a more desirable shape [13]. Or think of how vines can cling to vertical surfaces, and how the Venus Flytrap (*Dionaea muscipula*) can detect when prey has landed on its leaf [7].

In fact, in their natural environment, plants are continuously subjected to mechanical stimuli, from soil vibrations to rain, snow, and hail, to wind, and

I. Nisky et al. (Eds.): EuroHaptics 2020, LNCS 12272, pp. 149–157, 2020.
https://doi.org/10.1007/978-3-030-58147-3_17

contact with other nearby plants and animals [5]. With their natural environment in mind it is therefore not surprising that plants evolved to be able to detect mechanical stimulation. However, since at least the agricultural revolution, humans have been moving plants away from their natural habitat, in some specific cases even moving them indoors. We are moving plants indoors for decoration, health purposes (e.g., air quality, or mental health), and production of both decorative plants and for food production (e.g., indoor farming). In these indoor circumstances plants are no longer subjected to typical mechanical stimulation which has been shown to have clear effects on, for example, their growth rate [6,35,36].

In this context haptic technology could offer opportunities to provide stimulation to plants growing indoors in order to stimulate certain types of growing behaviors. In this sense, haptic technology becomes another tool, like LED lighting, or hydroponics systems, in allowing for controlled growth of plants [4]. In the remainder of this paper I will provide an overview of key terminologies and concepts related to plant touch perception. I will discuss examples of existing plant-technology interaction and highlight cases for haptics in particular. Finally, I will discuss application domains in more detail and will provide recommendations for the haptics community in order to kickstart research into 'plant haptics'.

2 Key Concepts in Plant Mechanosensing

Plants have evolved to be able to respond to mechanical stimuli that occur as a consequence of their growing environment [5]. Since at least the ancient Greeks, humans have known that mechanical stimuli affect plant growth [6]. Examples of how humans have used mechanical stimuli to modulate plant growth throughout history include the originally Chinese art of bonsai [13], where the binding of branches and trunks produces desirable shapes, *Mugifumi*, the Japanese practice of trampling wheat and barley seeds to improve plant growth [20], and studies by Darwin [11] on the movement of plants.

Since these long-known practices and early scientific studies much progress has been made regarding understanding the cellular mechanisms that underlie plant responses to mechanical stimuli, though much remains to be discovered [6]. The locus of plant touch perception is the cytoskeleton-plasma membrane-cell wall interface which subsequently integrates into molecular signaling specific to the mechanical stimulus, and signal transduction [24,35]. A description of the exact cellular and molecular mechanisms that enable mechanoreception in plants is beyond the scope of this paper. The interested reader is referred to review papers on this topic [10,24,28,35,36].

2.1 Terminology of Plant Responses to Mechanical Stimulation

In human perceptual sciences [26] as well as in engineering and computer science [18] the term 'haptic', which derives from the Greek word *haptikos* meaning 'able to come into contact with', is widely used. However, in plant biology a different Greek word is used to denote physical stimulation, namely *thígma* [7], meaning 'touch'. This term is used in different ways to denote specific classes of plant responses to mechanical stimuli. The term *thigmomorphogenesis*, coined by Jaffe [23], refers to the impact of mechanical stimuli on plant growth and development [22]. These are generally slow processes and the types of thigmomorphogenic sources and effects are varied and depend on the specific plant type. The greatest potential impact of haptic technology is in eliciting thigmomorphogenic effects because these constitute permanent changes to a plant's growth and development.

A different class of responses are *thigmotropic* responses. These responses refer to changes in plant growth that are related to the direction of the mechanical stimulation [7]. For example, roots may grow around physical barriers or grow towards the source of a vibration [14]. Thigmotropic effects are different from thigmomorphogenic effects in that the latter refer to structural changes in the plant's growth (e.g., thicker stems, stockier plants), while the former refers to adaptation of regular growth behavior. Thigmotropic responses can happen relatively quickly and provide another interesting application domain for haptic technology.

Finally, there are *thigmonastic* responses. These refer to a plant's responses to mechanical stimulation that is not related to the direction of the stimulation [7]. An example is the Venus Flytrap's leaves closing in response to stimulation of its mechanosensing trigger hairs located within the leave structure [7]. Thigmonastic effects occur rapidly, but are also momentary and do not result in structural changes. Some applications for haptic technology can be conceived of but they are more limited than for the other described responses.

2.2 Plant Responses to Mechanical Stimulation

Physical stimuli that plants are subjected to are varied and can be both internal and external [36]. The focus here is on external forces, because those can be more easily generated using haptic technology. Nevertheless, internal forces, including sensing changes in turgor in a plant's cell and self-loading on the vertical axis of the stem or due to fruit bearing under the influence of gravity [36] (*gravitropism* is the term reserved for gravity's influence on a plant's growth [10,28,36]), all affect plant growth and may interact with external forces.

Plants can be subjected to various forms of mechanical stimulation from precipitation, animals making contact with the plant, and contact with other plants. Under natural conditions wind may be the most common and persistent external force that affects plant growth [16,35]. Plants', in particular trees', acclimatization to wind conditions affects their branching (e.g., trees that are predominantly subjected to wind from a single direction, such as in coastal regions, show asymmetrical branching formations), stem, and even roots [16]. Stimulation of different plant organs such as leaf brushing, bending of the stem, mechanical stresses on the roots, and contact with reproductive organs all have effects on the plants' overall development and the development of each specific organ [5–8]. In general,

Fig. 1. A prototype with an Arduino Uno driving a Tectonic Elements TEAX25C05-8 transducer with a 250 Hz sine wave. The image shows results of approximately two weeks of continuous stimulation on the root orientation of pumpkin seeds (*Cucurbita maxima*). One root can be observed to orient towards the source of vibration. Root growth would be expected to be more diffuse without stimulation, though, note, this is a single-trial result for illustration purposes.

leaf brushing results in more compact plants, with thicker leaves, while bending of the stem in most cases results in shorter plants with shorter distances between branch nodes, an increase in width (i.e., radial growth) of the stem, and more flexible tissue [35]. These types of stimulation typically result in delay of flowering [35]. Forces exerted on a plants' leafs and stem may propagate to the roots, and, in trees at least, can result in a larger root mass [35]. In all, these adaptations to mechanical stimulation make sense in that they make plants more resilient to such stimulation in the future, while delayed flowering saves valuable resources. All of these thigmomorphogenic responses have a profound impact on plant development. Nevertheless, thigmonastic responses, especially stimulation of a plant's reproductive organs serves an important purpose as well. Flowering plants release pollen when physical stimulation by a pollinator is detected, a process referred to as buzz pollination [12].

Plants' response to mechanical stimulation may also trigger molecular and biochemical changes that serve as a defence against pests and fungi [29]. For example, stroking of a strawberry plant (*Fragaria ananassa*) [37] or thale cress (*Arabidopsis thaliana*) [3] resulted in increased resistance to a fungal pathogen. Pressure due to soundwaves (related to plant mechanosensing [36]) has been found to increase vitamin C and sugar, among other parameters, in tomato fruits (*Solanum lycopersicum*) [2], which is relevant to food production.

3 Plants, Touch, and Technology

Mechanical stimulation of plants results in overall more compact, stronger, and more resilient plants, with greener leaves (for a review of effects see [6]), and may affect plant food production [2] though this latter effect is less consistent [35]. Nevertheless, plants' general characteristic responses to touch result in ornamental and food producing plants with an appearance that is preferred by consumers and retailers, and that have benefits for growers in terms of production area, packaging, and transport [6]. In addition, using mechanical stimulation to activate a plants' defence mechanism has the potential to reduce the need for chemical pesticides [6,35].

Taken together, it is not surprising that researchers have created devices to automatically stimulate plants through brushing [27,32,38], or vibration (see [6] for several examples). However, these efforts have not been taken up widely due to technical limitations of the systems in question [6,35]. Interestingly, in the bioacoustics community researchers have been using contact speakers to generate vibrations detectable by plants [14,15]. Lab studies have shown that roots can orient towards the source of a vibration [15], and can even navigate towards this source [14]. Figure 1 shows that such setups can be relatively easily created with basic hardware.

In computer science, primarily human-computer interaction (HCI) research, plants are used for interaction but often only as input device [30] or as a display method where mechanical stimulation is used to move the plant [17]. Only in a few cases is attention paid to a plant's responses to these types of stimulation [25,33], but not as a central part of the designed system. Thus, there are opportunities for HCI designs to make plants' responses a more integral part of interactive systems (e.g., use changing growth patterns as a system's output).

4 Green Fingers, Untapped Potential

In nature plants respond to mechanical stimuli to adapt to their environment [35] and these responses result in an overall more desirable plant morphology [6]. Nevertheless, technical difficulties have held back developments for automation in horticulture [6,35], and there is only a nascent interest thus far in computer science to develop devices for haptic stimulation of plants [25,33]. Thus, there are several opportunities to investigate the role of haptic technology developed for humans [18] in providing mechanical stimulation to plants. The most straightforward application is leaf brushing with a mechanical stimulus. Sophisticated haptic devices might be able to deliver such stimuli with accurate control over the applied force and velocity of stimulation [18]. Using similar devices, or even off-the shelve components such as servo motors, stem bending might also be easily achieved. Air vortices produced by custom-built haptic devices [34] might also be used to provide brush-like stimulation and may be a less invasive method than using mechanical devices [35]. Recent advances in ultrasonic haptics [21] are of particular interest considering findings that demonstrate effects of ultrasound on plant development [1]. Ultrasonic haptic devices could be used for leaf, stem, root, and reproductive organ stimulation. As Fig. 1 demonstrates, applying vibrations to root structures is relatively easily achieved. Recent developments in wide-band vibrotactile actuators (e.g., Apple's Taptic Engine) might be of particular use here. In short, the haptics community is well-placed to provide novel methods and devices for applying mechanical stimuli to plants. Taking plants as an application area might result in new and unexpected opportunities for both haptics research as well as horticultural research. Here, I want to provide a few suggestions for directions in which to search for such opportunities.

First, we might consider the production of decorative and food-producing plants on a smaller scale. Indoor farming has seen quite an uptake in recent years, spurred on by developments of LED lights and hydroponic systems. With a global food system that is under severe pressure from human-made climate change, indoor farming might provide one route towards sustainable food security [4]. The use of haptic technology to provide mechanical stimulation to plants grown indoors can help produce more compact and resilient crops that require less space and less pesticides (for reviews see [6,35]). Thus, haptic technology might be another valuable tool in future food production through controlled-environment agriculture [4], especially considering that mechanical stimulation is a typical feature of plants' natural habitat [5]. Related to this, we might also consider food production in more futuristic scenarios. NASA has conducted experiments with food production in zero-G for decades and zero-G environments come with specific considerations for food production [9]. From this perspective, haptic technology could be applied in such environments to help reduce some of the unwanted effects of plant growth in zero-G.

Second, haptic technology in combination with observable plant responses to such stimulation could be used for educational purposes. Plants' responses to light and nutrients might be relatively well-known, but plants' responses to touch, less so. The demonstration of thigmomorphogenic, thigmotropic, and thigmonastic effects in biology lessons could be supported by haptic systems. Especially for thigmomorphogenic effects that typically take a longer time to develop, automated haptic systems to demonstrate such effects in the classroom might be fruitful.

Third, arts and design disciplines may benefit from haptic technology for the controlled mechanical stimulation of plants. Bio-mimetic design, for instance, could use haptic technology for the creation of, not just nature-inspired designs, but designs of which nature is an active, living part. Such ideas could even be extended to architecture, where haptic technology (e.g., robotic structures that can move and apply forces) could help shape living architectural structures. For an example, see the living bridge of Cherrapunji in India. Admittedly, such ideas are still somewhat speculative although research on 'cyborg botany' might suggest they are not too far off [31].

Fourth, as already hinted at, the haptics community might be of aid to research in biology, botany, and horticulture by providing state-of-the-art technology for the application and measurement of mechanical forces that can be applied to plants. Conversely, work on plant thigmo responses could also inspire haptics researchers, for example, in the design of plant-inspired haptic systems (see [19] for an example of plant-mimetic mechanosensors). Here, it is also important to stress the need for collaboration in all of the examples described. While the current paper aims to prompt the interest of the haptics community in plant thigmo responses, there is a large body of literature from the biological sciences that extends and adds nuance to the topics discussed here. The haptics community has a firm grasp on the technology necessary to create breakthrough innovations in haptics research, now it is time to see if they possess green fingers too.

References

1. Aladjadjiyan, A.: Physical factors for plant growth stimulation improve food quality. In: Food Production-approaches, Challenges and Tasks, vol. 270 (2012)
2. Altuntas, O., Ozkurt, H.: The assessment of tomato fruit quality parameters under different sound waves. J. Food Sci. Technol. **56**(4), 2186–2194 (2019)
3. Benikhlef, L., et al.: Perception of soft mechanical stress in arabidopsis leaves activates disease resistance. BMC Plant Biol. **13**(1), 133 (2013)
4. Benke, K., Tomkins, B.: Future food-production systems: vertical farming and controlled-environment agriculture. Sustainability: Sci. Pract. Policy **13**(1), 13–26 (2017)
5. Biddington, N.L.: The effects of mechanically-induced stress in plants–a review. Plant Growth Regul. **4**(2), 103–123 (1986)
6. Börnke, F., Rocksch, T.: Thigmomorphogenesis-control of plant growth by mechanical stimulation. Sci. Hortic. **234**, 344–353 (2018)
7. Braam, J.: In touch: plant responses to mechanical stimuli. New Phytol. **165**(2), 373–389 (2005)
8. Chehab, E.W., Eich, E., Braam, J.: Thigmomorphogenesis: a complex plant response to mechano-stimulation. J. Exp. Bot. **60**(1), 43–56 (2009)
9. Cooper, M., Douglas, G., Perchonok, M.: Developing the NASA food system for long-duration missions. J. Food Sci. **76**(2), R40–R48 (2011)
10. Coutand, C.: Mechanosensing and thigmomorphogenesis, a physiological and biomechanical point of view. Plant Sci. **179**(3), 168–182 (2010)
11. Darwin, C., Darwin, F.: The power of movement in plants. Appleton (1897)
12. De Luca, P.A., Vallejo-Marin, M.: What's the 'buzz' about? the ecology and evolutionary significance of buzz-pollination. Curr. Opin. Plant Biol. **16**(4), 429–435 (2013)
13. Elias, T.: History of the introduction and establishment of bonsai in the western world. In: Proceedings of the International Scholarly Symposium on Bonsai and Viewing Stones, pp. 19–104 (2005)
14. Gagliano, M., Grimonprez, M., Depczynski, M., Renton, M.: Tuned in: plant roots use sound to locate water. Oecologia **184**(1), 151–160 (2017)
15. Gagliano, M., Mancuso, S., Robert, D.: Towards understanding plant bioacoustics. Trends Plant Sci. **17**(6), 323–325 (2012)
16. Gardiner, B., Berry, P., Moulia, B.: Wind impacts on plant growth, mechanics and damage. Plant Sci. **245**, 94–118 (2016)
17. Hammerschmidt, J., Hermann, T., Walender, A., Krömker, N.: InfoPlant: multimodal augmentation of plants for enhanced human-computer interaction. In: 2015 6th IEEE International Conference on Cognitive Infocommunications (CogInfoCom), pp. 511–516. IEEE (2015)

18. Hayward, V., Astley, O.R., Cruz-Hernandez, M., Grant, D., Robles-De-La-Torre, G.: Haptic interfaces and devices. Sens. Rev. **24**(1), 16–29 (2004)
19. Huynh, T.P., Haick, H.: Learning from an intelligent mechanosensing system of plants. Adv. Mater. Technol. **4**(1), 1800464 (2019)
20. Iida, H.: Mugifumi, a beneficial farm work of adding mechanical stress by treading to wheat and barley seedlings. Frontiers in plant science **5**, 453 (2014)
21. Inoue, S., Makino, Y., Shinoda, H.: Active touch perception produced by airborne ultrasonic haptic hologram. In: 2015 IEEE World Haptics Conference (WHC), pp. 362–367. IEEE (2015)
22. Jaffe, M.: The involvement of callose and elicitors in ethylene production caused by mechanical perturbation. In: Fuchs, Y., Chalutz, E. (eds.) Ethylene, vol. 9, pp. 199–215. Springer, Dordrecht (1984)
23. Jaffe, M.J.: Thigmomorphogenesis: the response of plant growth and development to mechanical stimulation. Planta **114**(2), 143–157 (1973)
24. Jaffe, M.J., Leopold, A.C., Staples, R.C.: Thigmo responses in plants and fungi. Am. J. Bot. **89**(3), 375–382 (2002)
25. Kurihara, W., Nakano, A., Hada, H.: Botanical puppet: computer controlled shameplant. In: 2017 Nicograph International (NicoInt), pp. 68–71. IEEE (2017)
26. Lederman, S.J., Klatzky, R.L.: Haptic perception: a tutorial. Attention Percept. Psychophys. **71**(7), 1439–1459 (2009)
27. Morel, P., Crespel, L., Galopin, G., Moulia, B.: Effect of mechanical stimulation on the growth and branching of garden rose. Sci. Hortic. **135**, 59–64 (2012)
28. Moulia, B.: Plant biomechanics and mechanobiology are convergent paths to flourishing interdisciplinary research. J. Exp. Bot. **64**(15), 4617–4633 (2013)
29. Pillai, S.E., Patlavath, R.: Touch induced plant defense response. J. Plant Biol. Res. **4**(3), 113–118 (2015)
30. Poupyrev, I., Schoessler, P., Loh, J., Sato, M.: Botanicus interacticus: interactive plants technology. In: ACM SIGGRAPH 2012 Emerging Technologies, p. 4. ACM (2012)
31. Sareen, H., Maes, P.: Cyborg botany: Exploring in-planta cybernetic systems for interaction. In: Extended Abstracts of the 2019 CHI Conference on Human Factors in Computing Systems, p. LBW0237. ACM (2019)
32. Schnelle, M.A., McCraw, B.D., Schmoll, T.J.: A brushing apparatus for height control of bedding plants. HortTechnology **4**(3), 275–276 (1994)
33. Seo, J.H., Sungkajun, A., Suh, J.: Touchology: towards interactive plant design for children with autism and older adults in senior housing. In: Proceedings of the 33rd Annual ACM Conference Extended Abstracts on Human Factors in Computing Systems, pp. 893–898. ACM (2015)
34. Sodhi, R., Poupyrev, I., Glisson, M., Israr, A.: Aireal: interactive tactile experiences in free air. ACM Trans. Graph. (TOG) **32**(4), 1–10 (2013)
35. Sparke, M.A., Wünsche, J.N.: Mechanosensing of plants. Horticultural Rev. **47**, 43–83 (2020)
36. Telewski, F.W.: A unified hypothesis of mechanoperception in plants. Am. J. Bot. **93**(10), 1466–1476 (2006)
37. Tomas-Grau, R.H., Requena-Serra, F.J., Hael-Conrad, V., Martínez-Zamora, M.G., Guerrero-Molina, M.F., Díaz-Ricci, J.C.: Soft mechanical stimulation induces a defense response against botrytis cinerea in strawberry. Plant Cell Rep. **37**(2), 239–250 (2018)
38. Zhou, J., Wang, B., Zhu, L., Li, Y., Wang, Y.: A system for studying the effect of mechanical stress on the elongation behavior of immobilized plant cells. Colloids Surf. B **49**(2), 165–174 (2006)

The Impact of Control-Display Gain in Kinesthetic Search

Zhenxing Li(✉), Deepak Akkil, and Roope Raisamo

Faculty of Information Technology and Communication Sciences,
Tampere University, Tampere, Finland
Zhenxing.li@tuni.fi

Abstract. Kinesthetic interaction typically employs force-feedback devices for providing the kinesthetic input and feedback. However, the length of the mechanical arm limits the space that users can interact with. To overcome this challenge, a large control-display (CD) gain (>1) is often used to transfer a small movement of the arm to a large movement of the onscreen interaction point. Although a large gain is commonly used, its effects on task performance (e.g., task completion time and accuracy) and user experience in kinesthetic interaction remain unclear. In this study, we compared a large CD gain with the unit CD gain as the baseline in a task involving kinesthetic search. Our results showed that the large gain reduced task completion time at the cost of task accuracy. Two gains did not differ in their effects on perceived hand fatigue, naturalness, and pleasantness, but the large gain negatively influenced user confidence of successful task completion.

Keywords: Control-display gain · Force-feedback device · Kinesthetic search

1 Introduction

Kinesthetic interaction as a form of human-computer interaction (HCI) is based on applying force feedback to provide motion sensations in muscles, tendons, and joints [1]. There is an increasing number of kinesthetic applications in different fields, such as education [2], medical training and simulation [3].

Providing realistic force feedback requires dedicated devices such as haptic gloves [4], kinesthetic pens [5] or grounded force-feedback devices (e.g., Geomagic Touch [6]). Among them, force-feedback devices provide a reliable desktop interface with high-resolution forces (up to 1 kHz) [7]. A major limitation of force-feedback devices is that the length of the mechanical arm limits the interaction space [7]. A common solution is to scale a small motion of the mechanical arm to a larger motion of the onscreen haptic interaction point (HIP), i.e., employing a large control-display (CD) gain [8].

The concept of CD gain has been previously studied in the context of pointing devices such as the mouse, touchpad and handheld VR controllers. The results suggest that applying a high CD gain can help reduce task completion time [9, 10]. In the context of kinesthetic interactions, some studies suggested that the visual feedback provided by different CD gains can influence kinesthetic perception and sometimes

I. Nisky et al. (Eds.): EuroHaptics 2020, LNCS 12272, pp. 158–166, 2020.
https://doi.org/10.1007/978-3-030-58147-3_18

even override the perception available through force feedback [11–13]. Further, while using a force-feedback device, applying a large CD gain leads to a mismatch between hand motions and HIP motions, which thus could potentially influence the user's control of the HIP.

Previous studies have used different techniques to enable kinesthetic interactions in large virtual environments without directly using a large CD gain. Dominjon et al. [14] used the bubble technique which adjusts the HIP speed based on the relative positions of the HIP and its bubble to reach objects. Li et al. [15, 16] employed gaze modality to move the HIP for reaching remote targets. Both methods maintained the unit CD gain while touching objects.

Overall, there is an agreement that applying a large gain may influence kinesthetic interactions [14–16]. However, it is still not clear how different CD gains affect task measures such as task completion time, accuracy of interaction and user experience in real-world kinesthetic tasks. In order to fill this gap, we conducted an experiment involving kinesthetic search on a soft tissue. Kinesthetic search is a typical kinesthetic task we perform in the physical world. It requires the users to touch the object and move their fingers along the surface to detect textural and material abnormalities on or under the surface. In a computer-based kinesthetic search task, the user needs to move the HIP while applying appropriate inward force to detect anomalies and the precise control of the HIP is crucial for efficient and accurate interactions.

We evaluated two commonly used CD gains in kinesthetic search: a large CD gain (=3.25) determined by the size of the required virtual space was compared to the baseline unit CD gain (=1). We varied the types of the search area as an independent variable since the effects of the CD gain may be influenced by the interaction area.

We collected objective data (the search time, the number of lumps that the participants missed and the search pattern gathered from the movement data of the HIP) and subjective data (the perceived hand fatigue, naturalness, pleasantness and user confidence in finding all the lumps) to evaluate the two CD gains. The study focused on the below research questions in the context of kinesthetic search:

- Are there differences in the task efficiency and search accuracy using two gains?
- Are there differences in user experience using two gains?

The paper first introduces the experiment, following by the results and discussion.

2 Experiment

2.1 Selection of CD Gains

The explored soft tissue was a cuboid model (52 × 32 × 32 cm along the x-, y- and z-axes). The model was placed at the center of the virtual space and fully filled the screen of the display. The physical workspace of the force-feedback device used in the experiment was 16 × 12 × 12 cm [6].

The study compared two CD gains (*high* and *default*). The *high* gain was 3.25, determined by the ratio between the tissue size and the device workspace (i.e., 52/16). Thus, a 1 cm arm movement lead to a 3.25 cm HIP movement and the workspace was increased to 52 × 39 × 39 cm which could cover the dimension of the virtual tissue.

The *default* gain was 1 and thus the workspace was 16 × 12 × 12 cm. To explore the virtual space beyond this workspace, we employed gaze as the section mechanism to relocate the device workspace [16]. The user had to pull the mechanical arm backward to a reset position and gaze at the target area for 500 ms. The workspace would then lock to that area until the user repeated this process. Such a method allowed robust switching of the workspace and ensured that there were no accidental switches during the task. This selected mechanism is not relevant from the perspective of the experiment. All analyses (e.g., search time) pertained only to the period when the user touched the virtual tissue, avoiding any potential influence of this mechanism.

2.2 Experiment Design

A within-subject experiment was designed in a controlled laboratory setting. The task for the participants was to identify the number of lumps underneath a soft tissue.

We manipulated the types of the search area as an independent variable with two levels: four small areas or one large area (Fig. 1(B)). For the large area, the tissue (52 × 32 cm, along the x- and y-axes) was divided into four areas with the size 26 × 16 cm each. The trial of searching the large area included only one area (size = 26 × 16 = 416 cm^2), and four trials as a task group covered the area of the whole tissue. For the small areas, the tissue was divided into 16 small areas with the size 13 × 8 cm each. To make the search size of all trials consistent, one trial of searching the small areas consisted of four randomly selected areas out of the 16 possible options (size = 13 × 8 × 4 = 416 cm^2), and four trials as another task group covered the whole tissue.

The sizes of these areas were selected based on the required search time to avoid a very long experiment. Simultaneously, they were used to examine the effects of the two CD gains in practical applications. The size of each small area was selected so that it could be covered by both workspaces of the two gains. In contrast, the large area could be covered by the workspace using the *high* gain but was beyond the workspace using the *default* gain. The user needed to relocate the workspace four times to fully search the large area (see Fig. 1(A) as an example).

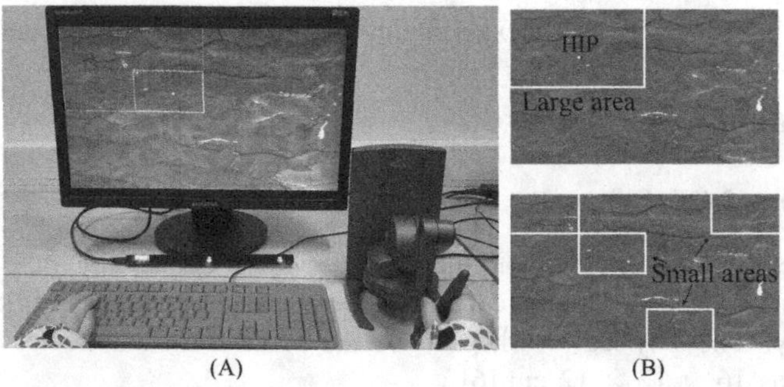

Fig. 1. (A) shows the experiment environment. The display shows an example of using the *default* CD gain to search a large area. The device workspace with the white boundary switches to the bottom right part of the large area where the user gazes at. (B) shows the area types.

The lump number for each trial was randomly selected from 1 to 4. For each task group with four trials, the total number were 10 (1 + 2 + 3 + 4 = 10). The lumps were sphere models. Since we are interested in examining user control by collecting the movement data of the HIP, all lumps were set as the same radius (0.3 cm) for simplicity. The lumps were randomly distributed (along the x- and y-axes) within the search areas, but placed at the fixed depth (1.5 cm) and were invisible to the participants.

Each participant needed to complete four task groups (2 gains × 2 types of the areas = 4 task groups) with 16 trials (4 task groups × 4 trials per group = 16 trials) and 40 lumps (4 task groups × 10 lumps per group = 40 lumps).

The haptics were developed using H3DAPI with OpenHaptics rendering system [17]. The stiffness of tissue and lumps were implemented by the linear spring law with different stiffness coefficients (tissue: 0.06 and lumps: 0.1) and the friction was implemented by the kinetic friction with the same friction coefficient (both: 0.01). The visual deformation was implemented by the Gauss function, linearly increased following the HIP depth. HIP was visualized as a sphere with 0.3 cm radius.

The participants were asked to input the number of lumps they found using a keyboard after each trial. The system checked and recorded the missing number and the search time. In addition, the system also logged HIP movement data along x-, y- and z-axes during the task. A 7-point Likert scale questionnaire was used to record the subjective data. User confidence were collected after each trial and other subjective data (hand fatigue, naturalness and pleasantness) were collected after each task group.

The participants signed an informed consent form and were asked to complete the tasks as accurately and quickly as possible. They were free to adopt their own strategy (e.g., horizontal and vertical searching) with a maximum of two full searches for each trial. In addition, no extra hand-rest equipment was provided. The order of the CD gains and the area types were counterbalanced among the participants.

2.3 Participants and Apparatus

24 participants were recruited from the local university community (16 women and 8 men), aged between 20 to 35 years (M = 26.17, SD = 4.26). Six participants had used a similar force-feedback device (1–2 times). An MSI GS63VR 7RF laptop was used as the host computer. We used a Samsung 245B monitor as the display, an EyeX [18] to track the gaze, a Touch X device [6] as the kinesthetic interface and a keyboard to input the participants' answer, shown in Fig. 1(A). We employed H3DAPI [17] for haptics and Tobii SDK [18] for accessing the eye tracker.

3 Results

3.1 Objective Data

We first conducted the Shapiro–Wilk Normality test that all data were not normally distributed (all $p < .001$). Thus, we used the 2 × 2 (gains × area types) aligned rank transform (ART) repeated-measures non-parametric ANOVA [19] for the analysis. The Wilcoxon signed-rank test was used for the post hoc analysis. Table 1 shows the overall ART ANOVA results. We focus our analysis on the main effect of CD gains and its significant interaction effect with the area types.

Table 1. Tests of within-subject effects on the objective data (significant values are in bold).

Sources	CD gains			Area types			Interaction effect		
	DF	F	Sig	DF	F	Sig	DF	F	Sig
Search time	1,23	42.07	**<.001**	1,23	5.68	**.026**	1,23	32.53	**<.001**
Missed lumps	1,23	16.33	**.001**	1,23	0.49	.491	1,23	1.89	.183
Covered area	1,23	62.83	**<.001**	1,23	66.58	**<.001**	1,23	78.83	**<.001**
Search depth	1,23	18.95	**<.001**	1,23	0.06	.803	1,23	0.25	.621

Search Time: We calculated the mean search time of four trials in each task group. The results showed that the *high* gain (M = 144.27, SD = 49.48) led to a shorter task completion time than the *default* gain (M = 209.93, SD = 63.86; Z = −4.200, $p < .001$). Figure 2(A) illustrates the interaction effect. In searching the large area, using the *high* gain (M = 112.48, SD = 44.59) led to approximately 47.7% shorter time than using the *default* gain (M = 215.26, SD = 74.22; Z = −4.286, $p < .001$). In searching small areas, using the *high* gain (M = 176.06, SD = 65.26) caused approximately 14.0% shorter time than using the *default* gain (M = 204.60, SD = 67.57; Z = −2.257, $p = .025$).

Missed Lumps: We calculated the sum of the missed lumps for each task group. Figure 2(B) shows that the participants using the *high* gain (M = 1.98, SD = 1.19) missed more lumps than using the *default* gain (M = 0.90, SD = 0.77; Z = −3.426, $p = .001$).

Covered Area: we calculated the proportion of the searched area based on the movement and the radius of HIP. Using the *high* gain (M = 83.29, SD = 4.96) caused searching a smaller area than using the *default* gain (M = 88.39, SD = 4.68; Z = −4.229, $p < .001$). Figure 2(C) shows that using the *high* gain (M = 78.02, SD = 7.37) led to searching a smaller area than using the *default* gain (M = 88.17, SD = 4.91; Z = −4.286, $p < .001$) in searching the large area. There was no difference in searching small areas. A participant's pattern for searching a large area is shown in Fig. 2(E) as an example.

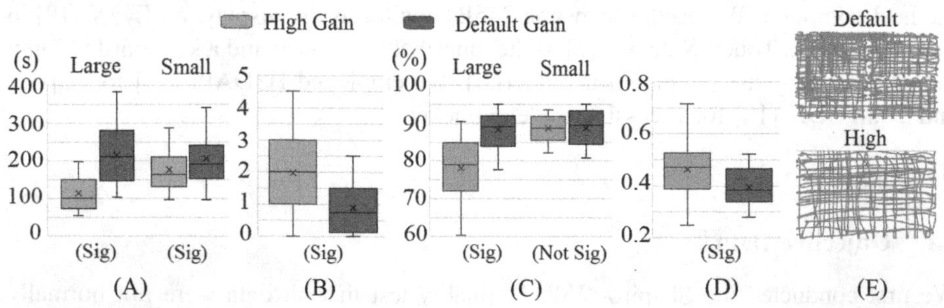

Fig. 2. (A) shows the search times based on the gains and the area types (i.e., large and small); (B) shows the number of the missed lumps based on two gains; (C) shows the area proportion the participants searched based on the gains and the area types; (D) shows the average absolute deviation of the HIP depth based on two gains; (E) shows a participant's pattern for searching a large area. The line in the boxplot is the median value and the cross mark is the mean value.

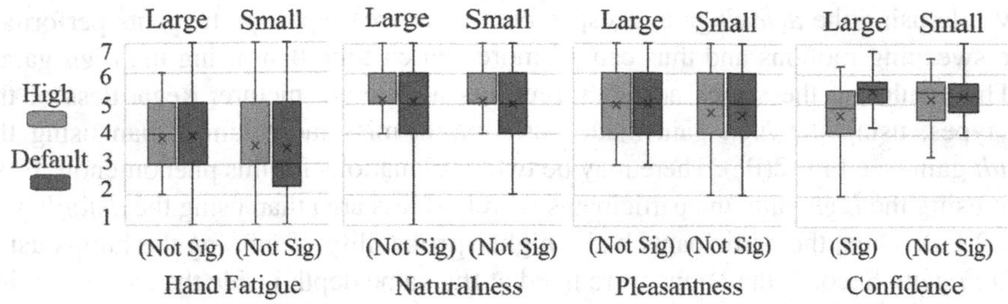

Fig. 3. Subjective results of the study (a higher value is better).

Search Depth: The HIP stability in the search depth may directly affect the search accuracy. To evaluate the stability, we calculated the average absolute deviation value of the HIP data (along z-axis) at the lump depth (1.5 cm) for both gains. Figure 2(D) shows that the *high* gain (M = 0.46, SD = 0.11) caused a lower stability of the HIP than the *default* gain (M = 0.39, SD = 0.07; Z = −3.857, $p < .001$) in the search depth.

3.2 Subjective Data

The data were analyzed with the Wilcoxon signed-rank test (Fig. 3). There were no statistically significant differences between two gains, in terms of perceived hand fatigue, naturalness, pleasantness, as well as user confidence in searching small areas. For the large area, using the *high* gain (M = 4.64, SD = 1.13) led to less confidence than using the *default* gain (M = 5.41, SD = 0.76; Z = −3.312, $p = .001$).

4 Discussion

We experimentally examined the effect of CD gains on kinesthetic search. The results show that CD gains and the area types have significant effects on task performance.

4.1 Differences in Task Completion Time and Search Accuracy

This study focused on the comparison of two different CD gains, where the movement of the device arm led to the different amount of HIP movement (1× and 3.25× respectively). Although the kinesthetic interaction involves complex hand behaviors and interaction feedback, our results show that a large gain increases the movement speed of HIP and thus reduces the task completion time, consistent with the common effect of the CD gain in the pointing tasks using the mouse [9].

However, the search time while using the *high* gain was influenced by the area types (Fig. 2(A)). It can be understood if we consider the search strategy used by our participants. Participants typically adopted a strategy that involved horizontal or vertical sweeping motions (Fig. 2(E)). Searching a large area easily enabled the participants to perform fewer sweeping motions. Searching multiple smaller areas made them perform numerous sweeping motions, potentially leading to longer task times.

While using the *default* gain, irrespective of the area types, participants performed more sweeping motions and thus caused more search time than using the *high* gain.

The results on the search accuracy presents a different picture. Regardless of the area types, using the *high* gain made participants miss more lumps than using the *default* gain (see Fig. 2(B)). There may be two explanations for this phenomenon. First, while using the *high* gain, the participants searched less area than using the *default* gain (Fig. 2(C)). Thus, the participants had a higher probability of missing the lumps using the *high* gain. Second, the lumps were fixed at the same depth inside the tissue. To find the lumps effectively, the participants had to maintain a constant depth of the HIP that could optimally touch the lumps while performing the sweeping motions. A more stable HIP depth presents better probability to find the lumps. Our result (Fig. 2(D)) demonstrated that using the *high* gain causes an increased variability in the HIP depth than using the *default* gain. Previous studies show that hand stability degrades under the stress of the force [20] and fatigue [21]. For the *high* gain, the stability issues may be amplified due to the scaling motion, and thus resulted in lower search accuracy.

4.2 Difference in User Experience

CD gain can potentially affect user experience, such as ease of use and pleasantness, in some HCI applications (e.g., [10]). Surprisingly, we did not find any difference between the two gain conditions in kinesthetic search, in terms of naturalness, pleasantness and hand fatigue. User confidence was influenced by two gains. Participants were generally less confident in finding all lumps while using the *high* gain, specifically while searching a large area. They likely had perceived the limited control over the HIP movement and were aware that they missed many areas. Using the *default* gain made participants more accurate in finding all lumps and subjectively more confident.

4.3 Limitations and Future Studies

This study has a few limitations. First, we examined two commonly used CD gains. Technically, the CD gain values that lie between them are rarely used due to the unsuitable workspace. Two levels (high and low) could sufficiently examine the general effect of the CD gain. However, a very large gain (i.e., the resulted workspace is much larger than the required space size) may cause different user performances (e.g., increase the task completion time, like [10]). Future work may examine this aspect.

Second, we used constant gains along x-, y- and z-axes. Dynamic gains were proposed for the pointing tasks (e.g., [8, 22]). However, their feasibilities for kinesthetic interaction are unknown. Dynamic gains (e.g., velocity-based) may lead to dynamic kinesthetic feedback and affect touch perception. Further, different CD gains could be potentially applied along the different axes. These should be studied further.

Third, the experiment involved a simple cuboid model with a flat surface. Practical applications may include models with irregular shapes and uneven surfaces (e.g., a heart model). The flat surface was a simple model that we could use to examine the effects of CD gains. Future work should test how results differ for complex models.

Fourth, we focused on kinesthetic search, a specific type of kinesthetic interaction. The CD gains may have different effects on different kinesthetic tasks, such as weight perception [11, 13]. Future work could examine CD gains in other kinesthetic tasks.

Fifth, this study included a short-term evaluation with new users. A prolonged usage or recruiting users such as medical professionals who are familiar with kinesthetic search may lead to different results. We propose these for the future research.

5 Conclusion

This study investigated the effects of CD gains on kinesthetic search. The experiment shows that a large gain improves task efficiency at the cost of user control and thus search accuracy. Our result experimentally demonstrates the significance to maintain the unit CD gain for accurate kinesthetic interaction. In addition, the findings of the study increase theoretical understanding of the CD gain effects on the task performance and user experience, which provide an experimental basis for designing new interaction techniques based on the CD gain for efficient and accurate kinesthetic interaction.

References

1. El Saddik, A., Orozco, M., Eid, M., Cha, J.: Haptics Technologies. Springer, Berlin (2011). https://doi.org/10.1007/978-3-642-22658-8
2. Grønbæk, K., Iversen, O.S., Kortbek, K.J., Nielsen, K.R., Aagaard, L.: Interactive floor support for kinesthetic interaction in children learning environments. In: Baranauskas, C., Palanque, P., Abascal, J., Barbosa, S.D.J. (eds.) INTERACT 2007. LNCS, vol. 4663, pp. 361–375. Springer, Heidelberg (2007). https://doi.org/10.1007/978-3-540-74800-7_32
3. Bielser, D., Gross, MH.: Interactive simulation of surgical cuts. In: PCCGA, pp. 116–442 (2000)
4. HaptX. https://www.haptx.com. Accessed 1 Dec 2019
5. Kamuro, S., Minamizawa, K., Tachi, S.: An ungrounded pen-shaped kinesthetic display: device construction and applications. In: WHC, pp. 557–567 (2011)
6. 3D Systems. https://www.3dsystems.com. Accessed 1 Dec 2019
7. Massie, T.H., Salisbury, J.K.: The PHANToM haptic interface. In: DSC, vol. 55, no. 1 (1994)
8. Argelaguet, F., Andújar, C.: A survey of 3D object selection techniques for virtual environments. Comput. Graph. **37**(3), 121–136 (2013)
9. Casiez, G., Vogel, D., Balakrishnan, R., Cockburn, A.: The impact of control-display gain on user performance in pointing tasks. J. CHI **23**(3), 215–250 (2008)
10. Kwon, S., Choi, E., Chung, M.K.: Effect of control-to-display gain and movement direction of information spaces on the usability of navigation on small touch-screen interfaces using tap-n-drag. Ind. Ergon. **41**(3), 322–330 (2011)
11. Dominjon, L., Richard, P., Fre, L.C., Richir, S.: Influence of control/display ratio on the perception of mass of manipulated objects in virtual environments. In: Proceeding of IEEE VR (2005)
12. Li, M., et al.: Evaluation of pseudo-haptic interactions with soft objects in virtual environments. PLoS One **11**(6), e0157681 (2016)

13. Samad, M., Gatti, E., Hermes, A., Benko, H., Parise, C.: Pseudo-haptic weight: changing the perceived weight of virtual objects by manipulating control-display ratio. In: CHI, Paper 320 (2019)
14. Dominjon, L., Lécuyer, A., Burkhardt, J.M., Barroso, G.A., Richir, S.: The "bubble" technique: interacting with large virtual environments using haptic devices with limited workspace. In: WHC, pp. 639–640 (2005)
15. Li, Z., Akkil, D., Raisamo, R.: Gaze augmented hand-based kinesthetic interaction: what you see is what you feel. IEEE Trans. Haptics **12**(2), 114–127 (2019)
16. Li, Z., Akkil, D., Raisamo, R.: Gaze-based kinaesthetic interaction for virtual reality. Interact. Comput. **32**, 17–32 (2020)
17. H3DAPI. http://www.h3dapi.org. Accessed 1 Dec 2019
18. TOBII. https://www.tobii.com/. Accessed 1 Dec 2019
19. Wobbrock, J.O., Findlater, L., Gergle, D., Higgins, J.J.: The aligned rank transform for nonparametric factorial analyses using only ANOVA procedures. In: CHI, pp. 143–146 (2011)
20. Borg, G., Sjöberg, H.: The variation of hand steadiness with physical stress. Motor Behav. **13**(2), 110–116 (1981)
21. Gates, D.H., Dingwell, J.B.: The effects of muscle fatigue and movement height on movement stability and variability. Exp. Brain Res. **209**(4), 525–536 (2011). https://doi.org/10.1007/s00221-011-2580-8
22. Wobbrock, J.O., Fogarty, J., Liu, S.Y., Kimuro, S., Harada, S.: The angle mouse: target-agnostic dynamic gain adjustment based on angular deviation. In: CHI, pp. 1401–1410 (2009)

The Arm's Blind Line: Anisotropic Distortion in Perceived Orientation of Stimuli on the Arm

Scinob Kuroki(✉)

Nippon Telegraph and Telephone Corporation, 3-1 Morinosato Wakamiya, Atsugi, Kanagawa, Japan
shinobu.kuroki.ub@hco.ntt.co.jp

Abstract. Given that mechanoreceptors are highly heterogeneously distributed and there is no direct sensory signal of the distribution, it must be challenging for the brain to identify stimuli in external space by remapping sensory inputs. Some previous studies reported perceptual distortion of tactile space, reflecting a difference in scales for different body parts. Here we report another example in which the orientation of stimuli perceived on the arm is rotated regionally, or even flipped. This illusion cannot be explained simply in terms of the resolution difference of mechanoreceptors.

Keywords: Psychophysics · Orientation perception · Motion perception · Arm · Braille · Reference frame

1 Introduction

Mechanoreceptors on the skin are heterogeneously distributed, and sampling of neural signal in the brain may differ depending on the body part. Given that fact, it may be challenging for the brain to robustly represent stimuli presented to different body sites. In this study, perception of orientation and direction of stimuli presented on the hand and on the arm was investigated. An example of perceptual distortion of tactile space at the peripheral, wherein orientation (trajectory) of the stimuli on the forearm appears to fool the responses of the receptors, is introduced.

Interesting discrepancies between a perceived spatial representation and a physical space have been reported. Perceived space of the stimuli (i.e., the distance between two points of contact) shrinks along the proximodistal axis [1–3] and perceived location (i.e., the exact location of contact) shifts toward anchor points such as the wrist and elbow [2, 4]. Still, the general understanding of haptic-space representation and how it changes across the body remains poorly understood. In particular, whether it can be ascribed to the somatotopic mapping (i.e., receptor distribution and receptive field size), remains under discussion.

Variation of the distribution density of receptors is not a problem that occurs only in regard to the skin. Another 2D sensor array, the retina, has a heterogeneous sensor distribution, and computation of most of the basic visual features differ between central and peripheral vision. For example, the signal detection threshold degrades from central vision towards peripheral vision; however, this degradation is relatively weak in detecting flickering and moving signal [5]. On the other hand, in the case of touch,

I. Nisky et al. (Eds.): EuroHaptics 2020, LNCS 12272, pp. 167–175, 2020.
https://doi.org/10.1007/978-3-030-58147-3_19

differences in perception of motion and orientation due to different stimulation sites have been sparingly studied [6].

In this study, the following two questions are addressed: (i) whether the direction of a simple moving stimulus that can be easily captured and tracked by eye (and presumably by hand) can be discriminated by the arm and, if not, (ii) how does the arm differ from the hand; that is, whether the difference can be ascribed to a difference in receptor distributions and in which reference frame the difference occurs.

2 Method

As shown in Fig. 1A, tactile stimuli were presented to subjects by a piezoelectric braille display (stimulator, hereafter) (Dot-view2, KGS, Japan) with an array of pins with diameter of 1.3 mm and inter-pin distance of 2.4 mm. Each pin can be switched independently to either the "on" position (maximum 0.7-mm normal displacement or less when damped by the contacting hand) or the "off" position (no displacement), and the status of the pins ("on" or "off") was updated every 100 ms. They touched the stimuli with the volar surface of their left hand or forearm (Fig. 1B). Their view of the display was occluded by a black cardboard plate, and the subjects wore earplugs to mask noise made by the stimulator.

In the experiment on direction judgment (Fig. 1C), as the stimulus, one dot was moved in one direction at 50 mm/s for two seconds. The dot moved every 0.1 to 0.2 s, and this variation was unavoidable due to a characteristic of the display. The direction of the dot movement was upward or downward to the right or left (LU, LD, RU, and RD in Fig. 1C). Note that a new starting point of the dot was chosen in every trial, and when the dot reached the edge of the stimulus area, it appeared from the opposite edge. The length of the trajectory was varied across trials, but in all trials, the dot was moved on the same trajectory for more than 5 cm (i.e., half the diagonal of the stimulus area). The stimuli were presented within 32 × 32 pins for the "hand" condition and "arm" condition, while they were presented within 8 × 8 pins for the "s_hand" condition and 45 × 32 pins for the "l_arm" condition. The dot stimuli moved in parallel to the diagonal of these stimulation areas. The subjects were asked to answer two two-alternative forced choices (2-AFCs): whether the stimulus moved upward or downward (Q1) and leftward or rightward (Q2). After each response, a feedback signal was sent to the subjects by beep sound. Ten subjects participated.

In the experiment on orientation description, the stimuli were 12 aligned dots, presented within a circular area of 32 pins in diameter (74.4 mm), and their orientation was one of eight possibilities (Fig. 1D). The dots were presented sequentially in one direction ["move + condition"], in the opposite direction ["move−"], in random order ["shuffle"], or presented all at once ["static"] within 2 s. Partially spatially overlapping with adjacent dots, each dot consisted of four to six "on" pins and appeared for 0.1 to 0.2 s. These variations were unavoidable due to a characteristic of the display. The subjects were asked to report the perceived orientation of the stimuli with respect to the stimulator surface by pressing two keys according to the response mapping (Fig. 1D) presented on a screen in front of them on the keyboard (e.g., the subject pressed '4' and 'R' when they perceived vertical orientation). Since whether the line stimuli were perceived symmetrically remained obscure, both edges of perceived shape (rather than one orientation) were recorded. The subjects did know that the presented stimuli passed the centre of the stimulus area, but they did not know that the stimuli were straight lines. No feedback signal was provided. Three different posture conditions were tested for different groups of ten subjects: subject's hands/arms oriented straight ahead ("normal"), rotated outward ("divergent"), or rotated inward ("convergent"), as shown in Fig. 4.

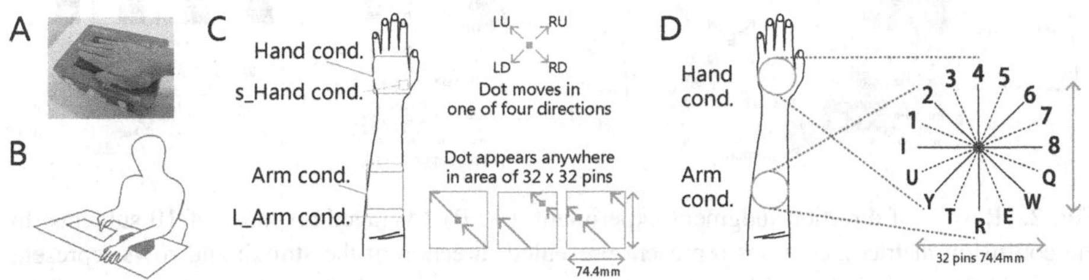

Fig. 1. (A) Braille-type stimulator. (B) View of the setup used in the experiments. The blue box represents the braille stimulator. (C) Subjects reported perceived direction of the stimulus trajectory in the direction-judgment experiment by pressing two keys (e.g., 'up' and 'left' for the illustrated stimuli at the bottom of the diagram). (D) A response map for the orientation-description experiment was presented on a screen in front of the subjects. The subjects reported perceived orientation (both edges) of the stimulus trajectory with respect to the stimulator surface by pressing two keys.

3 Results

3.1 Direction-Judgment Experiment

To investigate perceived direction through the skin, the moving-dot stimuli were presented obliquely on the volar surface of the hand (palm) and on the arm (forearm) of each subject by braille. When a dot reached the outer boundary of the presentation area, it reappeared at the opposite end of its line of motion and started to move in the same direction along that line (Fig. 1C). This repetitive motion could induce ambiguity of the direction of motion, although the direction could be easily reported visually (preliminary reports). Note that a dot moved on the same trajectory for a longer distance than the two-point discrimination threshold for the forearm [7].

As shown in Fig. 2A, averaged performances of ten subjects were above chance level under the "hand" condition, while they were slightly lower when the dot was moving to the upper right (RU). Meanwhile, under the "arm" condition, the observed pattern of responses differed dramatically from that under the "hand" condition. Contrary to intuition, the performance not only dropped but was biased in a particular direction. The subjects tended to report the direction from upper left to lower right instead of that from upper right to lower left, regardless of the physically presented stimuli (see schematic illustrations in Fig. 2A). Note that the observed patterns do not simply reflect response “key-pressing” bias, since the subjects pressed “L” or “R” and “U” or “D” at roughly equal probability.

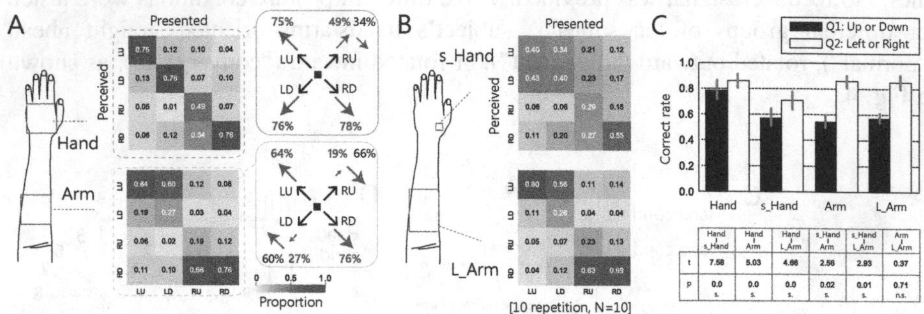

Fig. 2. Results of direction-judgment experiment. (A, B) Averaged response of 10 subjects. In the confusion matrices, columns represent presented direction of the stimuli and rows represent perceived direction. Diagonal lines represent correct responses. In the schematic pattern of the observed trend, where blue arrows represent correct responses and red ones represent incorrect responses. (C) Averaged correct rates of 10 subjects for each 2-AFC. Error bars represent 95% confidence intervals. The table lists the main effect group comparison (Ryan’s method, $\alpha = .05$) of ANOVA. (Color figure online)

Two remaining conditions were designed to test whether mechanoreceptor distribution or receptive-field size could explain this apparently odd anisotropic representation of perceived orientation on the arm. In the “s_hand” condition, the stimulus area was reduced one-quarter of that under the “arm” condition to compensate for the difference in two-point discrimination thresholds for the palm and the forearm [7]. Although the performance dropped, it did not show similar directional bias as that under the “arm” condition (e.g., the RU stimuli were perceived roughly equally as applied in all directions). Since previous literature reported that distances feel longer along the mediolateral axis than along the proximodistal axis of the arm [1–4], the same experiment was conducted under the “l_arm” condition, under which the stimuli area was elongated 1.4 times in the vertical direction only. Although the direction discrimination performance slightly improved, it showed a similar directional bias to that under the “arm” condition. Subjects’ reports (Fig. 2C) were entered into a two-way repeated ANOVA with four stimulus location and two 2-AFCs. The ANOVA results indicated that all main effects and interactions were significant ($F(3, 27) = 19.9$, $p < 0.0001$ for stimulus location; $F(1, 9) = 67.8$, $p < 0.0001$ for 2-AFC; $F(3, 27) = 32.2$, $p < 0.0001$ for interaction). Main-effect group comparison (Ryan’s method, $\alpha = .05$) revealed significant differences between all stimulus location pairs except the pair of the “arm” and “l_arm” conditions (Fig. 2C). These statistical tests suggest that neither the difference in two-point discrimination thresholds nor the difference in the tactile space can explain the difference between perceived orientation on the hand and that on the arm.

3.2 Orientation-Description Experiment

Since anisotropic distortion of perceived orientation on the arm was unexpectedly observed, the experimental task was changed, and this phenomenon was investigated in more detail. Subjects were asked directly to indicate the orientation of stimuli with respect to the stimulator's surface. The stimuli were dots aligned in one of eight possible orientations (Fig. 1D) with four different dot sequences presented: dots appear one by one in one direction ("move+" condition), in the other direction ("move−"), in random order ("shuffle"), and appear at once ("static").

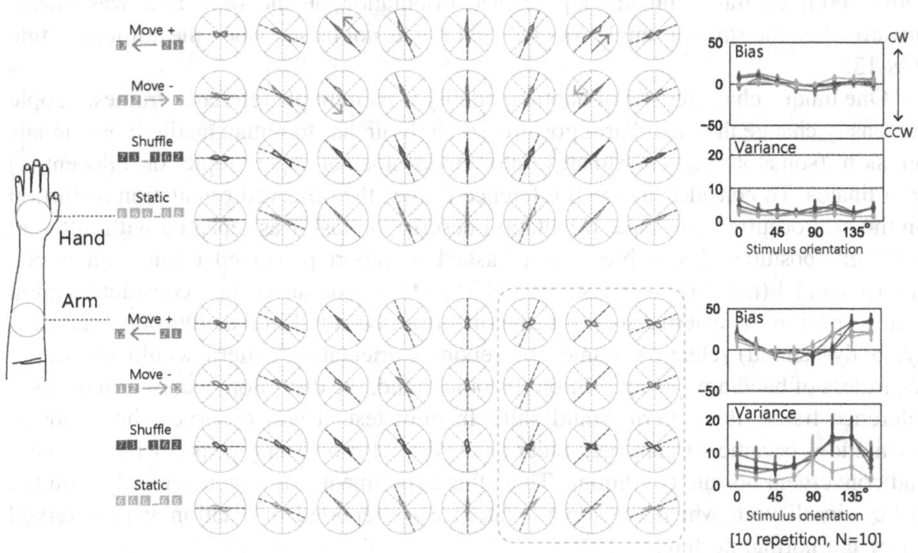

Fig. 3. Results of orientation-description experiment. Each column represents the results obtained with a stimulus presented at 0 to 157.5°. A stimulus at 0° was defined with respect to the front edge of the braille display placed parallel to the table. Black dashed lines represent orientation of the presented stimuli, and each coloured line represents the proportion of reported orientation under each condition. Bias represents the mean discrepancy between reported and veridical orientations of the stimuli, while variance represents variability for each stimulus. Error bars represent 95% CI. (Color figure online)

As shown in Fig. 3, ten subjects achieved good orientation-description performance when the aligned dot stimuli were presented on the hand: the coloured radar charts have well-defined peaks along the black dashed line. Calculated bias and variance are small regardless of stimulus orientation or dot sequence ("move ±," "shuffle," or "static"). Meanwhile, their performance degraded when the stimuli were presented on the arm: both bias and variance became large with stimuli presented around 135° (enclosed by the yellow dotted line in the figure). The intensity of bias was similar for different dot sequences, but variance was smaller under the static condition (purple line in the graph). Anisotropic distortion of perceived orientation on the arm was also observed in this experiment. Perceived orientation on the arm was biased inward (i.e., clockwise for left hand/arm) when the stimulus angle was 135°; however, it was not biased outward nor inward when the stimulus angle was 45°. Note that the 135° stimuli in this experiment and the RU and LD stimuli in the direction-judgment experiment were not identical, but they had the same orientation, and in both cases, the subjects could not properly report the orientation of the stimuli. In addition, particular distortion patterns of each stimulus depending on the area of stimulus presentation (e.g., the stimuli on the lower half of the forearm are more biased compared to those on the upper half) were not observed. Rather, the response patterns roughly kept a linear symmetric shape. According to our pilot test with a smaller number of subjects, this phenomenon may show "body-central symmetry": perceived orientation on the right arm was biased inward when the stimulus angle was 45°, but it was not biased when the stimulus angle was 135°.

One unique characteristic of haptic modality is its multiple reference frames. People can easily change their hand/arm posture, so the brain has to remap tactile input signals on skin (somatotopic) coordinates into environmental (spatiotopic or allocentric) coordinates. To consider in which reference frames the observed orientation distortion on the arm occurred, the same orientation-description task was repeated with different hand/arm postures. The subjects were asked to report perceived orientation in the environmental (not skin) reference frame. The stimulator stayed in a constant location with respect to the external world. If the distortion occurred in the environmental (i.e., eye/body-centred) reference frame, the reported orientation pattern would be similar regardless of hand/arm posture. If, on the other hand, the distortion occurred in the skin reference frame, the pattern would shift. In pilot test, it was observed that reported orientations became obscure (i.e., radar charts do not show sharp peaks) with divergent and convergent posture conditions. Thus, this experiment was conducted only with the static stimuli with which the lowest variance of reported orientation was observed under the normal posture.

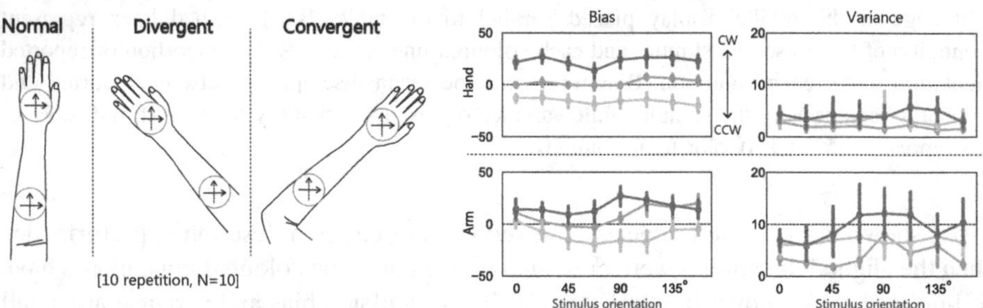

Fig. 4. Results of orientation-description experiment with varied posture. Hands/arms of subjects were oriented straight ahead ("normal," represented in red), rotated outward ("divergent," blue), or rotated inward ("convergent," green). The subjects were asked to report perceived orientation of the static stimuli with respect to the stimulator surface (environmental coordinate, represented as arrows in the figure). Note that the result obtained under the normal condition is a re-posting of the result presented in Fig. 3. (Color figure online)

Reported orientation was distorted in the "divergent" and "convergent" conditions even when the stimuli was presented on the hand (Fig. 4), and this finding is in line with that of the previous study that conducted an orientation-matching task with aluminium bars [8]. Observed variances in these conditions were higher than that in the "normal" condition, suggesting a higher level of task difficulty under the former conditions. According to the bias, the influence of the skin reference frame on perceived orientation was observed. In the divergent condition, the orientation in the environmental reference frame is shifted clockwise compared to that in the skin reference frame. Indeed, the reported orientation shifted clockwise. In contrast, the reported orientation shifted counter clockwise in the convergent condition. Bias for each posture condition was almost constant regardless of the stimulus orientation. On the other hand, when the stimuli were presented on the arm, bias of reported orientations varied according to the stimulus orientation in all posture conditions. The baselines (i.e., averaged performances across all orientations) differed according to posture condition, and this difference seems consistent with observed biases under the hand condition. Observed patterns of biases seem to be roughly consistent with the hypothesis that the distortion on the arm occurred in the skin coordinate, since it peaked around 90° under the divergent condition and around 0° under the convergent condition. Note that it remains unclear at this moment that whether the observed discrepancy with varied posture reflects the difference in discrepancy between the skin and environmental reference frames or reflects the difference induced by remapping difficulty and/or tightness of posture.

4 Discussion

Direction-discrimination performance and orientation-description performance at different body sites were measured. Reported orientations when the stimuli were presented on the arm were distorted in relation to those observed when the stimuli were presented on the hand (palm). In particular, inwardly inclined trajectory/shape is perceived more inwardly inclined. This distortion cannot be simply explained by the difference in receptor distribution, and shifted according to the skin reference frame. This study showed a clear example that the representation of simple stimuli is distinctly different when the stimuli are presented on different body sites. There might be a difference between central touch and peripheral touch in terms of computational processing.

The perceptual asymmetry of the mediolateral axis and proximodistal axis lines/motions on the arm has been reported. Jones et al. [4] presented moving stimuli by a three-by-three array on the volar surface of the forearm, and they reported that across-arm movement appears to be more easily recognized than along-arm movement. That result suggests that the edges of the arm may serve as landmarks for localizing cue. Closer-to-reality stimuli (in terms of resolution) were used in this study, and a similar trend was observed: subjects made more mistakes when presented with two alternative forced choices of direction between proximal or distal on the arm rather than those between medial or lateral (Fig. 2C). Note that the reported proximal-distal direction was even "flipped" in the present direction-discrimination experiment, and performance varied both under the movement and the static conditions in the present orientation-description experiment. These results seem difficult to fully understand in the context of previously introduced hypotheses on, for example, gravitation of anchor points, stretching of tactile space, receptive field shape, and the pixel model [1–4].

The perceptual asymmetries of inwardly and outwardly inclined lines/motions on the arm, on the other hand, have never been reported. Studies about distortion in haptic perception of parallelity on the hand might be relevant to the current findings, though the stimuli and task were not identical. Kappers, et al. [8] repeatedly and systematically investigated the distortion of perceived orientation by using a matching task of two oriented bars (a stable one for reference and a rotatable one for orientation matching). Their findings are consistent with the results presented here in the sense that the perceived orientation is not represented isotopically in all orientations and the distortion pattern changes according to hand posture. It may also be worthwhile considering the influence of dermatome difference. Mechanoreceptors on the hand/arms are distributed across multiple dermatomes, and the responses of each dermatome are projected to the brain through individual spinal segment. Though a previous study reported a minor effect of dermatomes during intensity discrimination task on the arm [9], the orientation (spatial relationship) might be calculated differently depending on whether the stimuli are presented within or across dermatomes. It seems that our inwardly and outwardly inclined stimuli were presented to the same degree over two dermatomes (C6 and T1) [10]; however, roll rotation of the arm actually non-uniformly stretches and rotates skin, and this uniformity remains unclear. In addition, the differences in cortical representation of each skin area (i.e., cortical magnification) may be related to the observed distortion on the arm, since the correlation with acuity of shape perception on the finger has been reported [11]. These are issues awaiting further investigation. It would be useful if the simple tasks used in this study could work as a probe to reveal the underlying remapping process of spatiotemporal perception by touch.

References

1. Miller, L.E., Longo, M.R., Saygin, A.P.: Mental body representations retain homuncular shape distortions: evidence from Weber's illusion. Conscious. Cogn. **40**, 17–25 (2016)
2. Cholewiak, R.W., Collins, A.A.: Vibrotactile localization on the arm: effects of place, space, and age. Percept. Psychophys. **65**(7), 1058–1077 (2003)
3. Fiori, F., Longo, M.R.: Tactile distance illusions reflect a coherent stretch of tactile space. Proc. Natl. Acad. Sci. **115**(6), 1238–1243 (2018)
4. Jones, L.A., Kunkel, J., Piateski, E.: Vibrotactile pattern recognition on the arm and back. Perception **38**(1), 52–68 (2009)
5. Nishida, S.: Advancement of motion psychophysics: review 2001–2010. J. Vis. **11**(5), 11 (2011)
6. Kuroki, S., Nishida, S.: Human tactile detection of within- and inter-finger spatiotemporal phase shifts of low-frequency vibrations. Sci. Rep. **8**(1), 1–10 (2018)
7. Weinstein, S.: Intensive and extensive aspects of tactile sensitivity as a function of body part, sex, and laterarity. In: Kenshalo, D.R. (ed.) The Skin Senses (1968)
8. Kappers, A.M.L., Viergever, R.F.: Hand orientation is insufficiently compensated for in haptic spatial perception. Exp. Brain Res. **173**(3), 407–414 (2006)

9. Shah, V.A., Casadio, M., Scheidt, R.A., Mrotek, L.A.: Spatial and temporal influences on discrimination of vibrotactile stimuli on the arm. Exp. Brain Res. **237**(8), 2075–2086 (2019). https://doi.org/10.1007/s00221-019-05564-5
10. Fardo, F., Finnerup, N.B., Haggard, P.: Organization of the thermal grill illusion by spinal segments. Ann. Neurol. **84**(3), 463–472 (2018)
11. Duncan, R.O., Boynton, G.M.: Tactile hyperacuity thresholds correlate with finger maps in primary somatosensory cortex (S1). Cereb. Cortex **17**(12), 2878–2891 (2007)

Evaluation of Changes in Perceived Intensity and Threshold of Moisture Sensation of Clothes Associated with Skin Moisture

Shanyi You(✉), Mai Shibahara, and Katsunari Sato

Nara Women's University, Kitauoyahigashicho, Nara, Japan
taz_yuu@cc.nara-wu.ac.jp

Abstract. Moisture sensation is an important determinant of clothing comfort. Conventional studies have attempted to elucidate the mechanism behind moisture sensation by using wet sample cloths that vary by water content. However, these studies did not consider the impact of the moisture levels of the skin that makes contact with the samples. In this study, we investigated changes in skin moisture sensation in terms of perceived strength and detection thresholds, based on contact with sample cloths, given various skin moisture conditions. In the first experiment, participants reported their perceived moisture levels for sample cloths that varied in water content and temperature, after making forearm contact with each sample cloth. The result showed that participants felt small amounts of moisture when skin moisture was increased. In the second experiment, participants' detection thresholds were evaluated using the staircase method, based on forearm-sample contact. The results showed that skin moisture did not affect the threshold of moisture sensation.

Keywords: Moisture sensation · Skin moisture · Perceived intensity · Threshold

1 Introduction

Clothing comfort is a complex issue. It can be influenced by several factors, including human physiology, human psychology, fabric characteristics, and climate [1]. For clothing that makes skin contact, moisture sensation is also an important factor in clothing comfort. However, the mechanism behind the perception of moisture sensation is not fully understood.

Conventional studies of moisture sensation have shown that the major factor affecting moisture sensation is changes in temperature or heat flow [2]. It has previously been found that moisture sensation varied for different parts of the skin [3]. Perceived humidity has been found to vary according to static and dynamic touch but to be independent of changes in pressure. Specifically, in the case of dynamic touch, participants rated stimuli that were dry and cold as being drier than similar stimuli with relatively higher temperatures [4]. Conversely, in the case of static touch, contact with a dry, cold cloth has been shown to lead to temperature changes in the skin that mirror

I. Nisky et al. (Eds.): EuroHaptics 2020, LNCS 12272, pp. 176–184, 2020.
https://doi.org/10.1007/978-3-030-58147-3_20

those that result from contact with a damp cloth. This represents a case of the perception of illusory moisture [5].

These and other conventional studies focused on the moisture level of the object in question (e.g., fabric) but did not consider effects attributable to the moisture level of the skin. The sensation of moisture in clothing is caused by the contact between the fabric and the skin. If the skin is wet to any degree, heat transfer between the fabric and the skin changes accordingly. Since heat transfer is an important factor in moisture sensation, the moisture level of the skin could also affect sensation.

In this study, we focused on the moisture level of the skin and evaluated changes in the moisture sensation by controlling the water content and temperature of the wet cloth. The change in the threshold of moisture sensation was measured as well. We expect that this study will contribute to the understanding of moisture perception and to improving the comfort of clothing that comes into direct contact with the skin.

2 Experiment 1: Perceived Intensity of Moisture Sensation

Perceived intensity of moisture sensation is the degree to which a person can feel moisture when the skin is in contact with a physical entity (e.g., a fabric). In order to confirm changes in the perceived intensity of moisture sensation, based on different skin moisture conditions, we had participants evaluate the strength of moisture sensation when their forearms touched sample cloths with varying temperatures and moisture levels.

2.1 Experimental Materials

The size of the sample cloths (cotton broad) was 6.5 cm^2. Each cloth was affixed to a 4 cm^2 Peltier element, for the purpose of temperature adjustment. A hotplate (Nissin Kogyo Co., Ltd., NHP-M30N) set at 33 °C was prepared for adjusting the skin temperature of the arm before it came into contact with the sample cloth. Skin moisture levels were measured by using a triplesense (MORITEX Co., Ltd., TR-3).

2.2 Experimental Conditions

The participants were 20 female university students. This experiment was approved by the Ethics Review Committee of Nara Women's University. The participants were informed of the relevant experimental procedure, and the experiment was conducted with each participant's consent. The test area in the experiment was the forearm segment of the dominant hand, 5 cm proximal to the wrist joint. The room temperature and environmental humidity were kept at 23 °C ± 0.5 °C and 50 ± 2%RH, respectively, throughout the experiment.

For each trial, the skin temperature was adjusted to one of two levels: 27 °C (below normal skin temperature) and 39 °C (above normal skin temperature). The moisture content of the sample cloth was adjusted to one of three levels: dry, low moisture, and high moisture. Using a dropper, the wet sample clothes were prepared with 100 µl and 300 µl of water for the low and high moisture conditions, respectively. During the experiment, skin moisture was adjusted using a wet towel.

Moisture sensation was evaluated using a four-point Likert scale, ranging from 0–3, as shown in Fig. 1. We selected the four-point scale, as previous experiments [5] have shown that participants can easily and precisely express their sensations of moisture, using this a scale of this length. We analyzed these values on an interval scale [6].

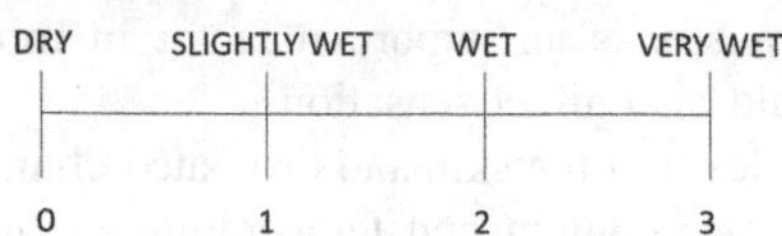

Fig. 1. Moisture sensation rating scale.

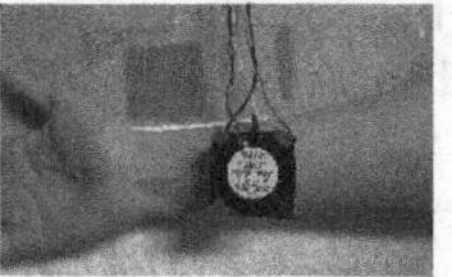

Fig. 2. Forearm experiment.

2.3 Experimental Procedure

Prior to the experiment, participants were asked to touch the dry and high-moisture sample cloths in order to ensure they had knowledge of the two ends (0 and 3) of the scale. Then, the initial skin moisture of participants' forearm was measured by the triplesense.

The participants closed their eyes during the experiment, so as not to see the sample cloth. First, the skin temperature was adjusted with the hot plate for 1 min. Next, the experimenter guided each participant's dominant hand to their side and brought the sample cloth into contact with the forearm for 4 s (Fig. 2). The forearm was then removed from the sample cloth and moisture sensation was evaluated. Afterwards, the arm was returned to the hot plate to restore each participant's skin temperature. In the above procedure, six conditions with different levels of water content and different temperatures were evaluated in random order for each participant. Each condition was repeated once.

Thereafter, a wet towel was placed on the arms of each participant for 5 min to adjust skin moisture. After that, the Triplesense was used to confirm the projected increase in skin moisture. This procedure was repeated six times (varied according to condition) to evaluate outcomes in each of the six conditions.

2.4 Results

Skin moisture levels before and after adjustment are shown in Fig. 3. We confirmed that skin moisture content was successfully increased for each participant, following adjustment.

Averages of the results of the four-step evaluations were calculated, based on moisture sensation ratings given before and after skin moisture adjustment, in each condition. Summaries of moisture sensation ratings are shown in Fig. 4.

It can be seen from Fig. 4 that, when the wet towel was not used (i.e., skin moisture was low), the moisture sensation rating was higher than when the wet towel was used (i.e., skin moisture was high). This indicates that, when the skin moisture was high, the moisture was perceived very little.

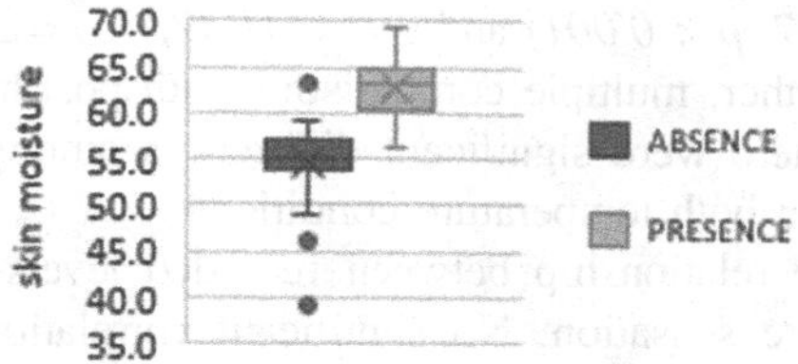

Fig. 3. The skin moisture content before and after the adjustment in Experiment 1.

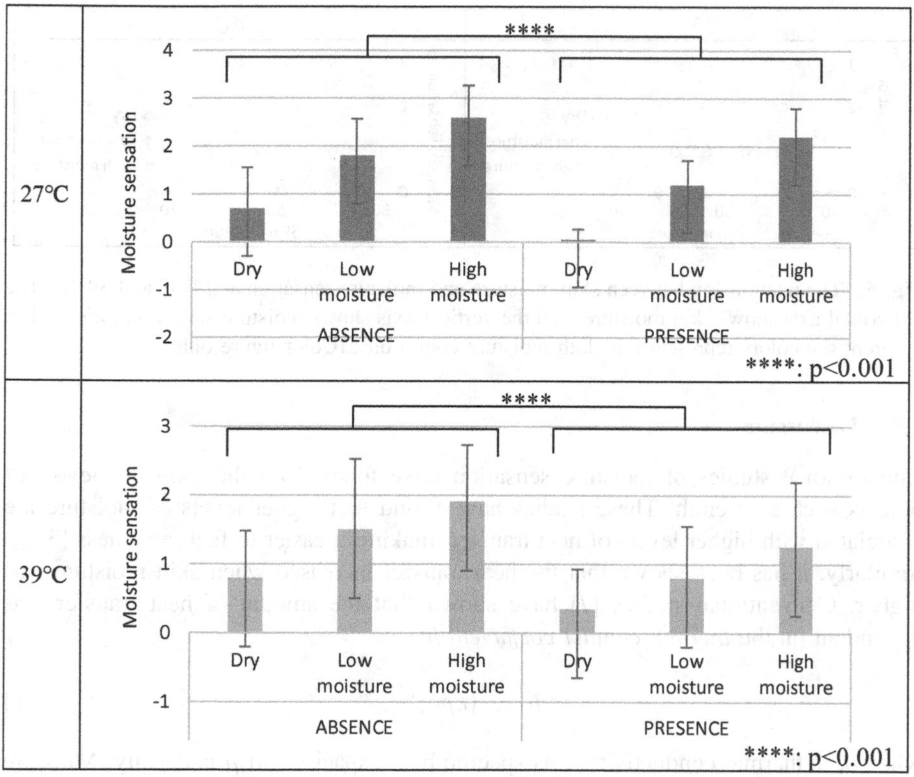

Fig. 4. Moisture sensation at 27 °C and 39 °C. The horizontal axis shows the moisture content of the sample cloth and the skin. The vertical axis shows the moisture sensation ratings. The errors bars represent standard deviations.

A 2 (temperature) × 3 (cloth moisture) × 2 (skin moisture) ANOVA, at $\alpha = 0.05$, revealed a significant difference according to skin moisture ($F\ (1, 19) = 26.11$, $p < 0.001$). In addition, the temperature × cloth moisture interaction was significant. Multiple comparisons showed no significant simple main effects in the dry and low-moisture conditions, but showed a significant effect in the high-moisture condition ($F\ (1, 57) = 17.17$, $p < 0.001$). Furthermore, there were significant differences within the

27 °C (F $(2, 76) = 84.77$, $p < 0.001$) and 39 °C (F $(2, 76) = 20.66$, $p < 0.001$) temperature conditions. Further, multiple comparisons with post-hoc tests, using Ryan's method, showed that there were significant differences among three cloth moisture conditions ($p < 0.001$) in both temperature conditions.

Figure 5 depicts the relationship between recorded levels of skin moisture and participant-rated moisture sensation. No significant correlation was found between these measures (27 °C: R(Dry) = −0.60, R(Low) = 0.11, R(High) = −0.33, 39 °C: R (Dry) = −0.28, R(Low) = −0.08, R(High) = 0.09).

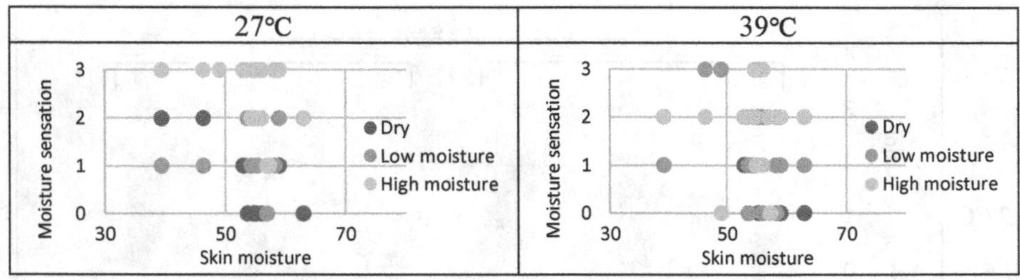

Fig. 5. The relationship between skin moisture and moisture sensation at 27 °C and 39 °C. The horizontal axis shows skin moisture, and the vertical axis shows moisture sensation ratings. The different dot colors represent the cloth moisture conditions. (Color figure online)

2.5 Discussion

Conventional studies of moisture sensation have focused on the moisture levels of objects such as a cloth. These studies have found that higher levels of moisture are associated with higher levels of heat transfer, making it easier to feel moistness [3, 5]. Similarly, it has been shown that the heat transfer increased when skin moisture was higher. Conventional studies [7] have shown that the amount of heat transfer was dependent on the *thermal contact coefficient* h:

$$h = (k\rho c)^{1/2} \tag{1}$$

where k is thermal conductivity, c is specific heat capacity and ρ is density. Moisture had a higher thermal contact coefficient than skin, when the movement of heat was larger. However, the current study (Fig. 4) showed that moisture sensation ratings decreased when skin moisture increased. This trend depended not on absolute skin moisture content but on relative content within individuals (Fig. 5).

Comprehensive judgments of moisture levels have been found to vary, based on friction and softness, in addition to temperature. Conventional studies have shown that contact with a cold object, with a high surface friction level, made people perceive the object as soft and moist [5, 8, 9]. In this experiment, skin moisture increased after the wet towel was applied. It is possible that, as the skin became softer, the sample cloth increased in perceived, relative hardness. It is presumed that this would have impaired individuals' ability to feel moistness. Additionally, when skin moisture levels increased, the number of active mechanoreceptors might have changed, correspondingly. Further research is required to test this assumption. In addition, skin sensations before and after contact with the sample cloth could be affected. It is possible that the amount of water on the skin's surface changed slightly when the moist stimulus touched the skin. If judgments of moisture levels are based on the difference in sensation between pre- and post-contact, it should be easier to perceive the change when the skin is dry, initially.

Changes in clothes' perceived moisture levels depended on the skin's moisture content. This is an important phenomenon not only for the design and development of comfortable clothing but also for psychophysical research. For example, since the skin temperature affects thermal and vibro-tactile stimulation, we adjusted the skin temperature before the experiment [10, 11]. Our results indicate that, during research measuring perceived moisture levels, it should be confirmed that the moisture content of the skin does not change during the experiment.

3 Experiment 2: The Threshold of the Moisture Sensation

The threshold of moisture sensation is the minimum water content required to illicit the feeling of moisture on the skin. We sought to confirm changes in the threshold of moisture sensation on the skin, based on different skin moisture conditions. As such we measured changes in sensation, based on cloths of varying moisture content coming into contact with skin with varying surface moisture levels.

3.1 Experimental Materials

The experiment materials were the same as those used in the previous experiment (Sect. 2.1).

3.2 Experiment Conditions

The participants were 20 female university students. This experiment was approved by the Ethics Review Committee of Nara Women's University. The participants were informed of the relevant experimental procedure, and the experiment was conducted with each participant's consent. The test area in the experiment was the forearm segment of the dominant hand, 5 cm proximal to the wrist joint. The room temperature and environmental humidity were kept at 23 °C ± 0.5 °C and 50 ± 2%RH, respectively, throughout the experiment.

As with the room temperature, the sample cloth was kept at 23 °C. The sample cloths, each differing in water content by 5 μl, were prepared in advance and stored in a container with a lid. During the experiment, skin moisture was adjusted by using a wet towel.

3.3 Experiment Procedure

The threshold was evaluated using the staircase method of psychophysical experiments. The participants were asked to close their eyes until the end of the session.

First, the sample cloth with a water content of 0 μl was applied. The forearm of the participant was placed in contact with a new sample cloth of 20 μl higher water content than once before. Then, the participant stated whether they felt moisture or not. Once moisture was perceived, a cloth 10 μl lower in water content was attached to the forearm. This was repeated until moisture was no longer perceived. Once a participant stopped perceiving moisture, a sample cloth 5 μl higher in water content was applied to the forearm repeatedly, until moisture was perceived. The level of moisture, associated with moisture perception, was then recorded. The duration of contact was 3 s. After the sample cloth was removed from the forearm, the participants reported again on their perception of any moisture. Then, the forearm was placed back on the hot plate to recover skin temperature.

The whole process was repeated four times, and the average value of the four repetitions was taken as the threshold. In order to prevent variation in the water content of the sample cloth, a set of the sample cloths was prepared again, after the evaluation was completed twice.

All of the participants' skin moisture levels were adjusted with a wet towel that stayed on the arm for 5 min. Thereafter, it was confirmed that the skin moisture increased. The same evaluation was performed four times. To maintain skin moisture levels, skin moisture was adjusted for 5 min again, after two evaluations.

3.4 Results

Skin moisture content before and after adjustment is shown in Fig. 6. We confirmed that skin moisture content was successfully increased for each participant, following adjustment.

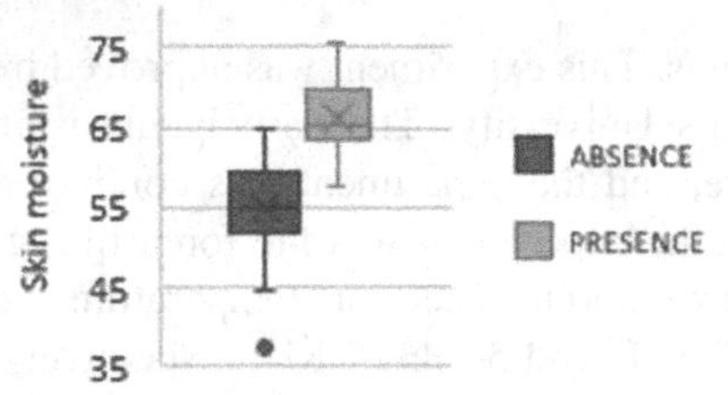

Fig. 6. The skin moisture content levels and after adjustment, in Experiment 2.

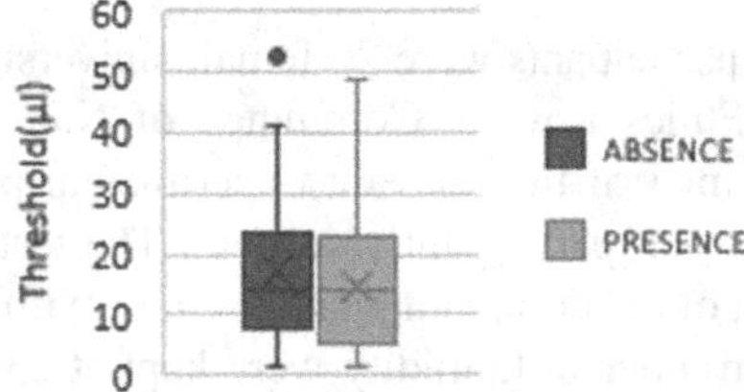

Fig. 7. The threshold of moisture sensation before and after the adjustment of skin moisture levels. The vertical axis represents the threshold value for moisture sensation.

Figure 7 is a graph summarizing the threshold values before and after the adjustment of skin moisture levels. As shown by the boxplot, the thresholds are almost the same regardless of the presence or absence of a wet towel (before and after the change in the skin moisture). The Wilcoxon signed-rank test revealed no significant difference in skin moisture ($p = 0.34$). It was found that skin moisture did not affect the threshold of the moisture sensation.

3.5 Discussion

Experiment 1, in Chapter 2, showed that the participants felt little moisture, when skin moisture was increased; when the level of the skin moisture was high, the threshold of the moisture sensation was presumed to be larger. However, the observed experimental result did not align with this hypothesis. It was found that skin moisture did not affect the threshold of moisture sensation.

We propose that the water remaining on the skin surface, after the removal of the wet towel, affected the observed threshold. When the skin moisture was increased by the wet towel, a small amount of water was left on the skin's surface. There is a possibility that this water decreased the threshold value of the skin. It is also possible that the moisture threshold (0–30 μl) was too low to be compared with the water content of the sample cloth (0, 100, 300 μl) in Experiment 1. We propose that the low water content of the sample cloth led to the diminished skin moisture effect. When an experiment on perceived intensity is performed with a sample cloth having a low level of water content, such as 30 μl, perceived moisture sensation is expected to be independent of variations in skin moisture. Further research is necessary to characterize the difference in results between the perceived intensity and detection threshold.

4 Conclusion

In this paper, we evaluated changes in perceived intensity and the threshold of the moisture sensation in different skin moisture conditions. The results showed that, when skin moisture was increased, participants felt a lower level of moisture. However, the threshold value of moisture sensation was not affected by skin moisture.

In future, we aim to investigate how the discrimination threshold changes with different temperatures and skin moisture conditions. Additionally, we will study how moisture sensation differs across body parts, such as how thresholds differ between the arm and the back.

References

1. Slater, K.: Human Comfort. Springfield, Ill. C.C. Thomas. USA (1985)
2. Koshiba, T., Tamura, T.: Factors governing the wet sensation of human skin. Jpn. Res. Assoc. Text. End-Uses **36**(1), 19–124 (1995)
3. Tamura, T.: A review of studies on regional differences of thermal and humidity sensitivity on human skin surface. Jpn. Soc. Sensory Eval. **11**(2), 81–88 (2007)

4. Shibahara, M., Sato, K.: Illusion of wetness by dynamic touch. IEEE Trans. Haptics **12**(4), 533–541 (2019)
5. Shibahara, M., Sato, K.: Illusion of moisture sensation of cloth by thermal control. Jpn. Res. Assoc. Text. End-Uses **56**(12), 951–958 (2015). (in Japanese)
6. Carifio, J., Perla, R.: Resolving the 50-year debate around using and misusing likert scales. Med. Educ. **42**, 1150–1152 (2008)
7. Ho, H., Jones, L.: Modeling the thermal responses of the skin surface during hand-object interactions. J. Biomech. Eng. **130**, 021005 (2008)
8. Tanaka, Y., Sukigara, S.: Evaluation of "shittori" characteristic for fabrics. J. Text. Eng. **54** (3), 75–81 (2008)
9. Okajima, T., Takeda, Y.: Tactile dryness of building materials. Trans. Architect. Inst. Jpn. **327**, 12–19 (1983)
10. Jones, L.A., Ho, H.-N.: Warm or cool, large or small? The challenge of thermal displays. Trans. Haptics **1**(1), 53–70 (2008)
11. Green, B.G.: The effect of skin temperature on vibrotactile sensitivity. Percept. Psychophys. **21**, 243–248 (1977)

The Effects of Simultaneous Multi-point Vibratory Stimulation on Kinesthetic Illusion

Keigo Ushiyama(✉), Satoshi Tanaka, Akifumi Takahashi, and Hiroyuki Kajimoto

The University of Electro-Communications, Chofu, Tokyo, Japan
{ushiyama,tanaka,a.takahashi,kajimoto}@kaji-lab.jp

Abstract. Kinesthetic sensation is important for improving presence and immersion in VR environments. However, presenting kinesthetic sensation usually requires a large space so as to avoid users colliding with objects or other users. One of the ways to tackle this issue is to use kinesthetic illusion, which is a way of presenting kinesthetic sensation without physical motion. However, realizing dynamic motion and fast movement remains difficult. Considering that multiple synergist muscles are usually involved in a movement such as walking or even simple arm movement, stimulating multiple synergist muscles might enhance the illusion. Thus, we investigated whether multi-point vibratory stimulation to multiple synergist muscles enhances induced kinesthetic illusions. We found that stimulating multiple synergist muscles created more vivid illusions. Additionally, we found that our method was effective for inducing steady illusions. We also calculated the contribution of each proposed stimulation point to the illusion.

Keywords: Kinesthetic illusion · Proprioception · Tendon vibration

1 Introduction

Presenting proprioception and kinesthetic sensation is important for improving presence and immersion in VR environments. Usually, kinesthetic sensation is presented using equipment that incorporates actual user motion such as walking. However, this requires a large space. Otherwise users might collide with objects and other people.

These issues can be resolved by inducing only kinesthetic sensation without physical motion. Kinesthetic illusions, which are illusions of the position and movement of one's own body and induced by stimulating proprioceptors such as muscle spindles [1], can achieve this goal.

Electronic supplementary material The online version of this chapter (https://doi.org/10.1007/978-3-030-58147-3_21) contains supplementary material, which is available to authorized users.

I. Nisky et al. (Eds.): EuroHaptics 2020, LNCS 12272, pp. 185–193, 2020.
https://doi.org/10.1007/978-3-030-58147-3_21

Kinesthetic illusions are often induced using about ~100-Hz tendon vibrations [4]. Although skin deformation [2] and electrical stimulation of tendons [5] can induce movement illusion, tendon vibration is more effective, considering that muscle spindles contribute a great deal to kinesthetic sensation [7]. The intensity of the illusion depends primarily on the vibration frequency and amplitude [8,10]. The preload force of the vibrator is also known to affect the threshold of vibration amplitude for eliciting the illusion [3].

However, the intensity of the illusion induced by tendon vibration is not enough for realizing dynamic and fast movement. One reason might be that the stimulation point is limited. Yaguchi et al. [14] reported that kinesthetic illusion was enhanced by stimulating two synergist muscles or tendons at both ends. Even though previous studies have examined stimulating one or two synergist muscles, actual movement involves many synergist muscles. Therefore, stimulating multiple synergist muscles can induce the more natural and large kinesthetic illusion.

In our previous report [13], we preliminarily confirmed that multi-point vibratory stimulation induced more steady illusion than the illusion induced by one or two points of stimulation. In this paper, we investigate the effects of multi-point vibration in more detail as well as the contribution of each vibration position to the illusion.

2 Methods

Twelve participants (10 men and 2 women; aged 21 to 25 years old; all right-handed) took part in the experiment. We presented vibration stimulation to seven positions on the left chest, upper arm, and forearm to induce the arm extension illusion, and Fig. 1 (left) shows the vibrator positions (1 to 7) over the same tendons that we tested in our previous report [13]. We hypothesized

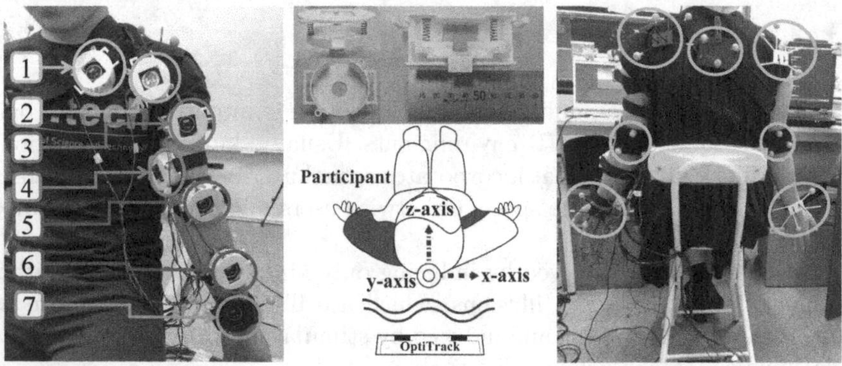

Fig. 1. (Left) Positions of vibrators. (Center) The design of the vibrator case, and the coordinate system (top view) used to record participants' movement. (Right) Positions of markers for measuring movement via optical tracking camera.

that each vibratory stimulation would contribute differently to the illusion and illusions induced by each vibration could be composited linearly as the vector model of the illusion [9,11].

Optical motion capture (OptiTrack V120:Duo) was used to measure participant movement. Figure 1 (right) shows seven positions on the neck, shoulders, elbows, and wrists where we placed retroreflective markers. The marker positions were recorded in a left-handed coordinate system, as shown in Fig. 1 (center).

2.1 Vibratory Stimulation

The vibrator (Acouve Lab VP 210) was hung on three springs (overall stiffness: 1.2 N/mm) in the vibrator case (Fig. 1, center) The bottom of the vibrator case was covered by a sponge to avoid pain, and the case was mounted with rubber bands and a supporter (Fig. 1, left). The preload force of vibrators was adjusted from 1.2 N to 2.4N by observing the displacement of the head contacting the skin. Based on a previous study [8], we set the vibration frequency to 70 Hz. The acceleration amplitude was adjusted to 90 m/s^2 with an accelerometer (Sparkfun LIS331). The input signal to the vibrators was generated with the same system as in our previous report [13].

2.2 Procedure

The experiments were carried out over two days by dividing the trials of presenting vibration in half so as to prevent participants' fatigue.

Participants were told the posture during the experiment and asked to wear the experiment devices. In particular, the vibrator cases were mounted on the target positions identified by touch. Each vibrator's acceleration amplitude and each OptiTrack marker were calibrated. After that, we measured the ability to express movement, and collected data of the vibration-induced illusions.

Measurement of the Ability to Express Movement. Preliminarily, we measured the ability to mirror the movement of the left arm with the right arm in order to screen out participants who cannot accurately evaluate the illusions by this method.

The experimenter moved the participant's left arm sinusoidally around the shoulder in two directions (flexion/extension and adduction/abduction) and participants mirrored the movement with their right arm. Six trials (three in each direction) were carried out randomly. The three trials in each direction included two trials of slow movement (about 3°/s) and one trial of fast movement (about 10°/s). The order of trials was different for each participant. The duration per one trial was 10 s.

Data Collection of Induced Illusions. We applied 127 vibration patterns (2^7-1), which included all combinations of the seven vibrators. Each vibration

Table 1. The evaluation scales of the illusion [6,12].

Questions	(1) minimum and (10) maximum
Vividness	(1) The illusion was not vivid at all (10) Perceived the illusion as if they were actually
Duration	(1) There was no illusion (10) the illusion evoked for stimulation
Magnitude	(1) The arm felt like it did not move very much (10) The arm felt like it moved as much as was possible

pattern was applied one time and the order of patterns differed across participants. Vibration was applied for 5 s with closed eyes. A 5-s interval divided each trial and a 1-min interval separated every 10 trials. Participants were alerted to the timing of the next trial via headphones, which also served to mask sound cues via white noise.

During the vibration, participants were asked to express the perceived illusion by their right arm. After the vibration, participants answered three questions on a scale of 1 to 10 (Table 1), based on previous studies [6,12]

Prior to data collection, participants became accustomed to the measurement procedure through a practice stage in which the five trials were carried out. The order of vibration patterns differed from those in the actual measurement.

2.3 Data Analysis

We calculated the angular velocity by dividing the angle difference between the initial and the end arm position by the vibration duration. The arm angle was calculated using the arm vector from shoulder position to wrist position measured by OptiTrack. The flexion/extension (y-z plain; extension is the positive direction) and adduction/abduction (x-z plain; abduction is the positive direction) directions were used for analysis.

3 Results

3.1 Measurement of the Ability to Express Movement

We calculated the error angle of right arm movement with respect to the left arm movement. In the flexion/extension direction, the average error was $1.92\pm 5.43°$. In the adduction/abduction direction, the average error was $-5.79\pm 3.91°$.

There was no participant who was not able to mirror both arms at all. Thus, we used the data of all participants for analyzing.

3.2 Data Collection of Induced Illusion

The 0.72% data (11 trials/1524 trials) was excluded from data analysis because tracking was lost during vibration. We analyzed the angular velocity of the right arm movement that expressed the illusory movement of the left arm.

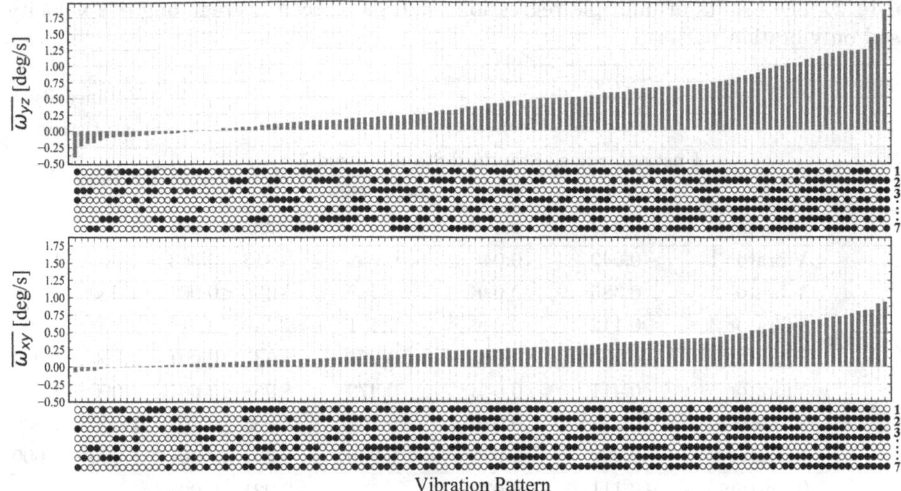

Fig. 2. (Above) Average ω_{yz} for each vibration pattern. (Below) Average ω_{xy} for each vibration pattern. vibration pattern corresponds to vibrators 1 through 7 (from the top).

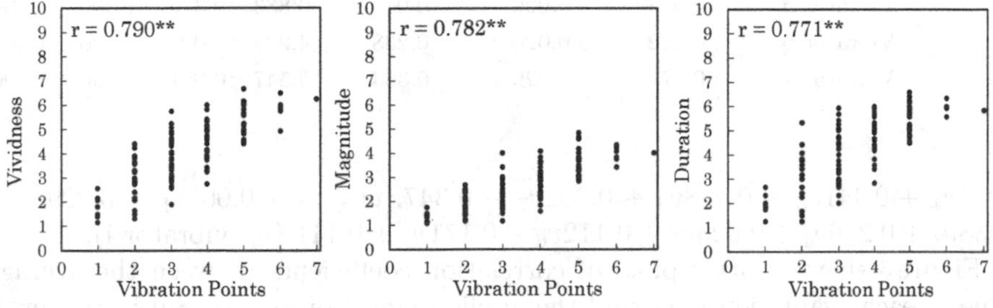

Fig. 3. Correlations (r) between vibration points and each subjective evaluation. **, $p < 0.01$

The angular velocity in the extension/flexion direction is represented by ω_{yz} and the angular velocity in the adduction/abduction direction is represented by ω_{xy}. Figure 2 shows the average angular velocity of each vibration pattern in each direction. Vertical vibration patterns on the horizontal axis indicate which vibrators used (1 to 7 from the top). The open circles indicate a vibrator was not used and closed circles indicate that it was.

A multiple regression analysis of the average angular velocity in each direction based on the vibration pattern (each vibrator was coded as ON = 1, OFF = 0) yielded the coefficients shown in Table 2. The regression equations for each angular velocity were statistically significant (ω_{yz} model: $F_{(7,119)} = 32.942$, $p < 0.001$, the adjusted $R^2 = 0.640$, ω_{xy} model: $F_{(7,119)} = 29.204$, $p < 0.001$, the adjusted $R^2 = 0.610$) and expressed as follows: $\omega_{yz} = -0.042v_1 + 0.385v_2 + 0.112v_3 +$

Table 2. The results of multiple regression analysis of each average angular velocity based on vibration patterns.

		Unstandardized coefficients	Standard error	Standardized coefficients	t	p	Collinearity statistics	
							Tolerance	VIF
ω_{yz} model	(Constant)	−0.347	0.067		−5.20	<0.001		
	Vibrator 1	−0.042	0.046	−0.049	−0.913	0.363	1.000	1.000
	Vibrator 2	0.385	0.046	0.449	8.396	<0.001	1.000	1.000
	Vibrator 3	0.112	0.046	0.131	2.440	0.016	1.000	1.000
	Vibrator 4	0.254	0.046	0.296	5.537	<0.001	1.000	1.000
	Vibrator 5	0.411	0.046	0.479	8.958	<0.001	1.000	1.000
	Vibrator 6	0.138	0.046	0.161	3.003	0.003	1.000	1.000
	Vibrator 7	0.292	0.046	0.340	6.361	<0.001	1.000	1.000
ω_{xy} model	(Constant)	−0.111	0.038		−2.921	0.004		
	Vibrator 1	−0.005	0.026	−0.010	−0.179	0.859	1.000	1.000
	Vibrator 2	0.229	0.026	0.489	8.788	<0.001	1.000	1.000
	Vibrator 3	0.038	0.026	0.080	1.441	0.152	1.000	1.000
	Vibrator 4	0.216	0.026	0.460	8.275	<0.001	1.000	1.000
	Vibrator 5	0.023	0.026	0.049	0.883	0.379	1.000	1.000
	Vibrator 6	0.112	0.026	0.238	4.288	<0.001	1.000	1.000
	Vibrator 7	0.171	0.026	0.364	6.547	<0.001	1.000	1.000

$0.254v_4 + 0.441v_5 + 0.138v_6 + 0.292v_7 - 0.347$, $\omega_{xy} = -0.005v_1 + 0.229v_2 + 0.038v_3 + 0.216v_4 + 0.023v_5 + 0.112v_6 + 0.171v_7 + 0.111$ (v_i: vibrator i).

Figure 3 shows scatter plots of correlation coefficients between the average value of each evaluation scale and the applied vibration points. Subjective evaluation values were averaged for each vibration pattern.

4 Discussion

4.1 Multiple Regression Analysis of Average Angular Velocity Based on Vibration Pattern

In ω_{yz} model (Table 2), vibrators 5, 2, and 7 had the highest standardized coefficients, in that order. Actually, vibrators 5 and 2 were always included in the higher order patterns in Fig. 2 (above). In ω_{xy} model, vibrators 2, 4 and 7 had the highest standardized coefficients in that order. Vibrators 2, 4 and 7 were always included in the higher order patterns in Fig. 2 (below).

It was common for vibrators 2 and 7 to have large effects in each model. The main difference was that vibrator 5 had the largest effect in the ω_{yz} model and vibrator 4 had a relatively small effect, while in the ω_{xy} model, the vibrator 4 had one of the largest effects. This can be understood by considering a vector model of muscle spindles [9,11], that is composed of the vectors of expected illusion directions and magnitudes when the muscle is stimulated. The coracobrachialis (vibrators 2 and 4), and the wrist flexors (vibrator 7) have vectors that point toward the compounded direction of extension and abduction, and the biceps brachii (vibrator 5) have a vector purely in the extension direction.

More interestingly, wrist flexors, which are not related to the motion of the shoulder joint directly, contributed a great deal to the illusion. We considered it is possibility because vibration applied to wrist flexors induced a motor image similar to what is experienced when the arm is moved by external force exerted on the hand. This image could have enhanced the kinesthetic illusion. The participants actually commented that they felt passive arm movements.

In Fig. 2, the minus value means that participants expressed flexion. This is because tonic vibration reflex (TVR) was evoked in the experiment accidentally. In some participants, the reflex was induced even for the vibration patterns that induced strong illusions in others. These data decreased the accuracy of the models of multiple regression analysis.

4.2 Relationship Between the Strength of the Illusion and the Points of Vibration

We found significantly positive correlations between each subjective evaluation scale and the vibration points (Fig. 3). This means that vivid and large kinesthetic illusions can be induced by multi-point vibratory stimulation. It also suggests that even though vivid illusions can be induced by strong vibration applied to single point, the same effect can be elicited by mild vibration distributed over several points.

4.3 Tendons and Muscles of the Stimulation Points

In this experiment, the tendons were not stimulated directly in all positions. Albeit the tendons of the coracobrachialis were located under the deltoid (vibrator 2) and biceps brachii (vibrator 4), each position was effective for illusion. This result indicates that the illusion was elicited by indirect vibration to the tendon, and the muscle spindles can be stimulated effectively through coracoid process (located in vibrator 2). In particular, the vibrator 4 contributed differently to the direction of the illusion than vibrator 5 did. Kinesthetic illusion can be induced even by stimulating muscle belly [3,4]. This implies that the deltoid contributed to the illusion induced by vibrator 2 and the biceps brachii also contributed to the illusion induced by vibrator 4.

5 Conclusion and Future Work

The purpose of the present study was to investigate the effect that increasing the number of vibratory stimuli has on the kinesthetic illusion. Vibrators were placed at seven positions on the synergist muscles around the chest, upper arm, and forearm, and all vibration combinations were tested. We found that multi-point vibration induced illusions steadily and found the optimized vibration pattern which evoke more rapid illusion than the illusion induced by seven vibratory stimuli (Fig. 2).

In this experiment, the average angular velocity of the illusion was about 2 deg/s at most. Thus, we think that the limit of vibration-induced illusions is suggested. In future, further analysis of the data needs to be performed, and combinations with other modalities should be investigated.

Acknowledgement. This research was supported by JSPS KAKENHI Grant Number JP18H04110.

References

1. Burke, D., Hagbarth, K.E., Löfstedt, L., Wallin, B.G.: The responses of human muscle spindle endings to vibration of non-contracting muscles. J. Physiol. **261**(3), 673–693 (1976)
2. Collins, D.F., Refshauge, K.M., Todd, G., Gandevia, S.C.: Cutaneous receptors contribute to kinesthesia at the index finger, elbow, and knee. J. Neurophysiol. **94**(3), 1699–1706 (2005)
3. Ferrari, F., Clemente, F., Cipriani, C.: The preload force affects the perception threshold of muscle vibration-induced movement illusions. Exp. Brain Res. **237**(1), 111–120 (2019)
4. Goodwin, G.M., Mccloskey, D.I., Matthews, P.B.: The contribution of muscle afferents to kinestesthesia shown by vibration induced illusions of movement and by the effects of paralysing joint afferents. Brain **95**(4), 705–748 (1972)
5. Kajimoto, H.: Illusion of motion induced by tendon electrical stimulation. In: World Haptics Conference, pp. 555–558 (2013)
6. Naito, E., Ehrsson, H.H., Geyer, S., Zilles, K., Roland, P.E.: Illusory arm movements activate cortical motor areas: a positron emission tomography study. J. Neurosci. **19**(14), 6134–6144 (1999)
7. Proske, U., Gandevia, S.C.: The proprioceptive senses: their roles in signaling body shape, body position and movement, and muscle force. Physiol. Rev. **92**, 1651–1697 (2012)
8. Roll, J.P., Vedel, J.P.: Kinaesthetic role of muscle afferents in man, studied by tendon vibration and microneurography. Exp. Brain Res. **47**(2), 177–190 (1982)
9. Roll, J.P., Albert, F., Thyrion, C., Ribot-Ciscar, E., Bergenheim, M., Mattei, B.: Inducing any virtual two-dimensional movement in humans by applying muscle tendon vibration. J. Neurophysiol. **101**(2), 816–23 (2009)
10. Schofield, J.S., Dawson, M.R., Carey, J.P., Hebert, J.S.: Characterizing the effects of amplitude, frequency and limb position on vibration induced movement illusions: implications in sensory-motor rehabilitation. Technol. Health Care **23**(2), 129–141 (2015)
11. Thyrion, C., Roll, J.P.: Predicting any arm movement feedback to induce three-dimensional illusory movements in humans. J. Neurophysiol. **104**(2), 949–959 (2010)
12. Tidoni, E., Fusco, G., Leonardis, D., Frisoli, A., Bergamasco, M., Aglioti, S.M.: Illusory movements induced by tendon vibration in right- and left-handed people. Exp. Brain Res. **233**(2), 375–383 (2014)

13. Ushiyama, K., Tanaka, S., Takahashi, A., Kajimoto, H.: Reinforcement of kinesthetic illusion by simultaneous multi-point vibratory stimulation. In: SIGGRAPH Asia 2019 Posters, pp. 1–2 (2019)
14. Yaguchi, H., Fukayama, O., Suzuki, T., Mabuchi, K.: Effect of simultaneous vibrations to two tendons on velocity of the induced illusory movement. In: Proceedings of IEEE International Conference of Engineering in Medicine and Biology Society (EMBC), pp. 5851–5853 (2010)

Isometric Force Matching Asymmetries Depend on the Position of the Left Hand Regardless of Handedness

Giulia Ballardini(✉) and Maura Casadio

Department of Informatics, Bioengineering, Robotics and Systems Engineering, University of Genoa, Genoa, Italy
giulia.ballardini@edu.unige.it

Abstract. Several studies highlighted differences in behavioral performance between the two hands, either due to hand dominance or to specialization of the brain hemisphere. In a previous study, right-handed individuals performed a bimanual isometric force-matching task with the arms in different configurations. There we found that the accuracy of the performance depended on the position of the left hand. Matching performance was worse when the left hand was in the lower position, regardless the symmetry of the arm configurations. In the present study, we tested the hypothesis that this effect is related to handedness, i.e., that in both right- and left-handed individuals the performance depends on the position of the non-dominant hand. Left-handed and age-matched right-handed participants were required to apply simultaneously the same amount of force in the upward direction, with the arms in symmetric or asymmetric configurations. No visual feedback of limb positions was provided. We found that for both groups the absolute and the signed (bias) difference of force between the sides depended on the position of the left hand. Thus, this role of the left arm was not determined by handedness, but likely by the specialization of the brain hemisphere. However, handedness influenced the performance: left-handers had a higher absolute error than right-handers in almost all conditions. No main effect of the left hand position was found for the variable error, but left-handers in most configurations had higher variable error when the left hand was in the lower position.

Keywords: Handedness · Bimanual task · Laterality

1 Introduction

About 89.5% of the world population prefers using the right hand [1] in the execution of various uni-manual motor tasks. This high number of right-handed individuals has led to a bias also in the study of sensorimotor abilities and motor control [2, 3]. In fact, the majority of researches on the upper limb focused only on right-handed individuals performing uni-manual tasks with their preferred arm, also called dominant arm. This approach reduces the experimental design complexity, but does not take into account the interaction between the two arms [4]. In addition, it limits the possibility to determine sensorimotor differences between left- and right-handed individuals (see also

I. Nisky et al. (Eds.): EuroHaptics 2020, LNCS 12272, pp. 194–202, 2020.
https://doi.org/10.1007/978-3-030-58147-3_22

[2] for a review) and the ability to determine whether the upper limb asymmetries were due to the hand dominance or to a specialization of the brain hemisphere unrelated to handedness. Actually, some tasks showed upper-limb behavioral asymmetries in left-handed individuals identical to the right-handers, suggesting that these observed effects were not determined by handedness, but likely by the specialization of the brain hemisphere. For example, this is the case of stiffness [5] and weight perception [6]. In other tasks, instead, the upper-limb behavioral asymmetries, such as target reaching accuracy [7] and finger pinch movement discrimination [4], were found mirrored with respect to right-handers, suggesting that the observed effects were due to handedness. To explain upper-limb asymmetries Goble et al. [3, 7] proposed the dichotomous model, based on study on right-handed individuals. According to this theory, the dominant arm, in bimanual reaching and position matching tasks, relies more on visual feedback, whereas the non-dominant arm relies more on proprioceptive feedback. This model could also explain results on the sense of effort [8] and grasp force [9] in bimanual tasks. However, to the best of our knowledge the handedness effect on behavioral asymmetries has not been evaluated in bimanual force control tasks with both hands are actively engaged toward a common goal.

In a previous study on right-handed individuals [10], we investigated the ability to simultaneously apply an equal amount of isometric force in the upward direction with the two arms either in symmetric or asymmetric configurations. There, we found that performance was not influenced by the symmetry of the arms configuration, but by the position of the left hand, indicating a leading role of the non-dominant limb on the bimanual performance of such task. We hypothesize that this effect is related to the participants' handedness and therefore that the results would be mirrored in left-handed individuals. For testing our hypothesis, in this study, we repeated the same experiment on a population of young left-handers and on an age-matched group of right-handers. Our hypothesis will be supported if in both populations the performance will depend on the position of the non-dominant hand. Conversely, if force performance will depend on the position of the left hand, results, contrary to our hypothesis, will indicate a hemispheric specialization in the brain, independent of handedness.

2 Materials and Methods

2.1 Experimental Set-up

The experimental set-up has been previously described in [10]. Briefly, we used a device composed of a base plane and two vertical bars, each with a metal linear guide, where a custom-made handle could slide or be fixed (Fig. 1a). In this experiment, the handles were locked in fixed positions by a mechanical block. The force exerted on the handle in the upward direction was measured by a micro load cell (CZL635, Phidgets Inc; full range scale of 5 kg, precision of 0.05%; Fig. 1a, detailed view). The force recoded by the load cells were sent to a DAQ board (NI USB-6008, National Instruments) connected to a laptop via USB. The control software was developed in LabVIEW (National Instruments). During the experiment the participants were sitting in front of the device (Fig. 1b), on such way that to move the handles at the top of the metal guide they had to completely extend their arms. They had to grasp the handles, maintaining their thumb and index fingers in contact with the bottom surface of the plates (Fig. 1b, detailed view) and push in the upward direction. The view of their hands, arms and shoulders was blocked by a black curtain attached to the device for the entire duration of the experiment. The instructions were displayed on a screen placed in the middle of the two vertical bars.

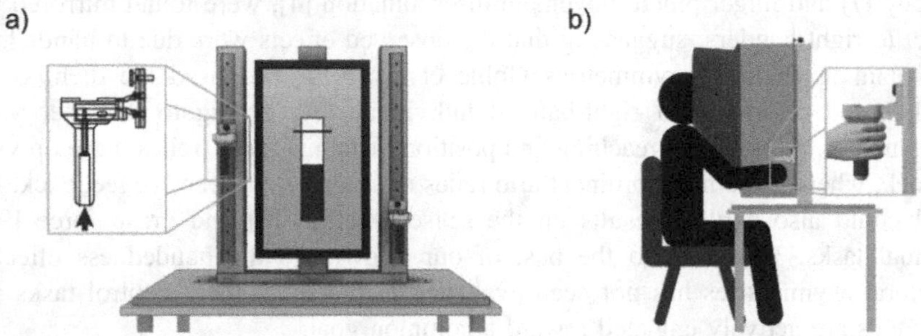

Fig. 1. (a) Rendering of the device. It is composed of a base plane and two vertical bars, each with a metal linear guide, where the handle could be fixed in different positions. Between the two bars there is a screen where a horizontal line indicates the target force that the participants have to match applying simultaneously the same amount of force with the two hands. The height of the blue bar in the screen is controlled by the sum of the force recorded by the load cells placed in the two handles, as shown in the detailed view. (b) Experimental set-up. Participants were sitting in front of the device grasping the two handles -as shown in the detailed view- locked in different configurations. A black curtain attached to the device prevented the visual feedback of the upper limbs. (Color figure online)

2.2 Protocol: Bimanual Isometric Force Matching Task

Participants were required to apply simultaneously the same amount of isometric force on both handles. In each trial the handles were placed in one of the four different configurations (HC; Fig. 2a) corresponding to all the possible combinations of two different heights, respectively 0.10 m (Down, D) and 0.30 m (Up, U) above the baseline position, i.e. handle in contact with the base plane. Accordingly, in two configurations the handles were in symmetric positions: both down (DD) or both up (UU) while in the other two they were in asymmetric positions: left down and right up (DU), and vice versa (UD). During each trial, participants did not receive any feedback of the force applied by each hand, but they could see on the device's screen the total force exerted as a vertical bar whose height was equal to the sum of the two forces. On the screen we also provided the target force to match, displayed as a horizontal line that has to be reached by the bar controlled by the bilateral force applied by the participants (Fig. 1a). Two different target force levels were presented: 9.8 N or 19.6 N (Fig. 2b). Each target force was presented five times for each hand configuration, in random order, for a total of 40 trials (4 hand configurations * 2 target forces * 5 repetitions). To complete each trial, participants had to communicate to the experimenter when they reached the required amount of force and to maintain it for 0.5 s (holding time interval).

There was no time constrain to complete each trial. If they attempted to apply the forces sequentially with the two hands an error message was provided and the trial was discarded.

A familiarization phase was performed before the task. During this phase, participants had the additional visual feedback of the force applied by each hand, displayed as two additional bars on each side of the main bar representing the total force. During this phase, they performed four trials, with different combinations of hand configuration and target force. Then, we asked if they correctly understood the task, otherwise they could extend the familiarization phase. The entire experiment lasted about 30 min; participants could rest anytime they needed, but they did not ask for any pause.

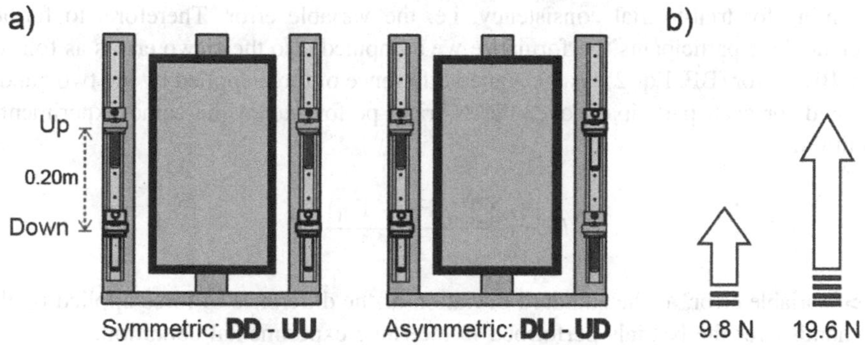

Fig. 2. Protocol of the bimanual isometric force matching task. The participants were asked to push upward the handles applying simultaneously equal isometric force with the two arms. (a) The handles could be placed in four configurations (two symmetric: DD, UU; two asymmetric: DU, UD). (b) We required to match one of the two possible levels of force: 9.8 N and 19.6 N. Each required target force has to be matched by the sum of the force applied to each handle.

2.3 Participants

36 participants (20 females, aged 23–33 yo) voluntarily participated to this study. Before starting the experiment, we evaluated the hand dominance by the 10-item Edinburgh Handedness Inventory (laterality quotient (LQ) score −100: 100) [11]. Based on this score, we divided participants in two age-matched groups: 13 left-handers (LQ score below −50; 8 F; 25 ± 3 yo (mean ± std); LQ score: −86 ± 14) and 23 right-handers (LQ score above 50; 12 F; 26 ± 2 yo; LQ score: 75 ± 14).

Inclusion criteria were: (i) no evidence or known history of neurological disease; (ii) normal joint range of motion and muscle strength; (iii) no problems of visual integrity that could not be corrected with glasses or contact lenses, as i.e. they could clearly see the feedback displayed on the device's screen. Each participant signed a consent form to participate in the study and to publish the results of this research. The research and the consent form were conformed to the ethical standards of the 1964 Declaration of Helsinki and approved by the local Ethical Committee.

2.4 Data Analysis

We focused on the difference of force applied by the two hands in the holding time interval. Our primary outcome was the Absolute Error (AE, Eq. 1), computed as the absolute value of the difference between the forces exerted by the left (F_L) and the right (F_R) arms, averaged for each participant over the N trials performed in the same experimental condition (i.e., same hand configuration and target force):

$$AE = \frac{\sum_{i=1}^{N} |F_{Li} - F_{Ri}|}{N}. \tag{1}$$

The absolute error could be influenced by two concurrent factors: (1) a systematic tendency to exert more force with one arm, i.e. the bias error, (2) a variable component accounting for trial-to-trial consistency, i.e. the variable error. Therefore, to further understand the participants' performance we computed also these two errors as follow:

> Bias Error (BE, Eq. 2), as the signed difference of force applied by the two hands, averaged for each participant over the N trials performed in the same experimental condition:

$$BE = \frac{\sum_{i=1}^{N} (F_{Li} - F_{Ri})}{N}. \tag{2}$$

> Variable error, as the standard deviation of the difference of force applied by the two hands over the N trials performed in the same experimental condition.

Statistical Analysis. Our primary goal was to investigate whether the ability to exert equal isometric forces with the two arms in different configurations was influenced by handedness. The secondary goal was to verify if the performance of left-handed participants depended on the symmetry of the hand configuration, the position of the left hand and the target force. Using IBM SPSS Statistics 25 (International Business Machines Corporation), we performed a repeated-measures ANOVA on the three indicators (absolute, bias and variable error) with one between-subjects factor: 'handedness' (2 levels: left- and right–handed participants) and with three within-subject factors: 'symmetry' (2 levels: symmetric HC and asymmetric HC), 'left hand position' (2 levels: up and down), and 'target force' (2 levels: 9.8 N and 19.6 N). We verified the normality of the data using Anderson-Darling test [12]. The null hypothesis was rejected for the absolute and the variable error, thus these data were corrected applying the fractional rank method [13]. We tested for the sphericity of the data using Mauchly's test, and it was verified for all indicators. We performed a post-hoc analysis (Tukey's method) to further investigate statistically significant effects. Statistical significance was set at the family-wise error rate of $\alpha = 0.05$ and applying Bonferroni correction for multiple comparison the threshold required for significance was set to $\alpha = 0.05/3 = 0.0167$.

3 Results

Left-Handed Participants Had Worse Performance in the Bimanual Force-Matching Task. The absolute error (Fig. 3a) was influenced by handedness (group effect: $F_{1,34} = 6.75$; $p = 0.014$). Specifically, left-handers performed the task with a higher difference of force between the sides than right-handers. However, this population effect for each participant could be due to the bias or to the variable error, as well as to their combination, regardless of handedness. Indeed, for both the bias and the variable errors (Fig. 3b and 3c), the handedness main factor did not reach the threshold of significance (bias and variable error: $p > 0.05$).

The Bimanual Performance was Influenced by the Position of the Left Hand Regardless of Handedness: also in Left-Handed Participants the Difference of Force Applied by the Two Hands Depended on the Position of the Left Hand and Not of the Non-dominant (Right) Hand. The absolute error significantly depended on the position of the left hand for both groups (left hand position effect: $F_{1,34} = 22.09$; $p < 0.001$), i.e. when the left hand was in the lower position the absolute error was higher. This could be explained by the bias error, which showed that participants of both groups tended to apply more force with the left hand when it was in the lower position (left hand position effect: $F_{1,34} = 16.44$; $p < 0.001$). However, this effect was more marked when the hands were in asymmetric configurations, i.e., the performance were not different in symmetric configurations (symmetry*left hand position interaction: $F_{1,34} = 16.00$; $p < 0.001$; post-hoc: DD-UU: $p = 0.759$; DU-UD: $p < 0.001$). As for the variable error, no effect of the left hand position was found in the overall population (left hand position effect: $p > 0.05$), while only the left-handers in most configurations (3 out of 4) had higher variable error when the left hand was in the lower position (left hand position*group interaction: $F_{1,34} = 11.86$; $p = 0.002$). For all the three indicators there were not significantly main effect of the symmetry of the arm configuration (symmetry effect: absolute, bias, and variable error: $p > 0.05$), consistently for both groups.

The Error was Influenced by the Required Total Amount of Force, Regardless of Handedness. The level of the target force had a significant - or close to significance - effect on all the three indicators (absolute-error: $F_{1,34} = 6.89$; $p = 0.013$; variable-error: $F_{1,34} = 11.86$; $p = 0.002$; bias-error: $F_{1,34} = 6.22$; $p = 0.018$), i.e. as expected these indicators were higher for higher target force, regardless of handedness, symmetry of the hand configuration and position of the left hand (all interactions for all indicators: $p > 0.05$).

All the above-mentioned results were confirmed also on the sex-matched subgroup.

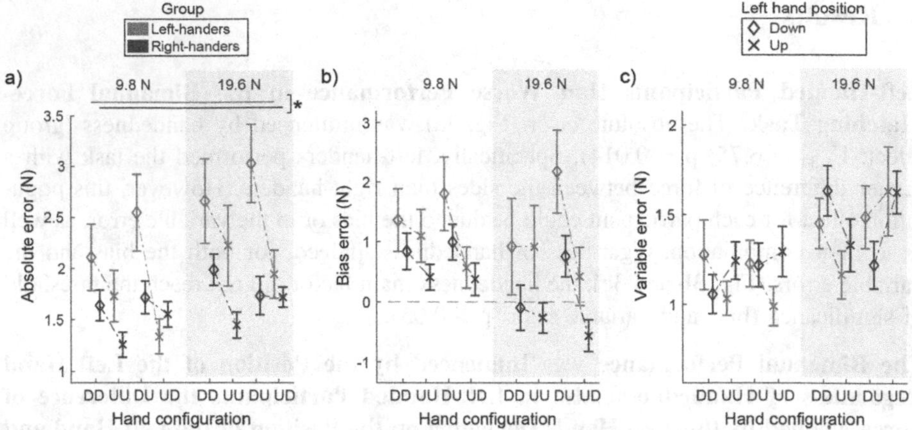

Fig. 3. Indicators of performance computed on the difference between the forces applied by the left and the right hand in terms of: (a) absolute error, (b) bias error and (c) variable error. Each indicator has been reported for the four hand configurations (symmetric: DD and UU; asymmetric: DU and UD, with D-down and U-up are the lower and the higher position respectively, and the first and the second letter are the position of the left and the right hand respectively). The left hand in 'U' position is represented by the 'x' symbol, while in the 'D' position by the 'diamond' symbol. Data are reported separately for each target force: white background indicates 9.8 N, light gray background 19.6 N. All the panels show the results (mean ± SE) separately for the left- and the right- handed population (in light blue and dark gray, respectively). (Color figure online)

4 Discussion

The Difference of Force Applied by the Two Hands in Term of Absolute Error was Influenced by Handedness in all the Experimental Conditions. Indeed, left-handed participants had higher absolute error than right-handed participants. This result supports the conclusions of previous studies, such as [4], suggesting that bimanual proprioception was less accurate in left-handed individuals. This study extends this finding to a bimanual isometric force matching task, where participants integrated the proprioceptive information from their arms positioned in symmetric or asymmetric configurations, not relying on visual feedback.

Bimanual Force Matching Performance Depended on the Position of the Left Hand Regardless the Handedness, i.e. the performance seemed to be influenced by the specialization of the brain hemisphere, evolving independently from handedness. The significant effect of the left hand position in right-handed individuals [11] was supported by the dichotomous model [3] observed in motor [7] and force tasks [8, 9]. This model suggests that during bimanual activities, the dominant right arm relies more on visual feedback, while the non-dominant left arm on proprioceptive feedback. Thus, in our experiment, where participants could not rely on visual feedback, the left arm might be advantaged and play a key role in solving the task. The present study on left-handed individuals extends this finding, suggesting that the observed asymmetry in bimanual force matching performance with different arm configurations could be due to a specialization of right hemisphere, evolving independently from handedness. This result is also supported by a study [14] on people with unilateral stroke, suggesting a specific contribution of the right hemisphere in controlling the production of bilateral force. Also other studies found a specialization of the right hemisphere in different but related tasks, such as controlling limb impedance for stabilizing limb position at the end of movement [15] and generating force for adapting to dynamic variation, such as unexpected perturbation [16]. Note that the signed difference between the forces applied by the two hands was higher and significant when the two arms were in asymmetric configurations. This was expected, since in this case the central nervous system has to apply different neural commands for each side of the body, accounting for the difference in arm configuration. However, humans have a universal tendency to perform coordinated bimanual movements, by activating homologous limb muscles in synchrony (e.g. [17]). The present results suggest that this tendency could be present also in bimanual isometric force matching tasks, explaining at least in part the more similar performance in symmetric configurations. We also found that the performance variability was partially influenced by handedness, since only left-handed individuals tended to have higher variability when the left hand was in the lower position than in the other configurations. A crucial role of the right hemisphere for variability of bilateral force control has been suggested in [14], but its interaction with handedness has not been extensively studied.

The Error is Influenced by the Required Total Amount of Force, Regardless of Handedness. The total amount of the requested force had a relevant effect on the performance, increasing the difference between the two hands and its variability. Further, the bias error highlighted that the left hand applied more force than the right for the lower target force, but this effect was decreased and even inverted for the higher target force, consistently with previous results in sequential [8, 9] and concurrent [10] matching task. The results of the present study extend the previous findings, highlighting that this effect was not influenced by the handedness.

Limitation and Future Directions. The results obtained in the right-handers were consistent with what reported [10] for both the absolute and the bias error. Instead, the variable errors in our young participants were lower and in a different relation to hand configurations. The difference was due to the older participants included in the previous study who had significantly higher variable errors, as found also in [18]. We plan to further investigate the influence of aging on this specific task in a future study.

Acknowledgement. This study was supported by Ministry of Science and Technology, Israel (Joint Israel-Italy lab in Biorobotics “Artificial somatosensation for humans and humanoids”).

References

1. Papadatou-Pastou, M., et al.: Human handedness: a meta-analysis (2020)
2. Elliott, D., Chua, R.: Manual asymmetries in goal-directed movement. In: Elliott, D., Roy, E. A. (eds.) Manual Asymmetries in Motor Performance, Boca Raton, CRC (1996)

3. Goble, D.J., Noble, B.C., Brown, S.H.: Proprioceptive target matching asymmetries in left-handed individuals. Exp. Brain Res. **197**(4), 403–408 (2009)
4. Han, J., et al.: Bimanual proprioceptive performance differs for right- and left-handed individuals. Neurosci. Lett. **542**, 37–41 (2013)
5. Leib, R., et al.: Force feedback delay affects perception of stiffness but not action, and the effect depends on the hand used but not on the handedness. J. Neurophysiol. **120**(2), 781–794 (2018)
6. Buckingham, G., et al.: Handedness, laterality and the size-weight illusion. Cortex **48**(10), 1342–1350 (2012)
7. Goble, D.J., Brown, S.H.: Upper limb asymmetries in the matching of proprioceptive versus visual targets. J. Neurophysiol. **99**(6), 3063–3074 (2008)
8. Scotland, S., et al.: Sense of effort revisited: relative contributions of sensory feedback and efferent copy. Neurosci. Lett. **561**, 208–212 (2014)
9. Mitchell, M., et al.: Upper limb asymmetry in the sense of effort is dependent on force level. Front. Psychol. **8**(APR), 1–8 (2017)
10. Ballardini, G., et al.: Interaction between position sense and force control in bimanual tasks. J. Neuroeng. Rehabil. **16**(1), 1–13 (2019)
11. Oldfield, R.C.: The assessment and analysis of handedness: the Edinburgh inventory. Neuropsychologia **9**(1), 97–113 (1971)
12. Anderson, T.W., Darling, D.A.: A test of goodness of fit. J. Am. Stat. Assoc. **49**(268), 765–769 (1954)
13. Friedman, M.: The use of ranks to avoid the assumption of normality implicit in the analysis of variance. J. Am. Stat. Assoc. **32**(200), 675–701 (1937)
14. Kang, N., Cauraugh, J.H.: Right hemisphere contributions to bilateral force control in chronic stroke: a preliminary report. J. Stroke Cerebrovasc. Dis. **27**(11), 3218–3223 (2018)
15. Mani, S., et al.: Contralesional motor deficits after unilateral stroke reflect hemisphere-specific control mechanisms. Brain **136**(4), 1288–1303 (2013)
16. Mitrovic, D., et al.: A computational model of limb impedance control based on principles of internal model uncertainty. PLoS ONE **5**(10), e13601 (2010)
17. Swinnen, S.P.: Intermanual coordination: from behavioural principles to neural-network interactions. Nat. Rev. Neurosci. **3**(5), 348 (2002)
18. Shim, J.K., et al.: Age-related changes in finger coordination in static prehension tasks. J. Appl. Physiol. **97**(1), 213–224 (2004)

Computational Model of a Pacinian Corpuscle for an Electrical Stimulus: Spike-Rate and Threshold Characteristics

Madhan Kumar Vasudevan(✉), Rahul Kumar Ray, and Manivannan Muniyandi

Touch Lab, Department of Applied Mechanics, Indian Institute of Technology Madras, Chennai 600036, India
madhan.kv@gmail.com, rahulraiecb@gmail.com, mani@iitm.ac.in

Abstract. Understanding the response of Pacinian Corpuscle (PC) for an electrical stimulus through a computational model can give better insight into the physiology. Although there are simpler models available in the literature, models simulating spike-rate and threshold characterizations are still missing. These characterizations may lead to the development of tactile displays combining both electrical and mechanical stimuli, especially high-frequency vibrations. We developed a PC model with equivalent circuits of the electrode-skin interface, PC's neurite, and the first Ranvier node. The input electrical stimulus is a current pulse with varying amplitude (0 to 2 mA) and varying frequency (5 Hz to 1600 Hz). The model is characterized initially for the frequency response, and then the spike-rate and threshold characteristics were simulated. The spike-rate traces for electrical stimuli show the phase-locking phenomenon similar to the mechanical stimuli responses of PC, however the plateau lengths are larger for the spike-rate traces with electrical stimuli compared to that of the mechanical stimuli. This is reflected as a large difference in the threshold characteristics for one and two impulses-per-cycle. Moreover, threshold characteristics are little influenced by the neural noise. This model can be extended to study the combination of electrical and mechanical stimuli.

Keywords: Electrical stimulation · Neurite · Ranvier node · Spike-rate · Threshold.

1 Introduction

The electrical stimulus applied over the skin elicits different sensations such as continuous or intermittent touch depending on the frequency, phase, and amplitude of the stimulus [14]. The variations in the electrical stimulus can induce the receptors of various modalities such as mechano-, chemo-, and thermoreceptors. These variations are then decoded by the CNS to perceive various aspects of touch and other sensations [15]. Pacinian corpuscle (PC) being the most sensitive mechanoreceptor in the human body is responsible for the sensation of

I. Nisky et al. (Eds.): EuroHaptics 2020, LNCS 12272, pp. 203–213, 2020.
https://doi.org/10.1007/978-3-030-58147-3_23

high-frequency vibrations. PC neurite and the axon can be selectively stimulated by an electrical stimulus to induce the sensation of high-frequency vibration. To understand the physiology of a PC for mechanical or electrical stimuli, it is necessary to understand the morphology of a PC in finer details [6,6,7,11,22,25] and the computational models [3,19,23,26,27].

The PC is made of an onion-like lamellar structure filled with interlamellar fluid [22]. The lamellar structure of the capsule acts as a mechanical band-pass filter [9,19], which is the main reason for the rapid adaptiveness of the PC for a mechanical stimulus. Although the lamellar structure of the PC helps it to be the most rapidly adaptive mechanoreceptor, the neurite membrane and inner core also contribute to its rapid adaptiveness [1]. Due to the presence of stretch-activated and voltage-activated ion channels (SAICs and VAICs) in PC neurite [21], the applied mechanical stimulus gets converted into electrical spikes. Since lamellae are attached to each other by tight junctions, fluid in the inner core is considered to be electrically isolated from that in the outer core [12]. Moreover, the outer core is electrically isolated from the transductive portion of the neurite [1].

The exact location of action potential generation is still in controversy. It was believed that the core of the PC is rigid [17,19], however, it was discovered later that the neurite of the PC contains SAICs and VAICs [1]. Models were developed, assuming that the action potential spikes from the first Ranvier node [14]. The controversy related to the site of initiation of action potential inside the PC capsule started from the discovery of the generator or local potential of PC inside the capsule [10]. Based on the majority of the evidence, it appears that the action potential at 1st Ranvier's node can be generated by the direct electrical stimulation at the nerve fiber [1].

1.1 Motivation and Objective

From the literature, it is clear that the characterization of a PC model for electrical stimuli in terms of spike-rate and threshold characteristics is still missing. Neural spikes encode different features of the stimulus which include amplitude, frequency and even the location of the stimulus over the skin [27]. Spike rate and threshold characteristics of a neural spike train help in understanding how the stimulus features are conveyed to the CNS. Such characterizations would be useful in comparing the physiological characteristics of a PC stimulated electrically, and that of a PC stimulated mechanically. Our objective in this work is to model and characterize the response of a PC for an electrical stimulus applied over the skin. We characterize our model based on frequency response, spike-rate, and threshold characteristics.

2 Method

The computational model of a Pacinian corpuscle excited by electrical stimuli is described in this section. This model is an extension of our previous modeling work (BMS model) [2–4] for an isolated PC excited by mechanical stimuli into a PC model excited by electrical stimuli. The model includes the electrode-skin interface, neurite, and the first Ranvier node of a PC. The simulation involves the application of electrical stimuli of various frequencies and the characterization of the model for frequency response, spike-rate and threshold characteristics.

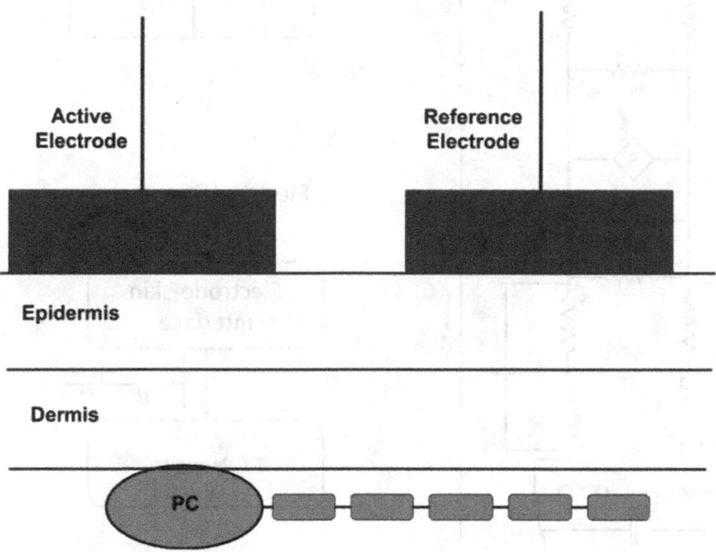

Fig. 1. Electrical stimulation of a Pacinian nerve through the skin surface. The stimulus current pulse is applied through a two-electrode electrical circuit. The axon of PC located deep within the skin gets excited by the current stimulus applied perpendicular to it. The frequency and amplitude of the current pulse can be varied for the measurement of a threshold of sensation.

2.1 Model Description

The electrode-skin interface, PC's neurite, and the first Ranvier node are modeled using electrical parameters available from Chan [8], Saadi et al. [24], Khorshid et al. [16] and Biswas et al. [2]. The conceptual depiction of the proposed model is shown in Fig. 1. The electrical current pulse is applied over the skin through active and reference electrodes. PC has a myelinated axon and it is located deep within the skin compared to other mechanoreceptors [6]. In our model, we assume that the axon of the PC is perpendicular to the application of an electrical stimulus. The applied electrical stimulus gets filtered by the skin layers, and then the filtered stimulus undergoes two-stage non-linear neural-spike generation. This spike generation process is modeled as an Adaptive Relaxation Pulse Frequency Modulator (ARPFM) which was introduced by Biswas et al. [4]. The ARPFM is similar to the integrate-and-fire neuron model except that

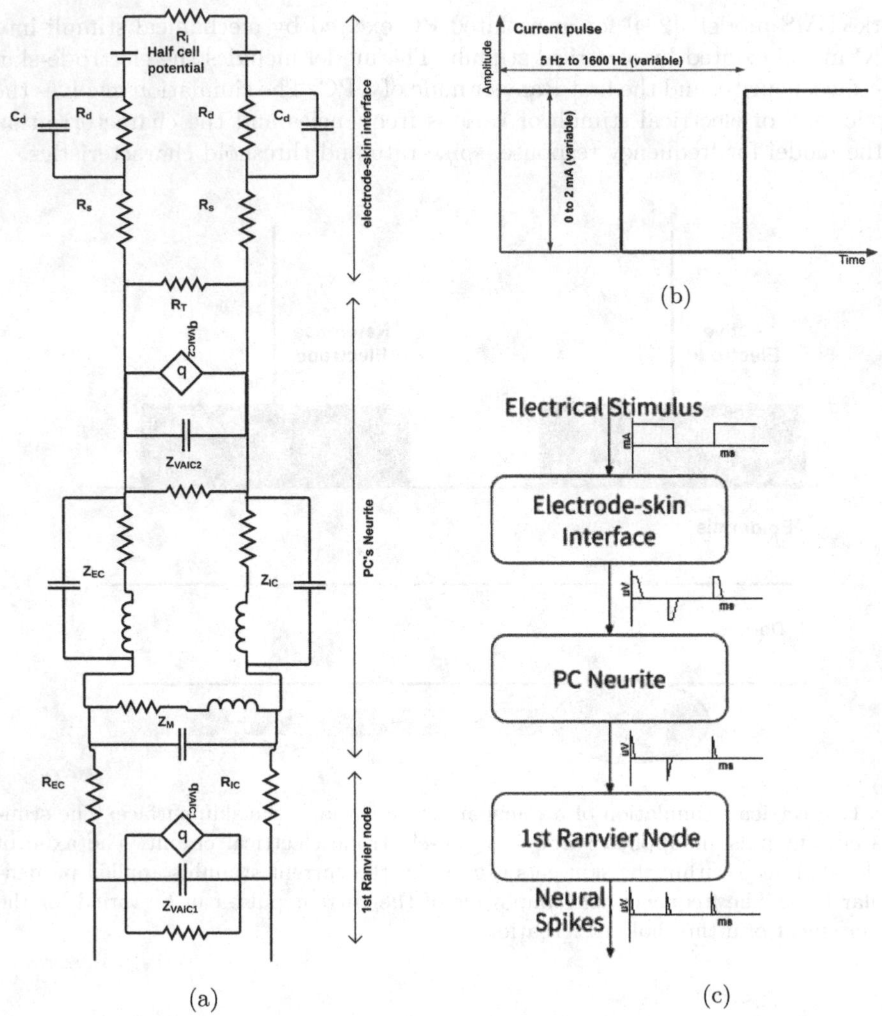

(a) (b) (c)

Fig. 2. (a) Equivalent circuit model of the electrode-skin interface, PC's neurite, and the first Ranvier node. The values of potential and impedance are adapted from [2,8, 16,24]. (b) Current pulse with variable amplitude and frequency, and with fixed duty cycle (50%). (c) Block diagram of a PC model with a representation of signals at each level.

the integrator is lossy, and the threshold is adaptive. The neural noise for the ARPFM stage is modeled as an additive random noise and multiplied with the adaptive threshold for the spike generation.

The proposed model of a PC for electrical stimulation is shown in Fig. 2. It consists of three stages, as shown in Fig. 2c, electrode-skin interface model, PC's neurite model, and the first Ranvier node model. Each of them is modeled with electrical components whose parametric values are adapted from Chan [8],

Saadi et al. [24], Khorshid et al. [16] and Biswas et al. [2]. The electrode-skin interface model, as shown in Fig. 2a (I), consists of tissue impedance R_T, skin resistance R_s, double layer impedance R_d, C_d, and half cell potential. Since there are two electrodes, active and reference, the same model is mimicked for both. That is, the electrical properties and parameters of both electrode-skin interfaces are assumed to be identical. The skin impedance measurement reviewed in Lu et al. [20] explains that the electrical impedance of the skin is mainly due to stratum corneum (the uppermost layer of the epidermis) and the impedance of the other layers is comparatively low. The impedance function is given as

$$K(s) = \frac{0.0259s + 1859}{0.01858s + 1} \quad (1)$$

The second stage of the model contains a circuit equivalent of the PC's neurite as shown in Fig. 2a (II). This stage introduces non-linearity in the generation of the receptor potential. It consists of a non-linear dependent charge source q_{VAIC2} and its impedance Z_{VAIC2}. The impedances of extra-cellular Z_{EC} and intra-cellular Z_{IC} matrices are also included in the model along with impedance of axolemma membrane Z_{M}. The PC's neurite parameters and the values of impedances are the same as given in Biswas et al. [2]. The output of the second stage is the receptor potential. It is assumed that the receptor potential from the second stage is the input to the third stage. The third stage of the model includes the equivalent circuit of the first Ranvier node, as shown in Fig. 2a (III). This circuit contains a non-linear dependent charge source q_{VAIC1} and the impedance parameter Z_{VAIC1}, both models the VAIC located in the first Ranvier node. The ARPFM threshold is actually a measure of the refractoriness of the VAICs to go for the next avalanche opening. In comparison to the model of VAIC, as ARPFM, the threshold is adaptive and amplified by the threshold amplification factor in the refractory period, as found in Loewenstein and Altamirano-Orrego [18]. It is observed that after the time (t) = 2.5 ms, the experimental data have an exponential decay with a time constant of 0.56 ms. The threshold amplification factor (TAF) considered for the ARPFM [4] is given as

$$TAF = 1 + (7.75 * t^{-0.16} * exp(-0.56t)) \quad (2)$$

The electrical stimulus is shown in Fig. 2b, which has variable amplitude from 0 to 2 mA and variable frequency from 5 Hz to 1600 Hz. The duty cycle of the applied electrical stimulus is always 50%.

Model Parameters: The model parameters for the PC model are same as given in [3,19] and also the parameters and approximations for the electrode-skin interface model are same as given [8,16,24].

R_i, R_d, R_s, R_T : Resistance of electrical stimulation circuit, double layer, skin and tissue, respectively

C_d : Double layer capacitance

$q_{\mathrm{VAIC2}}, q_{\mathrm{VAIC1}}$: Voltage activated ion channels in neurite (2) and 1st Ranvier node (1) are modeled as voltage dependent charge sources.

Z_{EC}, Z_{IC}, Z_M : Impedance of extracellular matrix, intracellular matrix and membrane of axolemma, respectively.

$K(s)$: Transfer function of electrode-skin interface model in Laplace domain.

Model Approximations: The following are the approximations assumed for the developed model,

a) We consider only one type of mechanoreceptor under the skin, the PC, although the Meissner is also a rapidly adaptive receptor.
b) The axon of a PC is perpendicular to the direction of the electrical stimulus.
c) The half cell potential of the electrode-skin interface model is assumed to be identical. Moreover, elements of the electrode-skin interface model for both the electrodes are identical, according to [8].
d) All the simulations are limited to only the vibrotactile stimulus with the frequency ranging from 5 Hz to 2000 Hz with 50% duty cycle.

3 Results and Discussion

The objective of this work is to develop a PC model for an electrical stimulus and characterize the model for frequency response, spike-rate and threshold characteristics.

3.1 Frequency Response

Figure 3a shows the bode plot of the electrode-skin-PC model, including the electrode-skin interface, PC's neurite, and the first Ranvier node. It may be observed that the overall model is a typical high-pass filter. The slope of the magnitude plot changes significantly; from 1 Hz to 10 Hz it is ≈10 dB/decade, from 10 Hz to 10 kHz it is ≈20 dB/decade and from 20 kHz the slope approaches zero. On the other hand, the phase difference is found to be within 90 to 45° for the frequencies from 1 Hz to 10 kHz, reaching its peak phase shift of ≈90°.

The bode plot shown here can be compared to that of the PC model with a mechanical stimulus [3]. Although the magnitude plot is almost the same as that of the PC model with a mechanical stimulus, the phase plot differs slightly for frequencies lesser than 5 Hz. Since the present model focuses on frequencies of more than 5 Hz, this small difference may be ignored. Also, at this low frequency other mechanoreceptors respond better than the PC [15].

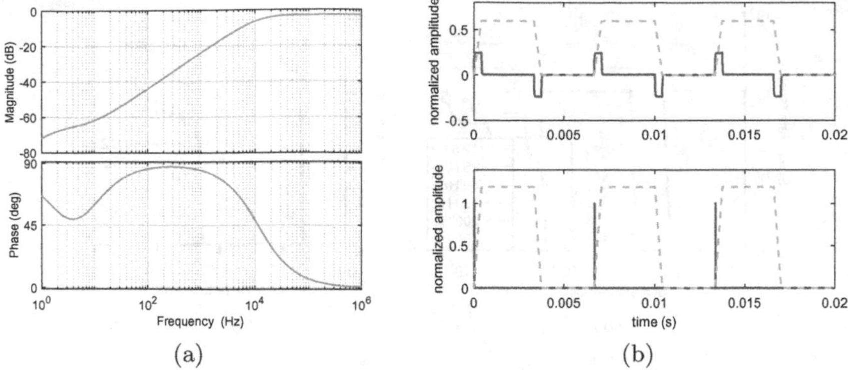

Fig. 3. (a) Magnitude response (top) and phase response (bottom) of the PC model for electrical stimuli. The magnitude in (a) indicates the gain in dB with respect to 1 mA electrical stimulus. (b) The response of the electrode-skin interface model (top), the spike response at the output of the first Ranvier node (bottom). The dashed lines in (b) indicate the applied electric stimulus as shown in Fig. 2b. All the amplitudes are normalized for the representation. The ordinate unit for dashed lines is mA and that for the solid lines is in μV.

3.2 Spike-Rate Characteristics

The spike response of the developed model is shown in Fig. 3b. The electrical stimulus that gets filtered by the electrode-skin interface model reaches the PC's neurite then the first Ranvier node to fire the neural spikes, as shown in Fig. 3b. The response shown here is the typical one impulse-per-cycle (ipc) response for the applied electrical stimulus.

The spike-rate versus stimulus amplitude plots for various stimulus frequencies are shown in Fig. 4a and 4c. We have also shown the spike-rate plots generated by a PC model for a mechanical stimulus in Fig. 4b and 4d (images adapted from Biswas et al. [4]), which are given adjacently for comparison. For each stimulus frequency of Fig. 4a, the spike-rate is zero until the stimulus reaches a certain threshold. Once the threshold is reached, it increases steeply and reaches a series of plateaus at spike-rates that are multiples of stimulus frequency. This phenomenon is known as phase-locking [5]. For instance, a 50 Hz stimulus gets plateaued during 50 *Spikes Per Second* (*sps*), 100 *sps*, 150 *sps* and 200 *sps* which are consistent with one, two, three and four impulse-per-cycles respectively according to Johnson [13] who recorded the population response from median nerves of 26 monkeys. This phase-locking, non-linear jumps, and plateau of the spike-rates depend on the stimulus frequency, as mentioned in [5]. The spike-rate gets saturated for 800 Hz and 1600 Hz at around 1000 *sps*.

Unlike the spike-rate plot for mechanical stimulus as given in Fig. 4b which contains short plateau lengths, the spike rate plot for electrical stimulus contains longer plateaus. These longer plateaus are reflected in the threshold characteristics as well which we are discussing subsequently in the next subsection.

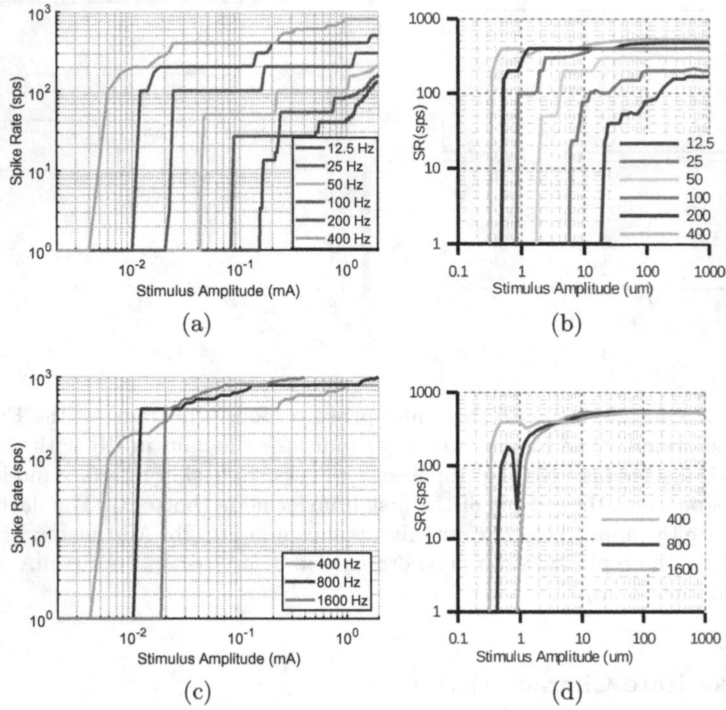

Fig. 4. (a) and (c) represents the Spike-rate (sps - number of spikes per second) characteristics plotted for the variation in stimulus amplitude (0 to 2 mA) with various constant stimulus frequencies. (b) and (d) represents the same for mechanical stimuli (image adapted from Biswas et al. [4]). Although the spike-rate characteristics for electrical stimuli exhibit a phase-locking phenomenon [5], the plateau lengths are larger compared to that of the characteristics in the mechanical stimuli model.

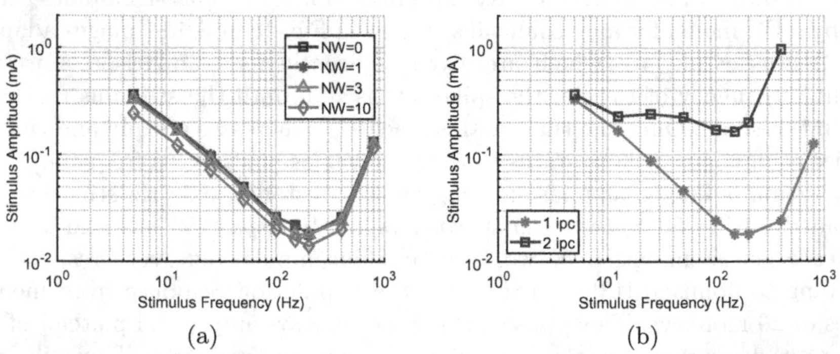

Fig. 5. (a) Stimulus amplitude versus stimulus frequency, for various noise weights. Except for NW = 10, we can observe that there is only little influence of noise weights in the threshold. (b) Minimum stimulus amplitude required to elicit one and two impulse-per-cycle (ipc) for varying stimulus frequency. This plot can be compared with the plataeu lengths of Fig. 4a to understand the shift in 2 ipc from 1 ipc curve here.

Physiologically, longer plateaus may be due to the various levels of saturation in the VAIC characteristics of the first Ranvier node induced by the electrical stimuli. Moreover, both the electrical and mechanical stimuli spike rate plots in Fig. 4 contain varying thresholds for each stimulus frequency to initiate spikes. This can be correlated with the increasing magnitude response of the model, as shown in Fig. 3a. Moreover, we can observe the saturation of spike-rate traces for the stimulus frequencies greater than 400 Hz as shown in Fig. 4d around 600 *sps*, whereas in Fig. 4c the saturation occurs only after 1000 *sps*.

3.3 Threshold Characteristics

The threshold characteristics simulated in this model are for one impulse-per-cycle (ipc) for various neural noises, which is shown in Fig. 5a. We can observe that the model response is not much influenced by the neural noise, except for noise weight (NW) of 10. The threshold characteristics simulated here may be considered as consistent with the well known psychophysical VPT curve [28] and the lowermost threshold is achieved for 200 Hz electrical stimulus. Moreover, the minimum stimulus amplitude required to elicit one and two impulse-per-cycle (ipc) for varying stimulus frequency is plotted and shown in Fig. 5b. We can observe that each trace of the threshold curve is unique and further away. This may be due to the longer plateaus as shown in Fig. 4.

3.4 Possible Extensions

In this paper, we assumed that the electrical stimulus reaches the first Ravier node only through the PC neurite, not directly. A special case can be considered for this model in which the applied electrical stimulus reaches both PC's neurite and the first Ranvier node simultaneously. In this case, input to the first Ranvier node will be the sum of the output from PC's neurite (receptor potentials) and the electrical stimulus itself, both acting together to reach the threshold in the Ranvier node. Although this special case is not explicitly shown in the model figures, it can also be simulated from the same model by simple addition, as mentioned here. The model can also be simulated for various duty cycles of electrical stimuli whereas in the present work we have considered a fixed duty cycle of 50%.

4 Summary

We developed a model for the response of Pacinian Corpuscle excited by an electrical stimulus. We adapted and combined the models of the electrode-skin interface, PC's neurite, and the first Ranvier node, and characterize them together for frequency response, spike-rate, and threshold characteristics. Although there are models known for the electrical stimulus to a PC, none of them explain in terms of the aforementioned characteristics. We have shown the frequency response using a bode plot, which shows constant gain after 10 kHz. The spike-rate plots for varying stimulus amplitudes were simulated and the plateaus were observed for each stimulus frequencies. The threshold characteristics were simulated for one and two impulses-per-cycle (ipc). Furthermore, for a 1 ipc, the threshold characteristics were simulated for various noise weights. We compared our results with the existing models of PC for mechanical stimulus. The future work may include developing a PC model 1) for the combination of electrical and mechanical stimuli (hybrid stimuli) and perform psychophysical experiments to validate the model responses, and 2) for two different electrical stimuli of varying amplitude, frequency, and phase applied in two different locations over the skin to elicit various sensations.

References

1. Bell, J., Bolanowski, S., Holmes, M.H.: The structure and function of Pacinian corpuscles: a review. Prog. Neurobiol. **42**(1), 79–128 (1994)
2. Biswas, A., Manivannan, M., Srinivasan, M.A.: A biomechanical model of Pacinian corpuscle & skin. In: 2013 Biomedical Sciences and Engineering Conference (BSEC), pp. 1–4. IEEE (2013)
3. Biswas, A., Manivannan, M., Srinivasan, M.A.: Multiscale layered biomechanical model of the Pacinian corpuscle. IEEE Trans. Haptics **8**(1), 31–42 (2015)
4. Biswas, A., Manivannan, M., Srinivasan, M.A.: Vibrotactile sensitivity threshold: nonlinear stochastic mechanotransduction model of the Pacinian corpuscle. IEEE Trans. Haptics **8**(1), 102–113 (2015)
5. Bolanowski Jr., S., Zwislocki, J.J.: Intensity and frequency characteristics of Pacinian corpuscles. I. Action potentials. J. Neurophys. **51**(4), 793–811 (1984)
6. Cauna, N., Mannan, G.: The structure of human digital Pacinian corpuscles (corpuscula lamellosa) and its functional significance. J. Anat. **92**(Pt 1), 1 (1958)
7. Cauna, N., Mannan, G.: Development and postnatal changes of digital Pacinian corpuscles in the human hand. J. Anat. **93**(Pt 3), 271 (1959)
8. Chan, A.Y.: Biomedical Device Technology: Principles and Design. Charles C Thomas Publisher, Springfield (2016)
9. Grandori, F., Pedotti, A.: A mathematical model of the Pacinian corpuscle. Biol. Cybern. **46**(1), 7–16 (1982). https://doi.org/10.1007/BF00335347
10. Gray, J.A.B., Sato, M.: Properties of the receptor potential in Pacinian corpuscles. J. Phys. **122**(3), 610–636 (1953)
11. Hubbard, S.: A study of rapid mechanical events in a mechanoreceptor. J. Phys. **141**(2), 198–218 (1958)
12. Ide, C., Hayashi, S.: Specializations of plasma membranes in PCS: implications for mechano-electric transduction. J. Neurocytol. **16**(6), 759–773 (1987)
13. Johnson, K.: Reconstruction of population response to a vibratory stimulus in quickly adapting mechanoreceptive afferent fiber population innervating glabrous skin of the monkey. J. Neurophysiol. **37**(1), 48–72 (1974)
14. Kajimoto, H., Kawakami, N., Maeda, T., Tachi, S.: Tactile feeling display using functional electrical stimulation. In: Proceedings of the 1999 ICAT, p. 133 (1999)
15. Kandel, E., Schwartz, J., Jessell, T.: Principles of Neural Science. Prentice-Hall International Edit. Elsevier, Amsterdam (1991)
16. Khorshid, A.E., Alquaydheb, I.N., Eltawil, A.M.: Electrode impedance modeling for channel characterization for intra-body communication. In: Fortino, G., Wang, Z. (eds.) Advances in Body Area Networks I. IT, pp. 253–266. Springer, Cham (2019). https://doi.org/10.1007/978-3-030-02819-0_19

17. Loewenstein, W.R.: The generation of electric activity in a nerve ending. Ann. N. Y. Acad. Sci. **81**(2), 367–387 (1959)
18. Loewenstein, W.R., Altamirano-Orrego, R.: The refractory state of the generator and propagated potentials in a Pacinian corpuscle. J. Gen. Physiol. **41**(4), 805–824 (1958)
19. Loewenstein, W., Skalak, R.: Mechanical transmission in a Pacinian corpuscle. An analysis and a theory. J. Physiol. **182**(2), 346–378 (1966)
20. Lu, F., et al.: Review of stratum corneum impedance measurement in non-invasive penetration application. Biosensors **8**(2), 31 (2018)
21. Pawson, L., Bolanowski, S.J.: Voltage-gated sodium channels are present on both the neural and capsular structures of Pacinian corpuscles. Somatosens. Mot. Res. **19**(3), 231–237 (2002)
22. Pease, D.C., Quilliam, T.A.: Electron microscopy of the Pacinian corpuscle. J. Cell Biol. **3**(3), 331–342 (1957)
23. Quindlen, J.C., Stolarski, H.K., Johnson, M.D., Barocas, V.H.: A multiphysics model of the Pacinian corpuscle. Integr. Biol. **8**(11), 1111–1125 (2016)
24. Saadi, H., Attari, M.: Electrode-gel-skin interface characterization and modeling for surface biopotential recording: impedance measurements and noise. In: 2013 2nd International Conference on Advances in Biomedical Engineering, pp. 49–52. IEEE (2013)
25. Spencer, P.S., Schaumburg, H.H.: An ultrastructural study of the inner core of the Pacinian corpuscle. J. Neurocytol. **2**(2), 217–235 (1973). https://doi.org/10.1007/BF01474721
26. Summers, I.R., Pitts-Yushchenko, S., Winlove, C.P.: Structure of the Pacinian corpuscle: insights provided by improved mechanical modeling. IEEE Trans. Haptics **11**(1), 146–150 (2018). https://doi.org/10.1109/TOH.2017.2769648
27. Vasudevan, M.K., Sadanand, V., Muniyandi, M., Srinivasan, M.A.: Coding source localization through inter-spike delay: modelling a cluster of Pacinian corpuscles using time-division multiplexing approach. Somatosens. Mot. Res. **37**(2), 63–73 (2020)
28. Verrillo, R.T., Fraioli, A.J., Smith, R.L.: Sensation magnitude of vibrotactile stimuli. Perc. Psychophys. **6**(6), 366–372 (1969). https://doi.org/10.3758/BF03212793

Haptic Technology

SwitchPaD: Active Lateral Force Feedback over a Large Area Based on Switching Resonant Modes

Heng Xu(✉), Michael A. Peshkin, and J. Edward Colgate

The Department of Mechanical Engineering, Northwestern University, Evanston, IL 60208-3111, USA
hengxu@u.northwestern.edu, {peshkin,colgate}@northwestern.edu

Abstract. We present a new device, the SwitchPaD, to generate an active lateral force on a bare fingertip over a large touch area. Like our previous device, the UltraShiver, the SwitchPaD uses synchronization of in-plane ultrasonic oscillation and out-of-plane electroadhesion to generate force. The UltraShiver, however, relied on a single longitudinal resonance to produce oscillations, resulting in an inconsistent force profile. The SwitchPaD switches between the first and the second longitudinal mode based on the finger position, resulting in a much more consistent force profile across the touch surface. Experiments are used to compare the performance of two different modal switching strategies. Results indicate that the SwitchPaD can generate 250 mN peak active lateral force over a large area, and that, with the proper switching strategy, the switch itself is imperceptible.

Keywords: Active lateral force feedback · Resonant modes switch · Ultrasonic oscillation · Electroadhesion

1 Introduction

Compared with friction modulation methods (ultrasonic friction modulation [1, 2], electroadhesion [3,4]), a surface providing active lateral force feedback may render a wider range of haptic effects. Included are effects such as potential well rendering [5] that require propulsive forces, and effects such as button click rendering [6] that require active forces on a stationary finger. In an effort to realize these benefits, a variety of interesting devices have been developed with the goal of providing active lateral force feedback. For instance, Gueorguiev et al. [7] proposed a traveling-wave based device that could provide 100 mN active lateral force with a pressing force of 0.5 N. Since devices employing traveling waves typically operate off resonance, a bulky actuator was required to generate strong lateral forces for haptic rendering. Our group has developed a range of resonant devices that can provide active lateral force, including the ShiverPaD [8] and eShiver [9]. In those devices (and a similar device in [10]), the touch surface

I. Nisky et al. (Eds.): EuroHaptics 2020, LNCS 12272, pp. 217–225, 2020.
https://doi.org/10.1007/978-3-030-58147-3_24

was oscillated by a voice coil actuator. Even though this approach generated a strong lateral force, the voice coil actuators were bulky, and the resonant frequencies were within the range of human hearing, resulting in noise.

The UltraShiver, for the first time, combined piezoelectric excitation of an ultrasonic resonance with electroadhesion to achieve both high forces and silent operation [11], but provided only a small touch area. This paper proposes a new haptic device, the SwitchPaD, that can achieve active lateral forces (±250mN) over a large area and in the inaudible region. It oscillates an electroadhesive surface in-plane by selectively exciting the first (22,390 Hz) or the second (53,320 Hz) resonant longitudinal mode based on the finger position. Experiments described in Sect. 5 were used to evaluate two different modal switching strategies and to investigate the ability of the SwitchPaD to provide consistent lateral force over a large area.

2 Background and Motivation

The UltraShiver consists of two piezoelectric actuators glued symmetrically on opposite sides of a sheet of anodized aluminum [11]. Forces are produced by synchronizing in-plane ultrasonic oscillation and out-of-plane electroadhesion. The former is tuned to the first longitudinal resonance of the UltraShiver. Even though the device shows good ability to control the lateral force, render convincing haptic effects, and localize them [6,11], it cannot provide a consistent magnitude of the lateral force over a large area due to the mode shape. The lateral force decreases to zero as the finger moves toward the nodal line, which is close to the center of the surface.

To remove this limitation, one possible solution is to take advantage of the second longitudinal resonance of the surface, because the anti-node line of the second mode lies at the same location as the node line of the first mode. In general, there are two methods: two-mode superposition and two-mode switching. In modal superposition, the first mode and the second mode are excited at the same time. Unfortunately, the combination of two modes produces a complicated spatiotemporal variation of the surface velocity, making it difficult to synchronize properly with electroadhesion. Additionally, due to the limitation of the power source and the PZT material, two modes cannot be excited fully at the same moment. They have to share the total power and the ability of the PZT material.

For these reasons, we chose to explore the modal switching strategy. In this strategy, when the finger slides from one end of the surface to the other, crossing the boundary where the lateral force generated by the second mode starts to be greater than that by the first mode, the resonance is switched from the first mode to the second. We call this approach the "SwitchPaD". The challenge here is how to design the SwitchPaD with optimal Q factors for each resonance so that it can generate high lateral force (greater than 125 mN peak), yet switch modes without perceptual artifact.

3 Method and Experiment

3.1 SwitchPaD Design

Structure. The structure of the SwitchPaD (as shown in Fig. 1 and Fig. 2) is similar to the UltraShiver (shown in [11]), and consists of two piezoelectric disc actuators and a sheet of anodized aluminum. The dimensions of the anodized aluminum are $104 \times 22 \times 1$ mm. The two soft piezos (SMD22T25R211WL, Steminc and Martins Inc, Miami, FL, USA) are 22 mm diameter $\times$ 0.25 mm thick. They are excited in-phase with one another to provide strong coupling to the longitudinal rather than flexural modes, and they are located 26 mm (= 104 mm/4) from the edge of the aluminum plate as a "sweet spot" for exciting both modes. This location ensures that the surface can generate enough lateral velocity in both modes and also attenuate quickly during the switch. This "sweet spot" was optimized via a series of FEA simulations and experiments which are beyond the scope of this paper.

Lateral Force Generation Principle. A detailed model of force generation with the UltraShiver is given in [11]. The SwitchPaD produces lateral force in the same way: as a result of friction being greater when electroadhesion is turned high than when it is turned low, and this effect being synchronized with in-plane oscillation. The direction and magnitude of the lateral force can be adjusted by varying the phase between the oscillation and the electroadhesion. The SwitchPaD is different, however, insofar as the resonant mode is switched depending on finger position. This ensures a high oscillation velocity wherever the finger may be on the touch surface.

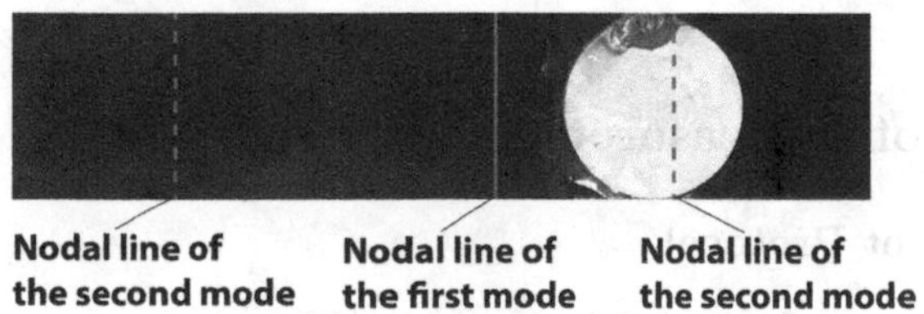

Fig. 1. Top view of the SwitchPaD.

3.2 Experiment Setup

Figure 2 shows the experiment setup, in which the SwitchPaD was mounted to mechanical ground (acrylic block) with four brass flexures, and the electrically grounded index finger of the dominant hand was constrained to move only up and down.

Two different sensors were used in this experiment setup, including a six-axis force sensor (ATI Nano 17 Force/Torque sensor) and a Laser Doppler Vibrometer

(LDV, IVS-500, Polytec, Inc). The force sensor was used to measure the lateral force on the surface and the LDV was used to measure the lateral velocity at the end of the surface (in Sect. 5.1) or at the side of the finger (in Sect. 5.2). In addition, a CCD sensor was used to track the finger position. The setup in Fig. 2 was placed on a linear rail controlled by a DC motor, which was used to move the SwitchPaD at a constant velocity (more details in Sect. 4.1 and 5.2).

The piezoelectric actuator voltage and the electroadhesive current were controlled with a custom voltage amplifier and a custom transconductance amplifier, respectively (more details were reported in [11,12]). All signals were recorded using a NI USB-6361 Multifunctional I/O Device with a 200 kHz sampling frequency.

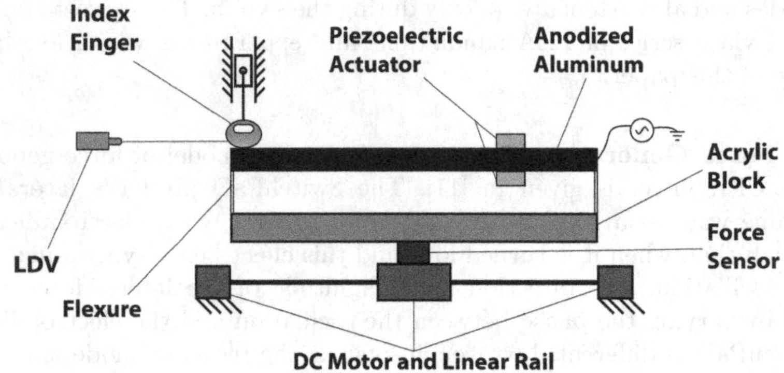

Fig. 2. Experiment platform.

4 Force Profile Measurement for Each Mode

4.1 Experiment Protocol

As a first experiment, the lateral force profile (force as a function of finger position) generated by each resonant mode was measured. The frequencies of the first and second modes were 22,390 Hz and 53,320 Hz, respectively. The input voltage of the piezoelectric actuator and electroadhesion were 30 V peak and 100 V peak with an offset of 200 V, respectively. The phase between the piezoelectric voltage and the electroadhesive voltage was set to generate peak lateral force (in Fig. 2).

During the experiments, the electrically grounded finger lightly touched the surface and kept a constant pressing force (0.3 N) as effectively as possible while the SwitchPaD was moved at 50 mm/s by the DC motor and the linear rail. The finger moved from the left end of the surface to the left side of the piezoelectric actuator. Each resonant mode was excited in one trial individually, and each trial was repeated ten times.

4.2 Results and Discussions

As shown in Fig. 4, both the first mode and the second mode can achieve around 300 mN active lateral force, but at different positions. For the first mode (blue curve in Fig. 4), the lateral force keeps a constant value (around 300 mN) from 0 mm to 25 mm and decreases to zero at 60 mm, which is consistent with the model prediction in [11]. For the second mode (red curve in Fig. 4), the lateral force is approximately constant from 25 mm to 65 mm. Thus, these results suggest that combining the first mode and the second mode is a promising method to provide a constant lateral force from 0 mm to 65 mm.

5 Resonant Mode Switch

Based on the results in Fig. 4, a switch located anywhere from 25 mm to 41 mm could be used to achieve an approximately constant lateral force (around 250 mN). In this section, the switch point was placed at 41 mm, and two different modal switching strategies were compared.

5.1 Mode Switch Strategy

When the finger slides from the left side to the right side and crosses the switch line (41 mm), the resonant mode is switched from the first mode to the second mode, vice versa. We explored two different switch strategies: instant and gradual.

For the instant switch, input voltage for the first mode is turned off at the same time input voltage for the second mode is turned on. Thus, there is only one resonant mode being driven at any instant, although there are ring-up (3.9 ms) and ring-down (6.9 ms) times for both modes due to their high Q values (first mode: 482, second mode: 651). When the LDV was used to measure the lateral velocity at the left end of the surface (in Fig. 2), a spike of the lateral velocity was found at the switch point (shown as the blue solid and dashed curves in Fig. 3(a)). The spike is due to the ring-up and ring-down times and is strongly perceived by the subject. This perceptual artifact makes the instant switch unacceptable for applications that require continuous force feedback on the surface.

To avoid the spike at the switch point, the idea of a gradual switch was investigated. In this strategy, the input voltage to the piezoelectric actuator at the first mode decreases gradually, and at the same time, the input voltage at the second mode increases gradually. This process takes around 100 ms, a time frame that was roughly optimized via simulation (see Fig. 3(b)). During this transition period, both resonant modes were excited, and electroadhesion was also operated at both frequencies with amplitude ramps that tracked those of the piezoelectric input voltages. The experiment results (red solid and dashed curves in Fig. 3(a)) show good agreement with the simulation results. More importantly, lateral force was preserved and the gradual switch could not be perceived.

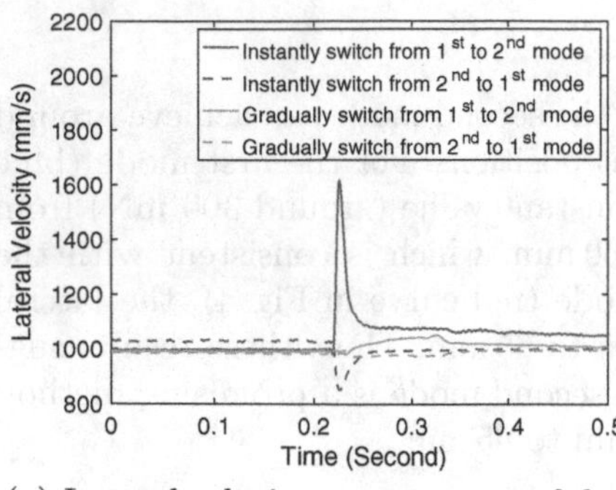

(a) Lateral velocity measurement of the instant switch and the gradual switch

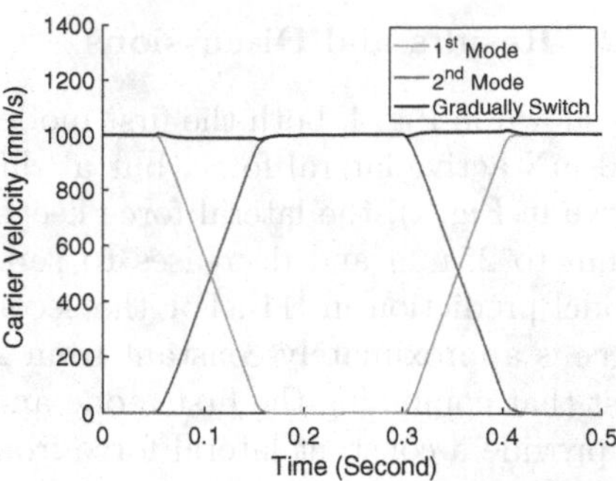

(b) Simulation of the gradual switch

Fig. 3. Lateral Velocity simulation and measurement during switches. All the lateral velocity simulations and measurements are at the left end of the surface. (Color figure online)

5.2 Experiment Protocol

This experiment is similar to that in Sect. 4.1, however, the LDV was used to measure the lateral velocity at the left side of the finger while the force sensor measured the lateral force on the surface. Experiments were performed with the two different switch strategies and repeated ten times.

5.3 Results and Discussions

The results are shown as force sensor measurements (in Fig. 4) and LDV measurements (in Fig. 5). Figure 4 shows that both the instant switch and the gradual switch can generate constant lateral force (around 250 mN) from 0 mm to 65 mm. The bandwidth of the force sensor, however, is below 50 Hz, so that it unable to detect rapid transients. LDV measurements instead were used to investigate performance up to 10 kHz.

Since the whole experiment setup was moved by a DC motor and a linear rail, which was operated via open-loop control of the velocity (more details in Sect. 4.1), the five trials in Fig. 5 do not align temporally. Figure 5(a) shows that, in every trial, there is a spike of the finger motion. This spike occurs when the finger crosses the mode switch line (41 mm). The peak velocity of this spike is around 35 mm/s with 10 ms width, which is above the human detection threshold of vibration [13,14]. In contrast to the instant switch, velocity spikes are essentially absent during the gradual switch. The lateral velocity varies within ±5 mm/s with 40 ms width, which cannot be perceived by subjects.

Thus, the instant switch strategy can only be used for specific haptic applications, such as button click rendering, in which a spatially discontinuous lateral force is acceptable. Since the gradual switch strategy can provide consistent lateral force over a large area, it can be used for more general haptic applications, such as shape rendering.

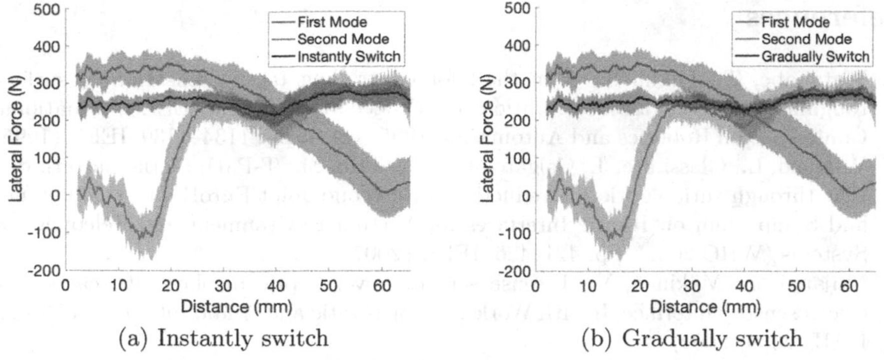

(a) Instantly switch (b) Gradually switch

Fig. 4. Lateral force measurement on the surface. The x-axis shows the distances from the left end of the surface. The solid curves and the shadows are the averages and standard deviations over ten trials at the corresponding experiment condition (the first mode, the second mode, the instant switch, and the gradual switch).

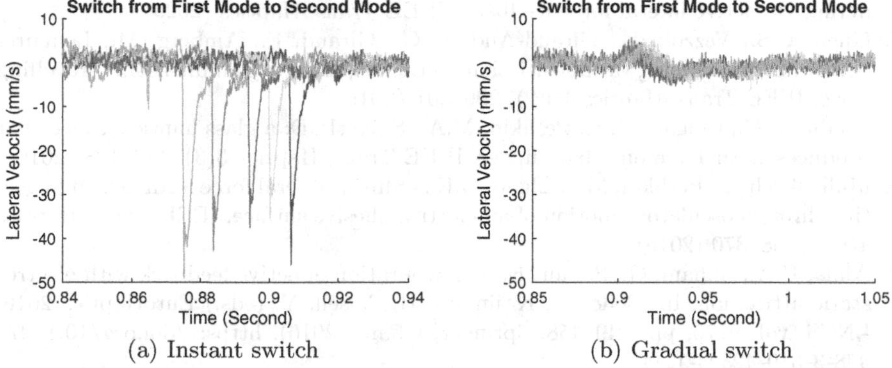

(a) Instant switch (b) Gradual switch

Fig. 5. Lateral velocity measurement at the left side of fingertip during instant switches and gradual switches. Each colored curve represents one trial. (Color figure online)

6 Conclusion

The SwitchPaD presented in this paper employs the first (22,390 Hz) and second (53,320 Hz) resonant longitudinal modes to achieve a strong active lateral force. By gradually switching between the first and second modes when the finger crosses a threshold position, the SwitchPaD can keep the lateral force consistent (250 mN peak) over a large area. This significantly improves the ability to render more general haptic effects compared to our previous device, the UltraShiver [11].

Acknowledgment. This material is based upon work supported by the National Science Foundation grants number IIS-1518602.

References

1. Watanabe, T., Fukui, S.: A method for controlling tactile sensation of surface roughness using ultrasonic vibration. In: Proceedings of 1995 IEEE International Conference on Robotics and Automation 1995, vol. 1, pp. 1134–1139. IEEE (1995)
2. Winfield, L., Glassmire, J., Colgate, J.E., Peshkin, M.: T-PaD: tactile pattern display through variable friction reduction. In: Second Joint EuroHaptics Conference and Symposium on Haptic Interfaces for Virtual Environment and Teleoperator Systems (WHC 2007), pp. 421–426. IEEE (2007)
3. Linjama, J., Mäkinen, V.: E-sense screen: novel haptic display with capacitive electrosensory interface. In: 4th Workshop for Haptic and Audio Interaction Design HAID 2009 (2009)
4. Shultz, C.D., Peshkin, M.A., Colgate, J.E.: Surface haptics via electroadhesion: expanding electrovibration with Johnsen and Rahbek. In: 2015 IEEE World Haptics Conference (WHC), pp. 57–62. IEEE (2015)
5. Osgouei, R.H., Kim, J.R., Choi, S.: Improving 3D shape recognition with electrostatic friction display. IEEE Trans. Haptics **10**(4), 533–544 (2017)
6. Xu, H., Klatzky, R.L., Peshkin, M.A., Colgate, E.: Localizable button click rendering via active lateral force feedback. IEEE Trans. Haptics (2020)
7. Ghenna, S., Vezzoli, E., Giraud-Audine, C., Giraud, F., Amberg, M., Lemaire-Semail, B.: Enhancing variable friction tactile display using an ultrasonic travelling wave. IEEE Trans. Haptics **10**(2), 296–301 (2016)
8. Chubb, E.C., Colgate, J.E., Peshkin, M.A.: ShiverPaD: a glass haptic surface that produces shear force on a bare finger. IEEE Trans. Haptics **3**(3), 189–198 (2010)
9. Mullenbach, J., Peshkin, M., Colgate, J.E.: eShiver: lateral force feedback on fingertips through oscillatory motion of an electroadhesive surface. IEEE Trans. Haptics **10**(3), 358–370 (2016)
10. Alma, U.A., Ilkhani, G., Samur, E.: On generation of active feedback with electrostatic attraction. In: Bello, F., Kajimoto, H., Visell, Y. (eds.) EuroHaptics 2016. LNCS, vol. 9775, pp. 449–458. Springer, Cham (2016). https://doi.org/10.1007/978-3-319-42324-1_44
11. Xu, H., Peshkin, M.A., Colgate, J.E.: UltraShiver: lateral force feedback on a bare fingertip via ultrasonic oscillation and electroadhesion. IEEE Trans. Haptics **12**(4), 497–507 (2019)
12. Shultz, C., Peshkin, M., Colgate, J.E.: The application of tactile, audible, and ultrasonic forces to human fingertips using broadband electroadhesion. IEEE Trans. Haptics **11**(2), 279–290 (2018)
13. Mountcastle, V.B.: The neural replication of sensory events in the somatic afferent system. In: Eccles, J.C. (ed.) Brain and Conscious Experience, vol. pp. 85–115. Springer, Heidelberg (1965). https://doi.org/10.1007/978-3-642-49168-9_4
14. Kandel, E.R., Schwartz, J.H., Jessell, T.M., Siegelbaum, S., Hudspeth, A.J.: Principles of Neural Science, vol. 4. McGraw-hill, New York (2000)

BY

Visuo-Haptic Display by Embedding Imperceptible Spatial Haptic Information into Projected Images

Yamato Miyatake[1], Takefumi Hiraki[1], Tomosuke Maeda[2], Daisuke Iwai[1](✉), and Kosuke Sato[1]

[1] Osaka University, Toyonaka, Osaka, Japan
{miyatake,hiraki,iwai,sato}@sens.sys.es.osaka-u.ac.jp
[2] Toyota Central R&D Labs., Inc., Nagakute, Aichi, Japan
tmaeda@mosk.tytlabs.co.jp

Abstract. Visuo-haptic augmented reality (AR) systems that represent visual and haptic sensations in a spatially and temporally consistent manner are used to improve the reality in AR applications. However, existing visual displays either cover the user's field-of-view or are limited to flat panels. In the present paper, we propose a novel projection-based AR system that can present consistent visuo-haptic sensations on a non-planar physical surface without inserting any visual display devices between a user and the surface. The core technical contribution is controlling wearable haptic displays using a pixel-level visible light communication projector. The projection system can embed spatial haptic information into each pixel, and the haptic displays vibrate according to the detected pixel information. We confirm that the proposed system can display visuo-haptic information with pixel-precise alignment with a delay of 85 ms. We can also employ the proposed system as a novel experimental platform to clarify the spatio-temporal perceptual characteristics of visual and haptic sensations. As a result of the conducted user studies, we revealed that the noticeable thresholds of visual-haptic asynchrony were about 100 ms (temporal) and 10 mm (spatial), respectively.

Keywords: Visuo-haptic display · High-speed projection

1 Introduction

Visuo-haptic displays that are used to provide visual and tactile sensations to users and maintain these two sensations consistent, both spatially and temporally, can promote natural and efficient user interaction in augmented reality

This work was supported by JST ACT-X Grant Number JPMJAX190O and JSPS KAKENHI Grant Number JP15H05925.

Electronic supplementary material The online version of this chapter (https://doi.org/10.1007/978-3-030-58147-3_25) contains supplementary material, which is available to authorized users.

I. Nisky et al. (Eds.): EuroHaptics 2020, LNCS 12272, pp. 226–234, 2020.
https://doi.org/10.1007/978-3-030-58147-3_25

(AR) applications. To facilitate effective visuo-haptic AR experiences, it is essential to ensure the spatial and temporal consistency of the visual and haptic sensations. Conventional systems allowed achieving the consistency by using optical combiners such as a half-mirror [10] or video see-through systems including a head-mounted display [2] to overlay visual information onto a haptic device. However, these systems require inserting visual display devices between a user and the haptic device, which constrains the field-of-view (FOV) and interaction space, and potentially deteriorates the user experiences. They also prevent multi-user interactions. A possible solution to this problem is to integrate a tactile panel into a flat panel display [1]. This enables a spatiotemporally consistent visuo-haptic display to facilitate multi-user interactions without covering their FOVs. However, such displays are limited to flat surfaces at the moment.

Another promising approach of visuo-haptic AR display to overcome the above-mentioned limitations is to combine a projection-based AR system for displaying visual information and a haptic display attached on a user's finger which is controlled by the luminance of each projected pixel [6,9]. This approach can be used to maintain the temporal and spatial consistency between the visual and haptic sensations while not being limited to flat surfaces owing to the projection mapping technology. However, potentially, the displayed image quality may be significantly degraded, as the luminance of the original image needs to be spatially modulated depending on the desired haptic information.

In this paper, we propose a visuo-haptic display based on the projection-based AR approach to provide both visual image and haptic control information. The proposed system controls a haptic display attached to a user's finger using temporal brightness information imperceptibly embedded in projected images using pixel-level visible light communication (PVLC) [4]. The embedded information varies with each pixel. We embed the temporal brightness pattern in a short period of each projector frame so that the modulation does not significantly affect the perceived luminance of the original projection image. Owing to the short and simple temporal pattern, the haptic feedback is presented with an unnoticeably short latency. We can design a visuo-haptic display with various surface shapes as the projection mapping technique can overlay images onto a non-planar physical surface. Multiple users can experience the system in which no visual display device needs to be inserted between the users and the surface.

We develop a prototype system comprising a high-speed projector that embeds spatially-varying haptic information into visual images based on VLC principle and a haptic display device that changes vibrations according to the obtained information. Through a system evaluation, we confirm if the proposed system can consistently represent visuo-haptic sensations. We can also use the system as a novel experimental platform to clarify the spatio-temporal perceptual characteristics of visual-haptic sensations. A user study is conducted to investigate whether the delay time and misalignment of visual-haptic asynchrony are within an acceptable range for user experiences.

2 Methods and Implementation

The proposed visuo-haptic AR display can represent haptic sensations corresponding to projected images when users touch and move the haptic display device on a projection surface. The system keeps the consistency of time and position between the visual and haptic sensations at a pixel level. Figure 1 shows the concept of the proposed system. The system comprises a projection system that can embed imperceptible information in each pixel of images and a haptic display device that can control a vibration.

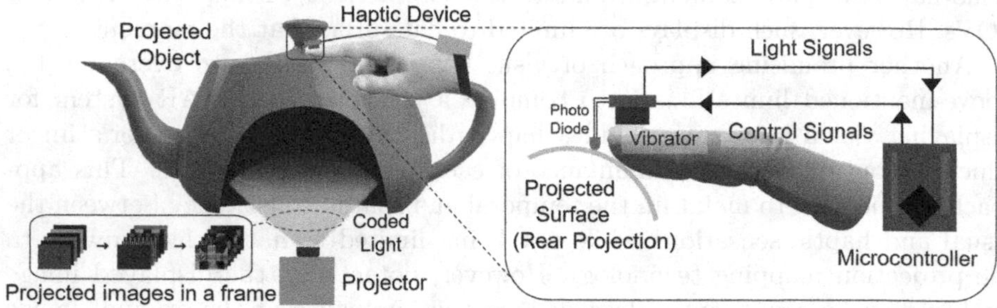

Fig. 1. Concept of the proposed system

2.1 Projection System

We utilize PVLC [4] for embedding haptic information into projected images using a DLP projector. When a projector projects an original image and its complement alternately at a high frequency, human eyes see only a uniform gray image owing to the perception characteristics of vision. Although human eyes cannot distinguish this imperceptible flicker, a photosensor can detect it and use it as signal information.

We employed a high-speed DLP projector development kit (DLP LightCrafter 4500, Texas Instruments) to project images using PVLC. We can control the projection of binary images using the specified software of the development kit. Each video frame consists of two segments. The first segment consists of 36 binary images that correspond to each bit of synchronization information and data for controlling haptic displays and takes 8.5 ms for projection. The second segment displays a full-color image for humans, which also compensates for the luminance nonuniformity caused in the first segment and takes 12 ms for projection. Thus, the time for projection in a frame is 20.5 ms, which means the frame-rate is 49 Hz. We embedded the 26 bits data on x and y coordinates ($x = 10$ bits, $y = 11$ bits) and the index number of the vibration information (5 bits) corresponding to a projected texture image using PVLC.

2.2 Haptic Display Device Controlled by PVLC

We developed a wearable haptic display controlled by PVLC. It comprises a receiver circuit with a photodiode (S2506-02, Hamamatsu Photonics), a controller circuit, a vibration actuator, and a Li-Po battery. The controller circuit has a microcontroller (Nucleo STM32F303K8, STMicroelectronics) and an audio module (DFR0534, DFRobot) for playing the audio corresponding to the vibration. The microcontroller is used to acquire the position and spatial haptic information by decoding the received signals to determine a type of vibration, and send it to the audio module that drives the vibration actuator. We use the linear resonant actuator (HAPTIC™ Reactor, ALPS ALPINE) as a vibration actuator. This actuator responds fast and has good frequency characteristics over the usable frequency band for haptic sensation. Therefore, the proposed haptic display device (hereinafter referred to as "Device HR") can present various haptic sensations.

Figure 2 provides an overview of the proposed system and the appearance of its user interface. We employed the data obtained from the LMT haptic texture database [8] as a source of projected images and the spatial haptic information. This database provides image files with textures and audio files with corresponding vibrations. We stored the vibration information in the audio module of Device HR in advance, and the device presents haptic feedback by playing the received index number of vibration information.

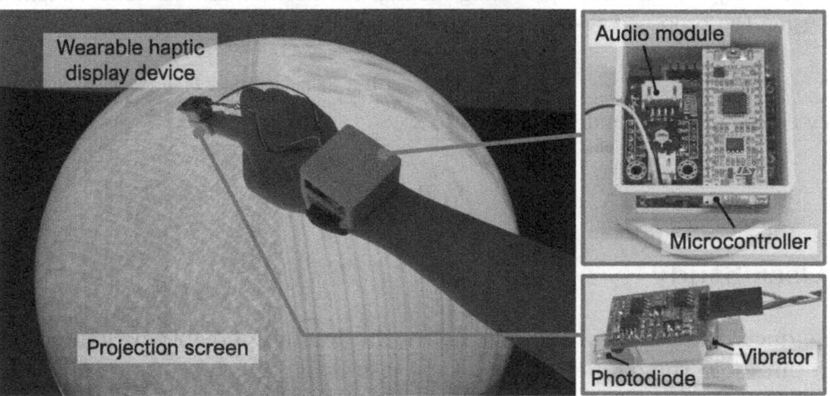

Fig. 2. Overview of the proposed system—the system comprises a projection system with a screen and a haptic display device controlled by the obtained light information

2.3 Latency Evaluation

The latency of the proposed system is defined as the duration from the time when the haptic display device is placed inside an area to the time when the device performs vibration. This latency (T_{late}) can be calculated as follows:

$$T_{late} = T_{wait} + T_{recv} + T_{vib} \quad (1)$$

where T_{wait} is the waiting time for synchronization, T_{recv} is the time to receive the data, and T_{vib} is the time to perform vibration using the actuator. According to the estimation of the previous work [3] and settings of the proposed system, we calculate T_{recv} equal to 8.5 ms and T_{wait} equal to 10.2 ms.

We measure T_{vib} by calculating the time from the moment when the microcontroller sends a control signal to the actuator of the two devices to the moment when the actuator is enabled. We project the sample image with the control information embedded for turning on the actuator in the left half of a projected image and that for turning off in the right half. We attach an acceleration sensor (KXR94-2050, Kionix) to the devices to detect vibration. We place the device at each on and off areas on the screen and conduct the measurement a 100 times using the microcontroller at each boundary of the area. As a result, the averaged values of T_{vib} are 66.5 ms for the Device HR.

Table 1 shows the values of T_{wait}, T_{recv}, T_{vib}, and T_{late} corresponding to each device. T_{vib} of the Device HR is a sum of latency values of the audio module and that of the actuator (HAPTIC™ Reactor) itself. We measure the latency of the audio module 100 times and the average value was 48.8 ms. Therefore, we can estimate that the latency of the HAPTIC™ Reactor is about $66.5 - 48.8 = 17.7$ ms.

Table 1. Delay time between providing the haptic information and visual information when using the proposed system—the evaluation was performed for Device HR

	T_{wait} [ms]	T_{recv} [ms]	T_{vib} [ms]	T_{late} [ms]
Device HR (with an audio module)	10.2	8.5	66.5	85.2

3 User Study

We conducted a user study to investigate the human perception characteristics of the threshold time of perception of the visual-haptic asynchrony and the misalignment tolerance of the visual-haptic registration accuracy of the proposed system. According to the previous studies, this threshold time was approximately 100 ms [7], and this misalignment tolerance was approximately 2 mm [5]. However, we could not simply apply these values to the proposed system. The visuo-haptic displays of the previous studies covered the user's view by visual displays from the user's fingers to which the haptic feedback was provided, while our system allows the user to see the finger directly. In the present user study, we performed the two experiments using the proposed system. The first experiment was focused on the evaluation of the threshold time of perception of a visual-haptic asynchrony, and the second was aimed to estimate the misalignment tolerance of the visual-haptic registration accuracy.

3.1 Setup for User Study

Figure 3 represents the experimental situation of the user study. We employed a tabletop display in which the proposed projection system was built-in. The screen size of the tabletop display was 0.77 m × 0.48 m, and its height was 0.92 m. We embedded the data of vibration control and the delay time into a projected image in each of the green and red areas separated by numbers. The haptic device turned on the vibration in the green area after the set delay time and turned off in the red area. We set the width of the moving area equal to 182 pixels in the projected image, and the resolution of the projected image was 1.22 pixels/mm in this setup; therefore, the actual width of the moving area was approximately 150 mm.

We implemented an alternative haptic display device for this study, which can present vibrations faster than Device HR by function limitation; this means it focuses on on/off of a constant vibration without the audio module. We used another actuator (LD14-002, Nidec Copal) in the device (hereinafter denoted as "Device LD"), and revealed the latency of Device LD is 34.6 ms by the same latency evaluation. Given that we could set the waiting time for sending a control signal to an actuator using the embedded data, the system was able to provide haptic feedback in the specific delay time (≥34.6 ms).

Fig. 3. Experimental situations considered in the user study, (a) appearance of the experiment related to the threshold time of perception of visual-haptic asynchrony, (b) appearance of the experiment related to the visual-haptic registration accuracy (Color figure online)

3.2 Participants and Experimental Methods

Ten participants (seven males and three females, aged from 21 to 24, all right-handed) volunteered to participate in the present user study. They were equipped with a haptic device on the index finger of their dominant hand. In the first experiment, we prepared 12 variations of the delay time from 50 to 160 ms at intervals of 10 ms (these delay times included the device-dependent delay time). The participants were instructed to move their index fingers from a red to a green area and answer whether they noticed the delay corresponding to the haptic sensation with respect to the moment when they visually identified the finger entering the green area. In the second experiment, we prepared 12 variations of misalignments from 0 (0 mm) to 26 px (21.32 mm). The participants were instructed to move their index fingers between the white boundary lines within each of the red and green areas as many times as they wanted. Then, they answered whether the timing of haptic feedback matched with crossing the red and green boundaries of the projected image. To influence the moving speed of the user's fingers, we displayed a reference movie in which a user was moving his/her finger at the speed of 150 mm/s during the experiment. The participants performed each procedure 12 times as the projected image had 12 areas in both the experiments. We defined the 10 different patterns of interconnections between providing haptic sensations with a delay time or a misalignment and the areas to cancel the order effects. The participants experienced these patterns in each experiment, thereby repeated each procedure 120 times in total.

3.3 Results and Discussion

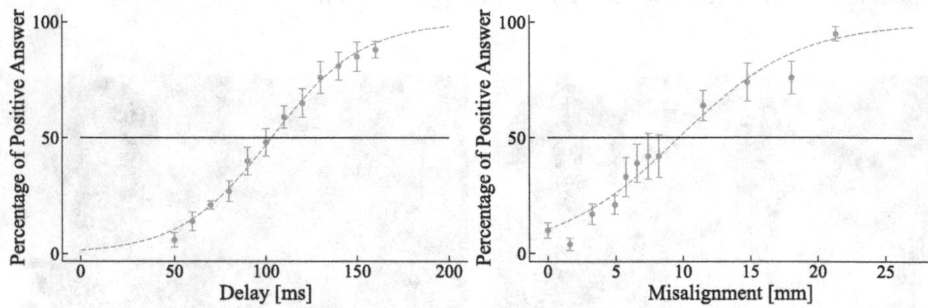

Fig. 4. Percentages of positive answers in the experiment for a threshold time of a visual-haptic asynchrony (left) and that for a visual-haptic regisration accuracy (right).

Figure 4 represents the averaged percentage of positive answers obtained in the first experiment. Herein, error bars represent the standard error of the means, and the curve is fitted using a sigmoid function defined as below:

$$y = \frac{1}{1 + \exp(-k(x - x_0))} \times 100 \tag{2}$$

In Figure 4 (left), we obtained the following values of parameters: $k = 0.04$ and $x_0 = 103$, as a result of the fitting and used these values in calculations. We identified the threshold time of perception of the visual-haptic asynchrony (T_{th}); It was set as the time at which users perceive the delay with a 50% possibility. The results indicated $T_{th} \approx 100$ ms, as presented in the fitted sigmoid curve. Similar results of T_{th} were reported in the previous research [7], that supported the result of this experiment. As the latency of Device HR is 85.2 ms, the proposed haptic displays meet the requirement of having the delay time such that users cannot perceive visual-haptic asynchrony with their eyes.

In Fig. 4 (right), we obtained the values of parameters: $k = 0.22$ and $x_0 = 9.7$, as a result of the fitting and used these values as parameters of the sigmoid function. We identified the misalignment tolerance of the visual-haptic registration accuracy (denoted as L_{th}) equal to the length perceived as a misalignment by half of the users. The results indicated $L_{th} \approx 10$ mm, as presented in the fitted sigmoid curve. The result indicates that the proposed system can provide haptic feedback to users without a perception of misalignment if it is kept within 10 mm, which can be considered as design criteria for visuo-haptic displays. Additionally, L_{th} (10 mm) is larger than the value (2 mm) reported in the previous research [5]. We can also conclude that the proposed system can extend the misalignment tolerance of the visual-haptic registration accuracy as the users can visually observe their finger with the haptic display in the system.

4 Conclusion

In the present paper, we proposed a novel visuo-haptic AR display that allowed eliminating visual-haptic asynchrony of the time and position perceived by the users. We implemented the projection system that could embed information into each pixel of images and the haptic display device that could control vibrations based on the obtained information. We conducted the user studies and revealed that the threshold of visual-haptic asynchrony obtained using the proposed visual-haptic display was about 100 ms for the time delay and about 10 mm for the position. From this result, we can conclude that the proposed display device can represent visual and haptic sensations in a synchronized manner as the system can represent them with the pixel-precise alignment at the delay of 85 ms. As future work, we will conduct similar user studies with various directions of the hand moving and with more participants to investigate the systematic thresholds. Furthermore, we will design a model to determine the vibration intensity and frequency of the haptic display based on haptic information corresponding to the texture image and the user's movements on a display.

References

1. Bau, O., et al.: TeslaTouch: electrovibration for touch surfaces. In: Proceedings of the UIST 2010, pp. 283–292 (2010)
2. Harders, M., et al.: Calibration, registration, and synchronization for high precision augmented reality haptics. IEEE TVCG **15**(1), 138–149 (2009)
3. Hiraki, T., et al.: Sensible Shadow: tactile feedback from your own shadow. In: Proceedings of the AH 2016, pp. 23:1–23:4 (2016)
4. Kimura, S., et al.: PVLC projector: image projection with imperceptible pixel-level metadata. In: Proceedings of the ACM SIGGRAPH 2008 Posters, p. 135:1, August 2008

5. Lee, C.-G., Oakley, I., Ryu, J.: Exploring the impact of visual-haptic registration accuracy in augmented reality. In: Isokoski, P., Springare, J. (eds.) EuroHaptics 2012. LNCS, vol. 7283, pp. 85–90. Springer, Heidelberg (2012). https://doi.org/10.1007/978-3-642-31404-9_15
6. Rekimoto, J.: SenseableRays: opto-haptic substitution for touch-enhanced interactive spaces. In: Proceedings of the CHI EA 2009, pp. 2519–2528, April 2009
7. Silva, J.M., et al.: Human perception of haptic-to-video and haptic-to-audio skew in multimedia applications. ACM TOMM **9**(2), 9:1–9:16 (2013)
8. Strese, M., et al.: A haptic texture database for tool-mediated texture recognition and classification. In: Proceedings of the HAVE 2014, pp. 118–123, November 2014
9. Uematsu, H., et al.: HALUX: projection-based interactive skin for digital sports. In: Proceedings of the ACM SIGGRAPH 2016 Emerging Technologies, pp. 10:1–10:2 (2016)
10. Wang, D., et al.: Analysis of registration accuracy for collocated haptic-visual display system. In: Proceedings of the HAPTICS 2008, pp. 303–310 (2008)

Manipulating the Perceived Directions of Wind by Visuo-Audio-Haptic Cross-Modal Effects

Kenichi Ito(✉), Yuki Ban, and Shin'ichi Warisawa

Graduate School of Frontier Sciences, The University of Tokyo, Chiba 2770882, Japan
itokenichi@lelab.t.u-tokyo.ac.jp, {ban,warisawa}@edu.k.u-tokyo.ac.jp

Abstract. Wind displays, which simulate the sensation of wind, have been known to enhance the immersion of virtual reality content. However, certain wind displays require an excessive number of wind sources to simulate wind from various directions. To realize wind displays with fewer wind sources, a method to manipulate the perceived directions of wind by audio-haptic cross-modal effects was proposed in our previous study. As the visuo-haptic cross-modal effect on perceived wind directions has not yet been quantitatively investigated, this study focuses on the effect of visual stimuli on the perception of wind direction. We present virtual images of flowing particles and three-dimensional sounds of wind as information to indicate wind directions and induce cross-modal effects in users. The user study has demonstrated that adding visual stimuli effectively improved the result corresponding to certain virtual wind directions. Our results suggest that perceived wind directions can be manipulated by both visuo-haptic and audio-haptic cross-modal effects.

Keywords: Wind display · Wind perception · Cross-modal.

1 Introduction

Improving immersion is necessary for most virtual reality (VR) content. An approach to ensure an immersive VR experience involves multisensory presentation. In this context, "wind displays" that simulate the sensation of wind for their users have become a popular topic of study. Heilig used wind along with odors and vibrations in Sensorama [1]. Moon et al. proposed WindCube [9], which simulates wind from several directions using 20 fans.

As many people experience the sensation of wind on their entire body every day, we can easily immerse ourselves in VR presentations with wind displays. The latter can improve immersion by faithfully reproducing the motion of objects in VR environments [5], self-motions [6,12], and climates [11].

Electronic supplementary material The online version of this chapter (https://doi.org/10.1007/978-3-030-58147-3_26) contains supplementary material, which is available to authorized users.

I. Nisky et al. (Eds.): EuroHaptics 2020, LNCS 12272, pp. 235–243, 2020.
https://doi.org/10.1007/978-3-030-58147-3_26

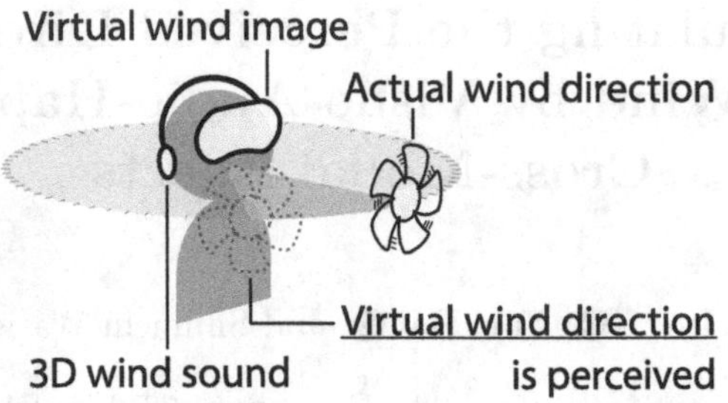

Fig. 1. Manipulation of perceived wind directions by multimodal stimuli

The reproduction of wind blowing from various directions is often crucial to simulate realistic wind. An array of fans [9] is the easiest and most applicable approach to this problem. However, the number of wind sources must match the number of desired directions of simulated wind. Therefore, wind display devices tend to be complicated and large under this implementation. VaiR [12] addresses this problem by implementing two rotatable bow-shaped frames that enable continuous change in wind directions. This approach could reduce the required number of wind sources, but requires actuators and mechanisms to move the device. We propose the presentation method of wind directions without an entire reproduction of physical wind by changing human perception.

Human perception of the directions of wind is investigated to design effective wind displays. Nakano et al. [10] demonstrated that the angles with respect to the human head corresponding to just noticeable differences (JND) in wind directions are approximately 4° in the front and rear regions and approximately 11° in the lateral region. Saito et al. [13] investigated the wind JND angles by presenting users with audio-visual stimuli. They reported that the JND angle values were much higher than those reported by Nakano et al. and suggested that the accuracy of wind perception was lowered by multisensory stimuli.

When we receive multisensory stimuli, different sensations are sometimes integrated with each other and our perception is altered. These phenomena are called cross-modal effects and they can alter the perception of physical stimuli. It has already been established that haptic sensations are altered by the visuo-haptic [7] and audio-haptic integrations [4]. Through cross-modal effects, we can provide rich tactile experiences without reproducing the stimulating physical phenomena completely faithfully.

We proposed a method to manipulate perceived wind directions by audio-haptic cross-modal effect [2] in order to simulate directional winds with simple hardware. We performed experiments that simultaneously presented wind from two fans and three-dimensional (3D) wind sounds, and concluded that the perceived wind directions could be changed by up to 67.12° by this effect.

It is suggested that congruent stimuli from two modalities strengthen the effect of cross-modal illusion on an incongruent stimulus from the other modality in tri-modal perception [14]. Therefore, we designed AlteredWind [3], which combines congruent visual and audio information about the wind direction to more effectively manipulate the perceived wind. We presented the audio-visual information through a head-mounted display (HMD) and headphones, as depicted in Fig. 1. Although we evaluated the perceived wind directions in a user study, there was only qualitative analysis and the sample size was small. In this study, we redesigned the experiment to quantitatively verify the visuo-audio-haptic cross-modal effects on wind direction perception. Our experiment compares the perceived wind directions across the combinations of different sensory modalities and placements of wind sources. Our result suggests new designs of wind displays utilizing cross-modal effects.

2 Implementing Visual, Audio, and Wind Presentation

In this paper, we define "virtual wind direction" as the direction of wind which is presented by multimodal stimuli and is different from the physical one. We implemented software for visual and audio presentations of the virtual wind directions and hardware for actual wind presentation.

2.1 Visual Presentation of Virtual Wind Directions

The virtual wind direction can be visually conveyed to a user by two primary methods—by suggesting the generation of wind or the existence of wind. The former effect can be realized by displaying a virtual image of a rotating fan. However, the wind approaching from behind the user cannot be expressed by this technique because the images would lie outside the field of vision. The latter technique uses images of particles being blown by the wind and images of flags or plants swaying in the wind. This technique can be used to convey wind originating from any direction. In this study, images of particles moving horizontally were used to verify the effects of visual information in a simple environment. We used an HMD (HTC Vive Pro) to display the image three-dimensionally and immersively. We programmed 1000 particles to emerge per second so that they were visible in the HMD along any flowing direction, as depicted in Fig. 2.

2.2 Audio Presentation of Virtual Wind Directions

We manipulated the perceived directions of wind by using 3D sounds recorded by a dummy head [2], which is a life-sized model of the human head with ears. The perception of a sound source can be localized around listeners when recorded sound is played binaurally. We installed a dummy head and a fan in an anechoic room and recorded the sound of the fan blowing wind against the dummy head from directions 30° apart from each other [2]. The sound was presented to users through noise-cancelling headphones (SONY WH-1000XM2). A-weighted sound pressure level (L_A) near the headphones measured with a sound level meter (ONO SOKKI LA-4350) was in the range of 46.8 dB–58.1 dB (it varied depending on the direction of the sound image). The direction of the audio information was always congruent with that of the visual one.

(a) Approaching from the front (b) Approaching from the right front

Fig. 2. Examples of the images of flowing particles

Fig. 3. A Plot of wind velocity

2.3 Proposed Device for Wind Presentation

We had designed a wind display device by placing a fan in front of the users and another behind them in our previous study [2]. In this experiment, we placed four fans (SANYODENKI San ACE 172) around the users' heads at 90-degree intervals. Each fan could be independently controlled by Arduino UNO connected to a computer. During the experiment, wind was generated from the fans at the front and the back or from those to the left and right. The four fans were affixed to camera monopods on a circular rail of 800 mm diameter and directed to the users' heads in about 300 mm ahead. In our previous study, results may have been affected by the participants' prior knowledge of the positions of fans. To ensure that the participants were not aware of the placement of the fans, we made the monopods detachable from the circular rail. The fans and the monopods were removed from the rail and hidden before the experiment and attached to the rail after the participants had worn the HMD.

We had continuously controlled the wind velocities corresponding to the two fans facing each other in the previous research [2] and confirmed that it was effective for manipulation of the perceived wind direction. "Continuously" here means that the wind from a particular fan was weakened as it moved further away from the virtual wind direction. We applied the same method in this study also and changed the wind velocities from 0.8 m/s to 2.0 m/s as shown in Fig. 3.

3 Experiment Regarding Perceived Wind Directions

3.1 Experiment Design

We conducted an experiment to verify the visuo-audio-haptic cross-modal effects on perceived wind directions. We presented the virtual wind directions which are not congruent with the actual wind directions by visual and audio information in this experiment. Six conditions were prepared by varying the existence of multimodal information (visual and audio, visual only, and audio only) and the placements of the fans (front-behind and left-right). Since we confirmed that the actual directions of wind are perceived without multimodal information [2], we omitted that condition in this user study. The experiment had a within-subjects design. For each condition, we presented wind coming from 12 virtual wind directions at intervals of 30° and each direction was repeated twice. Thus, there were a total of 144 presentations for each participant.

The Ethics Committee of the University of Tokyo approved the experiment (No. 19–170). Written informed consent was obtained from every participant. Each participant was instructed to sit in a chair and wear the HMD. As mentioned in Subsect. 2.3, the fans were attached after the participants wore the HMD, preventing them from being aware of the exact location (Fig. 4). To make the wind apparent on the head and neck of the participants, the lower ends of the fans were adjusted to match the height of the participants' shoulders. We asked the participants to stare toward the frontal direction marked in the VR view.

Each presentation consisted of a stimulation time of 12 s and an answering time of a few seconds. Considering the delay in starting and stopping the fans, the timing of wind presentation was advanced by 2 s compared with that of the other stimuli. The order of the presentations was randomized for each participant, and neither the participants nor the experimental staff was aware of the order beforehand. The participants responded with perceived wind direction of wind indicating the direction on the trackpad of a controller as shown in Fig. 5. Finally, they answered questionnaires about their experiences during the experiment.

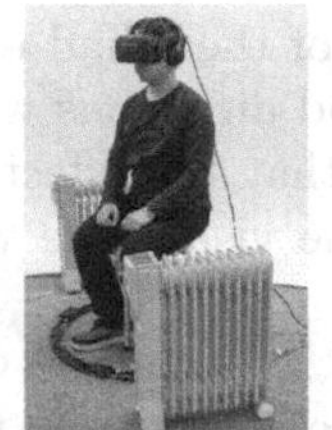

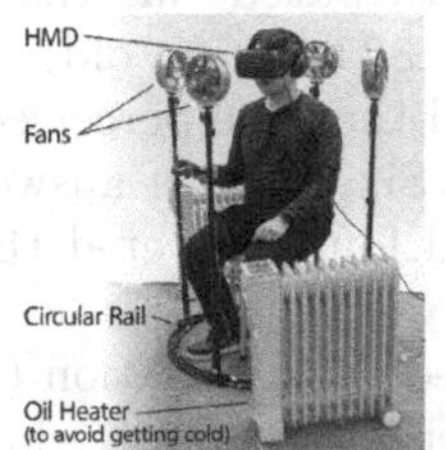

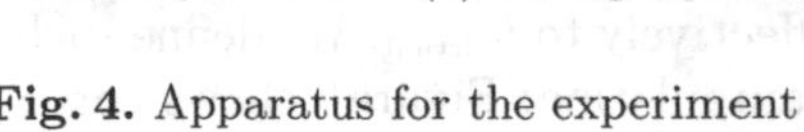
(a) Before experiment (b) During experiment

Fig. 4. Apparatus for the experiment

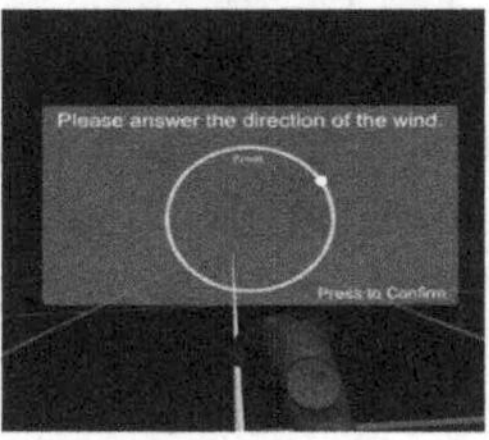

Fig. 5. Interface for answering directions

As the participants wore the HMD and the noise-cancelling headphones, fan operations could not be seen or heard by them. All the windows of the experiment room were closed and the air conditioner was turned off to ensure that there was no wind except the wind from the apparatus. Instead of air conditioning, two oil heaters which produce no airflow were used for warming. The apparatus was at least 1 m away from the walls so that they would not affect the airflow.

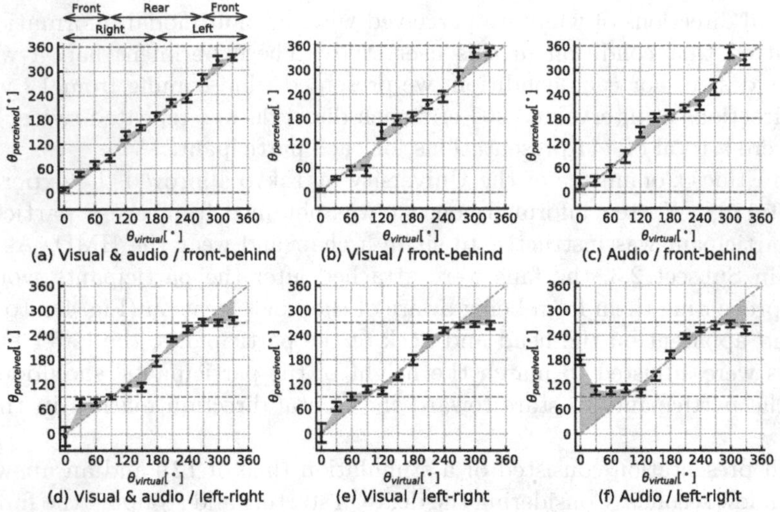

Fig. 6. Plots of the perceived wind directions ($\theta_{\text{perceived}} \pm$ standard error) and the virtual wind directions (θ_{virtual}). Diagonal dashed lines represents $\theta_{\text{virtual}} = \theta_{\text{perceived}}$. Horizontal dashed lines represents the actual directions of the fans.

3.2 Results

Twelve people of ages 22–49 participated in the experiment (7 men 23.9 ± 1.2 years old and 5 women 30.0 ± 11.0 years old). Two of them had ever researched haptics. Ten persons sensed that the perceived wind directions were affected by visual and audio stimuli. Six persons answered that the effect of the auditory stimuli was stronger and four answered that the visual one was stronger.

We calculated the average perceived wind direction ($\theta_{\text{perceived}}$) for every virtual wind direction (θ_{virtual}). The directions represent clockwise angles from the front. If $\theta_{\text{perceived}}$ was close to corresponding θ_{virtual}, we can judge that the perceived wind directions are manipulated effectively to θ_{virtual}. We define such conditions as "good performance" of the manipulation. Figure 6 shows correspondence between $\theta_{\text{perceived}}$ and θ_{virtual}. Data points near diagonal lines mean the good performance. Ones near horizontal lines mean a bad performance because it means that the actual wind directions are perceived instead of θ_{virtual}.

We performed statistical tests following the methods of directional statistics [8] for $\theta_{\text{perceived}}$ under three conditions corresponding to each fan placement. The Mardia–Watson–Wheeler test, which is a non-parametric test for two or more samples, was applied because Watson's U^2 test showed that some of the perceived directions do not arise from von Mises distribution. For directions with significant differences, post-hoc Mardia-Watson-Wheeler tests, with Hommel's improved Bonferroni procedure, were performed. The results have been tabulated in Table 1.

Table 1. Results of statistical tests on $\theta_{perceived}$ for each $\theta_{virtual}$. If there is a significant difference, the condition with $\theta_{perceived}$ closer to $\theta_{virtual}$ is indicated.

Fans	Pair	0°	30°	60°	90°	120°	150°	180°	210°	240°	270°	300°	330°
Front and Behind	All	*	*	*	*	n.s	†	n.s	n.s	*	n.s	n.s	n.s
	VA and V	n.s	n.s	n.s	VA*		n.s			n.s			
	VA and A	n.s	A*	A*	A*		VA†			VA†			
	V and A	V*	A*	n.s	n.s		n.s			n.s			
Left and Right	All	*	*	n.s	*	†	*	n.s	n.s	n.s	n.s	n.s	n.s
	VA and V	n.s	n.s		n.s	n.s	n.s						
	VA and A	n.s	n.s		VA*	VA*	A†						
	V and A	V†	V*		n.s	n.s	n.s						

VA: visual and audio, V: visual, A: audio, *: $p < .05$, †: $p < .1$, n.s.: $p \geq .1$

3.3 Discussion

We abbreviate visual and audio as VA, visual as V, and audio as A in the following discussion. Under the condition of the front-behind fans, the condition VA caused better performance of manipulation than the condition V when $\theta_{virtual}$ was 90°. The condition VA was marginally better than the condition A when ($\theta_{virtual} = 150, 240°$). On the other hand, the performance was better under the condition A than under the condition VA or V for several $\theta_{virtual}$. Under the condition of the left-right fans, the condition VA had a significantly ($\theta_{virtual} = 90°$) or marginally ($\theta_{virtual} = 120°$) better performance than the condition A. It was confirmed that the condition V was significantly ($\theta_{virtual} = 30°$) or marginally ($\theta_{virtual} = 0°$) better than the condition A. From these results, the combination of visual and auditory stimuli and the fans in a front-back position has shown overall better performance in the manipulation of the perceived wind directions.

Still, the performance of manipulation with the condition V or A was better than ones with the condition VA in several virtual wind directions. In these directions, the performance was already sufficient (the differences of $\theta_{perceived}$ and $\theta_{virtual}$ were at most 12°) with the condition V or A. Therefore, we conclude that the manipulation of the perceived wind directions is improved by using visuo-audio-haptic cross-modal effects under the condition that the perception cannot be sufficiently changed by using only visual or auditory stimuli.

For the left-right fan placement, the performance was worse than in front-behind one in condition VA and A, and there were little differences between condition VA and V. The reason may be that the participants did not localize correct sound images due to front-back confusions. Improvement of 3D sounds may enhance the cross-modal effects under left-right placements of fans.

Using more realistic images than the simple flowing particles could increase the effectiveness of visual stimuli. Further, the image of the particles may disturb the contents in practical applications. If the manipulation of the perceived wind can be realized with more diegetic visual information such as flags or swaying trees, it may be effectively used in practical applications using the wind display.

4 Conclusion

In this study, we proposed a method to manipulate the perceived directions of wind through visuo-audio-haptic cross-modal effects. We used virtual images of flowing particles as visual stimuli and 3D sounds of the wind as audio stimuli to make users perceive virtual wind directions. The user study demonstrated that combining visual and audio stimuli exerted significant effects than each stimulus alone corresponding to certain virtual directions. Combining the two modalities are considered to be effective when the perception cannot be sufficiently manipulated by only visual or auditory stimuli.

These results suggest that perceived wind directions can be altered by both visuo-haptic and audio-haptic cross-modal effects. Further improvements of the visual and audio stimuli used for manipulation should be considered in the future. The findings of this study can be applied to wind display technology to present various wind directions with limited equipment.

References

1. Heilig, M.L.: Sensorama simulator. US Patent 3050870, August 1962
2. Ito, K., Ban, Y., Warisawa, S.: Manipulation of the perceived direction of wind by cross-modal effects of wind and three-dimensional sound. In: 2019 IEEE World Haptics Conference (WHC), pp. 622–627, July 2019. https://doi.org/10.1109/WHC.2019.8816111
3. Ito, K., Ban, Y., Warisawa, S.: AlteredWind: manipulating perceived direction of the wind by cross-modal presentation of visual, audio and wind stimuli. In: SIGGRAPH Asia 2019 Emerging Technologies SA 2019, pp. 3–4. ACM, New York (2019). https://doi.org/10.1145/3355049.3360525
4. Jousmäki, V., Hari, R.: Parchment-skin illusion: sound-biased touch. Curr. Biol. **8**(6), R190–R191 (1998). https://doi.org/10.1016/S0960-9822(98)70120-4
5. Kojima, Y., Hashimoto, Y., Kajimoto, H.: A novel wearable device to present localized sensation of wind. In: Proceedings of the International Conference on Advances in Computer Enterntainment Technology ACE 2009, pp. 61–65. ACM, New York (2009). https://doi.org/10.1145/1690388.1690399
6. Kulkarni, S.D., Fisher, C.J., et al.: A full body steerable wind display for a locomotion interface. IEEE Trans. Vis. Comput. Graph. **21**(10), 1146–1159 (2015). https://doi.org/10.1109/TVCG.2015.2424862
7. Lécuyer, A.: Simulating haptic feedback using vision: a survey of research and applications of pseudo-haptic feedback. Presence: Teleoper. Virtual Environ. **18**(1), 39–53 (2009). https://doi.org/10.1162/pres.18.1.39
8. Mardia, K.V., Jupp, P.E.: Directional Statistics. Wiley, Hoboken (2000)
9. Moon, T., Kim, G.J.: Design and evaluation of a wind display for virtual reality. In: Proceedings of the ACM Symposium on Virtual Reality Software and Technology VRST 2004, pp. 122–128 (2004). https://doi.org/10.1145/1077534.1077558
10. Nakano, T., Yanagida, Y.: Conditions influencing perception of wind direction by the head. In: 2017 IEEE Virtual Reality (VR), pp. 229–230, March 2017. https://doi.org/10.1109/VR.2017.7892260

11. Ranasinghe, N., Jain, P., et al.: Ambiotherm: enhancing sense of presence in virtual reality by simulating real-world environmental conditions. In: Proceedings of the 2017 CHI Conference on Human Factors in Computing Systems CHI 2017, pp. 1731–1742 (2017). https://doi.org/10.1145/3025453.3025723
12. Rietzler, M., Plaumann, K., et al.: VaiR: simulating 3D airflows in virtual reality. In: Proceedings of the 2017 CHI Conference on Human Factors in Computing Systems CHI 2017, pp. 5669–5677 (2017). https://doi.org/10.1145/3025453.3026009
13. Saito, Y., Murosaki, Y., et al.: Measurement of wind direction perception characteristics with head mounted display. In: Entertainment Computing (EC2017), vol. 2017, pp. 138–144, September 2017
14. Wozny, D.R., Beierholm, U.R., Shams, L.: Human trimodal perception follows optimal statistical inference. J. Vis. **8**(3), 24–24 (2008). https://doi.org/10.1167/8.3.24

A 6-DoF Zero-Order Dynamic Deformable Tool for Haptic Interactions of Deformable and Dynamic Objects

Haiyang Ding[1(✉)] and Shoichi Hasegawa[2,3,4]

[1] Department of Computational Intelligence and Systems Science, Tokyo Institute of Technology, Tokyo, Japan
haiyang.d.aa@m.titech.ac.jp

[2] Precision and Intelligence Laboratory, Tokyo Institute of Technology, Tokyo, Japan
hase@pi.titech.ac.jp

[3] Independent Administrative Corporation, Tokyo, Japan

[4] Japan Science and Technology Agency (JST), Kawaguchi, Japan

Abstract. Continuous collision detetion is required for haptic interactions with thin and fast-moving objects. However, previous studies failed to eliminate the force artifacts caused by the tool's inertia. In this paper, we propose a multi-sphere proxy method for a 6-DoF deformable virtual tool with continuous collision detection. We use Zero-order Dynamics to avoid force artifacts caused by the tool's inertia. In addition, we eliminate the "tunneling" problem introduced by the use of Zero-order Dynamics. As a result, we support fast motions of both the tool and the virtual object with sphere-mesh-level contacts and real-time simulation. Stability is guaranteed via a position-based dynamics simulator and an optional multi-rate architecture.

Keywords: Haptic rendering · Continuous collision detection · Deformable objects

1 Introduction

Haptic interaction requires accurate collision detection in order to render the contact force between the virtual tool and the virtual environment. The "tunneling" problem, also called the pop-through effect, caused by using discrete collision detection can be avoided by using continuous collision detection (CCD).

Garre et al. [3] presented a deformable 6-DoF tool which was able to interact with deformable objects using CCD. However, they used bidirectional viscoelastic coupling between the device and the tool, which introduced force artifacts

Electronic supplementary material The online version of this chapter (https://doi.org/10.1007/978-3-030-58147-3_27) contains supplementary material, which is available to authorized users.

I. Nisky et al. (Eds.): EuroHaptics 2020, LNCS 12272, pp. 244–252, 2020.
https://doi.org/10.1007/978-3-030-58147-3_27

caused by the force of the tool's inertia. These artificial forces reduce rendering transparency, and fast motion input from the device may trigger stability problems when interacting with heavy objects.

To the contrary, the traditional proxy method [11] can directly manipulate the proxy position without force artifacts. This simulation method is called Zero-order Dynamics (ZoD) [6].

In this paper, we propose a multi-sphere proxy (MSP) model to achieve 6-DoF haptic interactions with ZoD. In addition, this method is also able to perform CCD by using the method from [2] to handle collisions between the proxy sphere and triangle meshes.

However, ZoD may induce the "tunneling" problem if the collisions are not identified during the constraint computation. In this paper, we perform an additional CCD to solve this problem. Therefore, constraints, such as deformation, can be performed correctly without causing the "tunneling" problem. We list our contributions below:

- An MSP model which enables 6-DoF haptic interactions with dynamic and deformable objects through sphere-mesh-level CCD contacts;
- A ZoD simulation of a deformable 6-DoF haptic tool which eliminates the force artifacts caused by the tool's inertia as well as the "tunneling" problem.

2 Background and Related Works

2.1 Haptic Rendering with Zero-Order Dynamics

ZoD [6] was proposed to describe the simulation of the traditional virtual proxy method [11]. This method uses a virtual finite-radius sphere, i.e., the proxy, to represent a 3-DoF virtual tool. When the device moves, the proxy tries to follow its path via iterative collision detections. If contact occurs, the proxy finds the closest position to the device constrained by contacts. If there is no contact, the proxy tracks the device perfectly. The time integration has no mass or velocity involved.

When considering the dynamics of the 6-DoF haptic tool, many studies use second-order dynamics [3,4] with virtual coupling [10], thus producing force artifacts. Meanwhile, the constraint-based [9] and configuration-based optimization [12] methods can be used to remove force artifacts. However, these methods are limited to rigid tools. Mitra et al. [6,7] used first-order dynamics but position constraints must be transferred into velocity constraints making contact forces difficult to compute.

2.2 Haptic Rendering with Continuous Collision Detection

The traditional virtual proxy method [11] checks just the collisions between the moving proxy and static virtual objects. Therefore, if a dynamic thin virtual object is moving quickly or deforming greatly during one time step and pass through the proxy position, the collision will not be detected.

Ding et al. [2] proposed a method utilizing triangle-proxy CCD and Proxy Pop-out processes to solve this problem. Triangle-proxy CCD computes the CCD between the triangle and the proxy by solving the coplanar condition between the moving triangle and the proxy center. After that, the Proxy Pop-out carries our iterative discrete collision detections between multiple triangular meshes and the proxy sphere. These processes are used to secure the start of the proxy to avoid the "tunneling" problem.

For the CCD used in 6-DoF haptic rendering, Ortega et al. [9] used a constraint-based god-object method for rigid-rigid haptic interactions. Garre et al. [3,4] can simulate deformation on both the tool and the object. Other haptic rendering approaches can be found in the survey [10].

3 Proposal

3.1 Multi-sphere Proxy Model for a Virtual Tool

As described in Sect. 2.2, CCD between the proxy and dynamic triangular models can be computed using a sphere proxy[2]. Here, we extend the method used in [2] via the MSP model.

The Oriented Particles (OP) [8] method uses ellipsoid particles to simulate solid deformations. We directly use the particles of the OP model as our MSP model. The particle of the virtual tool is called the tool particle, and the particle representing the device is called the device particle. Since we apply CCD and proxy method described in [2] for each tool particle, we use spherical particles instead of ellipsoids for better performance. However, since ellipsoids can be used to represent a more precise contact model compared to spheres, it will be interesting to conduct further research using ellipsoids.

Since the number of tool particles is related directly to both collision accuracy and simulation speed, collision accuracy and system efficiency should be balanced. In this paper, we use the same method as [8] to create the OP model. An example is shown in Fig. 3.

This method is similar to that employed in [1] and [12], which also utilize spheres. However, their methods cannot handle CCD contact between the sphere and the triangular mesh, and method in [1] cannot avoid force artifacts caused by the tool's inertia.

3.2 Six Degree-of-Freedom Haptic Rendering

Force and Torque Feedback. We calculate the force and torque feedback from the discrepancy between the virtual tool and the device. As shown in Fig. 1, we first calculate force and torque of each pair of tool and device particles. Only the particles that have collided are considered in the calculation. Next, we sum all the forces and all the torques, respectively. Finally, we divide each of these sums by the total number of tool particles. We use translational and rotational springs to adjust the feedback magnitude.

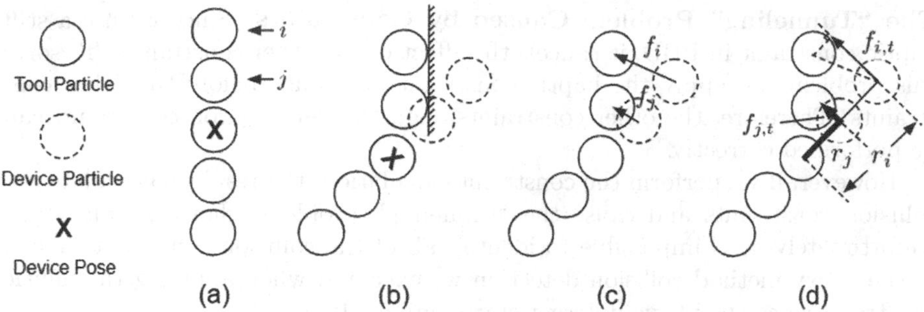

Fig. 1. Force and torque calculation of the multi-sphere Proxy model. (a) The start. (b) The deformable tool encounters a contact. (c) The force $\mathbf{f}_i$ and $\mathbf{f}_j$ and, (d) the torque $\mathbf{r}_i \times \mathbf{f}_{i,t}$ and $\mathbf{r}_j \times \mathbf{f}_{j,t}$ of tool particle i and j, respectively.

Haptic Contact Force Computation. The contact force applied to the virtual object is computed using the method in [2], which uses a distributed point-like contact force applied as an external force of PBD to the corresponding OP particle of the virtual object.

Multi-rate Architecture. To improve interaction stability, we use a multi-rate architecture. First, we simply synchronize all the tool particle positions from the physics thread to the haptic thread. The feedback force and torque are calculated in the haptic thread. However, the delay caused by synchronization may induce "dragging" forces. Therefore, we also synchronize the collision information (collision occurs or not) to eliminate the force artifacts when no contact occurs.

3.3 Zero-Order Dynamic Deformable Haptic Tool

As introduced in Sect. 2.2, the traditional proxy method [11] uses ZoD to avoid force artifacts. We achieve ZoD by manipulating the position of the tool particle directly with a position-based haptic constraint.

Haptic Constraint. The haptic constraint helps us to apply the haptic device input to the virtual tool by computing current position of the device particle as the goal position. The computation is as follows:

$$\mathbf{g}_i = Q_{dev}\mathbf{o}_i + \mathbf{d}_{dev}. \tag{1}$$

Here, $\mathbf{o}_i$ represents the tool particle's original barycentric coordinates before deformation, while Q_{dev} and $\mathbf{d}_{dev}$ represent for the device's orientation matrix (3×3) and position, respectively. The variable with a bold font refers to a 3×1 vector.

After the goal position is computed, we perform the traditional proxy method [11] to update tool particles to the goal positions, i.e., the positions of device particles (shown in Fig. 2(b)).

The "Tunneling" Problem Caused by Constraints. Since we use a stiff haptic constraint in PBD, it cancels the effect of the other constraints. To solve this problem, we apply the haptic constraint only once before the other constraints. Therefore, the other constraints, e.g., the deformation constraint, can be performed correctly.

However, if we perform the constraint calculation, the result may violate the collision constraints and cause the "tunneling" problem (shown in Fig. 2(c)). Unfortunately, it is impossible to identify all of the collision constraints ahead in the proxy method collision detection we executed when applying the haptic constraint since they have different starts and goals.

Solving the "Tunneling" Problem. To solve the "tunneling" problem caused by the constraint calculation, our proposal is to run an additional CCD, which is the same as the proxy method, for the constraint calculation (shown in Fig. 2(d)) to revise the particle positions. In order to do so, we record the tool particle positions before and after the PBD constraint calculation. After that, we perform the proxy method between the recorded start and end. The whole progress is shown in Fig. 2.

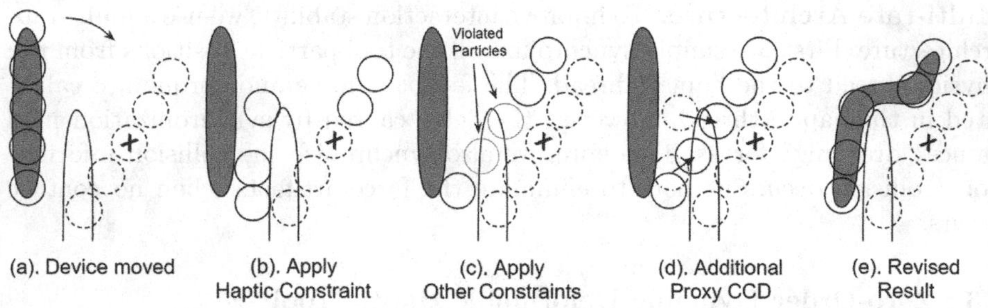

Fig. 2. Simulation of a zero-order dynamic deformable haptic tool

3.4 Simulation of a Zero-Order Dynamic Deformable Haptic Tool

Our simulation loop of the physics thread is executed as follows:

1. Physics thread start:
2. Simulate the virtual tool:
   (a) Perform triangle-CCD and Proxy Pop-out for tool particles; (Sect. 2.2)
   (b) Calculate and apply haptic constraint; (Sect. 3.3)
   (c) Perform the traditional proxy method for each tool particle;
   (d) Record current tool particle positions;
   (e) Perform and recored the result of other PBD constraint iterations;
   (f) Perform proxy CCD with the recorded start and end; (Sect. 3.3)

   (g) Update the tool particle position;
   (h) Calculate haptic contact forces; (Sect. 3.2)
3. Apply haptic contact forces as external forces to the contact object;
4. Simulate the other virtual objects.

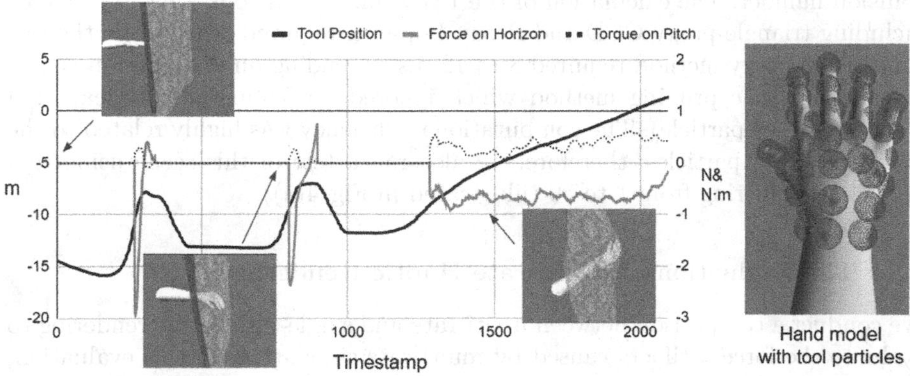

Fig. 3. Evaluation of interactions. Left: the recorded tool position, feedback force, and torque are presented with screenshots of a deformable hand interacting with a yellow curtain. Right: the hand model and its tool particles with the proxy radius. (Color figure online)

4 Evaluations

In this section, we will introduce two evaluations of our method. The evaluations were performed on an Intel i5-7300 4-core CPU PC without GPU. A Spidar-G6 was used as the haptic device. We also provide a supplementary video.

4.1 The Evaluation of Interactions and Efficiency

In this evaluation, we used a deformable 1 kg hand model [5] as the tool to interact with a double-faced, 8-m, 200 g curtain model. The curtain model contained 1089 vertices and 4096 triangles, while each vertex was modeled with one OP particle and connected with another eight nearby particles. The hand model had 3919 vertices and 3906 triangles with 37 OP particles (the placement of the tool particles can be further optimized). Only shape-matching and distance constraints were simulated for the two models. The PBD iteration counts were two and four for the curtain and the hand models, respectively. The feedback springs were both set to 20 N/m. No damping of virtual coupling was used.

The result is displayed in Fig. 3. A series of interactions were performed, including tapping and pressing, on the curtain model. From Fig. 3, we can see that the tool moved freely in the air without the force caused by the inertia. The system was stable even fast motions were inputted with strong impacts while the tool and the device deformation were performed correctly. The whole simulation took 2.5 s. The data was collected from the haptic thread which was running at 1kHz, while one time-stamp indicated one loop of the calculation.

The simulation ran at an average rate from 45 to 59 FPS, depended on the collision number. The calculation of the PBD constraints took 4 ms. The CCD, including triangle-proxy CCD and Proxy Pop-out, required 3 ms, while the two runs of the proxy method required 8 to 12 ms depending on the collision count. We used a basic pruning method which ignored far-off meshes (at least 1 m from every tool particle). The computational efficiency was highly related to the number of tool particles; therefore, we also ran a test of this scene using tool particles numbering from 1 to 50 (illustrated in Fig. 4(a)).

4.2 The Evaluation of Multi-rate Haptic Rendering

We conduct a comparison between multi-rate and single-rate haptic rendering to evaluate the force artifacts caused by multi-rate simulation. In this evaluation, the device slid along a virtual horizontal plane using the same hand model of the first evaluation. The results in Fig. 4(b) and (c) show that the horizontal "dragging" forces (forces on X and Z axis) arise only in the multi-rate case as the device moves. During our tests, the magnitude of the force artifacts was proportional to the computation time of the physics simulation.

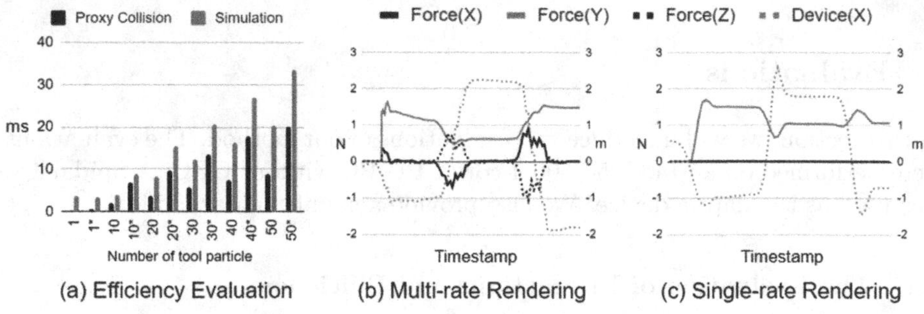

Fig. 4. Evaluation of efficiency (a) and Comparison Between Multi-Rate (b) and Single-Rate (c) Haptic Rendering. The efficiency evaluation utilized the average computation time for the simulation and proxy method using 1 to 50 tool particles. The label with "*" indicates collisions are occurred in the simulaiton while others are not.

5 Conclusion, Limitations, and Future Works

We propose a 6-DoF deformable tool that can interact with another deformable virtual object. The contact is handled between spherical proxies and triangular meshes. The force artifacts caused by the tool's inertia were eliminated, and tool deformation was performed without the "tunneling" problem. Our simulation is stable even with fast motion input using a large feedback spring. We also provide a multi-rate haptic rendering option which improves the stability but induces other force artifacts.

Compared to other studies, our system uses a less accurate contact tool model compared to the one in [3], in which the tool has 1441 triangular meshes for collision detection (not sure about the deformable object). The hand model we used in the evaluations does not have a rigid-body or joint constraint; therefore, the physics behavior of the hand model is less accurate than that used in [3]. Also, as we represent the tool with spherical particles, objects that are thinner than the gap between particles may result in penetrations. In the future, we plan to find solutions to cover this gap and use ellipsoids to handle accurate contacts.

Acknowledgment. This work was support by JSPS KAKENHI Grant Number 17H01774. Here, we are also grateful to the support of Springhead physical engine and the developers.

References

1. Cirio, G., Marchal, M., Otaduy, M.A., Lécuyer, A.: Six-oof haptic interaction with fluids, solids, and their transitions. In: 2013 World Haptics Conference (WHC), pp. 157–162. IEEE (2013)
2. Ding, H., Mitake, H., Hasegawa, S.: Continuous collision detection for virtual proxy haptic rendering of deformable triangular mesh models. IEEE Trans. Haptics **12**(4), 624–634 (2019)
3. Garre, C., Hernández, F., Gracia, A., Otaduy, M.A.: Interactive simulation of a deformable hand for haptic rendering. In: 2011 IEEE World Haptics Conference, pp. 239–244. IEEE (2011)
4. Garre, C., Otaduy, M.A.: Haptic rendering of objects with rigid and deformable parts. Comput. Graph. **34**(6), 689–697 (2010)
5. gotferdom: Free hand 3D model (2018). https://www.turbosquid.com/3d-models/hand-hdri-shader-3d-model-1311775. Accessed 25 Jan 2020
6. Mitra, P., Niemeyer, G.: Dynamic proxy objects in haptic simulations. In: 2004 IEEE Conference on Robotics, Automation and Mechatronics, vol. 2, pp. 1054–1059. IEEE (2004)
7. Mitra, P., Niemeyer, G.: Haptic simulation of manipulator collisions using dynamic proxies. Presence Teleoperators Virtual Environ. **16**(4), 367–384 (2007)
8. Müller, M., Chentanez, N.: Solid simulation with oriented particles. ACM Trans. Graph. **30**, 92 (2011)
9. Ortega, M., Redon, S., Coquillart, S.: A six degree-of-freedom god-object method for haptic display of rigid bodies. In: IEEE Virtual Reality Conference (VR 2006), pp. 191–198. IEEE (2006)
10. Otaduy, M.A., Garre, C., Lin, M.C.: Representations and algorithms for force-feedback display. Proc. IEEE **101**(9), 2068–2080 (2013)
11. Ruspini, D.C., Kolarov, K., Khatib, O.: The haptic display of complex graphical environments. In: Proceedings of the 24th Annual Conference on Computer Graphics and Interactive Techniques, pp. 345–352 (1997)

12. Wang, D., Tong, H., Shi, Y., Zhang, Y.: Interactive haptic simulation of tooth extraction by a constraint-based haptic rendering approach. In: 2015 IEEE International Conference on Robotics and Automation (ICRA), pp. 278–284. IEEE (2015)

Evaluating Ultrasonic Tactile Feedback Stimuli

Antti Sand[1(✉)], Ismo Rakkolainen[1], Veikko Surakka[1], Roope Raisamo[1], and Stephen Brewster[2]

[1] Tampere University, Tampere, Finland
antti.sand@tuni.fi
[2] University of Glasgow, Glasgow, UK

Abstract. Ultrasonic tactile stimulation can give the user contactless tactile feedback in a variety of human-computer interfaces. Parameters, such as duration, rhythm, and intensity, can be used to encode information into tactile sensation. The present aim was to investigate the differentiation of six ultrasonic tactile stimulations that were varied by form (i.e., square and circle) and timing (i.e., movement speed and duration, and the number of repetitions). Following a stimulus familiarization task participants (N = 16) were to identify the stimuli presented in the same order as in the familiarization phase. Overall, the results showed that it was significantly easier to identify stimuli that were rendered at a slower pace (i.e., longer duration) regardless of the number of repetitions. Thus, for ultrasonic haptics, rendering time was one important factor for easy identification.

Keywords: Ultrasonic haptics · Mid-air haptics · Stimuli design · User study

1 Introduction

Haptic feedback is commonly used in mobile phones, but so far the feedback has required physical contact with a device or the use of wearable actuators. Ultrasound tactile actuation [7] is a new approach for providing tactile feedback. It removes the limitation of contact and creates true mid-air haptic sensations.

Parameters such as duration, rhythm, and intensity, can be used to encode information into tactile sensation. Yet, it is unclear how to combine these for easy identification of stimuli. The ability of transducer arrays to rapidly update focal point location to create movement and shapes opens up new possibilities as to how much and what kind of information can be encoded into haptic feedback. However, this brings with it new open questions about how to best design haptic cues to convey information.

To better understand the technical and human limitations of ultrasonic haptic information transfer, we conducted a study to evaluate six different haptic feedback stimuli in terms of how quickly and accurately they could be identified and how well they would work as mobile phone notifications (e.g., receiving a phone call or a notification).

I. Nisky et al. (Eds.): EuroHaptics 2020, LNCS 12272, pp. 253–261, 2020.
https://doi.org/10.1007/978-3-030-58147-3_28

2 Related Work

Mid-air haptics can be made with a number of technologies that allow for tactile feedback on touchless interaction. Tactile sensations can be rendered to interactive spaces in 3D. Technologies such as pressurized air jets [16] or air vortex rings [18] can provide strong but rough feedback with some inherent time lag. Lasers [8,11] or electric arcs [17] are possible for very precise short range feedback.

Focused airborne acoustic air pressure produced by ultrasonic phased arrays [2,6,7] is particularly good at generating a range of tactile stimuli on the user's palm or fingertips. This technology allows for fast rendering of single points or multiple simultaneous ones for patterns such as volumetric shapes [9].

These inaudible sound waves (typically 40 kHz [7] or 70 kHz [6]) can be focused into a single location in space. As the human hand can not feel vibrations at 40 kHz, the emitted ultrasound is modulated at the focal point to a frequency of around 200 Hz. This frequency is detectable by mechanoreceptors sensitive to vibration and pressure [5]. At a focal point, the acoustic pressure becomes strong enough to slightly indent the human skin and stimulate the Pacinian corpuscles, thus generating a touch sensation.

For the most common hardware types, rendering multiple focal points simultaneously makes each point feel weaker as there are less transducers used for each point. The temporal resolution of touch perception is only a few milliseconds [10]. Hence, fast moving single points, through spatiotemporal modulation [4], can be used to create the sensation of an entire shape being rendered at once.

Tactons [1], or tactile icons, have been used to communicate messages non-visually to users through ultrasonic actuation [3] and have been used with immaterial [15] and virtual screens [14].

Previous work on identifiability of mid-air haptic shapes [13] found that users are better at identifying shapes with a single focal point or shapes organized in straight lines compared to circular shapes. However, there has been little research into the range of parameters that can be used in ultrasound haptics to find out what makes the most effective tactile cues. Therefore, we designed a study to find out what parameters can make the stimuli more identifiable.

3 Methods

Sixteen voluntary participants (11 male), aged 20 to 42 years (median age 29 years, SD 6.07) with normal sense of touch by their own report, took part in the study. Seven of them had no previous experience using ultrasonic haptics and the rest had experienced it once or multiple times.

3.1 Pattern Design

Preliminary testing indicated that stimuli using variation in duration and rhythm resulted in more reliable identification of six different haptic stimuli than variations in shape. Pretesting also suggested that a 3 × 3 cm stimulation area

Table 1. Haptic stimuli parameters used in the experiment.

Stimulus id	Total duration (ms)	Pulse duration (ms)	Repetitions	Breaks (ms)	Shape
1	400	400	1	0	Square
2	1100	400	2	300	Square
3	1800	400	3	300, 300	Square
4	1100	1100	1	0	Circle
5	500	100	2	300	Square
6	900	100	3	300, 300	Square

Fig. 1. Rendered shapes and repetitions. In stimuli 1 to 4, the focal point movement can be perceived, as shown by the arrow. Stimuli 5 and 6 use spatiotemporal modulation, making the entire shape concurrently perceivable.

provided stronger and clearer stimulus perception than larger areas and was therefore selected for the experiment.

Based on above the following six stimuli were designed. Each stimulus consisted of a rhythm formed by either one, two or three pulses, followed by a 300 ms break. Stimuli 1 to 3 had a 400 ms pulse duration, stimulus 4 had a single long pulse, and stimuli 5 and 6 had short, 100 ms pulses. Stimuli 5 and 6 used spatiotemporal modulation in which the focal point was rapidly and repeatedly updated to create a sensation of a single tactile shape.

Stimulus 4 had a circular shape, while the other stimuli had a square shape. It was also rendered in counterclockwise direction with the others rendered clockwise to see if the change in direction could be reliably noticed. Stimuli used in this experiment are described in Table 1 and visualized in Fig. 1.

3.2 Procedure

The UltraHaptics UHEV1 ultrasonic transducer array with 256 40 kHz transducers was used (see Fig. 1) to deliver haptic feedback.

Participants were to rest the wrist of their dominant hand comfortably on a foam pad with their palm facing down about 10 cm above the centre of the array. All participants used their right hand. They were instructed to keep their hand in the same position throughout the experiment. Near their left hand they had a keypad for input. No visual feedback was provided.

The experiment started with a written and verbal introduction and instructions, followed by a short training period, where the participants experienced each stimulus in a specific order (1 to 6), and repeated four times. They were instructed to try to remember the order number. In the experiment, the same stimuli were presented randomly, with each occurring three times (for a total 18 of stimuli). The participants wore headphones playing white noise to mask any audible noise from the array.

After each stimulus, a 500 ms audio sample was played as a cue for action. The participants were instructed to identify the stimulus by pressing a corresponding numeric key on the keypad in front of them as fast and accurately as possible.

Response times from the onset of the audio cue to the key press, as well as the key selected, were automatically logged in the database. Timing started after the stimulus had ended so that the stimulus duration would not affect the identification times. After an answer was given the experiment proceeded to the next stimulus and continued until all the stimuli were presented.

After the identification task, the participants were given the same stimuli again and they rated them on scales of *valence* (−4, unpleasant to 4, pleasant) and *arousal* (−4, calming to 4, arousing). They also rated the functionality of the stimuli as potential mobile phone notifications and incoming call notifications using a scale from 0 = does not suit at all to 7 = suits very well. At the end of the experiment, they were asked to freely describe their perceptions of the stimuli to see if the change in shape or drawing direction were detected.

The identification times were analyzed using one-way repeated measures analysis of variance (ANOVA). Bonferroni corrected t-tests at the .05 level were used for pairwise post hoc comparisons of the 15 pairs. The number of incorrect identifications, the valence and arousal ratings, and the stimulus suitability for mobile phone notifications ratings were first analyzed with Friedman tests and Bonferroni corrected Wilcoxon Signed Ranks tests were used for post hoc comparisons.

4 Results

One participant had trouble in sensing any of the stimuli. Therefore, this data was removed from the data set.

The ANOVA for the identification times showed a statistically significant effect of the haptic stimulus ($F(5,75) = 5.824$, $p < 0.01$). Post hoc tests showed that identification times were statistically significantly different between stimuli 1 and 3 ($MD = 949.193$, $p = 0.019$), 3 and 5 ($MD = 1235.229$, $p = 0.003$), and 3 and 6 ($MD = 551.813$, $p = 0.029$). Comparing just the duration, stimuli 2 and 3 were on average 500 ms and 700 ms quicker to identify than stimuli 5 and 6 respectively (See Fig. 2).

The Friedman test for the number of incorrect identifications showed a statistically significant effect of the haptic stimulus $X^2 = 18.673$, $p = 0.02$. Post hoc pairwise comparisons were not statistically significant.

Figure 2 shows the percentage of incorrect identifications made for each stimulus. Accuracy rate across all stimuli was 66%.

Three participants were able to identify each stimulus without errors. Stimulus 5 saw an 118% greater error rate compared with stimulus 2 and stimulus 6

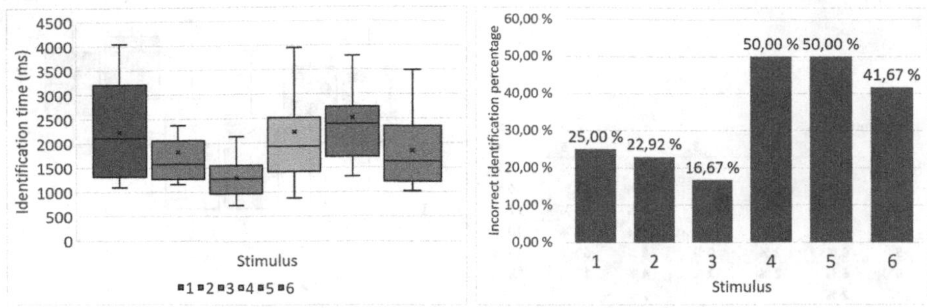

Fig. 2. Left: identification time distributions in ms from the start of the audio cue. Right: percentage of incorrect identifications for each stimulus.

saw an 150% increase compared with stimulus 3, which were those with the same rhythm. Stimulus 3 was the quickest and most accurate to identify. There were large differences in accuracy across the tested stimuli. When drawing a square shape at a lower speed for one, two or three repetitions, the accuracy was 78%. With stimulus 3 the accuracy rose to 83%, with 11 participants not mistaking once in all of the 48 identifications.

The confusion matrix for the stimuli is shown in Fig. 3. Correct identifications are on the main diagonal and the rhythmically similar options have a solid border. 100% correct identification would give a score of 48. Stimulus 3 is the best performing cue, with 5 the worst.

Stimulus 4 was often mistaken as stimulus 1, but not the other way around. Stimuli 5 and 6 were as likely to be confused with each other, as they were with those with the same rhythm, stimuli 2 and 3.

Ratings of stimulus suitability for mobile phone notifications (Fig. 3) showed a significant effect $X^2 = 28.114$, $p < 0.001$. Pairwise comparisons showed significant difference between stimuli 3 and 5 (Z = 3.307, p < 0.01), and 3 and 6 (Z = 3.311, p < 0.01). Stimulus 6 seems well suited for a general notification.

The ratings of arousal and stimulus suitability for a notification of an incoming phone call (Fig. 3) showed a statistically significant effect of the haptic stimulus $X^2 = 13.577$, $p = 0.019$. and $X^2 = 14.071$, $p = 0.015$, respectively. Post hoc pairwise comparisons were not statistically significant.

No statistically significant effect of stimulus were found on valence ($X^2 = 1.484$, $p = 0.915$). Overall, participants regarded all stimuli as quite pleasant with an average rating of 1.41 (SD 1.83).

Participants' descriptions of sensations failed to accurately identify differences in shape and direction, and the shapes were often incorrectly identified as being for example lines, crosses, vortexes or just random pokes. For the square shaped stimuli, there were 26 answers suggesting either a square shape, a line or a cross and 34 answers suggesting an oval, a vortex, a point, or a fluttering of points. For the circular shaped stimuli, there were 4 answers suggesting an oval, an infinity sign or a circle and 7 answers suggesting a linear movement or a

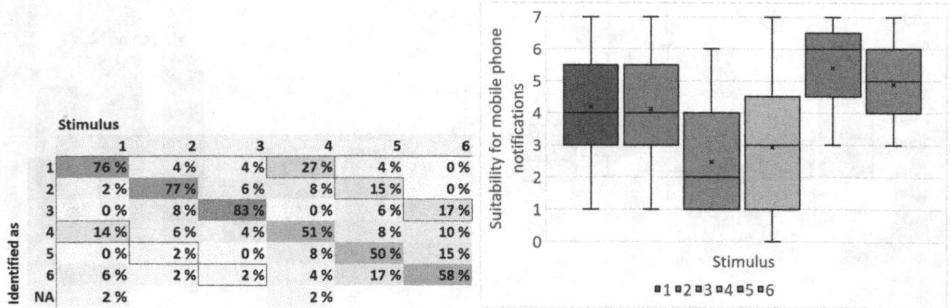

Identified as	Stimulus 1	2	3	4	5	6
1	76 %	4 %	4 %	27 %	4 %	0 %
2	2 %	77 %	6 %	8 %	15 %	0 %
3	0 %	8 %	83 %	0 %	6 %	17 %
4	14 %	6 %	4 %	51 %	8 %	10 %
5	0 %	2 %	0 %	8 %	50 %	15 %
6	6 %	2 %	2 %	4 %	17 %	58 %
NA	2 %			2 %		

Fig. 3. Left: confusion matrix for the stimuli. Stimuli (horizontal axis) and what they were identified as (vertical axis). Right: stimulus rating distribution on suitability for mobile phone notifications on a scale of 0 (does not suit at all) to 7 (suits very well).

fluttering of points. This suggests that shape and direction are not good options for cue design in this case.

5 Discussion

The results of the study showed that stimulus number 3 (See Table 1) was the fastest and most accurate to identify. It consisted of three 400 ms pulses forming a square pattern rendered at a slow rate. As the stimuli with lower rate were identified better than the ones with faster rate it seems that the duration had a greater effect than rhythm. However, as the experiment had multiple variables, isolation of the effect of one parameter is not viable.

The tested stimuli formed pairs of similarities. Stimuli 1 and 4 consisted of a single pulse, stimuli 2 and 5 had two pulses, and stimuli 3 and 6 had three pulses. In stimuli 1 to 3 the shape was drawn at a slower pace than in stimuli 5 and 6 so that the shape was completed in 400 ms.

The average identification time as well as the error rate of the stimuli 1 to 3 decreased as the number of pulses increased. A similar trend was seen for stimuli 5 and 6 so that adding the third pulse seemed to decrease the identification time and decrease the number of errors. Nevertheless, for the stimuli 4 to 6 there were considerably more errors than for the stimuli 1 to 3. Probably this is due to the very short duration of the pulse or due to the use of spatiotemporal modulation or both.

The results suggest that the longer duration of the stimulus led to a quicker and more accurate identification when the rhythm was two or three repetitions.

Palovuori *et al.* [12] reported that a 200 ms burst of ultrasound was an 'unmistakable' stimulus being functionally equivalent to a physical button click. However, in our experiment shortening the duration of the stimulus to 100 ms resulted in more incorrect identifications than prolonging the duration to 400 ms.

Stimulus 4, which had the longest duration to complete one shape had the highest number of errors and took a long time to identify. Increasing stimulus duration to 1100 ms did not make identification easier. However, it is unclear whether the shape drawn, or its drawing direction affected the identification. Considering that the participants couldn't reliably identify any shape or direction, it is assumed that it had very little effect, but further study is required.

Rutten *et al.* reported 44% identification rate when the participants were provided with visual representations of the stimuli [13]. Our stimuli formed pairs of similarity, thus identification by chance is 50%. However, some stimuli were considerably more accurate to identify. Our results seem to indicate that identifying any ultrasonic haptic shape is difficult, but that stimuli can be identified with the right parameters for which rendering time is a key factor.

6 Conclusions and Future Work

The present work reported an experiment that compared six different ultrasound haptic stimuli to see how fast and accurately they could be identified. Statistically significant differences were found for identification times, number of incorrect identifications and subjective ratings. With a rhythm of three repetitions, a 400 ms pulse duration made the cue quicker and easier to identify than with a pulse duration of 100 ms.

Further, using short pulse duration, when the shape was drawn rapidly and repeatedly, the identification time and errors increased. This could be either due to the pulse duration or due to the use of spatiotemporal modulation.

Subtle changes in shape or direction of draw seem to go unnoticed by the users. Building a vocabulary based on direction or complex shapes is not a viable method of information encoding. These results may help in designing easy to identify ultrasonic stimuli.

References

1. Brewster, S.A., Brown, L.M.: Non-visual information display using tactons. In: CHI 2004 Extended Abstracts on Human Factors in Computing Systems, CHI EA 2004. ACM, New York (2004)
2. Carter, T., Seah, S.A., Long, B., Drinkwater, B., Subramanian, S.: Ultrahaptics: multi-point mid-air haptic feedback for touch surfaces. In: Proceedings of the 26th Annual ACM Symposium on User Interface Software and Technology. UIST 2013. ACM, New York (2013)
3. Freeman, E., Brewster, S., Lantz, V.: Tactile feedback for above-device gesture interfaces: Adding touch to touchless interactions. In: Proceedings of the 16th International Conference on Multimodal Interaction, ICMI 2014. ACM, New York(2014)
4. Frier, W., Ablart, D., Chilles, J., Long, B., Giordano, M., Obrist, M., Subramanian, S.: Using spatiotemporal modulation to draw tactile patterns in mid-air. In: Prattichizzo, D., Shinoda, H., Tan, H.Z., Ruffaldi, E., Frisoli, A. (eds.) EuroHaptics 2018. LNCS, vol. 10893, pp. 270–281. Springer, Cham (2018). https://doi.org/10.1007/978-3-319-93445-7_24
5. Goldstein, E.B.: Sensation and Perception, 5th edn. Brooks/Cole Publishing Company, Pacific Grove (1999)

6. Ito, M., Wakuda, D., Inoue, S., Makino, Y., Shinoda, H.: High spatial resolution midair tactile display using 70 kHz ultrasound. In: Bello, F., Kajimoto, H., Visell, Y. (eds.) EuroHaptics 2016. LNCS, vol. 9774, pp. 57–67. Springer, Cham (2016). https://doi.org/10.1007/978-3-319-42321-0_6
7. Iwamoto, T., Tatezono, M., Shinoda, H.: Non-contact method for producing TACTILE sensation using airborne ultrasound. In: Ferre, M. (ed.) EuroHaptics 2008. LNCS, vol. 5024, pp. 504–513. Springer, Heidelberg (2008). https://doi.org/10.1007/978-3-540-69057-3_64
8. Jun, J.H., et al.: Laser-induced thermoelastic effects can evoke tactile sensations. Sci. Rep. **5**, 11016 (2015)
9. Long, B., Seah, S.A., Carter, T., Subramanian, S.: Rendering volumetric haptic shapes in mid-air using ultrasound. ACM Trans. Graph. **33**(6), 1–10 (2014)
10. Loomis, J.M.: Tactile pattern perception. Perception **10**(1), 5–27 (1981)
11. Ochiai, Y., Kumagai, K., Hoshi, T., Rekimoto, J., Hasegawa, S., Hayasaki, Y.: Fairy lights in femtoseconds: Aerial and volumetric graphics rendered by focused femtosecond laser combined with computational holographic fields. ACM Trans. Graph. **35**(2), 1–14 (2016)
12. Palovuori, K., Rakkolainen, I., Sand, A.: Bidirectional touch interaction for immaterial displays. In: Proceedings of the 18th International Academic MindTrek Conference: Media Business, Management, Content and Services. AcademicMindTrek 2014. ACM, New York (2014)
13. Rutten, I., Frier, W., Van den Bogaert, L., Geerts, D.: Invisible touch: how identifiable are mid-air haptic shapes? In: Extended Abstracts of the 2019 CHI Conference on Human Factors in Computing Systems. CHI EA 2019. ACM, New York (2019)
14. Sand, A., Rakkolainen, I., Isokoski, P., Kangas, J., Raisamo, R., Palovuori, K.: Head-mounted display with mid-air tactile feedback. In: Proceedings of the 21st ACM Symposium on Virtual Reality Software and Technology. VRST 2015, ACM, New York (2015)
15. Sand, A., Rakkolainen, I., Isokoski, P., Raisamo, R., Palovuori, K.: Light-weight immaterial particle displays with mid-air tactile feedback. In: 2015 IEEE International Symposium on Haptic, Audio and Visual Environments and Games (HAVE), October 2015
16. Sodhi, R., Poupyrev, I., Glisson, M., Israr, A.: Aireal: interactive tactile experiences in free air. ACM Trans. Graph. **32**(4), 1–10 (2013)
17. Spelmezan, D., Sahoo, D.R., Subramanian, S.: Sparkle: towards haptic hover-feedback with electric arcs. In: Proceedings of the 29th Annual Symposium on User Interface Software and Technology. UIST 2016 Adjunct. ACM, New York(2016)
18. Tsalamlal, M.Y., Issartel, P., Ouarti, N., Ammi, M.: Hair: haptic feedback with a mobile air jet. In: 2014 IEEE International Conference on Robotics and Automation (ICRA) (2014)

WeATaViX: WEarable Actuated TAngibles for VIrtual Reality eXperiences

Xavier de Tinguy[1(✉)], Thomas Howard[2], Claudio Pacchierotti[2], Maud Marchal[1], and Anatole Lécuyer[3]

[1] Univ Rennes, INSA, IRISA, Inria, CNRS, Rennes, France
xavier.de-tinguy@inria.fr
[2] CNRS, Univ Rennes, Inria, IRISA, Rennes, France
[3] Inria, Univ Rennes, CNRS, IRISA, Rennes, France

Abstract. This paper presents the design and evaluation of a wearable haptic interface for natural manipulation of tangible objects in Virtual Reality (VR). It proposes an interaction concept between encounter-type and tangible haptics. The actuated 1 degree-of-freedom interface brings a tangible object in and out of contact with a user's palm, rendering making and breaking of contact with, and allowing grasping and manipulation of virtual objects. Device performance tests show that changes in contact states can be rendered with delays as low as 50 ms, with additional improvements to contact synchronicity obtained through our proposed interaction technique. An exploratory user study in VR showed that our device can render compelling grasp and release interactions with static and slowly moving virtual objects, contributing to user immersion.

1 Introduction

Manipulation of objects in virtual reality (VR) commonly suffers from the absence of haptic sensations. As such, it is often unclear whether contact between one's virtual hand and virtual objects has been made, whether an object is properly grasped or not, and what the physical properties of the hand-object contact are. Conventional haptic interfaces for VR, be they grounded [9], body-grounded [11], or handheld [6] address this issue by applying forces to the user through an end-effector (e.g., a stylus), which mimics sensations of making and breaking contact as well as effects of mass, inertia, and collisions with the environment. However, such interactions are always mediated by the interface's end-effector, degrading the experience and preventing simultaneous manipulation

This project has received funding from the European Union's Horizon 2020 programme under grant agreement No 801413; project "H-Reality".
X. de Tinguy and T. Howard—Have contributed equally to this work.

Electronic supplementary material The online version of this chapter (https://doi.org/10.1007/978-3-030-58147-3_29) contains supplementary material, which is available to authorized users.

I. Nisky et al. (Eds.): EuroHaptics 2020, LNCS 12272, pp. 262–270, 2020.
https://doi.org/10.1007/978-3-030-58147-3_29

and exploration of the virtual object. Encounter-type haptic displays (ETHDs) solve the issue of rendering sensations of making and breaking contact, bringing their end-effector in contact with the user only when collisions with virtual objects occur [16]. Many types of grounded [7] and body-grounded ETHDs [10] exist. However, to the best of our knowledge, very few tackle the issue of grasping and manipulating objects. One work in this direction is that of [13], whose device allows grasping of the tangible end-effector, but presents the same issues as conventional haptic interfaces if the user wants to manipulate the grasped object. Passive haptics offers an alternative solution, superimposing virtual objects with similar tangible ones to create the illusion of truly manipulating virtual entities [5]. However, the number of required props for passive haptics increases with the complexity of the scene, making this approach unmanageable in rich virtual environments. Several different approaches aiming at rendering multiple virtual objects with few tangible ones exist to address this issue. They use reconfigurable or active tangible objects [4,12], augment passive props via wearable haptics [14,15] or use redirection techniques [1,8].

This paper presents a novel solution called "WeATaViX" at the interface between ETHDs and passive haptics, in the form of a wearable encounter-type device whose end-effector is a tangible object. It aims to provide physical presence for virtual objects while remaining as simple and unobtrusive as possible. The device is grounded on the back of the hand, secured to the skin via an ergonomic adhesive silicone layer. A servo motor moves a rigid link equipped with the tangible object towards and away from the user's palm. Unlike other wearable ETHDs (e.g., [3]), our end-effector aims at best fitting the shape properties of the virtual object, inherently solving shape rendering problems. With the device secured to the user's hand, the relative placement between the tangible and user's hand mimics that of their virtual counterparts. This paper presents our device, along with its dynamic analysis and a human-subject evaluation in VR.

2 The WeATaViX Haptic Interface

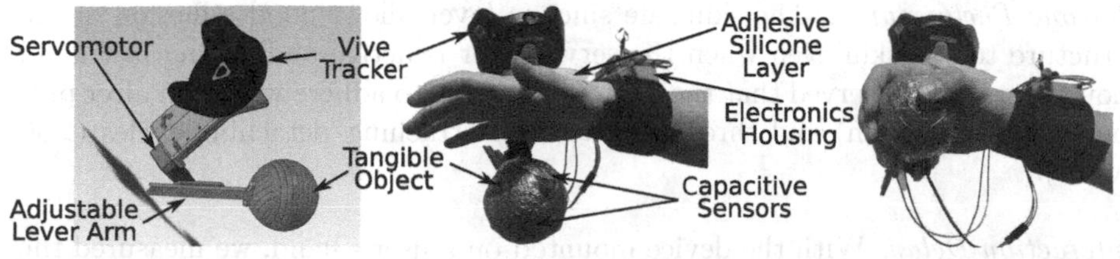

Fig. 1. Haptic device composed of a 3D-printed part anchored to an adhesive silicone layer attached to the hand. Two capacitive sensors cover the tangible, respectively facing the palm and the fingers during grasp closure.

2.1 Design and Description

A prototype of the device is shown in Fig. 1. It is composed of a 3D-printed structure to be placed on the back of the hand. Its profile is slightly curved to fit the shape of the hand. On the internal side, it is anchored in an adhesive silicone skin based on work by Chossat et al. [2], guaranteeing good adherence, comfort, and adaptability to different hand morphologies and skin properties. A HTC Vive Tracker can be attached on the external side. The distal side of the 3D-printed structure houses a HiTec HS-5065MG servomotor which controls the motion of a rigid link holding the tangible object. By moving the rigid link, the motor brings the tangible object towards or away from the user's palm. The tangible object is equipped with capacitive sensors to detect contacts with the hand. Further details are included in Fig. 1. A video of the device in action is available at https://youtu.be/JtcEYlwogpA. The device was designed with minimal weight as a target, weighing 85 g without the Vive tracker (185 g with tracker). Figure 2-A shows how the electronics are interconnected. Figure 2-B shows the VR setup. The HTC Vive tracker enables hand position tracking and, together with the capacitive sensors, animation of the user's hand avatar in VR.

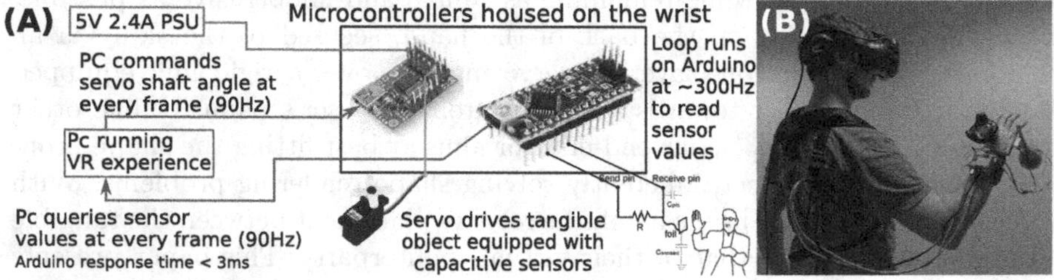

Fig. 2. (A) Schematic of the interconnected electronics structure for sensing and control. The capacitive sensing uses the Arduino CapacitiveSensor library. (B) VR setup.

2.2 Evaluation of the Device Performance

Silicone Performance. The skin-safe silicone layer allows good adhesion of the structure to the skin even when the servomotor is active and during fast hand movements. We observed that the device continues to adhere well even after prolonged use (>45 min) and throughout several attaching/detaching cycles (>30 cycles).

Interaction Delay. With the device mounted on a user's hand, we measured the delay between the command to engage the tangible and the contact detection on the palm, starting either far from the hand (servo shaft rotation of 80°), or very close (servo shaft rotation of 5°) to contact (0°). Over 100 trials, we measured a mean delay of 225 ms for the far position (SD 7.2 ms) and 49 ms for the near-grasp position (SD 8 ms). This leads us to estimate the fixed delay due to communication to be around 38 ms and the servo shaft rotation speed to be around the nominal rotation speed of 210 ms/90° despite the tangible object.

Influence of Motor Vibrations. The motor's motion sometimes induced transient vibrations of the Vive tracker which propagated to the hand avatar. To quantify this effect, a user wore the device with the palm facing downward against a fixed supporting structure. We recorded the tracker position while applying step motions to the servo shaft using the full range of movement of the tangible to maximise such vibrations. Over the course of 20 trials, we obtained a mean stabilisation time of 612 ms for the tracker, with induced positional errors up to 4.03 mm (SD 0.87 mm) and maximum angular errors of 2.76° (SD 0.43°).

3 Interaction Technique in VR

We implemented the simplest functional interaction technique for our device, with the aim of evaluating what functionalities and limitations are thus incurred. The rendering of an interaction between the user's hand and the tangible object uses a simple distance-based triggering paradigm. Whenever a virtual object comes close enough to the user's hand, the motor's shaft angle is driven proportionally to the virtual distance between the grasping location and the virtual object, moving the tangible object towards the user's palm. When the user's fingers touch the capacitive sensors through grasp closure, the object drifts to the predefined grasping location fitting a natural power grasp while the hand avatar is animated to envelop the virtual object. Upon release of the physical grasp, an invisible virtual proxy is released, followed by the virtual object 10 ms later. The tangible is immediately driven by smoothly interpolating the command position between that of the proxy and that of the virtual object over 100 ms (see Fig. 3). Although simple, this interaction technique elicited positive feedback from users.

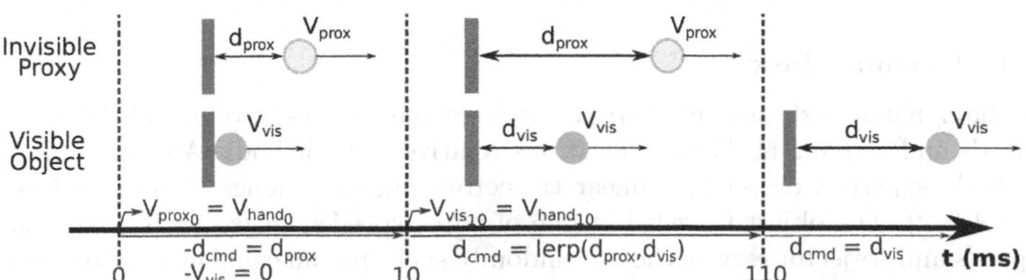

Fig. 3. Invisible proxy and visible object are released at 0 ms and 10 ms with the current hand speed v_{hand}. Their positions relative to the predefined grasping location in the virtual hand (orange) d_{prox}, d_{vis} are used to compute a smooth command d_{cmd}. (Color figure online)

4 User Study

We conducted a user study to evaluate our device and interaction technique in VR. We designed tasks covering a wide range of grasp and release interactions with different object speeds and positions relative to the user in order to determine the range of interactions supported by our device. 14 right-handed subjects (10 males, 4 females; ages 22–58 (M = 29)) participated in the study after providing written informed consent. Subjects wore the haptic device on their right hand, adjusted for their specific grasp. They viewed the virtual environment through a HTC Vive HMD, and held a Vive controller in their left hand to answer the experimental questions. We evaluated grasping and releasing in a static task where the virtual objects did not move, and in a dynamic task where the objects moved and had to be caught by the user. The virtual tasks lasted around 45 min per participant.

4.1 Static Task

Subjects stood facing a 10-cm-side cube (see Fig. 4-A) with the object appearing on one of its faces. They had to grasp and pick up the object using their dominant hand, after which they answered a first experimental question regarding synchronicity of the haptic and visual grasping interaction. They responded using a 5-point Likert scale ranging from "Not at all synchronous" (0) to "Totally synchronous" (5). In the second part of the interaction, they had to precisely place the object back onto a highlighted face of the cube. Upon releasing the object, they answered another question regarding synchronicity of the haptic and visual release using the same 5-point Likert scale. Trials were considered a failure if at any point, the subjects accidentally caused the object to drop. Subjects were instructed to minimize failures and task execution times. They began by performing 3 practice trials, then performed a total of 108 trials covering all combinations of grasping and releasing orientations with 3 repetitions each. After the trials, subjects filled out a questionnaire evaluating realism and ease of the interaction, device wearability and obtrusiveness, and task difficulty.

4.2 Dynamic Task

In the dynamic task, subjects were to catch virtual objects travelling at different speeds and arriving at different locations relative to their body. A cannon fired a single spherical object in a linear trajectory chosen amongst 7 options (see Fig. 4-B.2). The object travelled at one of three speeds: 1 m/s, 2 m/s or 3 m/s. Speeds and trajectories were chosen randomly such that an equal number of each speed was attempted for each trajectory and an equal number of attempts was made per trajectory. Subjects failed the trial if they failed to catch the sphere. After each catch, they were asked to rate synchronicity between the physical and virtual interaction, responding using the same 5-point Likert scale as in the static task. Subjects performed 3 practice trials, a total of 105 trials covering all combinations of object speeds and catching locations with 5 repetitions each, after which they filled out similar questionnaires to those from the static task.

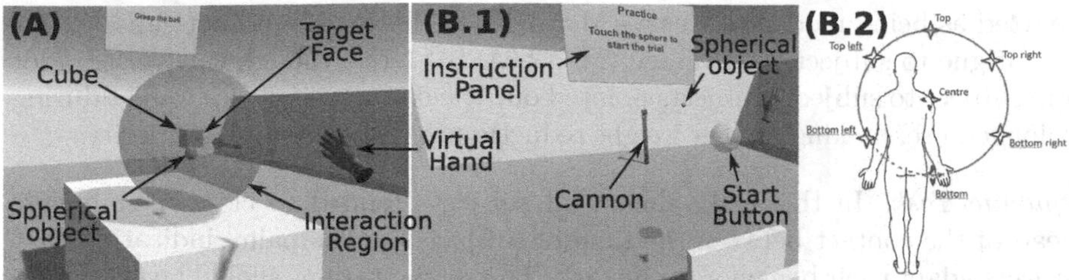

Fig. 4. Static (A) and dynamic (B.1) task environments. (B.2) The 7 catching positions.

4.3 Results and Discussion

Static Task. There was no visible effect of picking or placing position for all metrics, indicating that our device allowed similar task performances regardless of configuration, despite a single physical object approach and release direction. Task times were measured between the moment subjects entered the interaction region to the moment they respectively left it with the object in hand or completed the placing task. Picking task times were consistent within subjects, but variable between subjects (M = 2.51 s, SD 1.98 s). Placing task times followed a similar pattern (M = 2.31 s, SD 2.56 s). These task completion times indicate the device allows picking and placing interactions in reasonably short times. We measured the time between user's grasp closure and the contact between the tangible sphere and the palm of the hand. About half the population grasped the tangible object with the fingers first, while the other half waited for the physical object to collide with their palm for closing their grasp, which is an important consideration for the design of interaction techniques intended for assisting subjects during grasping. Subjects were consistently successful in both picking and placing tasks (M = 96.03%, SD 5.51% for picking; M = 86.24%, SD 9.04% for placing). Combined with the short task completion times, this is indicative of high adequacy of our device for grasping and releasing static virtual objects. We measured grasping positional error as the absolute distance separating the virtual object and the grasping position on the palm at the time of detected grasp closure. It appears all grasping orientations yielded similar errors (M = 2.6 cm, SD 1.1 cm). Subject's evaluation of picking synchronicity appears to positively correlate with grasping positional errors. Overall, subjects rated picking interactions as synchronous (M = 3.91, SD 0.86) and placing interactions as even more synchronous (M = 4.17, SD 0.92), but not significantly. Grasped objects were perceived as realistic (M = 4.14, SD 0.66), again reflecting adequacy of the device to manipulating static virtual objects. Users overall felt only moderately free in their movements (M = 3.71, SD 0.82). They reported device weight, motor vibrations and wiring as sources of obtrusiveness, rating the device as only moderately unobtrusive (M = 2.71, SD 0.73). Subjects reported high perceived virtual hand ownership (M = 4, SD 0.55), indicating that even with very rudimentary animation of the virtual hand our system is capable of maintaining immersion. The task was reported as being moderately easy (M = 3.42, SD 0.94) and did not cause excessive fatigue to subjects on average (M = 2.64, SD 1.15), though this varied a lot from subject to subject. Subjects pointed out the disturbing device weight during prolonged use, making further weight reductions as a future design priority.

Dynamic Task. In this task, almost all subjects tended to close their fingers ahead of the contact between the tangible object and the palm, indicating that subjects adapt their behavior to the task. These adaptations should be taken into account when designing interaction techniques supporting a wide range of tasks. Caught objects were perceived as moderately unrealistic (M = 2.93, SD 1.07), far below the perceived realism in the static task (means significantly different, $p < 0.001$, 2-sampled t-test). This is to be expected as the catching task amplified the perceptual effects of delays in our system. Subjects again felt moderately free in their movements (M = 3.43, SD 1.09) and rated device obtrusiveness similarly to that in the static task (M = 2.57, SD 0.76). Subjects reported a lower perceived virtual hand ownership than in the static condition (M = 3.07, SD 1; means significantly different, $p < 0.01$, paired t-test). Device limitations in the dynamic task combined with task difficulty seem to negatively affect the capacity of the device to ensure immersion. This highlights the need for improving the interaction technique if dynamic tasks are to be executed. Subjects reported that even on failed tasks, when the ball hit the hand and bounced off, the tactile feedback felt realistic. However, they also reported perceived delays in the haptic feedback which complicated the task and led to low perceived realism. Finally, the device required users to adapt their catching strategy, leading them to perform unnatural movements. This highlights a limitation of the current device and interaction technique when interacting with fast moving virtual objects.

5 Use Case

To showcase the adaptability of our device to multiple interactible virtual objects as well as the freedom of movement it provides, we designed a use case in which the user can freely roam about a virtual orchard, picking apples from trees, the ground, or tables, catching them as they fall, and even using them as ammunition in a game of "knock the cans" (see Fig. 5).

Fig. 5. Use case: (A) pick and throw an apple; (B) catch an apple falling from a branch.

6 Conclusion and Perspectives

We presented and evaluated a novel wearable haptic interface at the boundary between encounter-type displays and passive haptics. The device is grounded on the back of the hand thanks to an ergonomic adhesive silicone layer and uses a servomotor to bring a tangible object towards and away from a user's hand. We describe a simple interaction technique for VR allowing users to naturally grasp, manipulate and release objects while receiving compelling haptic feedback. Our device provides a simple and effective solution for tangible interaction with multiple virtual objects in large workspaces, with high adaptability to virtual environments. By grounding the device on the back of the user's hand, our system is unaffected by tracking issues incluencing conventional passive haptics. Furthermore, by mixing aspects from tangible haptics and encountered-type displays, our work opens perspectives towards ETHDs that provide the possibility of manipulation and object exploration through grasp closure. While our current solution only features a single fixed tangible, the ultimate goal will be to provide interactions with interchangeable or reconfigurable end-effectors in order to increase adaptability. Our device received positive feedback from users during its experimental validation, however several issues and limitations remain to be overcome. In the short term, it will be necessary to make our device at least partially wireless and more compact to increase portability and freedom of movement. Also, a reduction of the carried mass, introduction of mechanical damping elements between the servo and tracker, and improvements to the control law are avenues we wish to explore to overcome the issue of unwanted vibrations. Since in our current implementation the motor responds to the release of the grasp on the tangible object, the real object lags behind the virtual object. Our simple interaction technique compensates for this when interacting with static virtual objects but was shown to be less adequate for interactions with moving objects. We plan to explore both improvements to the control law as well as to the interaction technique (e.g. contact prediction) to make our device adaptable to a wider range of virtual interactions. Currently, the device only allows a single physical grasp position. Additional capacitive sensors should allow differentiation of grasps and thus a much wider range of object manipulations. Our interaction technique also only admits a single optimal virtual grasping location, thus providing various forms of grasping assistance to the user may improve usability. In the longer term, we wish to investigate using our device to apply force feedback towards or away from the hand, to simulate mass and inertia.

References

1. Azmandian, M., et al.: Haptic retargeting: dynamic repurposing of passive haptics for enhanced virtual reality experiences. In: Proceedings of ACM CHI, pp. 1968–1979 (2016)
2. Chossat, J.B., et al.: Soft wearable skin-stretch device for haptic feedback using twisted and coiled polymer actuators. IEEE Trans. Haptics **12**(4), 521–532 (2019)

3. Fang, H., et al.: An exoskeleton force feedback master finger distinguishing contact and non-contact mode. In: Proceedings of IEEE/ASME AIM, pp. 1059–1064 (2009)
4. He, Z., et al.: Robotic haptic proxies for collaborative virtual reality. arXiv preprint arXiv:1701.08879 (2017)
5. Insko, B.E., et al.: Passive haptics significantly enhances virtual environments. Ph.D. thesis, Univ. of North Carolina, USA (2001)
6. Kato, G., et al.: Hapsticks: a novel method to present vertical forces in tool-mediated interactions by a non-grounded rotation mechanism. In: Proceedings of IEEE World Haptics Conference, pp. 400–407 (2015)
7. Kim, Y., et al.: Encountered-type haptic display for large VR environment using per-plane reachability maps. Comput. Animat. Virtual World **29**(3–4), e1814 (2018)
8. Kohli, L.: Redirected touching: warping space to remap passive haptics. In: Proceedings of IEEE Symposium on 3D User Interfaces, pp. 129–130 (2010)
9. McNeely, W.A.: Robotic graphics: a new approach to force feedback for virtual reality. In: Proceedings of IEEE Virtual Reality Annual International Sympoisum, pp. 336–341 (1993)
10. Nakagawara, S., et al.: An encounter-type multi-fingered master hand using circuitous joints. In: Proceedings of IEEE ICRA, pp. 2667–2672 (2005)
11. Pacchierotti, C., et al.: Wearable haptic systems for the fingertip and the hand: taxonomy, review, and perspectives. IEEE Trans. Haptics **10**(4), 580–600 (2017)
12. Poupyrev, I., et al.: Actuation and tangible user interfaces: the vaucanson duck, robots, and shape displays. In: Proceedings of International Conference on Tangible and Embedded Interaction, pp. 205–212 (2007)
13. Ruffaldi, E.: Haptic rendering of juggling with encountered type interfaces. Presence **20**(5), 480–501 (2011)
14. Salazar, S.V., et al.: Altering the stiffness, friction, and shape perception of tangible objects in virtual reality using wearable haptics. IEEE Trans. Haptics **13**, 167–174 (2020)
15. de Tinguy, X., et al.: Enhancing the stiffness perception of tangible objects in mixed reality using wearable haptics. In: Proceedings of IEEE VR, pp. 81–90 (2018)
16. Yokokohji, Y., et al.: Designing an encountered-type haptic display for multiple fingertip contacts based on the observation of human grasping behaviors. Int. J. Robot. Res. **24**(9), 717–729 (2005)

Noncontact Thermal and Vibrotactile Display Using Focused Airborne Ultrasound

Takaaki Kamigaki(✉), Shun Suzuki, and Hiroyuki Shinoda

The University of Tokyo, 5-1-5 Kashiwanoha, Kashiwa-shi, Chiba-ken 277-8561, Japan
{kamigaki,suzuki}@hapis.k.u-tokyo.ac.jp, hiroyuki_shinoda@k.u-tokyo.ac.jp

Abstract. In a typical mid-air haptics system, focused airborne ultrasound provides vibrotactile sensations to localized areas on bare skin. Herein, a method for displaying heat sensations to hands where gloves are worn is proposed. The gloves employed in this study are commercially available gloves with sound absorption characteristics, such as cotton work gloves without any additional devices such as Peltier elements. The method proposed in this study can also provide vibrotactile sensations by changing the ultrasonic irradiation pattern. In this paper, we report basic experimental investigations on the proposed method. By performing thermal measurements, we evaluate the local heat generation on the surfaces of both the glove and the skin by focused airborne ultrasound irradiation. In addition, we performed perceptual experiments, thereby confirming that the proposed method produced both heat and vibrotactile sensations. Furthermore, these sensations were selectively provided to a certain extent by changing the ultrasonic irradiation pattern. These results validate the effectiveness of our method and its feasibility in mid-air haptics applications.

Keywords: Heat sensation · Vibrotactile sensation · Airborne ultrasound

1 Introduction

Recently, mid-air haptics technologies have attracted substantial interest because they can produce tactile sensations without any physical contact or the need for wearing any devices. Airborne ultrasound phased arrays (AUPAs) are one of the most practical devices in mid-air haptics. They can produce vibrotactile sensations on bare skin based on acoustic radiation pressure [1, 2]. The stimulus area, a focal point generated by AUPAs, can be down to wavelength, i.e., approximately 8.5 mm at 40 kHz (for typical AUPAs). It can be generated at an arbitrary position and controlled electronically. Complex stimulus patterns, such as the shapes of various objects based on multi-foci [3] and focal points with time-division [4], can be produced via the proper drive control of the transducers in AUPAs.

Most existing studies that employ AUPAs have developed applications that can only produce vibrotactile sensations. The realization of other types of haptic sensations using AUPAs, in addition to vibrotactile sensations, can expand the range of applications of AUPAs and contribute to the evolution of mid-air haptics technologies. In

I. Nisky et al. (Eds.): EuroHaptics 2020, LNCS 12272, pp. 271–278, 2020.
https://doi.org/10.1007/978-3-030-58147-3_30

this paper, we propose a method to produce heat sensations using AUPAs, in addition to producing vibrotactile sensations.

The proposed method requires users to wear gloves to produce heat sensations, whereas existing methods in mid-air haptics do not have such requirements. This is a limitation of the proposed method; however, the gloves that are required in the proposed method are ordinary ones, such as cotton work gloves that absorb ultrasound, without any additional devices such as Peltier elements. Additionally, vibrotactile sensations can be produced by changing ultrasonic irradiation patterns. We can find some practical applications where wearing gloves are acceptable. For example, at numerous factories, workers wear cotton work gloves while they work. Our method can be employed to prevent the inadvertent intrusion of workers' hands into dangerous zones by imparting a hot sensation as a danger alert. Our method can also be applied to surgery support by improving the surfaces of surgical gloves to absorb sound, where the glove remains disposable and battery-less.

Thermal displays in mid-air haptics, such as methods that employ infrared lasers [5] and thermal radiation [6], have been proposed. A method that uses lasers can also display a tactile sensation similar to a mechanical tap when an elastic medium is attached to the skin [7]. A generic comparison of the proposed method with the aforementioned laser-based methods is not straightforward; however, it is certain that the proposed method is the easiest, with no additional cost, with regard to application in a scenario where a worker wearing cotton gloves is already being aided by ultrasound mid-air haptics. A method for providing a cold sensation using AUPAs has been proposed; however, providing warm and hot sensations was beyond the scope of this method [8].

Herein, we report basic experimental investigations regarding the proposed method. We conducted two kinds of experiments: the first involves temperature measurements on the glove and the skin surface when the glove was exposed to ultrasound. The second is a perceptual experiment for confirming that our method provides heat and vibrotactile sensations and that it can display these sensations selectively by changing the ultrasonic irradiation patterns.

2 Proposed Method

Figure 1 shows an illustration of the proposed method. Both heat and vibrotactile sensations are produced at a focal point by AUPAs on a glove. Any type of glove can be used in this method, as long as it possesses sound absorption characteristics. A cotton work glove is used in this study. The production of heat sensations in this method is based on the phenomenon of sound absorption: heat is generated in the focal point on the glove owing to the absorption of sound, which in turn supplies heat to the skin at the contact points of the skin with the glove via the heat conduction. In addition, the generation of vibrotactile sensations utilizes the acoustic radiation pressure of the ultrasound on the glove.

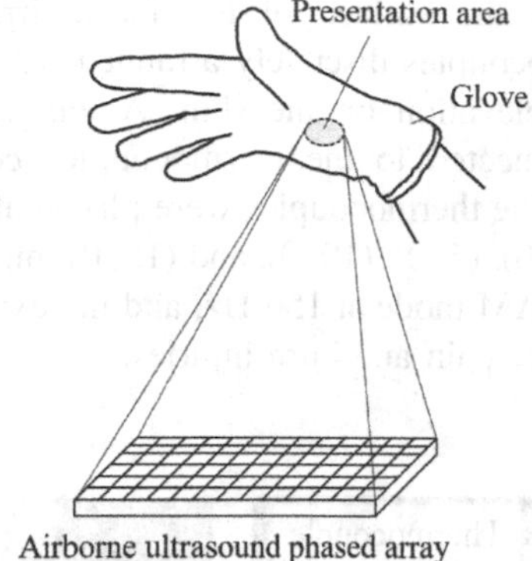

Fig. 1. Illustration of the proposed method

These sensations can be provided selectively by two modes of irradiation: static pressure (SP) mode where constant-amplitude ultrasound is irradiated and amplitude modulation (AM) mode where the ultrasound is modulated at 150 Hz. The acoustic absorption coefficient of the glove and the acoustic power at the focal point determine the temperature of the glove exposed to ultrasound. In SP mode, the ultrasound generates heat on the glove while inducing no vibrotactile sensations. In AM mode, the modulated ultrasound produces vibrotactile sensations. The modulation frequency of 150 Hz is selected so that the mechanoreceptors are excited efficiently [9]. The glove temperature also rises in AM mode. However, it is possible to adjust the amplitude and duration of the ultrasound to produce only vibrotactile sensations without heat sensation. The irradiation duration and amplitude should be adjusted to generate only vibrotactile sensations.

We confirmed the feasibility of the selective stimulation in the following experiments.

3 Experiments

3.1 Heat Generation on Glove Surface and Skin Surface

First, we measured the temperature elevation of a cotton work glove that is in contact with a human palm. The glove was exposed to ultrasound, and the temperature was measured both on the surface of the glove and the surface of the palm.

Figure 2 shows the experimental setup. We placed a fabric that was cut from a cotton work glove on an acrylic plate with square perforations of 50 mm side lengths. Subsequently, a hand was placed on the fabric to simulate the scenario of a user wearing the glove. We operated two 40 kHz AUPAs comprising 498 transducers (TA4010A1, NIPPON CERAMIC CO., LTD.) such that a focal point was generated on the fabric surface 200 mm above the surfaces of the AUPAs. The center of the focal point was considered as the origin point, (x, y) = (0, 0) mm, for the measurement. A thermography camera (OPTPI450O29T900, Optris) was employed to measure the temperature distribution on the surface of the fabric irradiated with ultrasound. Simultaneously, four thermocouples discretely arranged between the palm and fabric measured the temperature elevation on the skin. A temperature logger (SHTDL4-HiSpeed, Ymatic Inc.) connected to the thermocouples collected the data with a sampling period of 10 ms. The thermocouples were placed at 5 mm intervals in the x-direction, i.e., at (x, y) = (0, 0), (5, 0), (10, 0), and (15, 0) mm. The ultrasonic exposure patterns were SP mode and AM mode at 150 Hz, and the exposure time was restricted to 10 s in order not to induce pain and burn injuries.

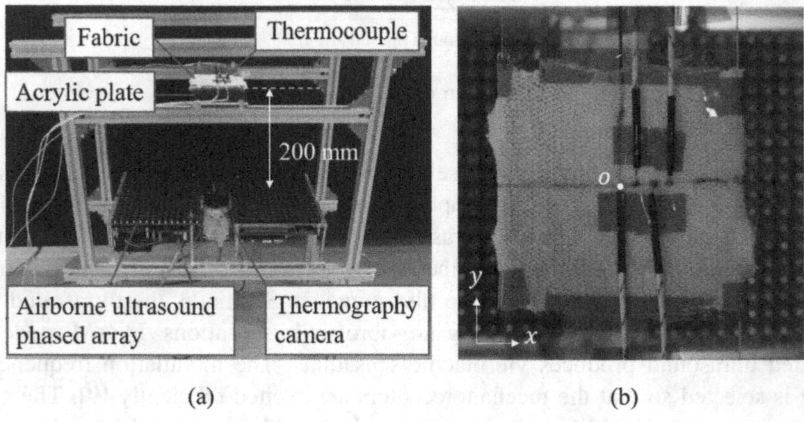

Fig. 2. Photograph of the experimental setup. (a) Complete view of the experimental setup, (b) enlarged view of thermocouples and fabric cut from a cotton work glove.

Figure 3 shows the images of the surface of the fabric that was exposed to the ultrasound over time, from which it can be clearly observed that the proposed method heated the localized area. The variation in the sizes of the focal points obtained for SP and AM modes can be attributed to distance deviation between the fabric and AUPA surfaces caused by the pressing force variation. Although some sidelobes were formed, the obtained temperature distribution is reasonable because the sound distribution on the surface with the focal point is formed in accordance with the sinc function [2]. The measured temperature was higher in the case of SP mode than that in the case of AM mode because the effective acoustic energy of SP mode is double of AM mode.

Figure 4 shows the measured results with regard to temperature changes on the skin. These results were obtained by matching the initial average temperatures for the two ultrasonic radiation patterns because the initial average temperatures differed by 1 °C for the two patterns (SP: 31.125 °C, AM: 30.125 °C). The temperature elevation was highest at the center of the focal point (x, y) = (0, 0) mm, and it decreased with an increase in the distance from the center of the focal point, for both radiation patterns. The temperature at (x, y) = (0, 15) mm did not increase for both radiation patterns. Thus, the proposed method heated the surface of the glove locally. Through comparison, it can be seen that the temperature in the case of SP mode was higher than that in

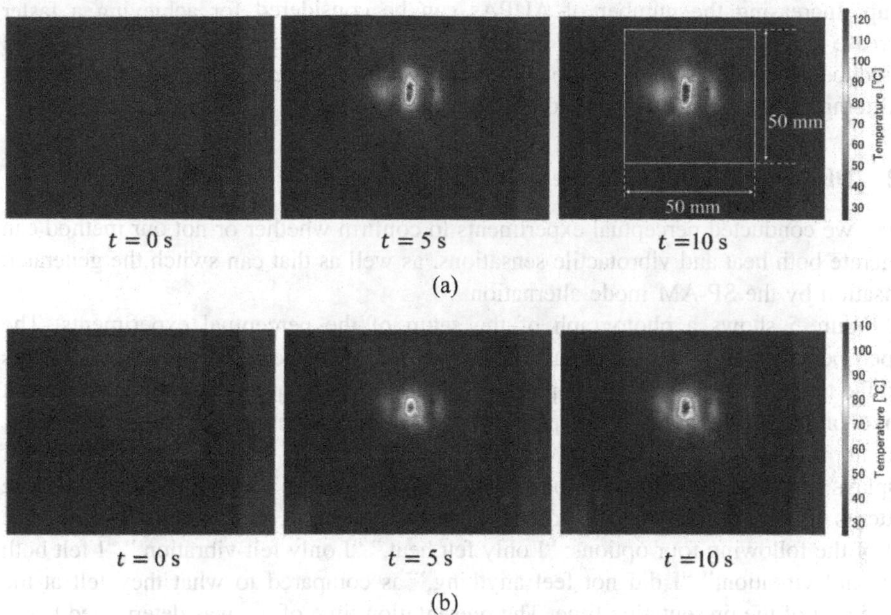

Fig. 3. Thermal images of observed area in the cases of (a) SP mode, (b) AM mode at 150 Hz.

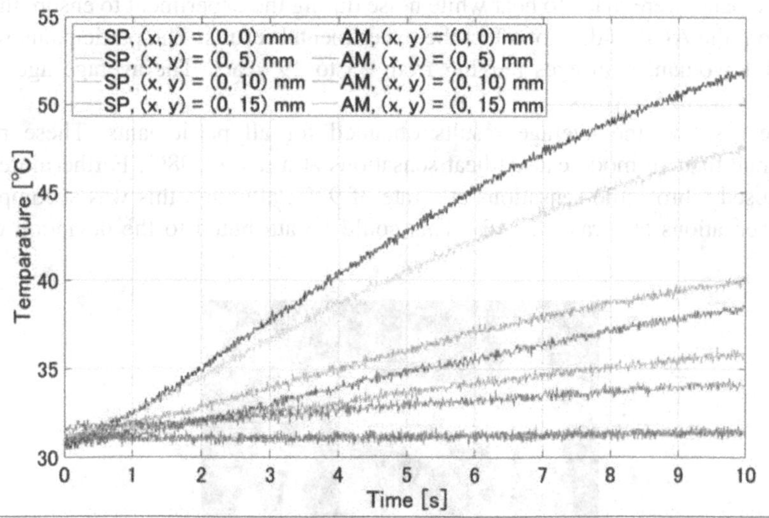

Fig. 4. Skin surface temperature measured by each thermocouple

the case of AM mode, at every measurement point. These characteristics were inconsistent with the result in Fig. 3. A temperature of 45 °C, which induces the sensation of pain [10], was achieved within 5.88 s at the earliest, at (x, y) = (0, 0) mm in the case of SP. Thus, the exposure time must be less than that in this setup. Increasing the number of AUPAs can be considered for achieving a faster increase in temperature because as the input acoustic energy increases, the time needed to induce pain will decrease. Although the relationship between the number of AUPAs and temperature elevation needs to be investigated, it is not considered in this study.

3.2 Distinguishing Vibrotactile and Heat Sensations

Next, we conducted perceptual experiments to confirm whether or not our method can generate both heat and vibrotactile sensations, as well as that can switch the generated sensation by the SP-AM mode alternation.

Figure 5 shows a photograph of the setup of the perceptual experiments. The experimental setup was the same as that shown in Fig. 2, except for the thermocouples and the temperature logger. Participants placed their hand, equipped with a cotton work glove, on the acrylic plate. We displayed three patterns: SP mode, AM mode (150 Hz), and no irradiation. Each pattern was displayed 10 times, and the total number of displays was 30 times. The order of displays was random. In each trial, one of the three patterns were presented for 5 s. After the presentation time, the participant chose one out of the following four options: "I only felt heat," "I only felt vibration," "I felt both heat and vibration," "I did not feel anything," as compared to what they felt at the beginning of the presentation time. The presentation time of 5 s was determined based on the result in Fig. 4, to ensure that the sensation of pain was not induced. Furthermore, the participants waited for 10 s after answering before the next trial was started. The participants were made to hear white noise during the experiment to ensure that the sound from the AUPAs did not affect the experimental results. The participants were 9 men and 1 woman, with ages ranging from 23 to 29 years. The average age of the participants was 25.4 years.

Table 1 shows the average results obtained for all participants. These results demonstrate that SP mode caused heat sensations at a rate of 98%. Furthermore, AM mode caused vibrotactile sensations at a rate of 97%, although this was accompanied by heat sensations at a rate of 61%. This could be attributed to the deviation of the

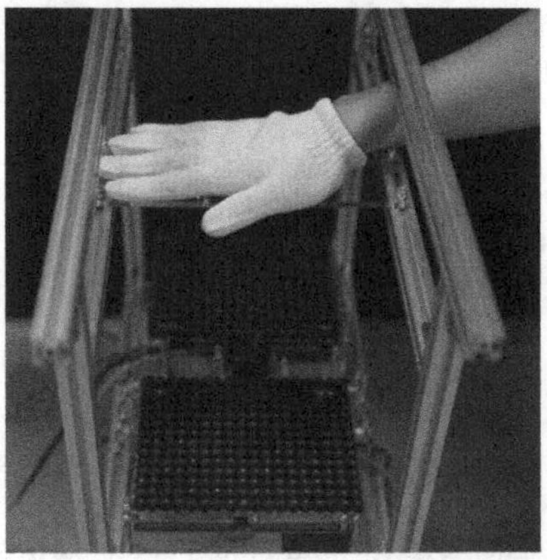

Fig. 5. Photograph of perceptual experiment

Table 1. Average results for 10 participants. The values in bracket are standard errors in each result.

Irradiation patterns	Answer			
	Heat only	Vibration only	Heat & vibration	None
SP mode	98% (1.3%)	0% (0%)	1% (0.9%)	1% (0.9%)
AM mode	3% (2.8%)	36% (7.9%)	61% (7.8%)	0% (0%)
No irradiation	0% (0%)	0% (0%)	0% (0%)	100% (0%)

initial temperature in each trial owing to residual heat from the previous trial. Thus, the proposed method has the possibility of providing only a vibrotactile sensation by adequate AM irradiation time.

In conclusion, the proposed method generated both heat and vibrotactile sensations. Additionally, these sensations were selectively generated to a certain extent by using different ultrasonic radiation patterns.

4 Discussion

The perceptual experiment shows that AM mode could provide vibrotactile sensation; however, this mechanism is not obvious. This mechanism has three possibilities, as follows: The first is that the acoustic radiation pressure of ultrasound passing through the mesh of the fabric stimulates skin. The second is that the acoustic radiation pressure vibrates the fabric, which taps skin. The third is that an elastic wave in fabric excited by irradiation ultrasound propagates to the skin. The actual mechanism needs further investigation.

The limit range of displaying heat sensation from the AUPAs is decided by acoustic energy at the focal point. The acoustic energy transmitted from AUPAs is inversely proportional to the square of the distance from AUPAs, assuming that energy attenuation in the air is ignored. The limit range in the setup of this paper is approximately 300 mm assuming that the temperature of the skin surface reaches 40 °C.

5 Conclusion

In this paper, we proposed a noncontact method for generating heat and vibrotactile sensations using focused airborne ultrasound. The proposed method provides heat sensations to a localized area by irradiating airborne ultrasound to the surface of a hand wearing a glove having sound absorption characteristics. This method also provides vibrotactile sensations by changing the ultrasonic irradiation pattern. It was confirmed that the proposed method locally provided both heat and vibrotactile sensations. In addition, our method generated these sensations selectively to a certain extent by varying the ultrasonic radiation pattern.

References

1. Iwamoto, T., Tatezono, M., Shinoda, H.: Non-contact method for producing tactile sensation using airborne ultrasound. In: Ferre, M. (ed.) EuroHaptics 2008. LNCS, vol. 5024, pp. 504–513. Springer, Heidelberg (2008). https://doi.org/10.1007/978-3-540-69057-3_64
2. Hoshi, T., Takahashi, M., Iwamoto, T., Shinoda, H.: Noncontact tactile display based on radiation pressure of airborne ultrasound. IEEE Trans. Haptics **3**, 155–165 (2010)
3. Carter, T., Seah, S.A., Long, B., Drinkwater, B., Subramanian, S.: UltraHaptics: multi-point mid-air haptic feedback for touch surfaces. In: Proceedings of the 26th Annual ACM Symposium on User Interface Software and Technology, pp. 505–514 (2013)
4. Korres, G., Eid, M.: Haptogram: ultrasonic point-cloud tactile stimulation. IEEE Access. **4**, 7758–7769 (2016). https://doi.org/10.1109/ACCESS.2016.2608835
5. Meyer, R.A., Walker, R.E., Mountcastle, V.B.: A laser stimulator for the study of cutaneous thermal and pain sensations. IEEE Trans. Biomed. Eng. BME **23**, 54–60 (1976). https://doi.org/10.1109/TBME.1976.324616
6. Saga, S.: HeatHapt thermal radiation-based haptic display. In: Kajimoto, H., Ando, H., Kyung, K.-U. (eds.) Haptic Interaction. LNEE, vol. 277, pp. 105–107. Springer, Tokyo (2015). https://doi.org/10.1007/978-4-431-55690-9_19
7. Lee, H., et al.: Mid-air tactile stimulation using laser-induced thermoelastic effects: the first study for indirect radiation. In: 2015 IEEE World Haptics Conference (WHC), pp. 374–380. IEEE (2015)
8. Nakajima, M., Hasegawa, K., Makino, Y., Shinoda, H.: Remotely displaying cooling sensation via ultrasound-driven air flow. In: 2018 IEEE Haptics Symposium (HAPTICS), pp. 340–343 (2018). https://doi.org/10.1109/HAPTICS.2018.8357198
9. Hasegawa, K., Shinoda, H.: Aerial vibrotactile display based on multiunit ultrasound phased array. IEEE Trans. Haptics **11**, 367–377 (2018)
10. Jones, L.A., Ho, H.-N.: Warm or cool, large or small? The challenge of thermal displays. IEEE Trans. Haptics **1**, 53–70 (2008). https://doi.org/10.1109/TOH.2008.2

KATIB: Haptic-Visual Guidance for Handwriting

Georgios Korres and Mohamad Eid(✉)

Engineering Division, New York University Abu Dhabi, Abu Dhabi, United Arab Emirates
{george.korres,mohamad.eid}@nyu.edu

Abstract. Haptic-visual guidance is shown to improve handwriting. This work presents a platform named KATIB (writer in Arabic) to support multi-stroke handwriting using haptic-visual guidance. A rotating neodymium magnet mounted onto a 2 DoF parallel robot underneath the writing surface is proposed to improve the fidelity of haptic guidance and mechanically decouple the stylus. The stylus utilizes stackable magnets to intensify magnetic forces. Full and partial haptic guidance methods are developed and evaluated. The current implementation demonstrates sufficient workspace of 80 mm × 60 mm and a perceivable (tangential) guidance force of 0.4 N. Magnetostatic analysis is conducted to study the effects of friction and tilting on the rendered force.

1 Introduction

Handwriting requires cognitive, visual-motor, and memory skills to master. Due to the complexity of human perception, a growing trend in the design of handwriting assistive technologies involves using multimodal interfaces [5,10]. Traditional assistive technologies for handwriting acquisition have focused on audio and visual modalities, but recently there has been a trend to exploit the haptic modality to further improve sensorimotor abilities [1,3,8,11].

In order to resemble real-life handwriting experience, haptic-visual feedback in touchscreen with stylus-based interaction are proposed [2]. A vibration motor is attached to the stylus to provide vibrotactile feedback based on the interaction with the touchscreen device [13]. Results showed that users benefit greatly from the vibrotactile feedback. Some combined vibrotactile feedback with visual guidance [9,15], and showed improved performance.

Although promising, existing methods for tactile and/or force feedback guidance are based on using mechanical attachments and do not address ergonomic

Supported by ADEK Award for Research Excellence (AARE) 2017 program under project AARE17-080.

Electronic supplementary material The online version of this chapter (https://doi.org/10.1007/978-3-030-58147-3_31) contains supplementary material, which is available to authorized users.

I. Nisky et al. (Eds.): EuroHaptics 2020, LNCS 12272, pp. 279–287, 2020.
https://doi.org/10.1007/978-3-030-58147-3_31

factors on handwriting performance (such as visual occlusion and flexibility in stylus grip and writing pressure). In this paper, we propose a multimodal system using contactless force feedback guidance based on magnetic forces that overcomes variations in grasping styles and applied pressure forces. Therefore, the contributions of this paper include the following: (1) proposing the hardware and software design of KATIB, a multimodal handwriting system with magnetic-based haptic guidance for improved ergonomics, (2) developing rendering algorithms for full and partial haptic guidance, and (3) characterization and technical evaluation of the proposed platform.

2 Related Work

Magnetically-driven haptic guidance is desirable since magnetic forces can be felt at the tip of the handwriting stylus without having mechanical attachment that may occlude the visual display. A common approach is to utilize an array of electromagnets to guide users to appropriate screen locations [16]. Actuated Workbench [7,12], Proactive Desk II [16], and Fingerflux [12] are tabletop systems that can make physical objects placed on the table move using an array of electromagnets. For instance, Fingerflux provided near-surface haptic feedback to guide the user's finger to appropriate locations on a touchscreen device [12]. However, this approach reported a drifting error of more than 10 mm which makes it unsuitable for handwriting tasks (literature suggests less than 3 mm error for handwriting tasks [4]).

An interesting approach is to attach a magnet or an electromagnet to the end effector of a two DoF motorized linkage mechanism to control the stylus, altogether placed underneath a writing surface (paper) to provide haptic guidance [6,14]. In the dePENd system, a computer controls the xy position of the magnet under the writing surface in order to move the pen and present haptic guidance [14]. As the linkage mechanism moves the magnet along a desired trajectory, it attracts the stylus through magnetic forces to move along the same trajectory. No visual guidance is provided. A recent study demonstrated a system to deliver dynamic guidance in drawing and sketching via an electromagnet placed underneath a pressure sensitive tablet [6]. The system allows the user to move the pen freely and renders pull back forces using a closed-loop time-free approach to minimize the error between the pen position and the desired trajectory.

These systems do not provide synchronized haptic-visual guidance for handwriting. Driven by previous findings that multimodal guidance is more effective for motor learning [10], KATIB provides mechanically decoupled, multimodal feedback system with synchronized haptic-visual guidance to support multi-stroke handwriting. Compared to electromagnetic-based guidance, the permanent magnet provides concentrated magnetic flux and thus higher fidelity of haptic guidance. Furthermore, larger haptic guidance forces are achievable with permanent magnet, compared to electromagnet. Finally, the heating effects of the electromagnet weakens the magnetic flux and thus reduces the haptic guidance force over usage.

3 KATIB System

A schematic diagram of the KATIB system design is shown in Fig. 1(a). The system comprises a 3D printed pen-like stylus that hosts cylindrical 5×5 [mm] (or more) and one cylindrical 2×2 [mm] vertically polarized N42 neodymium magnets that are stacked on top of each other forming the stylus tip, a low-cost 4 mm thick resistive touchscreen 640×480 pixels display to provide visual guidance and capture interactions between the writing surface and the stylus, a N42 neodymium magnet underneath the screen that is attached to the end-effector of a 2 DoF parallel manipulator for moving the magnet along a particular trajectory and provide magnetic haptic guidance, and a board computer that runs an application to provide synchronized audio, visual, and haptic guidance for the learner. A snapshot of the system prototype is shown in Fig. 1(b).

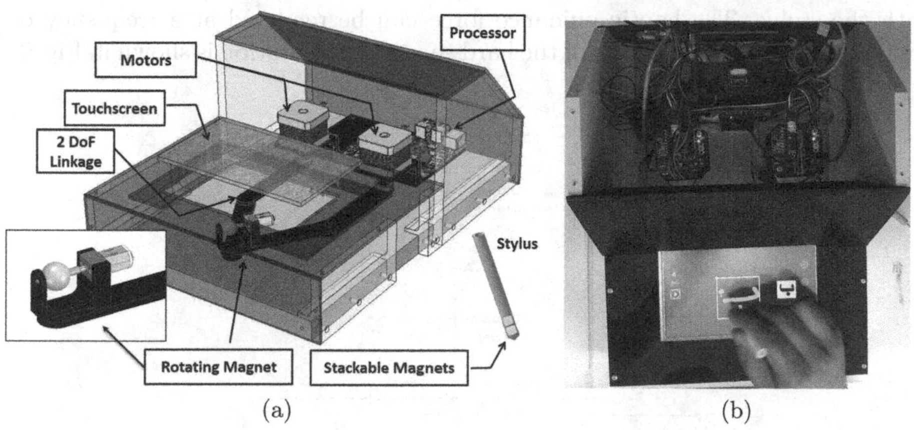

Fig. 1. (a) KATIB system design, (b) KATIB prototype.

3.1 Magnetic Force Acting on the Stylus

The magnetic force due to a non uniform magnetic field can be calculated through the following equation:

$$\mathbf{F} = \nabla(\mathbf{m} \cdot \mathbf{B}), \ \mathbf{m} = \frac{1}{\mu_0}\mathbf{B}_r V \tag{1}$$

whereas $\mathbf{m}$ is the magnetic moment vector. In case of a permanent magnet the magnetic moment can be expressed through the residual flux density of the magnet B_r. The residual flux density of the magnet is usually provided by the manufacturer and V is the volume of the magnet. The values of the parameters above, as well as the derivation of the magnetic flux distribution $\mathbf{B}$, are discussed in the magnetostatic analysis in Sect. 4.2

3.2 Hardware Implementation

Katib sytem was designed around a Raspberry Pi Model B+ single board computer which is running Linux software. It drives two NEMA17 stepper motors which are equipped with a 5:1 reduction planetary gearbox. The stepper motors are PID controlled through the uStepper driver platform which is based on an ATMEL ATMEGA328 MCU, the Trinamic TMC5130 Motor Driver and the Broadcom AEAT8800-Q24 Hall effect encoder. The motors can operate at a maximum update rate of 1800 Hz. The rotation of the end effector is achieved by the use of a custom solution comprised of a 15k RPM micro DC motor equipped with a 300:1 reduction gearbox and a quadrature hall effect encoder which is driven from a Texas Instruments DRV8838 driver and an ATMEL ATMEGA328. The Rasberry PI is also connected to a 640×480 TFT display which is driven by a generic HDMI-TFT driver. The display is equipped with a resistive touchscreen driven by the MICROCHIP AR1100 touchscreen controller to detect contact with the stylus. The haptic guidance force can be rendered at a frequency of 1800 Hz. A schematic diagram of the hardware implementation is shown in Fig. 2.

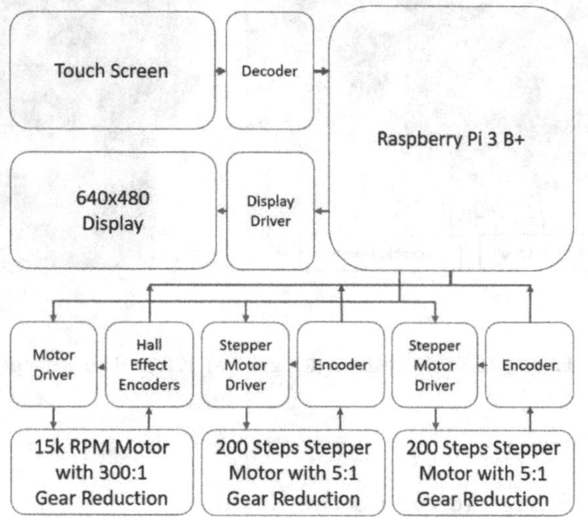

Fig. 2. Hardware implementation.

3.3 Graphical User Interface Design

As shown in Fig. 3, KATIB provides two interfaces: an instructor window and a learner window. The instructor window enables teachers to record a handwriting task, assign it to one or more learners, and examine the learners performance. On the other hand, the learner window provides learners with a list of handwriting

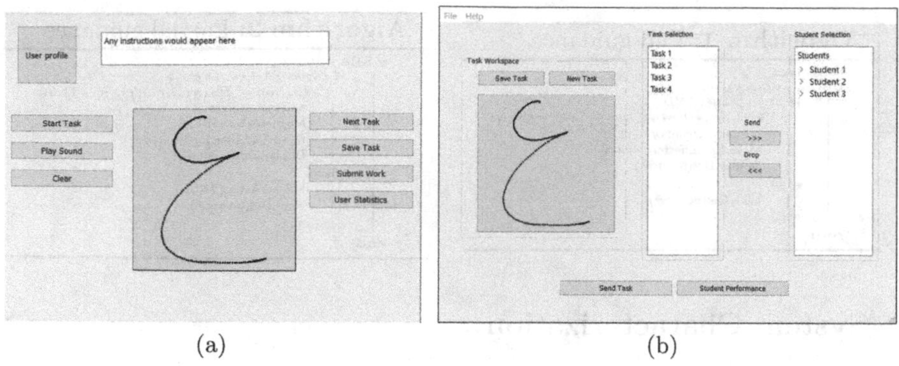

Fig. 3. (a) Learner interface, (b) Instructor interface.

tasks to exercise and record. The recorded handwriting tasks are evaluated and a report about the learner's performance is sent back to the instructor.

3.4 Haptic-Visual Guidance

Visual guidance is implemented by showing a visual trace of the entire handwriting task as well as a highlighted visual target for the immediate next move along the handwriting trajectory. In multi-stroke tasks, a flashing visual dot of different color is displayed to show the starting point of the subsequent stroke. As for haptic guidance, full and partial haptic-visual guidance methods are developed for KATIB system. In the full haptic guidance method, the system leads the movement by providing visual feedback about the next point to move to along the trajectory and applies a maximum force to move the stylus to the desired position. Once the user is at the desired position, the next point is identified and haptic-visual guidance is provided for the next point. The full guidance method is detailed in Algorithm 1. Figure 5 demonstrates high fidelity of full haptic guidance where the average root mean square (RMS) error is 2.73 mm.

Partial haptic-visual guidance is user-led. The KATIB system provides visual feedback for the next point to move the stylus to and provide force guidance only when the user deviates significantly from the desired trajectory. The magnetic force is switched on or off by rotating the magnet by 180° from its original position. If the user completes the entire handwriting task within the trajectory error threshold, no haptic feedback is applied. The partial haptic-visual guidance method is shown in Algorithm 2.

Algorithm 1: Full guidance

```
initialization;
while poitnts#>0 do
    if Stylus_activeAND
      ||Tip_Loc − Point_Loc||L2 < ε then
        points# =points# − 1;
        Move_End_Effector(points#);
        Display_Dot(points#);
    else
        blink_Dot(points#);
    end
end
```

Algorithm 2: Partial guidance

```
while poitnts#>0 do
    if StylusIsActive AND
      ||Tip_Loc − Point_Loc||L2 < ε then
        points# =points# − 1;
        MagnetActive(FALSE);
        Move_End_Effector(points#);
        Display_Dot(points#);
    else
        MagnetActive(TRUE);
        blink_Dot(points#);
    end
end
```

4 System Characterization

4.1 Haptic Guidance Workspace

The work space of the 2DOF parallel manipulator can be calculated by solving the forward kinematics problem. The stepper motors can be driven in up to 16 microsteps and since the planetary gearbox has a 5:1 reduction rate, the final number of steps per motor shaft revolution is 16000 steps. A Matlab scrip was written to solve the kinematic problem for the specified step resolution for a range between −45 to 45° per motor and 100 mm as the length of each parallel manipulator arm. The measured workspace for the device is 80 mm (width) by 60 mm (height), which is sufficient for most handwriting tasks.

4.2 Magnetostatic Analysis

In order to calculate the magnetic force acting on the stylus according to Eq. 1, we need to derive the magnetic field distribution for the specific configuration (magnet, glass screen, stylus magnet) under a steady current, which in this case is equal to zero. The magnetostatic equations are derived from Maxwell's equations under the assumption of either fixed or moving charges with a steady current. In such cases, Maxwell's equations can split into two pairs of equations: two equations describing the electric field (electrostatics) and two equations describing the magnetic field (magnetostatics).

The ANSYS Magnetostatic Analysis module was utilized to solve the magnetic field distribution for the different translational and tilt cases. The geometry consisted of a stack of 2 cylindrical magnets for the stylus(5×5 mm and 2×2 mm respectively) and one cylindrical (5×5 mm) magnet for the end effector, separated by a surface of glass material with 4 mm thickness ($\mu_{r/glass} = 5$). The structure was enclosed in a volume of simulated air ($\mu_{r/air} = 1$). Neodymium N42 (NdFeB) was used to simulate the magnets with a residual induction of 1300 mT and a coercive force of 955 KA/m. The geometry was meshed using a "sphere of influence", a heavily refined spherical mesh volume encapsulating the region whereas the magnets where interacting, to ensure a stable and accurate approximation. The final mesh comprised about 250k elements. Finally, the simulation was repeated for the different translation and tilt configurations of the stylus and end effector. The tilt axis was defined by the tangency between the base of the stylus magnet and the glass surface.

Effects of Stylus Displacement/Tilt on Guidance Force. As the magnet moves to the next position along the trajectory, it attracts the stylus to move into the same direction. The normal and tangential attraction forces are analyzed as function of the horizontal distance between the magnet and the stylus. The effects of displacement is simulated as shown in Fig. 4 (a). The results, shown in Fig. 4 (b), demonstrate a peak tangential force of 0.43 N from 2 mm to 5 mm displacement, which produces a perceivable haptic guidance. Furthermore, the effects of tilting the stylus on the amplitude of the attraction force is also studied. As shown in Fig. 4 (c), tilting the stylus up to 40° from the normal direction would results in a negligible tangential force.

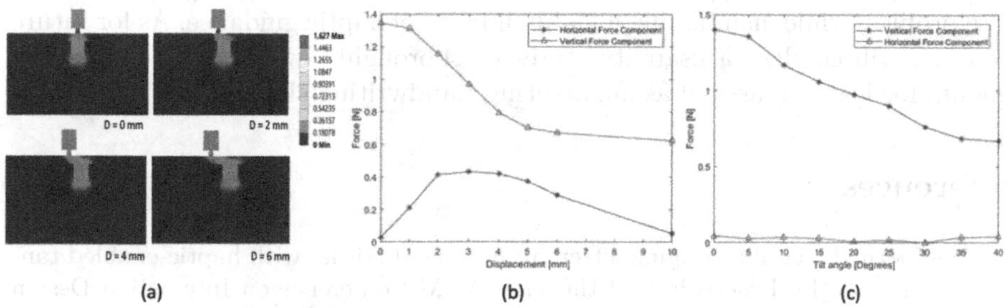

Fig. 4. (a) Magnetic interaction modeling, (b) magnetic forces against displacement, (c) magnetic forces against tilt angle.

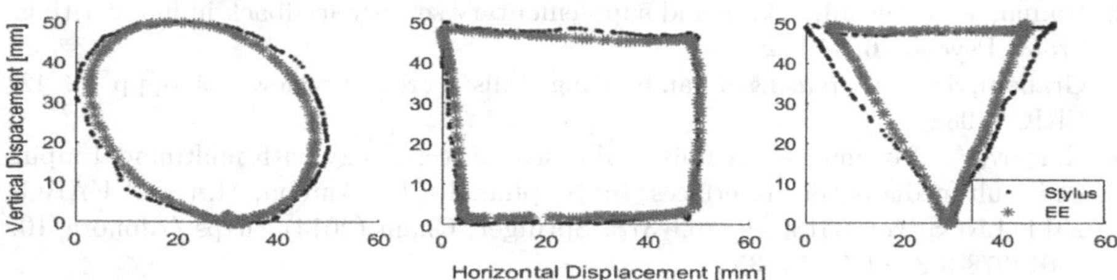

Fig. 5. Sample handwriting tasks with stylus drifting along the writing surface (EE for end-effector).

Friction Effects on Stylus. The effects of friction between the stylus and the handwriting surface is examined. The stylus freely drifts along the handwriting trajectory due to the attraction force by the moving magnet. The magnet and stylus positions are recorded for three different handwriting tasks and plotted in Fig. 5 where differences in traces are mainly due to friction effects. Since the static friction f_s is larger than the dynamic friction f_d, it must be true that the tangential force Fx is larger than the static friction force fs because the stylus is moving. Given the results from Fig. 4(b) showing that the tangential force maximizes at around 0.43 N at 3.5 mm away from the magnet implies that the static friction must be less than 0.43 N to be able to move the stylus (see Fig. 5). It must be noted that this is an extreme case and in reality the user is holding the stylus and applying some forces along the desired trajectory. Also, while the user is grasping the stylus there is always a tilt angle that reduces further the effects of the friction.

5 Conclusion

In this paper, we proposed a haptic-visual guidance system that utilizes magnetic forces for haptic guidance to support handwriting. By using magnetic forces, the system improves the ergonomics of handwriting (support for grasping styles and force profiles) while maintaining a highly fidelity of haptic guidance. As for future work, we will conduct a usability study to thoroughly evaluate the ergonomic benefits for learning as well as augmenting handwriting skills.

References

1. Asselborn, T., et al.: Bringing letters to life: handwriting with haptic-enabled tangible robots. In: Proceedings of the 17th ACM Conference on Interaction Design and Children, pp. 219–230. ACM (2018)
2. Cho, Y., Bianchi, A., Marquardt, N., Bianchi-Berthouze, N.: Realpen: providing realism in handwriting tasks on touch surfaces using auditory-tactile feedback. In: Proceedings of the 29th Annual Symposium on User Interface Software and Technology, pp. 195–205. ACM (2016)
3. Danna, J., Velay, J.L.: Basic and supplementary sensory feedback in handwriting. Front. Psychol. **6**, 169 (2015)
4. Graham, S.: Measurement of handwriting skills: a critical review. vol. 8, pp. 32–42. ERIC (1982)
5. Karpov, A., Ronzhin, A.: A universal assistive technology with multimodal input and multimedia output interfaces. In: Stephanidis, C., Antona, M. (eds.) UAHCI 2014. LNCS, vol. 8513, pp. 369–378. Springer, Cham (2014). https://doi.org/10.1007/978-3-319-07437-5_35
6. Langerak, T., Zarate, J., Vechev, V., Panozzo, D., Hilliges, O.: A demonstration on dynamic drawing guidance via electromagnetic haptic feedback. In: The Adjunct Publication of the 32nd Annual ACM Symposium on User Interface Software and Technology, pp. 110–112. ACM (2019)
7. Pangaro, G., Maynes-Aminzade, D., Ishii, H.: The actuated workbench: computer-controlled actuation in tabletop tangible interfaces. In: Proceedings of the 15th annual ACM symposium on User interface software and technology, pp. 181–190. ACM (2002)
8. Park, W., Korres, G., Moonesinghe, T., Eid, M.: Investigating haptic guidance methods for teaching children handwriting skills. IEEE Trans. Haptics **12**, 461–469 (2019)
9. Portillo, O., Avizzano, C.A., Raspolli, M., Bergamasco, M.: Haptic desktop for assisted handwriting and drawing. In: ROMAN 2005. IEEE International Workshop on Robot and Human Interactive Communication, pp. 512–517. IEEE (2005)

10. Sigrist, R., Rauter, G., Riener, R., Wolf, P.: Augmented visual, auditory, haptic, and multimodal feedback in motor learning: a review. Psychon. Bull. Rev. **20**(1), 21–53 (2013)
11. Teranishi, A., Korres, G., Park, W., Eid, M.: Combining full and partial haptic guidance improves handwriting skills development. IEEE Trans. Haptics **11**(4), 509–517 (2018)
12. Weiss, M., Wacharamanotham, C., Voelker, S., Borchers, J.: Fingerflux: near-surface haptic feedback on tabletops. In: Proceedings of the 24th Annual ACM Symposium on User Interface Software and Technology, pp. 615–620. ACM (2011)
13. Withana, A., Kondo, M., Makino, Y., Kakehi, G., Sugimoto, M., Inami, M.: Impact: immersive haptic stylus to enable direct touch and manipulation for surface computing. Comput. Entertainment (CIE) **8**(2), 9 (2010)
14. Yamaoka, J., Kakehi, Y.: depend: augmented handwriting system using ferromagnetism of a ballpoint pen. In: Proceedings of the 26th Annual ACM Symposium on User Interface Software and Technology, pp. 203–210. ACM (2013)
15. Yang, X.D., Bischof, W.F., Boulanger, P.: Validating the performance of haptic motor skill training. In: 2008 Symposium on Haptic Interfaces for Virtual Environment and Teleoperator Systems, pp. 129–135. IEEE (2008)
16. Yoshida, S., Noma, H., Hosaka, K.: Proactive desk ii: development of a new multi-object haptic display using a linear induction motor. In: IEEE Virtual Reality Conference (VR 2006), pp. 269–272. IEEE (2006)

ThermalTex: A Two-Modal Tactile Display for Delivering Surface Texture and Thermal Information

X. Guo[1(✉)], Y. Zhang[1,2], W. Wei[1], W. Xu[3], and D. Wang[1,2]

[1] State Key Lab of Virtual Reality Technology and Systems, Beihang University, Beijing 100083, China
hapticwang@buaa.edu.cn

[2] Beijing Advanced Innovation Center for Biomedical Engineering, Beihang University, Beijing 100083, China

[3] Department of Mechanical Engineering, The University of Auckland, Auckland 1142, New Zealand

Abstract. We present a two-modal surface display that can simultaneously display thermal and texture stimuli to the user's fingers. The texture is generated by electrovibration effect on a flexible film with a high thermal conductivity. The temperature of the film is controlled by Peltier element attached with a water-cooling heatsink. The performance of the prototyped device is evaluated by the temperature response to the step signal and the maximum electrostatic force of the display. The application of the display is demonstrated in a virtual reality environment where users can feel the two modal haptic property of the virtual object.

Keywords: Haptics · Electrovibration · Temperature · Thermal stimuli

1 Introduction

Human tactile perception systems contain diverse types of receptors, such as mechanoreceptors and thermoreceptors [1]. These receptors work in an integrated manner to perceive a haptic sensation [2]. For example, when human explores the tactile properties on the surface of real object, the texture and temperature are always perceived synchronously. This fact appears indicating that the simultaneous presentation of texture and temperature information is beneficial for the reproduction of surface characteristics.

Some researchers have tried to add temperature feedback to surface haptic feedback devices by combining pin-array and the thermal feedback [3–5]. The pin-array stimuli

---

Electronic supplementary material The online version of this chapter (https://doi.org/10.1007/978-3-030-58147-3_32) contains supplementary material, which is available to authorized users.

I. Nisky et al. (Eds.): EuroHaptics 2020, LNCS 12272, pp. 288–296, 2020.
https://doi.org/10.1007/978-3-030-58147-3_32

were actuated by piezoelectric bimorphs and the thermal feedback was created by a Peltier thermoelectric element. To improve the quality of teleoperation, Gallo et al. [6] presented a flexible tactile display delivering pattern and thermal stimuli. The pattern was realized by a hybrid electromagnetic-pneumatic actuation and the temperature is controlled by a Peltier element. More recently, Strese et al. [7] presented a tactile mouse to display the hardness, friction, warmth and roughness of virtual objects. The hardness, friction and roughness were actuated by the electromagnet, servo and voice coil actuator, and the thermal stimuli were created by a Peltier element.

We aimed to develop a new tactile interface, which can provide electrovibration and thermal stimuli. The electrovibration stimuli was generated on a high thermal conductivity film. The Peltier element provides the thermal stimuli through heating and cooling the film. Furthermore, we evaluated the performance of the prototyped device by measuring temperature change ability and the maximum electrostatic force. Finally, a special demo was designed to show the proposed display ability in presenting surface properties.

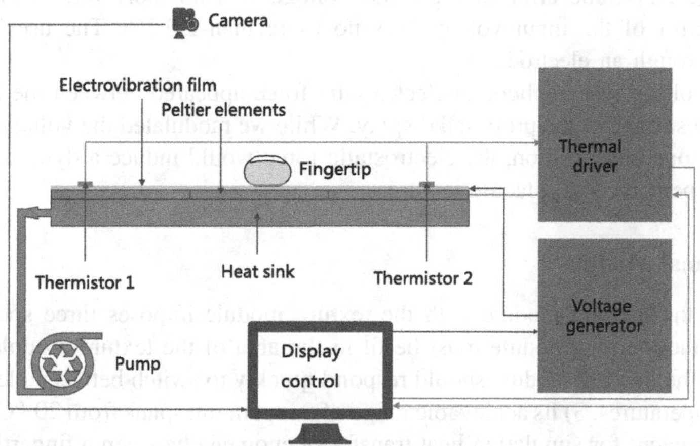

Fig. 1. The proposed tactile display: the system diagram

2 Design

The proposed display consists of two parts: a texture module and a thermal module. These modules work independently and are readily integrated. The texture module generates electrovibration effect by a flexible film. The thermal module controls the temperature by the Peltier effect. The display design was shown in Fig. 1.

2.1 Texture Module

In order to efficiently control the temperature of the texture module, the electrovibration actuator in the texture module needs to be thin and has high thermal conductivity. Therefore, we designed a thin electrovibration film with a high thermal conductivity.

Generally, the electrovibration actuator composed of three layers, including the base layer, conductive layer and insulating layer. Nano-carbon copper foil (NCF) is very thin and has high thermal conductivity, so we used it as the base layer and conductive layer. The NCF consists of a base layer of Nano-carbon and a copper layer, which are about 0.15 mm. The insulating layer is a key part of generating electrovibration stimulus, which needs to have high electrical resistance and is easy to coat on the copper layer. We coated a layer of Polyimide (PI) on top of the copper layer as the insulator, which is roughly 2 μm and has 0.12 w/(m.k) thermal conductivity. Totally, Electric vibrating film length 120 mm, width 80 mm, thickness 0.15 mm. Totally, electrovibration film is 120 × 80 × 0.15 mm.

A voltage controller was used to generate a periodic voltage applying to the conductive layer. This controller can output square wave, 0–350vpp and 10–5000 Hz voltage. The amplitude error of the input voltage was no more than ±4% and the frequency error of the input voltage was no more than ±0.2%. The user's wrist is grounded through an electrode.

As the voltage was applied, an electrostatic force appeared between the finger and the touching surface of the proposed display. While we modulated the voltage based on the time or touching position, the electrostatic force would induce a dynamic friction, which then perceived as a tactile texture.

2.2 Thermal Module

Combining the thermal module with the texture module imposes three specific constraints: 1) the thermal module must be fit in the area of the texture module (8 cm * 12 cm). 2) The thermal module should respond quickly to switch between elevated and reduced temperatures. 3) Its achievable range of temperature spans from 20 °C to 35 °C, which is sufficient for simulating heat transfer phenomena between a fingertip.

The Peltier element acted as a thermal actuator because it is only electrically driven and responds quickly to switch between elevating and reducing temperatures. In order to estimate the power density of the Peltier element, the maximum heat flux Q generated when a hand touched the entire haptic module area can be derived from two semi-infinite body models:

$$Q = \frac{T_e - T_d}{R_{e-d}} \tag{1}$$

where T_e is the environmental temperature, T_d is the targeted temperature of the device, and R_{e-d} is the thermal contact resistance between the environment and the device. The value of R_{e-d} at a contact force of approximately 2 N can be approximated as a function of the thermal conductivity of the tactile surface material (κ_d), shown as follows:

$$R_{e-d} = \frac{0.37 + \kappa_d}{1870 \times \kappa_d} \tag{2}$$

The resting temperature of the skin of the hand often varies between 25 °C and 36 °C [8]. We chose an average value and assumed the hand temperature is 30 °C. When the texture module is kept at 20 °C, the derived maximum heat flux is 6.1 kW/m^2. The whole area of the contact area is 8*12 cm^2. The maximum heat pumping capacity is 58.2 W.

We choose six 40 * 40 * 4.6 mm Peltier elements (TEC1-12704T125, Beijing Huimao Refrigeration Equipment Co., Ltd., China) with a maximum cooling capacity of 40.1 W was selected. The total area of the six Peltier elements is fit with the tactile contact area and the total heat cooling capacity is about 240 W, which is four times larger than the estimated values because the maximum cooling capacity of Peltier element is obtained at a temperature difference between the two plates of the Peltier element of 0 °C. Thus, the real maximum Qc depends on the heatsink performance. Since performance of Peltier thermoelectric module depends, we designed a water-cooling system to increase their performance. The water-cooling system consists of a pump, a water tank and a water-cooling block that is attached to the heat generation side of Peltier elements a piece of thermally conductive silicone.

As our device have a narrow working temperature range from 18 °C to 38 °C, and need a fast-transient response to simulate the finger contact with a material, we chose two NTC thermistor temperature sensors for our system. The thermistor is a Φ1.6 × 4.5 mm cylinder with a thermal sensitivity of ±0.1 °C (between 0 °C and 70 °C).

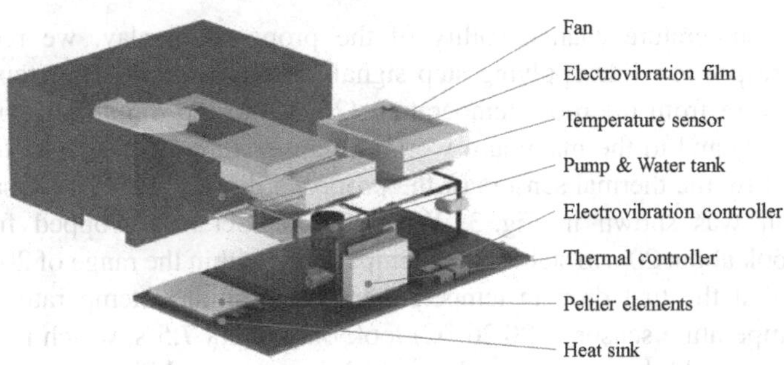

Fig. 2. The components of the ThermalTex

3 Fabrication and Assembly

The electrovibration film was fabricated by a molding process, e.g. the NCF was spin-coated with a fluorine PI, and then baked at 200 °C for 2 h.

The six Peltier elements are arranged as shown in Fig. 2 to form an 8 * 12 cm plane. The bottom surface of the Peltier elements was glued onto the heatsink and the top surface was glued onto the electrovibration film by the thermal grease. The thermal grease could reduce the thermal resistance between the Peltier elements, the heatsink and electrovibration films. Two NTC thermistor temperature sensor was attached to the top surface of the electrovibration, aiming to directly measure the temperature of the electrovibration film. In order to accurately measure the temperature of the electrovibration film, the temperature sensor needs to be closely attached to the top surface of the electrovibration film. We use thermal grease to wrap the temperature sensor and fix it on the top surface of the electrovibration film with adhesive tape.

4 Performance Measurements

The ThermalTex depends on the temperature and the electrostatic force applying to the fingertip. Different tactile sensations can be created by controlling the amplitude and frequency of electrostatic force and the variation of the temperature. Therefore, it is essential to know capacity of the proposed display in variation of the temperature and the generating electrostatic force.

We conducted two experiments with ThermalTex. In the first experiment, we measured temperature change ability. The maximum electrostatic force at different temperature was measured in the second experiment.

4.1 Temperature Response

To evaluate temperature change ability of the proposed display, we recorded the temperature response while applying step signals of temperature. The step signals of temperature were from the room temperature (25 °C) to the minimum working temperature (20 °C) and to the maximum working temperature (35 °C). The temperature was measured by the thermal sensors of the proposed display with 10 Hz sample rate.

The result was shown in Fig. 3. When the temperature dropped from 25 °C to 20 °C, it took about 20 s to stabilize the temperature within the range of 20 ± 0.5 °C. We noticed that the first drop in temperature to a minimum (temperature sensor 1: 20.52 °C, temperature sensor 2: 20.26 °C) took only about 7.5 s, which indicated that the temperature could decrease faster, but it took longer to stabilize to the target value. When the temperature rose from 25 °C to 35 °C, there was a similar trend. It took 30 s to stabilize within 35 ± 0.5 °C, and the first rise to the maximum value (temperature sensor 1: 35.93 °C, temperature sensor 2: 36.25 °C) took only about 9 s.

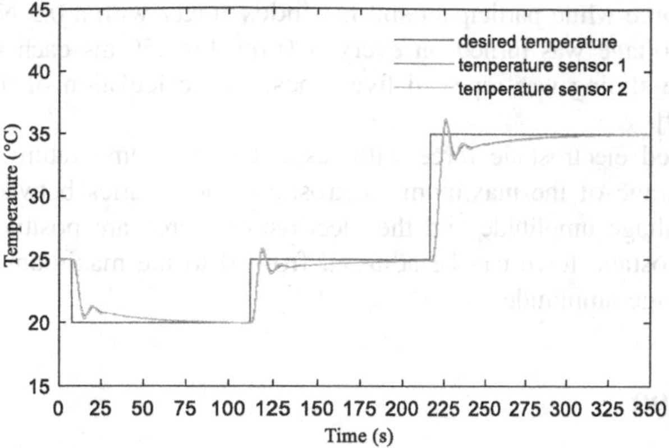

Fig. 3. The behavior of the proposed display for temperature steps

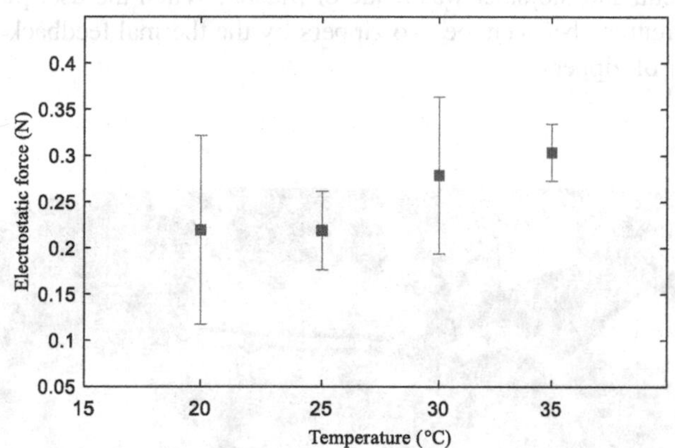

Fig. 4. The electrostatic force was estimated while applying the voltage amplitude of 350 Vpp at 140 Hz

4.2 Electrostatic Force

We used an apparatus to measure the maximum electrostatic force generating on the proposed display. Detail information of the apparatus can be seen in [9, 10]. The maximum electrostatic force occurred while applying the maximum amplitude of the voltage (350 Vpp square voltage). Thus, we choose the 350 Vpp amplitude of applying voltage. The frequency is 140 Hz, which located in the sensitive range of the electrovibration display [11, 12]. Four temperature levels (20, 25, 30, 35 °C) cover the working range of the proposed display.

A participant joined this experiment. To decrease the hydration level of the fingertip, talcum powder was used to dry the fingertip. We recorded the normal force and the tangential force while participant slid her index finger with a 0.5 N normal force. The applying voltage was turned on every 500 ms for 250 ms each time. For each temperature, the sliding was repeated five times. The calculation of the electrostatic force refer in [9].

The estimated electrostatic force with respect to the temperature was shown in Fig. 4. The average of the maximum electrostatic force varies between 0.22 N and 0.30 N. The voltage amplitude and the electrostatic force are positive related [13]. Thus, the electrostatic force can be adjusted from 0 to the maximum value by controlling the voltage amplitude.

5 Application

The ThermalTex can be applied to many situations, such as teleoperation, remote palpation and virtual shopping, etc. Figure 5 demonstrates an application in virtual shopping. In this scenario, we set up two kinds of clothes with zipper, one of which was made of metal and the other was made of plastic. When the user pulls the zipper, he/she can differentiate between the two zippers by the thermal feedback while feel the tactile sensation of zipper.

Fig. 5. Application of the proposed display. Users can see the virtual products by the VR headset and perceive the tactile sensations of the zipper in the virtual store by the proposed display

6 Conclusion

The paper presented a new tactile interface to provide feedback of surface properties of fine texture and temperature. The performance of the thermal and the texture feedback was assessed by the temperature response to the step signals and the maximum electrostatic force of the display, respectively. The results show that the temperature can change rapidly, but it takes about 30 s to stabilize to the target value. The maximum electrostatic force varies between 0.22 N and 0.30 N, when the normal force of 0.5 N was applied by the sliding finger. Potential application was demonstrated by a virtual shopping scenario.

The performance evaluation of the ThermalTex shows two challenging issues to be addressed. The temperature response time is slow and needs to improve for the requirements of real-time display. Furthermore, there is a large variation in the electrostatic force at 20 and 30 °C, and the reason needs to be explored in the future.

Acknowledgment. This research is supported by the National Key Research and Development Plan under Grant No. 2017YFB1002803, and the National Natural Science Foundation of China under the grant No. 61532003.

References

1. Toney, A., Dunne, L., Thomas, B.H., Ashdown, S.P.: A shoulder pad insert vibrotactile display. In: Seventh Proceedings of the ieee International Symposium on Wearable Computers, pp. 35–44 (2003)
2. Lederman, S.J., Klatzky, R.L.: Haptic perception: a tutorial. Attention Percept. Psychophysics **71**, 1439–1459 (2009)
3. Yang, G.H., Kyung, K.U., Srinivasan, M.A., Kwon, D.S.: Development of Quantitative Tactile Display Device to Provide Both Pin-Array-Type Tactile Feedback and Thermal Feedback. IEEE Computer Society, Los Alamitos (2007)
4. Yang, G.H., Kyung, K.U., Srinivasan, M.A., Kwon, D.S.: Quantitative tactile display device with pin-array type tactile feedback and thermal feedback. In: 2006 IEEE International Conference on Robotics and Automation, Orlando, FL, USA, pp. 3917–3922 (2006)
5. Hribar, V.E., Pawluk, D.T.V.: A Tactile-Thermal Display for Haptic Exploration of Virtual Paintings. Association for Computing Machinery, Dundee (2011)
6. Gallo, S., Son, C., Lee, H.J., Bleuler, H., Cho, I.J.: A flexible multimodal tactile display for delivering shape and material information. Sens. Actuators, A Phys. **236**, 180–189 (2015)
7. Strese, M., Hassen, R., Noll, A., Steinbach, E.: A tactile computer mouse for the display of surface material properties. IEEE Trans. Haptics **12**, 18–33 (2019)
8. Jones, L.A., Ho, H.: Warm or cool, large or small? the challenge of thermal displays. IEEE Trans. Haptics **1**, 53–70 (2008)
9. Guo, X., Zhang, Y., Wang, D., Lu, L., Jiao, J., Xu, W.: The effect of applied normal force on the electrovibration. IEEE Trans. Haptics **12**, 571–580 (2019)
10. Guo, X., Zhang, Y., Wang, D., Lu, L., Jiao, J., Xu, W.: Correlation between electrovibration perception magnitude and the normal force applied by finger. In: Prattichizzo, D., Shinoda, H., Tan, Hong Z., Ruffaldi, E., Frisoli, A. (eds.) EuroHaptics 2018. LNCS, vol. 10893, pp. 91–101. Springer, Cham (2018). https://doi.org/10.1007/978-3-319-93445-7_9
11. Guo, X., Zhang, Y., Wang, D., Jiao, J.: Absolute and discrimination thresholds of a flexible texture display. In: 2017 IEEE World Haptics Conference (WHC), pp. 269–274 (2017)

12. Vardar, Y., Güçlü, B., Basdogan, C.: Effect of waveform in haptic perception of electrovibration on touchscreens. In: Haptics: Perception, Devices, Control, and Applications: 10th International Conference EuroHaptics 2016, London, UK, pp. 190–203 (2016)
13. Shultz, C.D., Peshkin, M.A., Colgate, J.E.: Surface haptics via electroadhesion: expanding electrovibration with Johnsen and Rahbek. In: IEEE World Haptics Conference (WHC), Evanston, IL, United states, pp. 57–62 (2015)

Can Stiffness Sensations Be Rendered in Virtual Reality Using Mid-air Ultrasound Haptic Technologies?

M. Marchal[1,2(✉)], G. Gallagher[3], A. Lécuyer[3], and C. Pacchierotti[4]

[1] Univ. Rennes, INSA, IRISA, Inria, CNRS, Rennes, France
maud.marchal@irisa.fr
[2] IUF, Paris, France
[3] Inria, Univ. Rennes, IRISA, Rennes, France
[4] CNRS, Univ. Rennes, Inria, IRISA, Rennes, France

Abstract. Mid-air haptics technologies convey haptic sensations without any direct contact between the user and the interface. A popular example of this technology is focused ultrasound. It works by modulating the phase of an array of ultrasound emitters so as to generate focused points of oscillating high pressure, which in turn elicit haptic sensations on the user's skin. Whilst using focused ultrasound to convey haptic sensations is becoming increasingly popular in Virtual Reality (VR), few studies have been conducted into understanding how to render virtual object properties. In this paper, we evaluate the capability of focused ultrasound arrays to simulate varying stiffness sensations in VR. We carry out a user study enrolling 20 participants, showing that focused ultrasound haptics can well provide the sensation of interacting with objects of different stiffnesses. Finally, we propose four representative VR use cases to show the potential of rendering stiffness sensations using this mid-air haptics.

1 Introduction

Focused airborne ultrasound is nowadays the most mature and popular technology able to provide mid-air haptics. Arrays of ultrasonic transducers can produce phase-shifted acoustic waves which constructively interfere at points in space called focal points and destructively interfere elsewhere, conveying haptic sensations by varying acoustic radiation pressure on the skin. Focused ultrasounds have been already employed in several applications of Virtual Reality (VR) [4,5,9,10]. However, despite the recent popularity of mid-air ultrasound

This project has received funding from the European Union's Horizon 2020 programme under grant agreement No 801413; project "H-Reality".

Electronic supplementary material The online version of this chapter (https://doi.org/10.1007/978-3-030-58147-3_33) contains supplementary material, which is available to authorized users.

I. Nisky et al. (Eds.): EuroHaptics 2020, LNCS 12272, pp. 297–306, 2020.
https://doi.org/10.1007/978-3-030-58147-3_33

technologies, to the best of our knowledge, no study has analyzed if and to what extent ultrasound haptic arrays can provide effective stiffness sensations. Most work using this technology in VR has been in fact dedicated to the rendering of shapes [6,9] and textures [1].

This work studies the capability of mid-air ultrasound haptics of rendering stiffness sensations when interacting with virtual objects. More specifically, we aim at identifying the differential threshold for stiffness perception when using a focused ultrasound array to render objects in VR. Of course, it is important to highlight that we are not rendering force feedback as if the user was interacting with a real piston. Our objective is to understand whether we can elicit/simulate stiffness sensations using focused ultrasound arrays. Our paper comprises a perceptual evaluation as well as four VR use cases (see Fig. 3), where we show the potential of our approach as an alternative to contact haptic feedback [2,11] in VR scenarios.

2 Perceptual Evaluation

2.1 Experimental Setup

To validate focused ultrasound as a tool to provide stiffness sensations in VR, we prepared an experimental setup enabling participants to interact with 1-D stiffnesses. The setup is shown in Fig. 1. The virtual environment was composed of a virtual piston placed on a black table. The real environment was composed of an Ultrahaptics STRATOS platform, which is a commercial focused ultrasound array. It comprises a 16×16 planar array of transducers emitting 40 kHz ultrasound in an upward direction. The virtual environment was shown to the participant through an HTC Vive VR headset. A HTC Vive Tracker was attached to the dominant wrist of the participants to track the motion of their hands, and a virtual hand avatar mimicked this motion in the virtual environment. Finally, an HTC Vive controller was held by the participant in their non-dominant hand to answer the in-screen perceptual questions, and a pair of noise-canceling headphones avoided potential effects due to auditory cues arising from the device operation.

2.2 Haptic Rendering

The task consisted in comparing the stiffness of two virtual pistons. Each virtual piston was modeled as a 1-D spring following Hooke's law. Whenever a user enters in contact with the piston, the system simulates a spring-like feedback, where the pressure commanded by the Ultrahaptics device is defined by $p = k(z_0 - z) + p_0$ if the user contacts the piston, 0 otherwise. k is the simulated stiffness of the piston (in Pa/m, sound pressure over displacement), z the current altitude of the piston, $z_0 = 30$ cm its resting position, $\Delta z = z_0 - z$ its current compression, and $p_0 = 146.87$ dB SPL (441 Pa) the absolute detection threshold we registered at 30 cm (when $\Delta z = 0$). The piston is fully compressed

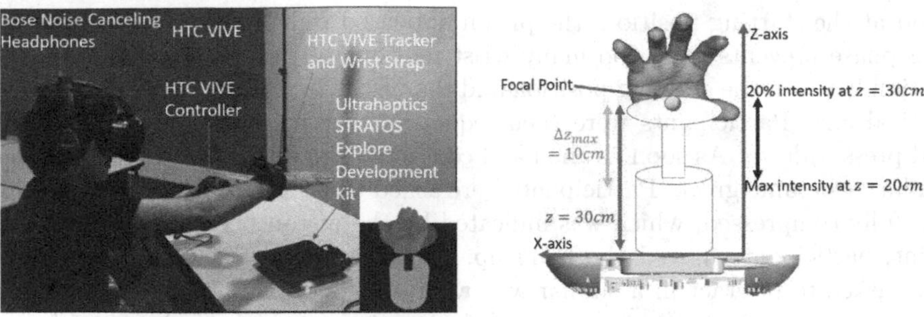

Fig. 1. (Left) Setup: subjects interact with the Ultrahaptics interface while wearing a HTC Vive display. The dominant hand is tracked using a Vive Tracker while the other hand holds a Vive Controller to answer the questions. The virtual scene with the piston is shown as an inset. (Right) Visual representation of the focal point and its relation with the displacement of the piston.

at z = 20 cm ($\Delta z_{\max}$ = 10 cm). The Ultrahaptics device generates localized pressure at a designated focal point. We rendered this point at the centroid of the upper plate of the piston (see Fig. 1). When the user interacts with the piston, this point results at the center of the user's palm as well as the center of the ultrasound array. As soon as the user contacts the piston at its resting position (z = 30 cm, Δz = 0 cm), the device starts generating a pressure on the palm. This pressure increases as the user presses the piston down, reaching its maximum when the piston is fully compressed (z = 20 cm, $\Delta z_{\max}$ = 10 cm). The STRATOS platform can provide a maximum of 163.35 dB SPL at z = 20 cm and p_0 thus represents 15% of the maximum power. We rendered the focal point with the Ultrahaptics device using spatiotemporal modulation (STM), introduced by Kappus and Long [8]. In STM, focal points are generated with a fixed frequency (usually the maximum achievable by the device, i.e., 40 kHz). Since this frequency is very high, it poses significantly fewer constraints on the temporal evolution of the peak intensity and focal point position. Frier et al. [3] have investigated the trade-off between pattern repetition rate in STM and perceived intensity. A study on human's detection of focal points and basic shapes rendered via focused ultrasound stimuli can be found in [4].

2.3 Experimental Procedure

Participants were first required to fill out a pre-experiment questionnaire. Then, they were asked to wear the HTC Vive headset, tracker, controller (see Fig. 1). Participants had to compare two pistons with different rendered stiffness, modeled by a 1D spring law, as detailed in Sect. 2.2. At the beginning of each interaction, the virtual environment only showed a transparent hand, marking the target hand position for starting the task. Before the start of the first interaction, the virtual hand of the participant was calibrated to ensure it matched the transparent hand along the three different axes. Once participants placed their hand at the starting position, the piston appeared right below it. This calibration phase prevents from too many wrist motions of the user's hand since the motion between the starting position and the piston is straightforward along the vertical axis. Participants were then requested to touch the top of the piston and press it down. As soon as the hand contacted the upper plate of the piston, the latter became green. Participants were asked to press onto the piston until it was fully compressed, which was indicated by the piston becoming red. At this point, participants moved the hand up, releasing the piston. After that, they were asked to interact in a similar way with a second piston. After this second interaction, participants were finally asked to judge *which* of the two pistons felt *stiffer*. One piston served as a reference, displaying a reference stiffness k_{ref}, while the other piston displayed a variable stiffness k_{test}. After preliminary testings, we considered 6 values of test stiffness k_{test} to be compared with 3 values of reference stiffness k_{ref}.

The three stiffness values of the reference piston were:

- $k_{ref,1} = 7358$ Pa/m (155.39 dB SPL when $\Delta z = z_{\max}$, which is 40% of the device power range).
- $k_{ref,2} = 13242$ Pa/m (158.91 dB SPL when $\Delta z = z_{\max}$, 60% of the range),
- $k_{ref,3} = 19126$ Pa/m (161.41 dB SPL when $\Delta z = z_{\max}$, 80% of the range),

The six values of the test piston were: +5884 Pa/m (20% of the device power range when $\Delta z = z_{\max}$), +2942 Pa/m (10%), +1471 Pa/m (5%), −5884 Pa/m, −2942 Pa/m, −1471 Pa/m with respect to the reference stiffness.

2.4 Conditions and Experimental Design

Two conditions are considered in our experimental design:

- **C1** is the difference of stiffness between the reference piston and the test piston, $|k_{ref} - k_{test}|$.
- **C2** corresponds to a binary variable, which is true if the piston perceived as the stiffest is indeed the one rendered with a higher stiffness constant.

The order of presentation of the two pistons was counterbalanced to avoid any order effect: every couple of pistons was presented in all orders. The starting reference was also alternated to ensure that fatigue did not influence the last block. Thus, participants were presented with 90 trials per reference stiffness (270 in total), divided in 5 blocks of 6 trials in a randomized order for each block. The experiment lasted approximately 40 min, with breaks between the blocks.

2.5 Participants and Collected Data

Twenty participants (16 males, 4 females) took part to the experiment, all of whom were self-identified right-handed. 18 of them had previous experience with haptic interfaces. All were naive with respect to the study objectives. The age range of the participants was between 21 and 29 years (M = 24).

For each couple of pistons, we collected as an objective measure the participant's answer. This answer corresponds to the piston (first or second) which was reported by the participant as the stiffest. The measure was then collected as a true discovery rate if the answer corresponds to the stiffest value rendered. Participants also completed a subjective questionnaire. The first set of questions was asked three times, after the 5 blocks dedicated to one reference stiffness, using a 7-item Likert scale: (Q1): I felt confident when choosing the response after each interaction; (Q2): After the experiment, I felt tired; (Q3): The task was easy; (Q4): It felt like pressing a real piston. Then, at the end of the experiment, we asked them to answer two open questions: (Q5): Would you describe what you felt as stiffness? If not, please attempt to describe it; (Q6): Do you any have any further comment or suggestion?

3 Results

Reference stiffness: $k_{ref,1}$. Answers to the questionnaire regarding confidence (Q1) ranged from 2 (nearly very unconfident) to 7 (very confident) out of 7, with a mean of 4.75 and standard deviation (SD) 1.2. Regarding fatigue (Q2), answers ranged from 1 (not fatigued) to 6 (moderately fatigued) out of 7, with a mean of 4.1 (SD = 1.3). When the user was asked how easy the task was (Q3), answers ranged from 4 (slightly easy) to 7 (very easy) out of 7, with a mean of 5.45 (SD = 0.9). The reported realness of the piston (Q4) ranged from 1 (not real at all) to 6 (moderately real) out of 7, with a mean of 3.6 (SD = 1.5). Figure 2a shows the psychometric curve as well as the mean and standard deviation for each comparison piston. We obtained a JND value of 20% using a 75% threshold, along with a Point of Subjective Equality (PSE) of 2.16%.

Reference stiffness: $k_{ref,2}$. Answers to the questionnaire regarding confidence (Q1) ranged from 2 to 6 out of 7, with a mean of 4 and SD 1.5. Regarding fatigue (Q2), answers ranged from 1 to 6 out of 7, with a mean of 4.25 and SD 1.5. When the user was asked how easy the task was (Q3), answers ranged from 3 to 7 out of 7, with a mean of 5.05 and SD 1.3. The reported realness of the piston (Q4) ranged from 1 to 6 out of 7, with a mean of 4 and SD 1.6. Figure 2b shows the psychometric curve as well as the mean and standard deviation for each comparison piston. Under this reference, a 75% differential threshold of 32% was obtained with a PSE of 3.65%.

Reference stiffness: $k_{ref,3}$. Answers to the questionnaire regarding confidence (Q1) ranged from 2 to 7 out of 7, with a mean of 4.1 and SD 1.3. Regarding fatigue (Q2), answers ranged from 2 to 6 out of 7, with a mean of 4.3 and SD 1.4. When the user was asked how easy the task was (Q3), answers ranged from 2 to 7 out of 7, with a mean of 5.25 and SD 1.4. The reported realness of the piston (Q4) ranged from 2 to 6 out of 7, with a mean of 3.94 and SD 1.5. Figure 2c shows the psychometric curve as well as the mean and standard deviation for each comparison piston. Differently from the other curves, this time we were not

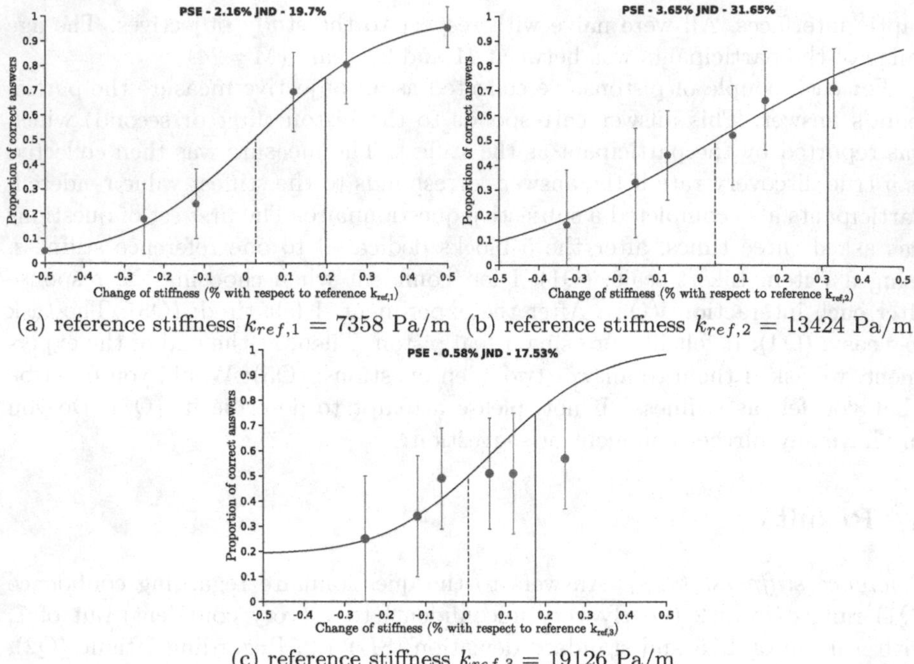

(a) reference stiffness $k_{ref,1} = 7358$ Pa/m (b) reference stiffness $k_{ref,2} = 13424$ Pa/m

(c) reference stiffness $k_{ref,3} = 19126$ Pa/m

Fig. 2. Psychometric curves for the three reference stiffness values, fitting a cumulative Gaussian to the data. We plot the proportion of correct answers in function of the percentage increase in stiffness with respect to the reference one $k_{ref,1}$. The vertical dashed and solid lines represent the PSE and the 75% differential threshold. Error bars represent standard deviation. (Color figure online)

able to reach proportions of correct answers close to 1 on the right-hand side of the curve. This result could be due to the fact that the considered reference stiffness $k_{ref,3}$ requires pressures close to the device maximum, i.e., 161.41 dB SPL when $\Delta z = z_{\max}$, which is the 80% of the device power range. For this reason, it was not possible to test large increments. Another explanation for this behavior could be the presence of refractions and artifacts generated by the acoustic waves, which become more intense as the peak pressure increases and interfere with the overall perception of stiffness. This latter point is supported by how users described the haptic sensation over the three reference conditions. In fact, while during experiments on reference stiffness $k_{ref,1}$ and $k_{ref,2}$ users most often reported to feel a "circular shape", during experiments on $k_{ref,3}$ users started to report feeling "lines" or "bars". The focal point generated by the device should remain circular at all intensity levels. For this reason, we evaluated the psychometric curve only taking into account the stiffness intensities for which users reported feeling a circular shape (blue points in Fig. 2c). Under this reference, a 75% differential threshold of 18% was obtained with a PSE of 0.58%.

Post-experiment Questionnaire. All users were able to detect that the force increased over the displacement. However, only 48% of the participants were able to feel that the minimum pressure they felt when they first interacted with the piston was always the same (146.86 dB SPL, 15% of the full range). When asked if the sensation they felt reassembled stiffness (Q5), 80% of users said that it did. The remaining 20% could not express what they felt, but still recognized an increase in force. When asked to describe the sensations they felt over the duration of all 60 pistons, answers ranged from "feeling a real piston" to feeling a "stream", "circular air flows", and "some kind of resistance".

4 Use Cases in Virtual Reality

We demonstrate the viability of rendering stiffness sensations using ultrasound focused arrays through four use cases in Virtual Reality, shown in Fig. 3[1].

The first use case (see Fig. 3a) represents a scene at a carnival fair. It is composed of a stand at a carnival fair, featuring a pump, a release button, and a balloon to be inflated. Users are asked to inflate a balloon by repeatedly pressing on the pump. Every time the pump is pressed, it becomes a little stiffer to render the increased pressure inside the balloon. The second use case (see Fig. 3b) is composed of a small piano placed on a table. Piano keys are generally weighted having a higher stiffness for the lower register and a lower stiffness for the higher register. We render four different octaves, each having variable degrees of stiffness. Users are able to select a different set of octaves by pressing a button next to the piano. The third use case (see Fig. 3c) is composed of a hospital room with a virtual patient lying upon a bed. A 2-cm-wide area on the patient's stomach was rendered stiffer than the rest. Users are instructed to palpate the users stomach and indicate where they feel the stiffer region. The fourth use case (see Fig. 3d) is composed of four blocks that need to be pressed in a certain sequence to open a door containing a treasure chest. Each block has a different stiffness. To access the treasure, users must press the blocks in order of stiffness, from the lowest to the highest. On top of the door, there are four lights, that indicate the progress of the task.

5 Discussion and Conclusions

Ultrasound haptics is considered a very promising technology, as it is able to convey compelling haptic sensations without any direct contact between the user and the interface. However, as only recently ultrasound arrays have become available, very few works have studied the type of haptic sensations we can render with this technology. This work evaluates whether it is possible to render stiffness sensations in Virtual Reality using haptic feedback generated by ultrasound focused arrays. To calculate the JND and the PSE for this type of stiffness sensation, we carried out a human subject study enrolling 20 participants. Subjects

[1] A video is available at https://youtu.be/sJKYV1nLIY.

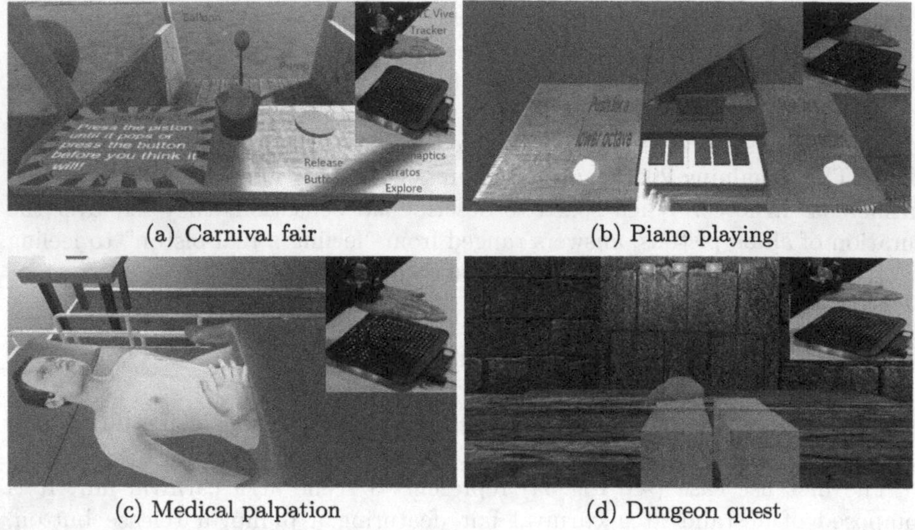

(a) Carnival fair (b) Piano playing

(c) Medical palpation (d) Dungeon quest

Fig. 3. We implemented four use cases in Virtual Reality. We render different stiffness sensations using the ultrasound stimuli generated by our Ultrahaptics interface. (Color figure online)

were asked to compare the perceived stiffness of multiple virtual pistons, whose stiffness was rendered by an Ultrahaptics device via ultrasound haptic stimuli. In the literature, researchers have shown that the JND for stiffness discrimination can range from 8 to 23% [7,12]. Jones and Hunter [7] have reported an average JND of 23% for participants comparing the stiffness of springs simulated using two servo-controlled electromagnetic linear motors. Each motor was coupled to one wrist of the subject. Tan et al. [12] calculated the JND of stiffness for a task which required grasping two plates with the thumb and index fingers and squeezing them along a linear track. A force which resisted the squeeze, simulating different levels of stiffness, was generated by an electromechanical system. When subjects had to squeeze the plates always for a fixed displacement, the JND registered was of 8%; on the other hand, when the displacement was randomized from trial to trial, the JND was of 22%. Of course, all these works rendered stiffness by providing kinesthetic feedback.

In our study, we found JND of 17%, 31%, and 19% for the three reference stiffness values 7358 Pa/m, 13242 Pa/m, 19126 Pa/m (sound pressure over displacement), respectively. The subjective questionnaires show that most subjects indeed identified the provided haptic sensations as stiffness. These results prove that it is indeed possible to simulate stiffness sensations using ultrasound haptic feedback in VR. Four use cases showed the potential and viability of our approach in immersive VR applications. Despite these promising results, our study has some limitations. First, it is important to stress that the haptic sensations rendered by ultrasound arrays is of course different than the haptic sensations usually felt when pressing a piston. For this reason, our objective is to *simulate* stiffness sensations. Another drawback is that the behavior we registered when commanding pressures higher than 162.43 dB SPL (90% of the maximum power of the device). The circular focal point started to feel like something different (a "line", a "bar") and the stiffness recognition rate significantly degraded. This is an issue we plan to address in the future, studying what happens from an acoustics point of view and understanding what it means in terms of human perception.

References

1. Beattie, D., Georgiou, O., Harwood, A., Clark, R., Long, B., Carter, T.: Mid-air haptic textures from graphics. In: Proceedings of the IEEE World Haptics (WiP Paper) (2019)
2. De Tinguy, X., Pacchierotti, C., Marchal, M., Lécuyer, A.: Enhancing the stiffness perception of tangible objects in mixed reality using wearable haptics. In: Proceedings of the IEEE Conference on Virtual Reality and 3D User Interfaces (VR), pp. 81–90 (2018)
3. Frier, W., et al.: Using spatiotemporal modulation to draw tactile patterns in mid-air. In: Prattichizzo, D., Shinoda, H., Tan, H.Z., Ruffaldi, E., Frisoli, A. (eds.) EuroHaptics 2018. LNCS, vol. 10893, pp. 270–281. Springer, Cham (2018). https://doi.org/10.1007/978-3-319-93445-7_24
4. Howard, T., Gallagher, G., Lécuyer, A., Pacchierotti, C., Marchal, M.: Investigating the recognition of local shapes using mid-air ultrasound haptics. In: Proceedings of IEEE World Haptics, pp. 503–508 (2019)
5. Howard, T., Marchal, M., Lécuyer, A., Pacchierotti, C.: Pumah: pan-tilt ultrasound mid-air haptics for larger interaction workspace in virtual reality. IEEE Trans. Haptics **13**, 38–44 (2020)
6. Inoue, S., Makino, Y., Shinoda, H.: Designing stationary airborne ultrasonic 3D tactile object. In: Proceedings of the IEEE/SICE International Symposium System Integration, pp. 159–162 (2014)
7. Jones, L.A., Hunter, I.W.: A perceptual analysis of stiffness. Exp. Brain Res. **79**(1), 150–156 (1990)
8. Kappus, B., Long, B.: Spatiotemporal modulation for mid-air haptic feedback from an ultrasonic phased array. J. Acoust. Soc. Am. **143**(3), 1836–1836 (2018)
9. Long, B., Seah, S.A., Carter, T., Subramanian, S.: Rendering volumetric haptic shapes in mid-air using ultrasound. ACM Trans. Graph. **33**(6), 181 (2014)
10. Makino, Y., Furuyama, Y., Inoue, S., Shinoda, H.: Haptoclone (haptic-optical clone) for mutual tele-environment by real-time 3D image transfer with midair force feedback. In: Proceedings of the ACM CHI, pp. 1980–1990 (2016)
11. Salazar, D.S.V., Pacchierotti, C., De Tinguy, X., Maciel, A., Marchal, M.: Altering the stiffness, friction, and shape perception of tangible objects in virtual reality using wearable haptics. IEEE Trans. Haptics **13**, 167–174 (2020)
12. Tan, H., Durlach, N., Beauregard, G., Srinivasan, M.: Manual discrimination of compliance using active pinch grasp: the roles of force and work cues. Percept. Psychophys. **57**, 495–510 (1995)

Midair Haptic Presentation Using Concave Reflector

Kentaro Ariga[1(✉)], Masahiro Fujiwara[1,2], Yasutoshi Makino[1,2], and Hiroyuki Shinoda[1,2]

[1] Graduate School of Information Science and Technology, The University of Tokyo, Tokyo, Japan
ariga@hapis.k.u-tokyo.ac.jp

[2] Graduate School of Frontier Sciences, The University of Tokyo, Chiba, Japan
Masahiro_Fujiwara@ipc.i.u-tokyo.ac.jp,
{yasutoshi_makino,hiroyuki_shinoda}@k.u-tokyo.ac.jp

Abstract. An airborne ultrasound tactile display (AUTD) can focus on an arbitrary position by controlling the phase shift and amplitude of each transducer, and provide a tactile stimulus on the human body without direct contact. However, beyond a distance from the phased array, it cannot secure a focal size comparable to the wavelength and the displayed pressure pattern blurs. In this study, we propose a method with a concave reflector to focus at a farther position and present a tactile sensation without enlarging the array. By appropriately designing the focal length of the concave reflector, the focal point can be formed at a distant position using a mirror formula while keeping the focus size non-extended. We conducted two experiments, the results of which show that the proposed method is valid and that a workspace of several square centimeters can be achieved.

Keywords: Airborne ultrasound haptics · Concave reflector · Mirror formula

1 Introduction

An airborne ultrasound tactile display (AUTD) [4,9] can present a tactile stimulus on the surface of a human body without direct contact. An AUTD creates an ultrasound focus at an arbitrary position within the workspace by controlling the phase shift and amplitude of the output emission of each ultrasonic transducer. However, it has difficulty forming a focus far from the phased array and presenting a tactile sensation [10]. Only up to a distance comparable to the aperture of the phased array can an AUTD secure a focal size comparable to the wavelength, and the focus diameter increases as the focus moves farther from the phased array. To tackle this problem, multiple AUTDs were used in previous studies to enlarge the aperture of the phased array or place them near the focus [6,13].

I. Nisky et al. (Eds.): EuroHaptics 2020, LNCS 12272, pp. 307–315, 2020.
https://doi.org/10.1007/978-3-030-58147-3_34

We propose a method for forming a focus and providing a tactile sensation farther from an AUTD without enlarging the phased array or placing the devices near the focal point. There are several approaches which provide comparable workspace extension without enlarging the phased array [1,5]. In the proposed method, ultrasound waves emitted from an AUTD are reflected and focused using a concave reflector. A technique of reflecting ultrasound waves in order to focus them, has been applied in other studies [7,8]. The technique has also been applied in a previous study in haptics [11], through which the phased array can present a tactile stimulus at an arbitrary position within the workspace, but a focus-formable distance has not been extended owing to the planar reflections. In the proposed method, it is possible to generate a focus farther from the AUTD and provide a tactile sensation while maintaining the ability to steer the focus electrically, by applying a mirror formula [3]. A mirror formula has been applied in another study so as to reconstruct the geometry of arbitrary reflectors using ultrasound phased arrays [12].

The theory supporting the proposed method and its advantages are described in Sect. 2. In Sect. 3, two experiments conducted in this area are described. First, we investigated whether a focus is formed when applying the proposed method and if a sufficient sound pressure that can provide a tactile sensation is obtained. Next, we experimentally confirmed the range of the workspace by focusing at certain points.

2 Proposed Method

2.1 Mirror Formula

The procedure to create a focal point at an arbitrary position using a concave reflector is illustrated in Fig. 1. To simplify the following equation, let us define the z-axis as shown in Fig. 1, where f is the focal length of the concave reflector. We assumed a paraxial approximation system and ignored any aberrations. When an acoustic source is placed at $A = (a_x, a_y, a_z)$, the reflected wave is concentrated and forms a focus at $B = (b_x, b_y, b_z)$ using the following mirror formula,

$$\frac{1}{a_z} + \frac{1}{b_z} = \frac{1}{f}, \tag{1}$$

as well as the magnification formulae,

$$b_x = -\frac{b_z}{a_z} a_x, \tag{2}$$

$$b_y = -\frac{b_z}{a_z} a_y. \tag{3}$$

Therefore, a focus can be created at (b_x, b_y, b_z) by driving the AUTD to reproduce the sound wave emitted from the image sound source placed at (a_x, a_y, a_z), as derived from Eqs. (1)–(3).

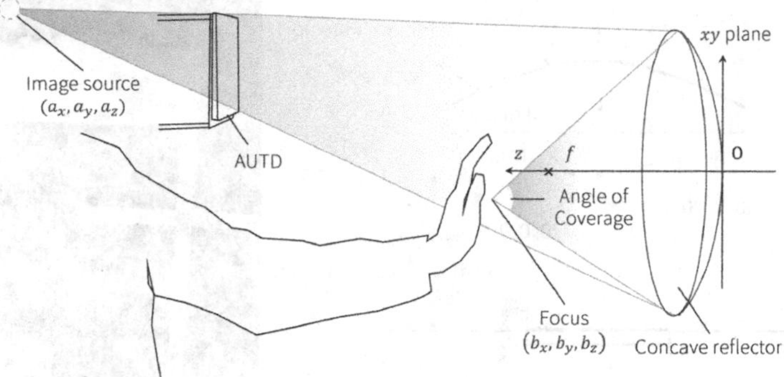

Fig. 1. The process to form a focal point using a concave reflector. AUTD reproduces the propagating waves as if they were emitted from the image sound source, which are reflected by the reflector and concentrated at the focal point.

2.2 Contributions by Concave Reflector

The focus can be formed farther away from the phased array because the angle of coverage can be changed arbitrarily through an appropriate design of the concave reflector. In the case where focal points are formed by direct incident waves or reflected waves from a planar reflector, the angle of coverage is uniquely determined for each focal position, and becomes narrower as the distance from the AUTD increases. As a result, the focal diameter lengthens and a tactile sensation cannot be obtained; in addition, the phased array has to be enlarged to increase the angle of coverage. By contrast, concave reflectors can have various focal lengths. If we appropriately design the focal length and use the mirror formula, the angle of coverage can be increased even at a position far from the AUTD. Therefore, the diameter of the focal point can be smaller, which produces a tactile stimulus, without enlarging the phased array.

3 Experiment

3.1 Implementation

Figure 2 shows the experiment setup. An AUTD is composed of 249 transducers. A T4010A1 (developed by Nippon Ceramic Co., Ltd.) was employed as the transducer. The T4010A1 emits a 40-kHz ultrasound at 121.5 dB in sound pressure level (SPL) at a distance of 30 cm. Each ultrasonic transducer was driven by the full power of the AUTD. The phase shift of each transducer was calculated and driven using the wavelength $\lambda = 8.5$ mm at a sound speed of $c = 340$ m. The concave mirror is a symmetrical parabolic dish shape, the diameter of which is 400 mm and the focal length is 180 mm. The distance between the AUTD and the reflector is 669 mm. A standard microphone (Brüel & Kjær 4138-A-015) was moved using a 1-axis motorized stage. The sound pressure was estimated by calculating the absolute value of the 40 kHz component of DFT.

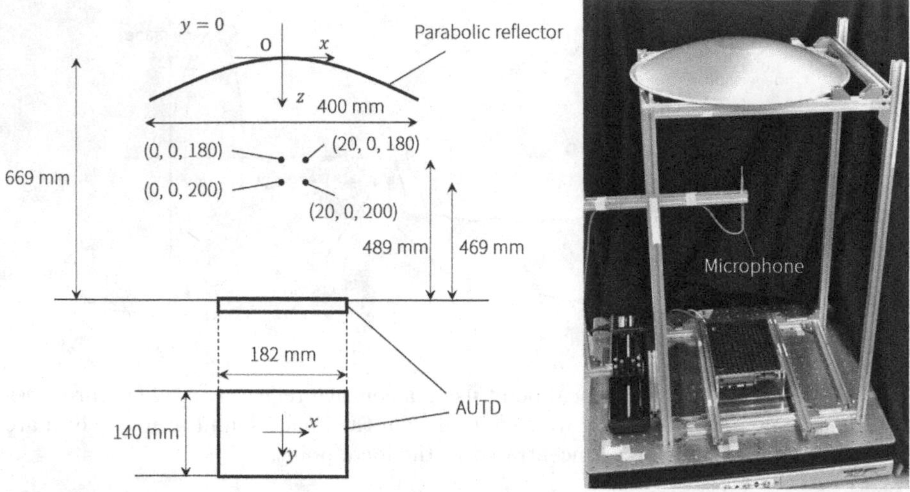

Fig. 2. Geometry of the proposed system. The position (0, 0, 180 mm) is the focus of the reflector.

3.2 Experiment 1: Validation of Mirror Formula

In Experiment 1, the sound pressure distribution around the focus was measured along the x-axis at the depth of designed focal position, and it was confirmed whether the focal point was formed by the mirror formula. The focal points were formed at (0, 0, 180 mm), (20, 0, 180 mm), and (20, 0, 200 mm), where the distances from the AUTD were 489 and 469 mm. As a reference experiment, the sound pressure distribution was also measured, where a focal point was directly formed at (0, 0, 200 mm) without using a reflector. The microphone was moved within the range of $-100\,\text{mm} \leqq x \leqq 100\,\text{mm}$, where it was fixed vertically upward when using the reflector and vertically downward when not using it, in order to steer the microphone toward the direction of sound wave arrival.

The results of Experiment 1 are shown in Fig. 3. As shown in this figure, it can be confirmed that each designed focus can be presented by the concave mirror. The maximum sound pressure at each focal position with the reflector is 6.34×10^3, 5.43×10^3, and 5.69×10^3 Pa, or 170.0, 168.7, and 169.1 dB SPL, respectively, which is sufficient to provide a tactile stimulus [2].

By contrast, the maximum sound pressure at (0, 0, 200 mm), without the reflector is 1.49×10^3 Pa, or 157.4 dB SPL. Furthermore, the focal diameter without the concave reflector is more than twice that with the reflector. Figure 3 shows that proposed method achieves a higher focal sound pressure and a smaller focal diameter than the conventional method, and that the proposed method can present tactile stimuli.

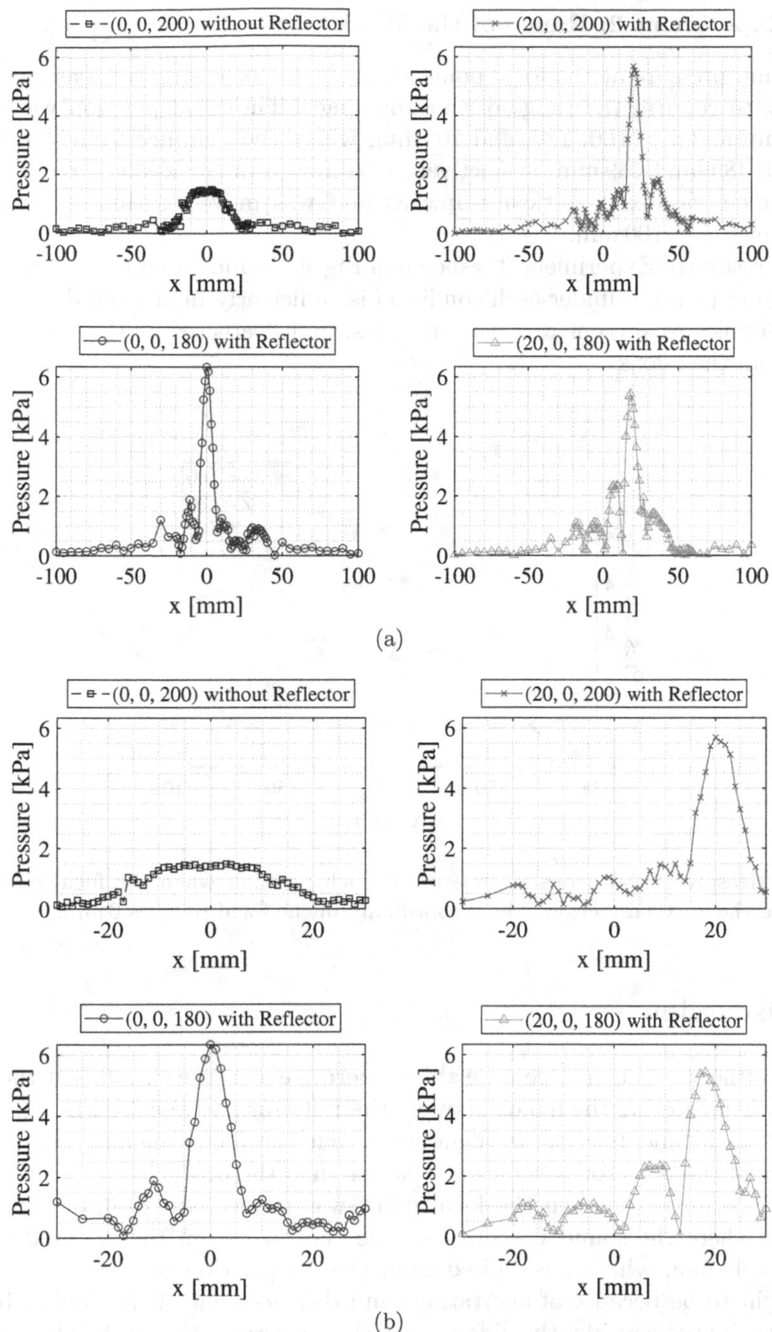

Fig. 3. Measured acoustic pressure distribution in the x-axis direction around each focal point. The position of the focus is at $(x, y, z) = (0, 0, 180\,\text{mm})$, $(20, 0, 180\,\text{mm})$, and $(20, 0, 200\,\text{mm})$ using the reflector, and at $(0, 0, 200\,\text{mm})$ without the reflector. (b) shows an enlarged view of (a).

3.3 Experiment 2: Range of the Workspace

The sound pressure at the focal point when the focus was moved was measured in order to confirm the range of the workspace. The focal position was set at $y = 0\,\text{mm}$ and $z = 160$, 180, and 200 mm, where the distance from the AUTD was 509, 489, and 469 mm, respectively, and moved in the x-axis direction. The microphone was fixed vertically upward and was moved within the range of $-100\,\text{mm} \leqq x \leqq 100\,\text{mm}$.

The results of Experiment 2 is shown in Fig. 4. As indicated in this figure, the focal sound pressure under each condition is sufficiently high around $x = 0\,\text{mm}$, which enables to present a tactile stimulus, and decreases as the focus moves away from the z-axis.

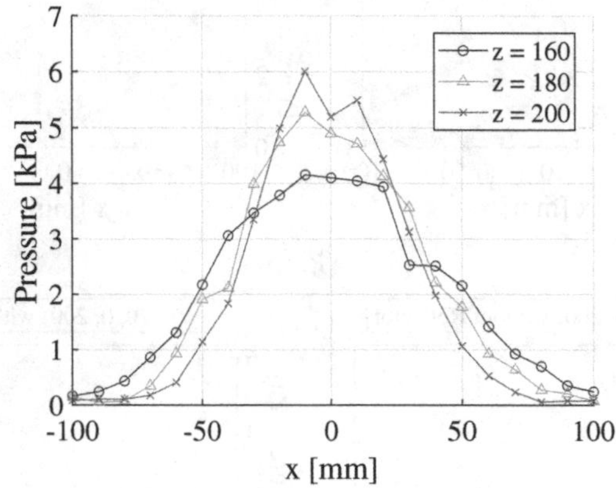

Fig. 4. Measured focal acoustic pressure at each position when the focal point was moved in the x-axis direction. The y coordinate of the focal point is 0 mm.

4 Discussion

As the authors' comment, we were able to perceive a tactile stimulus at the focal points when entering the hands between the reflector and the AUTD. The hand is an obstacle shielding the incident wave, but the focus remains because the aperture of the reflector is sufficiently larger than the obstacle.

In Experiment 1, when the focal point was formed at $(20, 0, 180\,\text{mm})$, the position where the sound pressure became the highest in the x-axis direction was $x = 18\,\text{mm}$, which was shifted from the set position of $x = 20\,\text{mm}$. This is thought to be because of aberrations and displacement of the device. In this study, we did not consider the difference in the acoustic path length, which causes a decrease in the focal sound pressure. Therefore, the focal sound pressure can be increased if we optimize the phase shift of each transducer.

The results of Experiment 1 showed that a focal point can be formed at an arbitrary point through the proposed method. Figure 5 shows an example

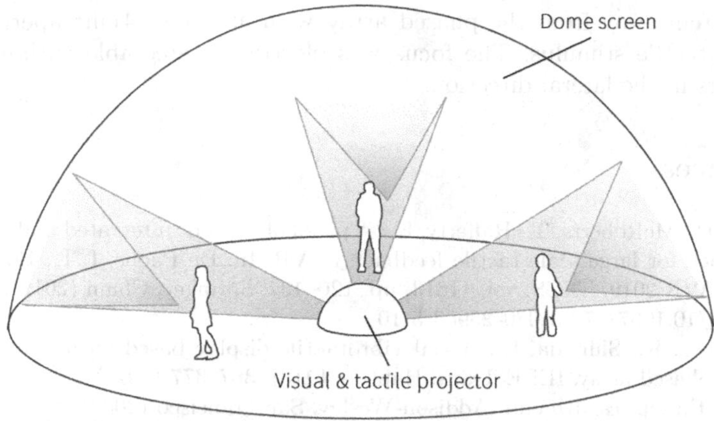

Fig. 5. Application example of the proposed method.

application of this system, in which an image and a tactile sensation are simultaneously presented in a large dome screen by a visual and tactile projector.

Although the proposed method theoretically forms a focal point at an arbitrary position, there are cases where a focal point cannot be formed depending on the position of the phased array and the concave reflector, as well as the aperture of the reflector. For example, in Experiment 2, when the focal point was set to (0, 0, 240 mm), the position of the image sound source formed by the AUTD was (0, 0, 720 mm) according to the mirror formula, which is only 51 mm below the AUTD. However, most of the emitted sound waves are not reflected in this case because when simply considering the geometric acoustic model, a concave reflector of approximately 2.57×10^3 mm in diameter is required if the position and focal length of the reflector are fixed. In fact, the focal sound pressure at (0, 0, 240 mm) was 1.20×10^3 Pa. Moreover, when the irradiation direction of the incident wave is sufficiently shifted from the direction toward the concave mirror, a focal point can be formed at an unexpected position. This is because the grating lobe generated focuses at unintended positions accompanying with the main focus. In fact, in Experiment 2, when the focal point was set at a position sufficiently far from the z-axis, the focus formed by the grating lobe was confirmed. In addition, in this paper we did not consider the attenuation during propagation, by which the amplitude of sound waves exponentially decreases with distance. Therefore, if the propagation distance is too long, the focal sound pressure can be attenuated significantly [2,4].

5 Conclusion

We proposed a method for focusing on an arbitrary point farther from the AUTD using a concave mirror. Our experiments showed that the proposed method can form a focus whose diameter is approximately the wavelength of ultrasound at a point 49 cm away from the phased array with an $18 \times 14\,\text{cm}^2$ aperture and present a tactile stimulus. The focus was electrically steerable within several centimeters in the lateral direction.

References

1. Brice, D., McRoberts, T., Rafferty, K.: A proof of concept integrated multi-systems approach for large scale tactile feedback in VR. In: De Paolis, L.T., Bourdot, P. (eds.) AVR 2019. LNCS, vol. 11613, pp. 120–137. Springer, Cham (2019). https://doi.org/10.1007/978-3-030-25965-5_10
2. Hasegawa, K., Shinoda, H.: Aerial vibrotactile display based on multiunit ultrasound phased array. IEEE Trans. Haptics **11**(3), 367–377 (2018)
3. Hecht, E.: Optics, 4th edn. Addison-Wesley, San Francisco (2002)
4. Hoshi, T., Takahashi, M., Iwamoto, T., Shinoda, H.: Noncontact tactile display based on radiation pressure of airborne ultrasound. IEEE Trans. Haptics **3**(3), 155–165 (2010)
5. Howard, T., Marchal, M., Lécuyer, A., Pacchierotti, C.: PUMAH: pan-tilt ultrasound mid-air haptics for larger interaction workspace in virtual reality. IEEE Trans. Haptics **13**, 38–44 (2019)
6. Inoue, S., Makino, Y., Shinoda, H.: Active touch perception produced by airborne ultrasonic haptic hologram. In: 2015 IEEE World Haptics Conference (WHC), pp. 362–367. IEEE (2015)
7. Ito, Y.: Linearly convergent aerial ultrasonic source providing a variable incident angle and acoustic radiation force by standing-wave ultrasonic field. Jpn. J. Appl. Phys. **48**(7S), 07GM11 (2009)
8. Ito, Y.: High-intensity aerial ultrasonic source with a stripe-mode vibrating plate for improving convergence capability. Acoust. Sci. Technol. **36**(3), 216–224 (2015)
9. Iwamoto, T., Tatezono, M., Shinoda, H.: Non-contact method for producing tactile sensation using airborne ultrasound. In: Ferre, M. (ed.) EuroHaptics 2008. LNCS, vol. 5024, pp. 504–513. Springer, Heidelberg (2008). https://doi.org/10.1007/978-3-540-69057-3_64
10. Korres, G., Eid, M.: Haptogram: ultrasonic point-cloud tactile stimulation. IEEE Access **4**, 7758–7769 (2016)
11. Monnai, Y., Hasegawa, K., Fujiwara, M., Yoshino, K., Inoue, S., Shinoda, H.: HaptoMime: mid-air haptic interaction with a floating virtual screen. In: Proceedings of the 27th Annual ACM Symposium on User Interface Software and Technology, pp. 663–667 (2014)
12. Rodriguez-Molares, A., Løvstakken, L., Ekroll, I.K., Torp, H.: Reconstruction of specular reflectors by iterative image source localization. J. Acoust. Soc. Am. **138**(3), 1365–1378 (2015)
13. Suzuki, S., Takahashi, R., Nakajima, M., Hasegawa, K., Makino, Y., Shinoda, H.: Midair haptic display to human upper body. In: 2018 57th Annual Conference of the Society of Instrument and Control Engineers of Japan (SICE), pp. 848–853. IEEE (2018)

Movement-Free Virtual Reality Interface Using Kinesthetic Illusion Induced by Tendon Vibration

Satoshi Tanaka(✉), Keigo Ushiyama, Akifumi Takahashi, and Hiroyuki Kajimoto

The University of Electro-Communications, 1-5-1 Chofugaoka, Chofu, Tokyo, Japan
{tanaka,ushiyama,a.takahashi,kajimoto}@kaji-lab.jp

Abstract. In current virtual reality (VR) systems, the physical movement of the body is required, which creates problems of safety, cost, and accessibility. To solve those problems, we propose a system that fixes a user's body, detects force when a user tries to move, and generates the sensation of movement using kinesthetic illusion caused by tendon vibration. We implemented a system limited to simple motion, and conducted an experiment to evaluate operability, body ownership, and agency. Although we could not statistically verify the effect of kinesthetic illusion, the results suggested that it may be possible that kinesthetic illusion could increase ownership and decrease agency.

Keywords: Kinesthesia · Tendon vibration · Virtual reality

1 Introduction

An interface for manipulating avatars is an essential component of virtual reality (VR). In science fiction works, VR is often described as a system in which users can subjectively move freely while their physical body is lying on a bed. However, present VR systems reflect the movement of the physical body to the avatar using devices such as position tracking controllers.

In VR, the movement of the physical body causes various problems. For example, the risk of collision and injury; the requirement for a large space and equipment; and the problem that people with limited mobility cannot use VR. The development of a VR system that does not require physical movement would lead to the solution of such problems of safety, cost, and accessibility.

A brain-computer interface (BCI) can be used to implement a movement-free VR system. However, BCI has technical difficulties, such as the requirement for expensive devices, and invasive means (e.g., anesthesia) are required to block motor commands.

Electronic supplementary material The online version of this chapter (https://doi.org/10.1007/978-3-030-58147-3_35) contains supplementary material, which is available to authorized users.

I. Nisky et al. (Eds.): EuroHaptics 2020, LNCS 12272, pp. 316–324, 2020.
https://doi.org/10.1007/978-3-030-58147-3_35

By contrast, a simpler method may exist. Even if the user's body is physically fixed, by altering user's kinesthesia (sense of body movement), the user may be able to feel moving as if not fixed. Fortunately, our kinesthesia can be modulated relatively easily. For example, the sense of motion can be generated even if the body is not actually moving if vibration stimulus is applied to the tendon [6]. This phenomenon is called kinesthetic illusion.

In this study, we propose a system that detects the force exerted by a user's body fixed on a rigid frame to control an avatar, and presents kinesthetic illusion using tendon vibration. In this paper, We implement a system limited to one degree-of-freedom motion as a proof of concept, and evaluate its operability and effects on body ownership and agency.

2 Related Work

When vibration is applied to the tendon of a muscle, kinesthetic illusion, in which the vibrated muscle is stretched, is generated [6]. This phenomenon is considered to be caused by the activation of the muscle spindle by vibration [6], and it has been reported that the nerve fires at a frequency that agrees with the vibration frequency up to a certain vibration frequency [4,15].

Research has also been conducted on the presentation of complex kinesthetic illusions using this phenomenon. Albert et al. [1] converted recorded nerve firing patterns to vibration and presented kinesthetic illusions that reproduced the movement in the recording. Additionally, Thyrion et al. [16] successfully presented three-dimensional kinesthetic illusion by predicting nerve firing the movement trajectory and converting it into vibration.

This phenomenon has also been applied to VR and human-computer interaction. For example, Hagimori et al. [7] combined tendon vibration with a visual stimulus using a head-mounted display (HMD) to make small physical movements perceived as large movements in VR. Barsotti et al. [3] proposed a system that combined kinesthetic feedback using tendon vibration with a BCI based on motor imagery.

Among the above techniques, in the BCI-based approach, the user does not need to move physically at all. However, in those BCI systems, the user has to perform motor imagery, rather than trying to move physically. This may cause a sense of unnaturalness in VR applications.

Other methods exist to modulate kinesthesia in addition to tendon vibration. Okabe et al. [13] reported that the illusion of finger movement was generated by presenting the flow field of tactile sensation according to the shearing force of the fingertip. Similarly, Heo et al. [8] generated the illusion of bending an object using vibrotactile stimuli according to the force applied to the object.

The modulation of a kinesthetic sensation can also result from visual stimuli alone. Pseudo-haptic feedback proposed by Lecuyer et al. [10] is a technique to present a sense of force only using visual stimuli. However, it was reported that a force-sensing stationary device, similar to our study, was perceived like moving in their experiment [10].

As a method similar to our study, Mochizuki et al. [12] proposed a VR interface that does not require physical movement by measuring the torque exerted by joints of the physically fixed user. However, their current system used visual stimulus only, and techniques that directly alter kinesthesia were not tested.

3 System

3.1 Operating Principle

The operation of the proposed system is shown in Fig. 1. In this system, the user's body (arms, legs, etc.) is fixed to a rigid frame. When the user tries to move the body, a force sensor attached at a fixing position detects the force exerted by the user (Fig. 1a). In response to the detected force, the computer simulates the user's intended motion and renders a virtual arm on the HMD (Fig. 1b). Simultaneously, a vibration stimulus is applied to the user's tendon in response to the simulated motion to present kinesthetic illusion (Fig. 1c). Thus, vision and kinesthesia are presented as if the user is moving the body, despite the user's body being fixed.

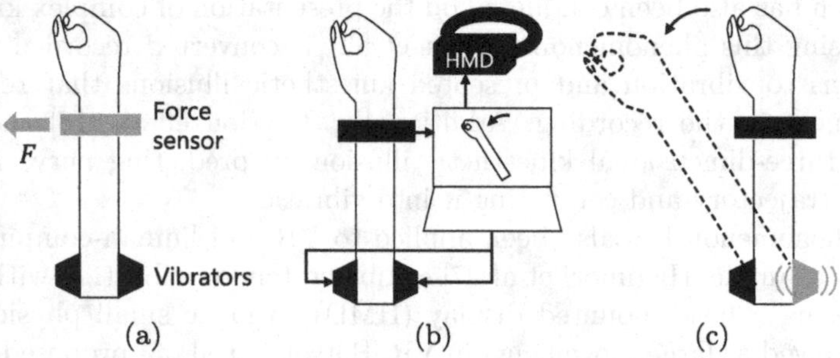

Fig. 1. Operation of the system (a) user tries to move and exerts a force. (b) Computer simulates movement and controls the HMD and vibrators. (c) Vibration induces illusory movement.

3.2 Implementation

As a proof of concept of the proposed method, we implemented a system limited to one degree-of-freedom motion. Although this technique may be applicable to the motion of various joints, we chose the extension and flexion of the forearm following the experiment of Roll et al. [14].

System Configuration. The system detects force using a load cell attached to the wrist fixing component, and connected to the PC through the front-end circuit. The game engine Unity (Unity Technologies) was used for the simulation and rendering, and visual stimuli were presented using an HMD Vive Pro (HTC Corporation) via an external graphics processing unit. The vibration waveform was generated using a waveform generator circuit, amplified using an audio amplifier (MUSE M50), and presented using two vibrators (Vp210, Acouve Laboratory). The vibrators were attached near the right elbow using an elastic fabric supporter, to vibrate distal tendons of the biceps brachii (BB) and triceps brachii (TB) muscles. To make the initial position of the virtual arm coincide with the physical arm, a position tracking device (Vive Tracker, HTC Corporation) was attached to the frame. Additionally, to enhance the ownership of the arm, the actual movement of the user's fingers was captured and reproduced in the display using Leap Motion (Ultraleap).

Algorithms. The angular velocity of the virtual elbow joint was proportional to the force applied to the load cell; that is, when the force applied to the load cell was F [N] (assuming $F = 0$ at system startup), the commanded value of the angular velocity was $\omega_{\text{command}} = 50F$ [deg/s]. The angle of the elbow joint θ [deg] was obtained by integrating the angular velocity, but was limited to -45 to 45°. For the angle and angular velocity, the extension direction was positive. Vibration waveforms were generated by frequency modulating a sine wave between 0 and 100 Hz. Although the amplitude changed according to the frequency because of the frequency response of the system, and kinesthetic illusion is likely to be diminished in lower frequencies, a simple linear mapping was used for the sake of simplicity. This algorithm is based on the knowledge that nerve firing corresponds to the vibration frequency [4,15] and its applications [1,16]. The vibration frequencies f_{BB} [Hz] (for BB) and f_{TB} [Hz] (for TB) were

$$f_{\text{BB}} = \begin{cases} 0 & (4\omega < 0) \\ 4\omega & (0 \leq 4\omega < 100) \\ 100 & \text{(otherwise)} \end{cases} \tag{1}$$

$$f_{\text{TB}} = \begin{cases} 0 & (-4\omega < 0) \\ -4\omega & (0 \leq -4\omega < 100) \\ 100 & \text{(otherwise)} \end{cases} \tag{2}$$

where ω [deg/s] was the angular velocity of the virtual elbow joint ($\omega = 0$ when the angle θ reaches positive or negative limit). However, the coefficients used in the above algorithms were set empirically and not determined theoretically.

4 Experiment

To evaluate the operability, body ownership, and agency of the avatar using the proposed system, we conducted an experiment in which the participants used the system under the following three conditions.

Tendon All elements of the proposed method were incorporated.

None Only visual stimulus was used, without vibration stimulus.

Tactile Only cutaneous cues were presented using high-frequency vibration that was unlikely to cause kinesthetic illusion. Based on the method of Bark et al. [2], an amplitude-modulated sine wave of 250 Hz was used for the vibration waveform, and the amplitudes of the waveform input to the BB and TB vibrators were $A_{\mathrm{BB}} = 0.1 \times 10^{\theta/45}$ and $A_{\mathrm{TB}} = 0.1 \times 10^{-\theta/45}$ (from 0 to 1), where θ [deg] is the virtual elbow angle.

4.1 Methods

Initially, 13 laboratory members that specialize in VR and/or haptics participated in the experiment, but because malfunctions of Leap Motion occurred during experiments involving three of the participants, their data were excluded from subsequent analyses. Finally, data from 10 participants (22 to 25 years old, average 23.6 years old, all right-handed, one female, nine male) were analyzed.

First, the acceleration amplitude of the BB vibrator attached to the participant's arm driven with a 100 Hz sine wave was adjusted to approximately 130 m/s^2 (which was determined to stably evoke kinesthetic illusion), using an accelerometer (LIS331HH, STMicroelectronics). We also presented 100 Hz sine wave stimuli and confirmed orally that the kinesthetic illusion occured in both the extension and flexion directions. When the illusion was not sufficiently obtained, the position of the vibrators was re-adjusted until the illusion occurred.

The participants then performed a task to evaluate the operability of the system based on Fitts' law [5,11] for each condition. The task consisted of 50 trials. In each trial, the participants controlled the virtual forearm and kept the angle aligned with the target for 1 s. The center position of the target was randomly generated from $-30°$ to $30°$ and the width was randomly generated from $5°$ to $15°$. White noise was presented using the built-in headphones of the HMD during task execution.

After the task for each single condition was complete, the participants answered the questions shown in Table 1 using the seven-point Likert scale (-3: totally disagree, $+3$: totally agree), to evaluate body ownership and agency. This questionnaire was a modified version of the questionnaire used in the study on rubber hand illusion by Kalkert et al. [9] and consisted of four categories: Ownership, Ownership Control, Agency, and Agency Control. The questions were translated into Japanese.

To cancel the order effect, the order of the conditions was counterbalanced as much as possible. However, due to the aforementioned malfunctions, among $3! = 6$ permutations, orders of conditions None-Tendon-Tactile and Tactile-Tendon-None was used only once, and the other orders were used twice.

4.2 Results

Operability Evaluation Using Fitts' Law. In the task results for each participant and each condition, the equation of the modified Fitts' law $MT = a +$

Table 1. Questionnaire for evaluating ownership and agency (based on Kalckert et al. [9])

Category	Question
Ownership	I felt as if I was looking at my own hand
	I felt as if the displayed hand was part of my body
	I felt as if the displayed hand was my hand
Ownership control	It seems as if I had more than one right hand
	It felt as if I had no longer a right hand, as if my right hand had disappeared
	I felt as if my real hand was turning like computer-generated image
Agency	I felt as if I could cause movements of the displayed hand
	I felt as if I could control movements of the displayed hand
	The displayed hand was obeying my will and I can make it move just like I want it
Agency control	I felt as if the displayed hand was controlling my will
	It seemed as if the displayed hand had a will of its own
	I felt as if the displayed hand was controlling me

$b \log_2(A/W + 1)$ [11] was fitted, where MT [s] is the time required for the trial, A [deg] is the difference between the angle at the start of the trial and the target angle, and W [deg] is the width of the target. Additionally, the index of performance IP was calculated [11]. Figure 2 compares the average IP for all participants under each condition. The repeated measures analysis of variance showed no significant differences between the conditions ($p = .898$). As a result of multiple comparisons using the Bonferroni correction, no significant differences were found ($p = 1.000$ for all pairs). Additionally, the correlation coefficient r of the fitting ranged from 0.139 to 0.851 with an average of 0.532.

Questionnaires. In the same manner as [9], the answers to questions belonging to the same category were averaged, and four scores (Ownership, Ownership Control, Agency, and Agency Control) were calculated (Fig. 3) and further analyses were done using these scores. The Wilcoxon signed-rank test between Ownership and Ownership Control, Agency and Agency Control was performed in each condition and significant differences in both Ownership-Ownership Control and Agency-Agency Control were found in all conditions ($p < .05$). Also, the Friedman test was performed between conditions for each score, but there was no significant difference ($p = .393$ for Ownership, $p = .087$ for Ownership Control, $p = .607$ for Agency, $p = .098$ for Agency Control).

4.3 Discussion

Operability. There was no significant difference in IP between the conditions, and as shown in Fig. 2, the average IP for the Tactile and Tendon conditions were almost the same or slightly lower than that of the None condition. Hence, we consider that operability was not improved by presenting kinesthetic illusion in the proposed system.

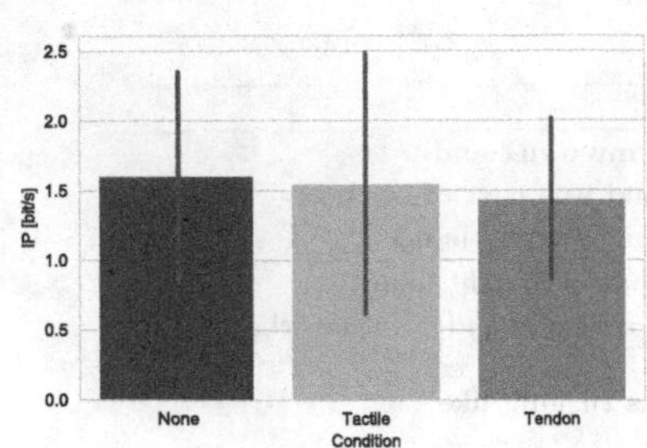

Fig. 2. Comparison of the average IP by condition (error bars indicate the standard deviation)

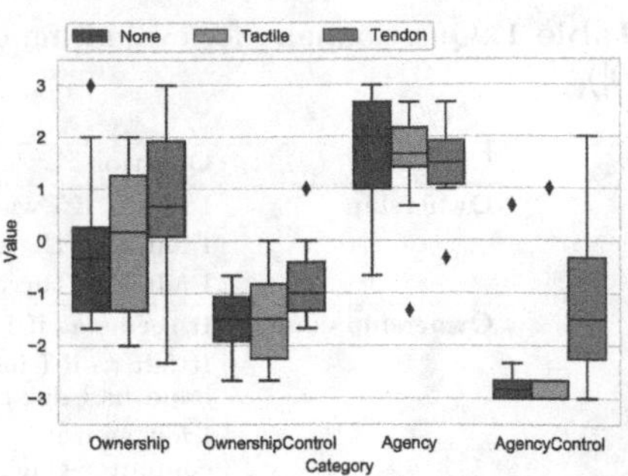

Fig. 3. Comparison of the questionnaire scores by condition

Body Ownership and Agency. For the results of the questionnaire, there was no significant difference in scores between the conditions, possibly due to large variability between participants. However, in Fig. 3, ownership tended to increase in the Tendon condition in comparison with other conditions. Therefore, while we cannot conclude that tendon vibration was effective in the movement-free VR interface, it may contribute to the generation of ownership.

Additionally, in Fig. 3, a small decrease in Agency and an increase in Agency Control in the Tendon condition were observed. In fact, some participants' comments suggested a lack of agency, such as "a feeling of being moved by others" for the Tendon condition. We consider that the loss of agency was because movement simulation and the method for the vibration presentation were imperfect, and the kinesthetic illusion was different from the intended motion of the user.

5 Conclusion

In this paper, we proposed and implemented a VR system that requires no physical body movement by detecting the force exerted by the user's fixed body and presenting kinesthetic illusion. As a result of the experiment, an improvement in operability caused by kinesthetic illusion was not confirmed. Additionally, although not statistically verified, the results suggest that the kinesthetic illusion may lead to an improvement of ownership and decrease of agency. As future work, a more precise verification of usefulness and improvement of the presentation method are required.

Acknowledgements. This research was supported by JSPS KAKENHI Grant Number JP18H04110.

References

1. Albert, F., Bergenheim, M., Ribot-Ciscar, E., Roll, J.P.: The la afferent feedback of a given movement evokes the illusion of the same movement when returned to the subject via muscle tendon vibration. Exp. Brain Res. **172**(2), 163–174 (2006). https://doi.org/10.1007/s00221-005-0325-2
2. Bark, K., Wheeler, J.W., Premakumar, S., Cutkosky, M.R.: Comparison of skin stretch and vibrotactile stimulation for feedback of proprioceptive information. In: 2008 Symposium on Haptic Interfaces for Virtual Environment and Teleoperator Systems, pp. 71–78. IEEE, March 2008. https://doi.org/10.1109/HAPTICS.2008.4479916
3. Barsotti, M., Leonardis, D., Vanello, N., Bergamasco, M., Frisoli, A.: Effects of continuous kinaesthetic feedback based on tendon vibration on motor imagery BCI performance. IEEE Trans. Neural Syst. Rehabil. Eng. **26**(1), 105–114 (2018). https://doi.org/10.1109/TNSRE.2017.2739244
4. Burke, D., Hagbarth, K.E., Löfstedt, L., Wallin, B.G.: The responses of human muscle spindle endings to vibration of non-contracting muscles. J. Physiol. **261**(3), 673–693 (1976). https://doi.org/10.1113/jphysiol.1976.sp011580
5. Fitts, P.M.: The information capacity of the human motor system in controlling the amplitude of movement. J. Exp. Psychol. **47**(6), 381–391 (1954). https://doi.org/10.1037/h0055392
6. Goodwin, G.M., McCloskey, D.I., Matthews, P.B.: The contribution of muscle afferents to kinaesthesia shown by vibration induced illusions of movement and by the effects of paralysing joint afferents. Brain: J. Neurol. **95**(4), 705–748 (1972). https://doi.org/10.1093/brain/95.4.705
7. Hagimori, D., Isoyama, N., Yoshimoto, S., Sakata, N., Kiyokawa, K.: Combining tendon vibration and visual stimulation enhances kinesthetic illusions. In: 2019 International Conference on Cyberworlds (CW), pp. 128–134. IEEE, October 2019. https://doi.org/10.1109/CW.2019.00029
8. Heo, S., Lee, J., Wigdor, D.: PseudoBend: Producing haptic illusions of stretching, bending, and twisting using grain vibrations. In: UIST 2019 - Proceedings of the 32nd Annual ACM Symposium on User Interface Software and Technology, pp. 803–813. ACM Press, New York (2019). https://doi.org/10.1145/3332165.3347941
9. Kalckert, A., Ehrsson, H.H.: The moving rubber hand illusion revisited: comparing movements and visuotactile stimulation to induce illusory ownership. Conscious. Cogn. **26**(1), 117–132 (2014). https://doi.org/10.1016/j.concog.2014.02.003
10. Lecuyer, A., Coquillart, S., Kheddar, A., Richard, P., Coiffet, P.: Pseudo-haptic feedback: can isometric input devices simulate force feedback? In: Proceedings IEEE Virtual Reality 2000 (Cat. No. 00CB37048), pp. 83–90. IEEE Computer Society (2000). https://doi.org/10.1109/VR.2000.840369
11. MacKenzie, I.S.: Fitts' law as a research and design tool in human-computer interaction. Hum.-Comput. Interact. **7**(1), 91–139 (1992). https://doi.org/10.1207/s15327051hci0701_3
12. Mochizuki, N., Nakamura, S.: Motion-Less VR: full-Body immersive VR interface without real body motion. In: The 24th Annual Conference of the Virtual Reality Society of Japan, pp. 6B–09 (2019, in Japanese)
13. Okabe, H., Fukushima, S., Sato, M., Kajimoto, H.: Fingertip Slip Illusion with an Electrocutaneous Display. In: International Conference on Artificial Reality and Telexistence, pp. 10–14 (2011)

14. Roll, J.P., Vedel, J.P.: Kinaesthetic role of muscle afferents in man, studied by tendon vibration and microneurography. Exp. Brain Res. **47**(2), 177–190 (1982). https://doi.org/10.1007/BF00239377
15. Roll, J.P., Vedel, J.P., Ribot, E.: Alteration of proprioceptive messages induced by tendon vibration in man: a microneurographic study. Exp. Brain Res. **76**(1), 213–222 (1989). https://doi.org/10.1007/BF00253639
16. Thyrion, C., Roll, J.P.: Predicting any arm movement feedback to induce three-dimensional illusory movements in humans. J. Neurophysiol. **104**(2), 949–959 (2010). https://doi.org/10.1152/jn.00025.2010

Haptic Display Using Fishing Rod

Daiki Naito(✉) and Hiroyuki Kajimoto(✉)

The University of Electro-Communications, 1-5-1 Chofugaoka, Chofu, Tokyo, Japan
{naito,kajimoto}@kaji-lab.jp

Abstract. This paper proposes a grounded haptic display that can widely present a force sensation using a fishing rod. The proposed haptic display incorporates a fishing rod to present the force sensation at the user's fingertip using a thread being reeled from the tip of the fishing rod. The multidirectional force sensation is presented by driving the base supporting the fishing rod using a pan-tilt mechanism, and wending the thread using the reel. The fishing rod is lightweight and moves at high speed. Additionally, it is possible to widely present the force by bending the fishing rod. This device requires conversion from the position of the user's fingertip and magnitude of the force to the necessary winding force of the thread and the posture of the fishing rod. This study developed a control method and evaluated it by considering a case wherein the force to be presented is exerted in the vertical direction.

Keywords: Virtual reality · Human interface · Haptic device · Fishing rod

1 Introduction

In the field of virtual reality, many studies are developing sensory presentation devices that reproduce sensations felt in the real world. Presenting a force sensation is an important element with regard to improving the perception of objects in virtual space [1].

A haptic display is a system that accurately tracks the movement of the user and presents the required force to their fingertips and tools [2]. Users can perceive the sensation of touching virtual objects created by a computer using the haptic display [3–5]. To present a force in an arbitrary direction and at an arbitrary point in space, the first basic structure to be considered is a structure using the robot arm of a serial link such as PHANToM [4]. One drawback of this type is that, to realize a large workspace, a rigid bar with the same length as the workspace moves within the space, which increases the risk and requires a motor with a large output at the base to generate a force at the end of the rigid bar. Another possible structure is the use of thread traction, as in the case of

Electronic supplementary material The online version of this chapter (https://doi.org/10.1007/978-3-030-58147-3_36) contains supplementary material, which is available to authorized users.

I. Nisky et al. (Eds.): EuroHaptics 2020, LNCS 12272, pp. 325–333, 2020.
https://doi.org/10.1007/978-3-030-58147-3_36

SPIDAR [5]. This device type has a problem that numerous winding mechanisms must be installed to cover the entire workspace. Hence, a large-scale device is required to realize a wide workspace.

To this end, we propose a new type of haptic display that can present a force sensation to the fingertips using a fishing rod shown in Fig. 1. The force sensation is presented to the finger by attaching a thread to the finger and controlling the winding force of the thread. The base of the fishing rod has a pan-tilt mechanism. Because the fishing rod is lightweight, even a motor with a relatively low output can control its attitude at high speed. Although there is a restriction whereby the pulling direction of the thread cannot be set in a perfectly arbitrary manner (for example, a downward force cannot be generated), this might serve as a technique with a wide workspace, in the sense of range.

This paper presents the configuration of the proposed system and the development and evaluation of a control method in the case wherein the force direction is vertical.

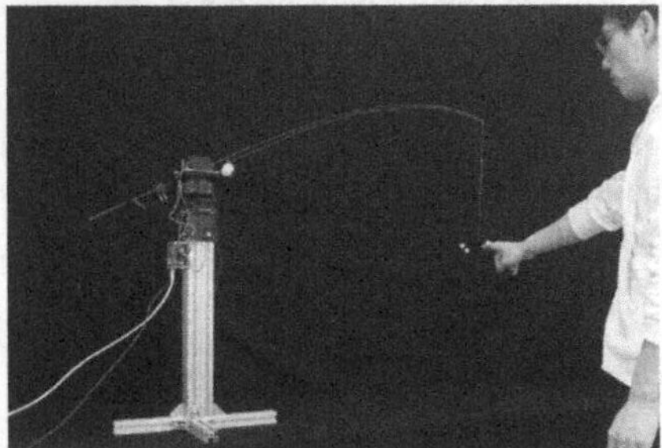
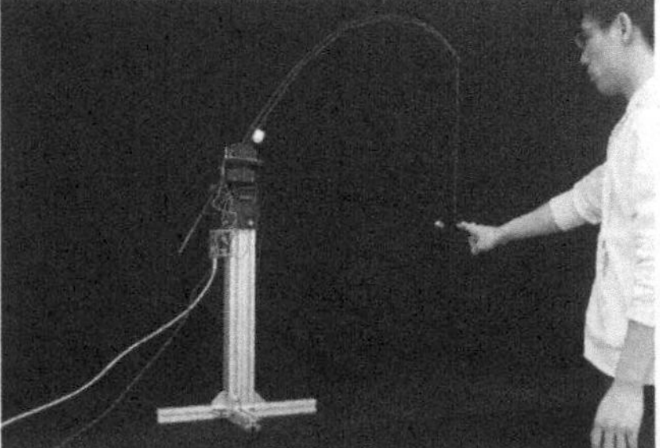

Fig. 1. Overview of device: (left) fishing rod is not bent; (right) fishing rod is bent.

2 Related Work

Here, the haptic displays using a thread are described. The thread-based haptic displays can be approximately divided into wearable type and grounded type displays.

The wearable type uses a technique for presenting the force sensation by mounting the device on the user's body. Hirose developed HapticGEAR [6] and Hosseini developed a similar device using twisted-thread actuators [7]. In this wearable type of haptic displays, the range wherein the force sensation can be exerted is wide, and the user can use it freely. However, wearing the device is time-consuming and imposes a burden to the user owing to the weight of the device. Additionally, the reaction force is inevitably generated in the mounting part when the force sensation is presented.

In contrast, a grounded thread-based display type, such as SPIDAR [5], fixes the device to the wall or floor. The device essentially comprises a cube type frame and four motors at the vertices, and exerts the force sensation to the finger using winding threads. The grounded type can generate a strong and accurate force because the device is grounded and not affected by the reaction force. However, there is a problem whereby the range of the force to be presented is limited within the cube. Additionally, although this device type can apply a multi-directional force using multiple threads, the operating range of the fingertip is more limited owing to interference amongst the threads. SPIDAR has been under development for a long time and comes in various types [8]. SPIDAR-S [9], which presents a force sensation to the user's finger with a single thread, and SPIDAR-MF [10], which corresponds to wrist twisting using a rotatable frame structure, have been proposed to solve the workspace problem. However, SPIDAR-S does not assume a multiple degree-of-freedom (DoF) force sensation, and SPIDAR-MF focuses on preventing the interference of threads corresponding to multiple fingers.

Based on these considerations, a haptic device using a thread currently has three requirements: suppressing the generation of the force to a position except for the intended position; preventing the interference between the body and the threads; presenting a force sensation within a wide range. The proposed haptic device satisfies these three requirements at the cost of sacrificing the accuracy of force direction. The problem of the repulsive force has been solved by developing the device as a grounded type. The interference between the body and the threads is prevented by using one thread from the fishing rod. Finally, the presentation of the force sensation within a wider range is realized using a method for driving the base of the fishing rod.

3 Device Configuration

The proposed device is shown in Fig. 1. A thread coming out of a fishing rod is attached to the user's fingertip end-effector. To wind up the thread, a DC motor (Maxon, 135079, 10 W) is attached to the reel and controlled using PWM control. To present the multi-directional force using the fishing rod, two servomotors (JX Servo, C 70 Digital Servo, 92 W) are mounted onto the base supporting the fishing rod, and a pan-tilt mechanism is adopted. The fingertip position is detected by an optical motion capturing system (OptiTrack, V 120: Trio) installed around the device, and optical markers (retroreflective balls) are attached to the fingertip end-effector. With the abovementioned device configuration, the winding of the thread attached to the end-effector and the posture of the fishing rod are controlled to present the force sensation.

Three effects are expected when using the fishing rod as a haptic display. First, compared with the case of multi-threads, the calculation of the force is simpler because the tension direction of the thread matches the direction of the force presented to the fingertip. Second, because the fishing rod is lightweight, it is possible to control the posture of the fishing rod at high speed, even when using a motor with relatively low torque. Third, the fishing rod can innately present a strong force to the fingertip. Although the necessary torque is calculated from the length of the fishing rod and thread tension, it can be suppressed when a strong force is required because the bending of the fishing rod has the effect of equivalently shortening the length of the fishing rod such as right figure of Fig. 1. In this case, however, the accuracy of the force direction is reduced.

4 Control Method

In this study, we considered a control method for presenting a directional force. Accordingly, we limited the investigation to the case of a vertical force; that is, the thread attached to the fingertip always exerts a straight upward pull. The optical motion capturing system detects the position of the fingertip and the proposed device vertically pulls the thread attached to the fingertip by controlling the winding of the thread and the posture of the fishing rod.

In this case, the fingertip position becomes the input information of the control system. When introducing this device into a virtual reality application, it is necessary to assess whether it touches the object in the virtual space or not, and to change the presenting magnitude of the force. Additionally, it is important to change the magnitude of the force to express the softness suitable to a virtual object. In other words, the direction and magnitude of the force presented to the fingertip are the targets that should be realized by controlling the device.

In the case of a serial link type haptic device, and particularly with regard to the control of an impedance device type such as PHANToM [4], the position of the end effector is the input, as in the present situation, while the direction and magnitude of the presented force are the targets to be realized. Because a conventional serial link haptic device does not deform the link mechanism itself, it is easy to measure the position of the end effector by carrying out forward kinematics calculations to obtain the torque required by the motor through inverse dynamics calculations. In our case, it is currently difficult to derive a calculation formula that considers the bending of the fishing rod in relation to the force presented at the fingertip using the proposed device.

Thus, we initially attempted to perform feedback control using the difference between the target direction of the force and the current direction of the thread such as Fig. 2(a). The optical marker was not only attached to the fingertip end-effector, but also to the thread (Fig. 2(b)), and the direction of the thread was obtained.

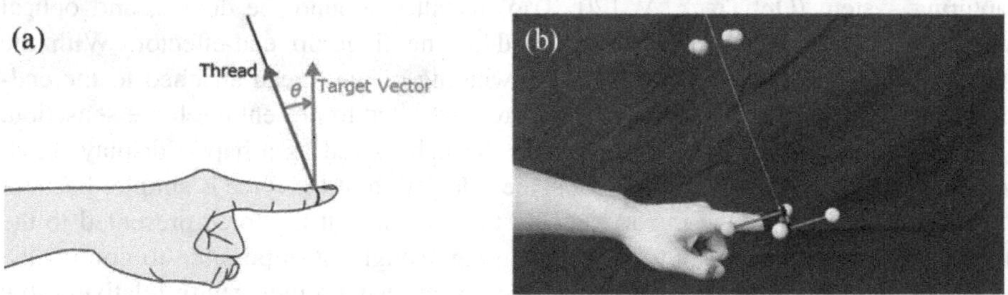

Fig. 2. Overview of feedback control; (a) consideration: (b) actual situation.

However, when the device was actually operated through feedback control, the problem of thread vibration occurred. One of the causes for this is that the sampling rate of the optical tracking was 120 Hz, which is close to the resonant frequency of the thread itself and increased the control difficulty. Additionally, the resonant frequency of the thread dynamically changed with the change of its length and tension.

Based on the abovementioned trial, we considered that it is currently difficult to control the fishing rod using feedback control. Thus, we decided to control the fishing rod using feedforward control. As mentioned above, the measurement information in the proposed device is the fingertip position, and the target values are the magnitude and direction of the force presented at the fingertip. This time, the target direction of the force was fixed to be vertical. Moreover, it is desirable that the winding force of the thread and the angle of the fishing rod are obtained as output information. To this end, it was necessary to investigate the relationship between the position of the fingertip and the winding force of the thread, the posture of the fishing rod, and the force generated by the winding force of the thread and posture of the fishing rod. In the next section, we present the measurement and fitting of the relationship between the input and output information.

5 Measurement

5.1 Measurement Method

In the measurement, the relationship between position x of the fingertip and the duty ratio command value d to the motor used for winding the thread, the attitude θ of the fishing rod, and the force f generated by the winding force of the thread and the attitude of the fishing rod were measured. The measurement environment is shown in Fig. 3. We attached a 1.2-kg load instead of a finger. The position x was the distance from the device to the load. The angle of the fishing rod is defined as 0° when it is parallel to the ground. To measure the force of the thread pulling the load, the load was placed onto an electronic balance. The weight of the load was recorded by starting the electronic balance before the measurement, and the traction force was calculated from the change of the electronic balance when the load was pulled by the thread.

The load was placed at 22.5 cm, 45.0 cm, 56.25 cm, 67.5 cm, 78.75 cm, and 90.0 cm with the device position as the origin. The duty ratio of the motor winding the thread at each position was changed by 10%, from 10% to 100%. Then, the angle of the fishing rod was adjusted such that the thread pulled the load exactly in the vertical direction. Additionally, because the motor of the reel started to rotate when the duty ratio was 5% or more, measurements were also made when the duty ratio was 5% at

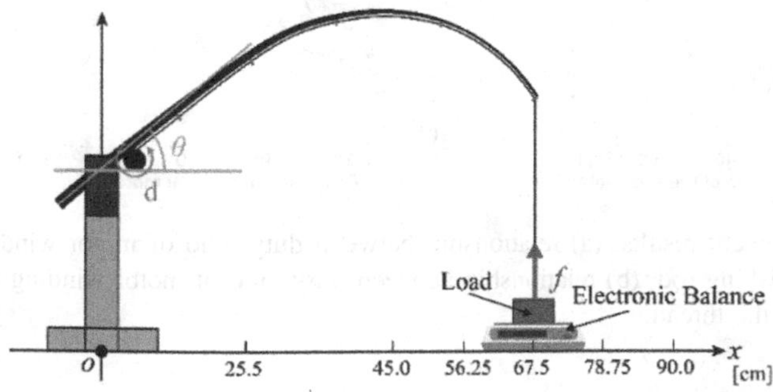

Fig. 3. Measurement environment.

each position. The measurement of the fishing rod's angle and the traction force to the 11 values of the duty ratio was repeated five times while the position was fixed, and the mean and the standard deviation were calculated.

5.2 Measurement Result

The measured means and the standard deviations were plotted in Fig. 4. The relationship amongst location x of the load, duty ratio d of the motor controlling the winding of the thread, and angle θ in the pitch axis direction of the fishing rod is shown in Fig. 4 (a). Figure 4 (b) shows the relationship amongst location x of the load, the duty ratio d of the reel motor, and the force f of the thread pulling the weight.

In Fig. 4(a) and (b), only one data point existed when the load was set to 90 cm. The total length of the fishing rod used in this study was 90 cm. Additionally, the duty ratio of 5% was the minimum duty ratio at which the motor of the reel started to rotate (the fishing rod began to bend).

The measurement data obtained when the load was placed at 22.5 cm and 78.75 cm was smaller in the number of plots than the data obtained when the load was placed from 45.0 cm to 67.5 cm. When the load was set to 78.75 cm, the posture of the fishing rod for pulling in the vertical direction was not observed with a duty ratio higher than 40%. Moreover, when the load was installed at 22.5 cm and the duty ratio exceeded 20%, the fishing rod began to bend not only in the pitch axis direction, but also in the roll axis direction. Therefore, the duty ratio over 30% was not measured.

According to Fig. 4 (a), the fishing rod's angle increases as the position of the load approaches the device. Similarly, in Fig. 4 (b), the traction force also increases. Increasing the fishing rod's angle is the same as increasing its whip. Therefore, it can be confirmed that the traction force also increases, when the whip of the fishing rod increases.

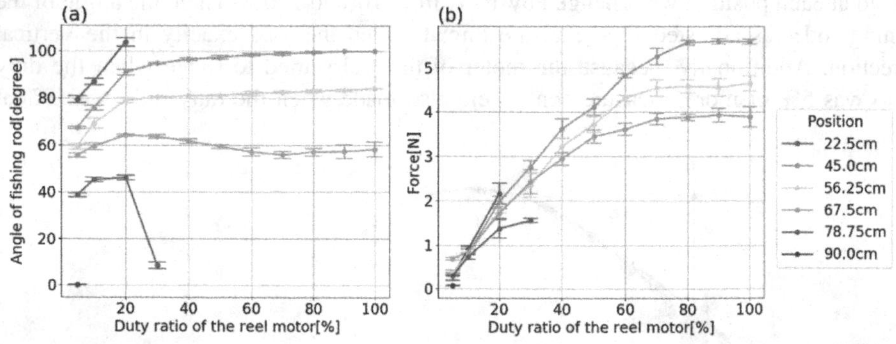

Fig. 4. Measurement results: (a) relationship between duty ratio of motor winding the thread and attitude of fishing rod; (b) relationship between duty ratio of motor winding the thread and pulling force of the thread.

In the proposed feedforward control, the position x of the fingertip and the magnitude f of the presented force are given, and the duty ratio d required by the motor winding the thread and the angle θ of the fishing rod must be obtained. From Fig. 4(b), the duty ratio d of the motor winding the thread was derived by inputting the position x of the fingertip and the strength f of the force to be presented. Then, based on Fig. 4 (a), the posture θ of the fishing rod was determined from the position x of the fingertip and the duty ratio d of the motor.

To realize the abovementioned procedure, the regression equation obtained by the least squares method was derived from the measurement data presented in Fig. 4. The regression curves based on the regression equation are shown for several locations in Fig. 5 and Fig. 6. Figure 5 shows the regression curve obtained from x and f to d, using a second order polynomial, while Fig. 6 shows the regression curve obtained from x and d to θ using a first order polynomial, which indicates that the regression curve adequately represents the data obtained by measurement.

Our control scheme was confirmed by performing actual feedforward control. The amount of force was fixed and the fingertip was moved. As shown in Fig. 7, the thread always pulled the fingertip straight upward, even when the fingertip was moved, which indicates that the desired feedforward control was realized.

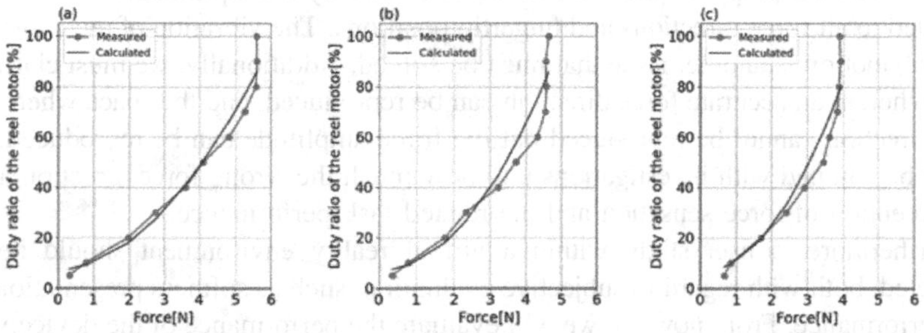

Fig. 5. Comparison of regression curves with measurements of thread pulling force and force winding the thread: (a) x = 45.0 cm; (b) x = 56.25 cm; (c) x = 67.5 cm.

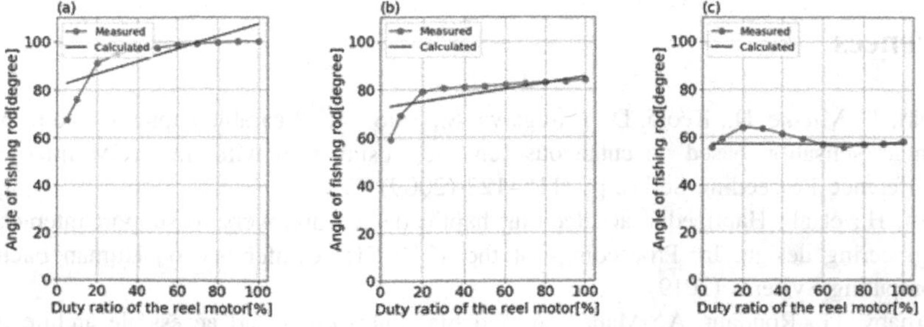

Fig. 6. Comparison of regression curve with measurements for force winding the thread and posture of fishing rod: (a) x = 45.0 cm; (b) x = 56.25 cm; (c) x = 67.5 cm.

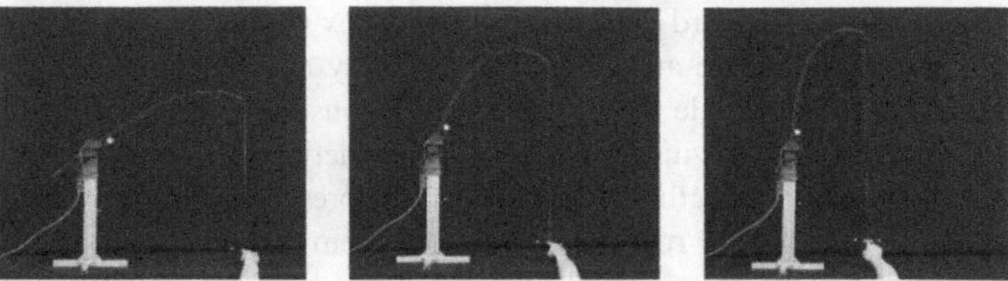

Fig. 7. Validation of control; the fingertip is continuously pulled in vertical direction with the same amount of force.

6 Conclusion

This paper proposes a haptic device using a fishing rod and describes the proposed control method that pulls the fingertip in the vertical direction. The relationship between the data was investigated by measuring the position of the fingertip, winding force of the thread, posture of the fishing rod, and force of the thread pulling the fingertip. The regression curves were obtained based on the measured data and incorporated into the control of the device. Thus, it was always possible to pull the fingertip in the vertical direction, even if the fingertip was moved.

However, the proposed control method is obviously a simplification, and must be expanded to all force directions and fingertip positions. The vibration of the fishing rod by quick motion is another issue that must be solved. Additionally, we must clarify the space wherein an accurate force direction can be reproduced, and the space wherein the force direction cannot be reproduced but the force amplitude can be reproduced. This should be coupled with investigations into how much the wrong force direction affects the perception of force sensation and the related task performance.

Furthermore, a user study within a virtual reality environment should be also conducted, both with regard to subjective evaluation, such as softness presentation, and task performance. From now on, we will evaluate the performance of the device with a focus on the user studies.

Acknowledgment. This research was supported by JSPS KAKENHI Grant Number JP18H04110.

References

1. Aoki, T., Mitake, H., Keoki, D., Hasegawa, S., Sato, M.: Wearable haptic device to present contact sensation based on cutaneous sensation using thin wire. In: ACM International Conference Proceeding Series, pp. 115–122 (2009)
2. Seifi, H., et al.: Haptipedia: accelerating haptic device discovery to support interaction & engineering design. In: Proceedings of the 2019 CHI Conference on Human Factors in Computing Systems (2019)
3. Barnaby, G., Roudaut, A.: Mantis: A scalable, lightweight and accessible architecture to build multiform force feedback systems. In: UIST 2019-Proceedings of the 32nd Annual ACM Symposium on User Interface Software and Technology, pp. 937–948 (2019)

4. Jarillo-Silva, A., Domínguez-Ramírez, O.A., Parra-Vega, V., Ordaz-Oliver, J.P.: PHANToM OMNI haptic device: Kinematic and manipulability. In: CERMA 2009-Electronics Robotics and Automotive Mechanics Conference, pp. 193–198. IEEE (2009)
5. Sato, M., Hirata, Y., Kawarada, H.: Space interface device for artificial reality—SPIDAR. Syst. Comput. Jpn. **23**, 44–54 (1992)
6. Hirose, M., et al.: HapticGEAR: the development of a wearable force display system for immersive projection displays. In: Proceedings-Virtual Reality Annual International Symposium (2001)
7. Hosseini, M., Meattini, R., Palli, G., Melchiorri, C.: Development of sEMG-driven assistive devices based on twisted string actuation. In: 2017 3rd International Conference on Control, Automation and Robotics (ICCAR), pp. 115–120. IEEE (2017)
8. Sato, M.: Development of string-based force display: SPIDAR. In: 8th International Conference on Virtual Systems and Multimedia (VSMM) (2002)
9. Ma, S., Toshima, M., Honda, K., Akahane, K., Sato, M.: SPIDAR-S: a haptic interface for mobile devices. In: Chen, Y., Christie, M., Tan, W. (eds.) SG 2015. LNCS, vol. 9317, pp. 203–206. Springer, Cham (2017). https://doi.org/10.1007/978-3-319-53838-9_18
10. Liu, L., Miyake, S., Akahane, K., Sato, M.: Development of string-based multi-finger haptic interface SPIDAR-MF. In: Proceedings of 23rd International Conference on Artificial Reality and Telexistence (ICAT), pp. 67–71 IEEE (2013)

Confinement of Vibrotactile Stimuli in Periodically Supported Plates

Ayoub Ben Dhiab(✉) and Charles Hudin

CEA, LIST, 91191 Gif-sur-Yvette, France
{ayoub.ben-dhiab,charles.hudin}@cea.fr

Abstract. For multitouch and multiuser interactions on a touch surface, providing a local vibrotactile feedback is essential. Usually, vibration propagation impedes this localization. Previous work showed that narrow strip-shaped plates could allow the confinement of vibrotactile stimuli to the actuated area. Adding to this principle, periodically supported plates also provide a non-propagative effect at low frequencies. Using both geometrical properties, we provide a device allowing a multi-touch interaction through an array of piezoelectric actuator. Experimental validation show that vibrations are well confined on top of actuated areas with vibration amplitude over 2 μm.

Keywords: Surface haptics · Confinement · Vibrotactile stimuli · Narrow plate · Ribbed plate · Non-propagating · Evanescent wave · Vibrations

1 Introduction

Wave propagation in today's surface haptic interfaces limits the user to a single point of interaction whereas most exploratory procedures benefit from the use of multiple fingers to properly process surface information [7]. Solving this problem requires to localize vibrations. There are several methods related to technologies that allow localized haptic feedback on continuous surfaces. For spatial localization of vibrations, i.e., the creation of local deformation at a point or area of the plate, there are two techniques: *Time-Reversal* [5] and *Modal Superimposition* [4]. For these two techniques, the wavelength of the vibrations to obtain a local deformation at the centimeter or millimeter scale necessarily involves high frequencies. The *Time-Reversal Wave Focusing* technique uses the propagation of elastic waves to generate constructive interference at a given time and position on the plate. The acceleration peak created by the focusing of the waves causes the ejection of a static or moving fingerpulp, giving a haptic feedback smaller than the fingertip. However, this approach needs a calibration procedure and is subjected to external parameters (e.g. temperature, finger interaction), which can impede the tactile stimulation. As for *Modal Superimposition*, it uses a truncated modal decomposition to focus a deformation shape and control its position. The high frequencies produced by an array of piezoelectric actuators

I. Nisky et al. (Eds.): EuroHaptics 2020, LNCS 12272, pp. 334–342, 2020.
https://doi.org/10.1007/978-3-030-58147-3_37

allow for the appearance of a tactile lubrication phenomenon, i.e. a variation in friction that is only felt when the finger is moved [10]. Both techniques rely on propagation of high frequency waves. During user interaction, the non-linear behavior of the finger significantly modifies the vibrational behavior of the plate. Such modifications are particularly striking for high frequencies thus leading to many complications (incl. flexural waves scattering, mode damping, mode translation). In order to avoid these limitations and provide a haptic feedback both to static and moving fingers, the *Inverse-Filter* approach [8] proposes to dynamically control low-frequency waves, which are less impacted by finger interaction, to provide the user with the desired stimuli only at the contact points. Nonetheless, this technique is based on (consequent) computation, signal processing and calibration procedures. As this technique uses waves within the tactile frequency range (0–1 kHz), it induces a global movement of the plate and therefore limits the interaction to five control points. Generally, all the previously mentioned methods depend on wave propagation and thus, are in need of calculations and signal processing in order to achieve localization. Therefore, we have investigated in our previous work [1] another method that overcomes such requirements by relying on geometry features and wave evanescence instead. We have shown that it is possible to obtain localized deformations above the actuators for low frequency signals in narrow thin plates (1D). A particularity of this technique is that it allows the spatial localization of low-frequency stimuli i.e. a local deformation of the plate even if we use low-frequencies with long wavelength. In addition, finger interaction does not attenuate low-frequency evanescent waves. No matter the surface in contact, i.e. finger, hand, arm, foot... the behavior of the vibrating plate is not impacted and a localized stimuli can be provided. Although interesting, this approach is limited to narrow plates. However, for a rich multitouch interaction, extending this approach to an arbitrary 2D plate becomes necessary and is the focus of this paper. Because waves tend to directly propagate in 2D plates, the confinement i.e. the suppression of propagative waves in favor of evanescence, was not easily achieved. In order to do so, we used a theoretical result raised by Mead [6]. He states that a periodically supported beam exhibits a non-propagating low-frequency band up to the first resonant mode of isolated segments. Meaning in our case that if we have a periodically bounded plate and if each formed areas have each a waveguide geometry, a cut-off frequency exists and we can have the spatial localization of low frequency stimuli. Therefore, in this paper, we verify this assumption experimentally with a novel setup and discuss the various possibilities it opens up.

2 Principle

2.1 Confinement by the Ribs

In this section we explain how vibrations can be confined along both axis of a ribbed plate depicted in Fig. 1. On the x-axis, we find ourselves solving a problem of an infinite beam supported periodically (Fig. 1). Regularly supported beams can be seen as a set of individual supported beam portions applying

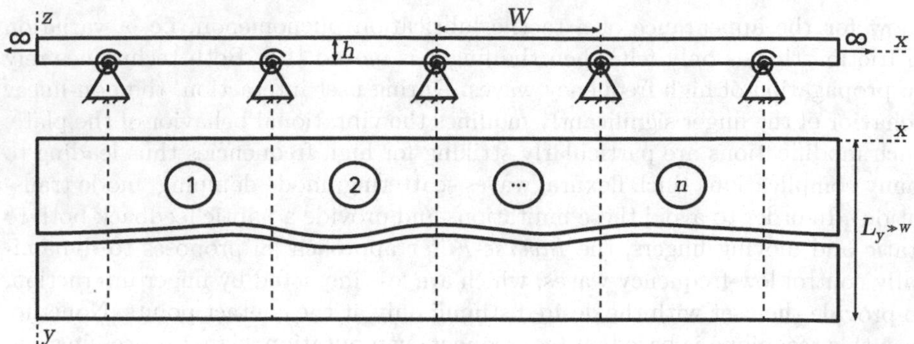

Fig. 1. Plate bounded by a set of equidistant ribs forming n areas which theoretically behaves like narrow-plate bounded on their longer edges to a rigid frame.

bending moments (represented by rotation spring in Fig. 1) to their neighboring beam portions. In such systems, vibrations propagate from one portion to another through bending moments produced at the junction between two portions. Mead [6] has shown that for frequencies bellow the first resonance of an individual portion, the propagation from one portion to another attenuates exponentially. This means that in our system, the cut off frequency f_1 that marks the transition between propagative and evanescent waves along the waveguide dimension, also corresponds to the transition from propagative to evanescent wave in the transverse dimension. This enables the extension of the waveguide approach describes in [1] to 2D surfaces regularly supported by thin ribs.

2.2 Evanescence and Waveguide Geometry

The localization along the y-axis of low frequencies is implied by the waveguide geometry of the propagation medium. In such medium, it was shown that bending waves cannot propagate below the cut-off frequency of the first propagation mode. This non-propagation effect allows the confinement of vibrotactile stimuli and thus can be used for multitouch haptic interactions. If we suppose a perfect bounding of the ribs to a rigid frame, each of the n areas possesses their first cut-off frequency equal to [1]:

$$f_1 = \frac{\pi}{4\sqrt{3}} \frac{h}{W^2} \sqrt{\frac{E}{\rho(1-\nu^2)}} \tag{1}$$

with ρ the mass density, E the Young's modulus, ν the Poisson's ratio, h the plate thickness, W the distance between ribs. When an area is submitted to a vibration of frequency $f < f_1$, waves propagating along the y-axis and the x-axis are evanescent and dies exponentially with the distance.

3 Experimental Validation

3.1 Apparatus

To verify the spatial localization provided by our method, we have set up a system designed to allow a user to freely explore a surface. The size of the device thus corresponds to the size of an adult hand, and the number of areas formed corresponds to the number of fingers. The only requirement associated with the implementation of this method is having a plate correctly bounded periodically to a rigid frame. If each created area have a waveguide geometry, a cut-off frequency exists and a localized vibration can be obtained.

Mechanical Components. The apparatus consists in a glass plate, measuring $150 \times 130 \times 0.5\,\mathrm{mm}^3$, bounded on a set of 6 equidistant ribs of width 4 mm forming 5 areas. In order to provide haptic feedbacks, piezoelectric actuators (muRata 7BB-20-3) of circular geometry, composed of two part: a plate of diameter 20 mm, thickness of 0.1 mm; and a piezoelectric transducer of diameter 12.8 mm, thickness of 0.11 mm, were glued on the bottom side of the plate between each ribs. The out of plane displacement of the surface is measured by a laser vibrometer (Polytec MLV-100/OFV-5000) mounted on a motorized 3 axis platform. This setup is illustrated on Fig. 2.

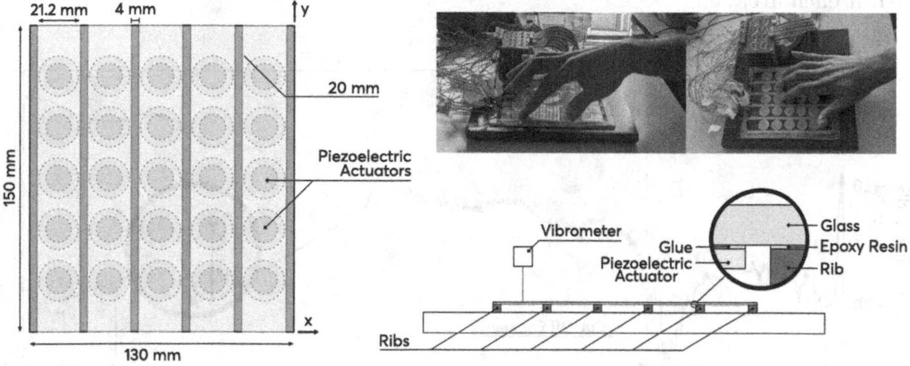

Fig. 2. Experimental setup. A glass plate is bounded onto a set of 6 equidistant ribs with epoxy resin. The actuators are glued to the bottom of the plate, 5 per area between each rib, for a total of 25 actuators.

Driving Components. Because vibration localization is induced by the propagation medium geometry, no specific calculation nor wave control implementation are needed. Signal amplification for each actuators was assured by a Piezo Haptics Driver DRV8862 from Texas Instrument. A voltage output module NI-9264 was used to send requested analog signals to each driver for amplification.

Communication with the NI-9264 voltage output module and signal design was made with Python using the PyDAQmx package.

3.2 System Frequency Response Function (FRF)

The frequency response function or FRF (ratio of the displacement at a given point to the voltage applied to an actuator in the frequency domain) for the actuator was measured by sending a linear sweep signal of 40 V amplitude going from 0 Hz to 5 kHz sampled at 10 kHz and is represented on Fig. 3. We search with this FRF frequencies where the amplitude at the center of the actuator (in light gold) presents noticeable differences with the amplitude outside the actuator area (in dark blue). We can notice that, from 0 to 1.2 kHz, vibration amplitudes inside the actuator area are 30 dB higher in average than outside the actuator area which shows that elastic waves do not propagate throughout the plate for vibration frequency in this range. The FRFs at other points were also measured and show the same behavior, though with some differences. We supposed in Sect. 2 that clamping conditions were perfect thus having the same cut-off frequency for every area. But the FRF in other areas actually present different cut-off frequencies meaning that the clamping quality of each rib was not assured. The boundary condition being made with epoxy resin we can assume that eventual air bubbles or gaps could have been left during the mounting process thus giving slight behavior differences between each area and locally within each area.

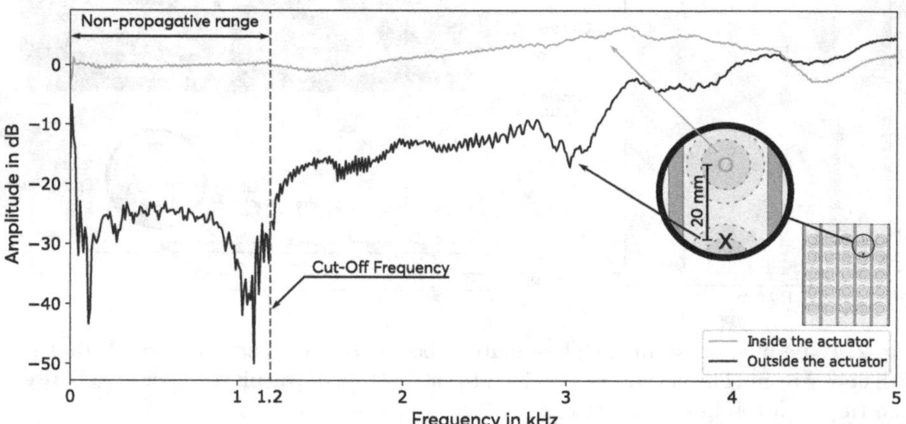

Fig. 3. System Frequency Response Function (FRF) $H_{dB} = 20\log_{10}|H|$ at the center of the activated actuator (in light gold) and 20 mm away outside the actuator inside the same area (in dark blue). (Color figure online)

3.3 Stimuli Confinement

To illustrate the multitouch capability of our device we mapped plate amplitudes when submitted to signals sent with multiple piezoelectric actuators. We chose a 250 Hz (tactile sensitivity peak) 5 cycles burst excitation and a driving voltage amplitude of a 100 V for every actuator used. Figure 4 shows the maximum displacement throughout the plate when using 1, 2, 3, 5, 6 and 7 actuators at the same time. We can notice that waves are confined above activated actuators with amplitudes reaching 3 µm inside the actuator area. Outside the actuated area, waves propagating along the y-axis are evanescent and die exponentially. Along the x-axis, vibration propagation is impeded by the clamped ribs surrounding the actuator. The energy representation with a logarithmic scale of Fig. 5 shows certain areas of interest. Apart from localized energy area above each actuator, we can note leaks with a 15 dB amplitude difference between vibrations inside actuator areas and leaked vibrations while other areas present a 30 dB difference. This leak is actually due to clamping imperfection caused during the prototype assembly. Another area which is less visible (squared in green in Fig. 5), asses the evanescent behavior along the x-axis due to the set of supports. This important loss of energy (around 25 dB loss) comes from the evanescence behavior and the width of the ribs which, by providing strong bending moments, reduce the transmitted energy. Thinner ribs would still provide a localized effect but with a reduced energy loss.

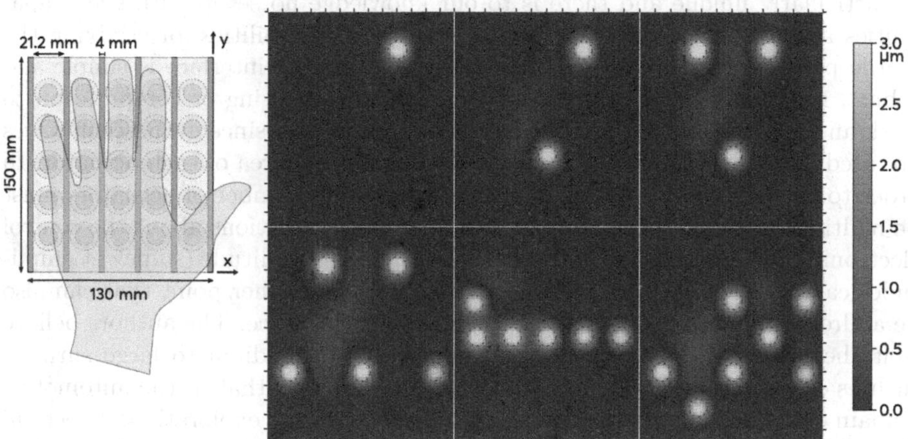

Fig. 4. Maximum out of plane displacement in μm of the plate for 1, 2, 3, 5, 6 and 7 actuators activated at same time. Vibration sources are localized and reach a vibration amplitude above 3 µm while the rest of the plate stays at rest.

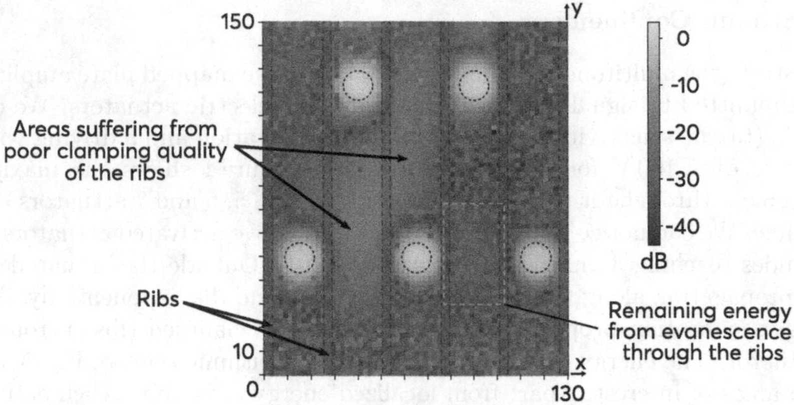

Fig. 5. Energy representation with a logarithmic scale of the plate submitted to 5 activated actuators. Most of the energy is localized above each piezoelectric actuator.

4 Discussion and Perspective

The device shown here can provide localized vibrotactile feedback in several areas of a continuous plate. The confinement is achieved along both axis by low frequency evanescence phenomenon: one based on the geometry of the propagation medium; the other based on the periodically placed supports. This system is particularly unique and there is to our knowledge no setup with such capabilities and we believe that our device offers new possibilities for studying the tactile perception and to develop new interactions. The interface is simple and robust, i.e. it does not require any special signal processing or computation to work and is not impacted by user interaction. However, since evanescent waves are used, the localization of stimuli is only done in the area of each actuator. In order to increase the resolution of such a device, the number of actuators must be multiplied as well as the number of ribs. It raises questions about the control electronics, the power consumption and the available amplitude (sufficient amplitudes cannot be guaranteed with small actuators). Another point that can also be addressed is concerning possible use case of such device. The authors believe that the method in question is easily applicable to medium to large surfaces such as car dashboards. Breitschaft and al. [2] indicates that in the automotive domain four tasks must be performed by tactile interiors: exploration, detection, identification and usage. These tasks are already achievable as is if the method is applied to a thin touch screen (<2 mm). The user can explore the surface freely with one or more fingers or by sliding or tapping or even by placing his hand entirely on the surface without changing its behavior. The detection of the localized stimulus and its identification is ensured by the shape and frequency content of the signal used, which can also be varied according to the position of the fingers if one wants to create areas with different vibratory behaviors. Finally the confirmation action is done by the selected zone by sending a signal with a strong frequency dynamic to simulate a button (this function can be improved by adding a force sensor). Future works will focus on two axes, one dealing with the possibility of thinner ribs and the other on an interface use case, in particular the localization of a vibratory targets on a flat continuous surface.

5 Conclusion

At low frequency, bounded ribbed-plates allow for the localization of vibrotactile stimuli. A piezoelectric actuator bounded to such medium mostly produce evanescent waves which are confined to the actuator covered area as long as the driving frequency is lower than the first cut-off frequency of our system. No control strategies nor signal processing are needed, directly sending the desired signal to the desired actuator suffice in order to obtain a localized vibrotactile stimuli. Multitouch and multiuser interactions can therefore be implemented with ease. Application such as tactile fingerspelling [3], tactile keyboard, tactile memory games and applications using exploratory procedures on a plates can be implemented here. Although the presence of actuators underneath the plate limits applications to surfaces where transparency is not necessary, recent breakthroughs shows that transparent piezoelectric actuators are at hand [9] making our technology compatible with screen surfaces.

Acknowledgment. The revised version of this article benefited from insightful comments from Dr. Sabrina Panëels.

References

1. Ben Dhiab, A., Hudin, C.: Confinement of vibrotactile stimuli in narrow plates: principle and effect of finger loading. IEEE Trans. Haptics (2020)
2. Breitschaft, S.J., Clarke, S., Carbon, C.C.: A theoretical framework of haptic processing in automotive user interfaces and its implications on design and engineering. Front. Psychol. **10**, 1470 (2019)
3. Duvernoy, B., Topp, S., Hayward, V.: "HaptiComm", a haptic communicator device for deafblind communication. In: Kajimoto, H., Lee, D., Kim, S.-Y., Konyo, M., Kyung, K.-U. (eds.) AsiaHaptics 2018. LNEE, vol. 535, pp. 112–115. Springer, Singapore (2019). https://doi.org/10.1007/978-981-13-3194-7_26
4. Enferad, E., Giraud-Audine, C., Giraud, F., Amberg, M., Semail, B.L.: Generating controlled localized stimulations on haptic displays by modal superimposition. J. Sound Vibr. **449**, 196–213 (2019)
5. Hudin, C., Lozada, J., Hayward, V.: Localized tactile feedback on a transparent surface through time-reversal wave focusing. IEEE Tran. Haptics **8**, 188–198 (2015)
6. Mead, D.J.: Free wave propagation in periodically supported, infinite beams. J. Sound Vibr. **11**, 181–197 (1970)
7. Morash, V.S., Pensky, A.E.C., Miele, J.A.: Effects of using multiple hands and fingers on haptic performance. Perception **42**, 759–777 (2013)
8. Pantera, L., Hudin, C.: Multitouch vibrotactile feedback on a tactile screen by the inverse filter technique. IEEE Trans. Haptics (2020)
9. Qiu, C., et al.: Transparent ferroelectric crystals with ultrahigh piezoelectricity. Nature **577**, 350–354 (2020)
10. Wiertlewski, M., Fenton Friesen, R., Colgate, J.E.: Partial squeeze film levitation modulates fingertip friction. PNAS **113**, 9210–9215 (2016)

2MoTac: Simulation of Button Click by Superposition of Two Ultrasonic Plate Waves

Pierre Garcia[1,2(✉)], Frédéric Giraud[1], Betty Lemaire-Semail[1], Matthieu Rupin[2], and Michel Amberg[1]

[1] Univ. Lille, Centrale Lille, Arts Et Metiers Institute of Technology, Yncrea Hauts-de France, ULR 2697 L2EP, 59000 Lille, France
{pierre.garcia,frederic.giraud, betty.lemaire-semail,michel.amberg}@univ-lille.fr

[2] Hap2u, 20 Rue du Tour de l'Eau, 38400 Saint-Martin-d'Hères, France
{pierre.garcia,matthieu.rupin}@hap2u.net

Abstract. Recent studies have shown that a button click sensation could be simulated thanks to ultrasonic vibrations. In this context, a travelling wave may enhance the simulation because it creates internal lateral stresses that are released during the stimulation. However, this solution is difficult to integrate on plates. We present 2MoTac, a method which superpose a longitudinal and a bending mode simultaneously on a plate, in order to create a pseudo-travelling wave. We present the design, and a psychophysical study to deduce the optimal ratio between the bending and longitudinal mode amplitudes, in terms of detection threshold and robustness.

Keywords: Haptic display · Tactile perception · Ultrasonic vibration · Vibration mode · Keyclick · Button click

1 Introduction

Most of new human-machine interaction devices now rely on touch screens. They have become so cost effective that physical buttons and knobs are removed and replaced by virtual buttons displayed on the flat and hard surface of the display panel in vending machines, car dashboards, and so on. However, touch screens do not involve the sense of touch in the interaction with the machine: they do not provide information through this channel, unlike physical buttons for instance. Therefore, sight is predominant for whom wants to interact with the aforementioned machines, leading to many disadvantages: elderly and visually-impaired individuals struggle to use touch screens, and drivers must look away from the road, which is a major safety issue in automobile. Therefore, technological improvements to touch screens are needed, in order to involve touch when interacting with a machine through its touch screen.

I. Nisky et al. (Eds.): EuroHaptics 2020, LNCS 12272, pp. 343–352, 2020.
https://doi.org/10.1007/978-3-030-58147-3_38

To create the illusion of touching a button on a panel, lateral displacement [1] or vibrotactile stimulation [11] can be used. The low frequency vibration of the touch screen directly stimulates the skin mechanoreceptors; because this vibration is difficult to contain, [3] identifies the frequency response of the touch screen (FRF) and then uses the frequency that maximises the vibration displacement at a specific point. In [7], the authors invert the FRF and can control the vibration at three different positions where actuators are located. However, using vibrotactile stimulation to create the simulation of a button click has the disadvantage to produce audible noise. To cope with this issue, [10] uses the rapid vibration of a standing wave at ultrasonic frequency, to reduce the internal stresses inside the finger pulp due to the friction that appears when pressing the touchscreen. More robust results are obtained if a net tangential force is produced, by using electroadhesion force synchronized on ultrasonic lateral displacement [15] or by using a travelling wave [6]. These solutions however seem difficult to be integrated on plates, without obstructing the view by actuators.

This paper introduces a new concept of tactile stimulation that uses two orthogonal modes to simulate a button click. The elliptical motion of the particles in contact with the fingertip creates a lateral force as if a travelling wave was used. After presenting the technological principle that derives from [2], a study has been conducted in order to define the ratio between the two modes that optimizes the feeling of a single click when reversing the lateral force direction.

2 Presentation of the Two Modes Stimulation

2.1 Theoretical Background

We consider a plate, which length is L, thickness b and height H. Its vibration can result in the superposition of different types of modes: the longitudinal modes are characterized by the out-of plane translation of the plate's cross-sections while bending modes are characterized by an in-plane translation and an out-of plane rotation of the cross-section, as described in Fig. 1.

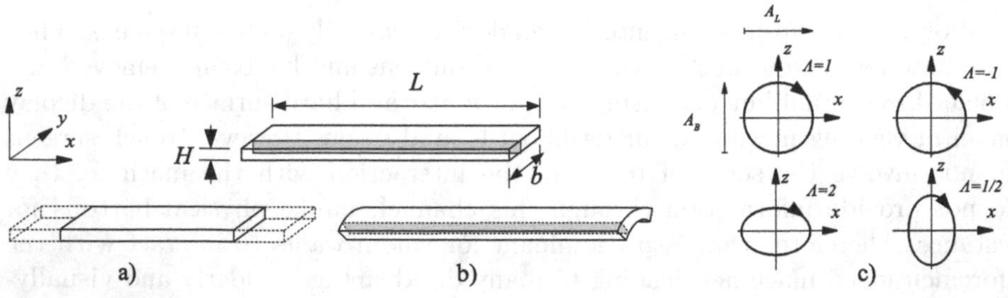

Fig. 1. Vibration mode a) Longitudinal, b) bending and c) Elliptical motion in the middle of the top surface for several values of Λ

When the two modes are produced at the same vibrating frequency, then the point in the middle of the top surface undergoes a displacement which has two components:

- along x, due to the longitudinal mode, denoted $a_L(t)$,
- along z, due to the bending mode, denoted $a_B(t)$.

In this paper, the two modes are energized at the same frequency f, with a phase shift of $\pm\pi/2$. We then write:

$$a_L(t) = \frac{A_L}{2}\sin(2\pi f t) \qquad a_B(t) = \frac{A_B}{2}\cos(2\pi f t) \tag{1}$$

with A_L and A_B the vibration amplitude of each mode (peak-peak); if $A_B < 0$, the phase shift is $\pi/2$. Hence, by combining the two vibration modes, the points at the top surface follow an elliptical motion, which is able to create a net lateral force, as described in [4]. By changing the ratio $\Lambda = \frac{A_L}{A_B}$, the shape of the elliptical trajectory is modified, as described in Fig. 1c), promoting the normal displacement over the longitudinal one for $\Lambda < 1$ and vice versa for $\Lambda > 1$. In the remaining of this paper, we introduce the maximum vibration speed $U_L = 2\pi f A_L$ and $U_B = 2\pi f A_B$ because they combine frequency and displacement.

2.2 Related Work

To produce a button click sensation, [10] uses a rapid change in friction between the fingertip and a plate by increasing or decreasing ultrasonic vibration. The detection threshold of the click is found to be lower if the friction is decreased compared with a friction increased. A fraction of the elastic energy stored in the finger pulp is due to some tangential contact forces that appear during touch. When it is suddenly released, it creates a stimulus that is detectable by the skin's mechanoreceptors. However, the detection thresholds of this click are higher compared with those measured to detect a friction change in presence of a lateral motion of the fingertip [12]. Therefore, [6] suggests to use a travelling wave instead of a standing wave. Indeed, the results show a lower perceptual threshold for travelling waves, and participants press less times per trial and exert smaller normal force on the surface in order to perceive click effect.

Travelling wave can increase the tangential contact forces due to the elliptical motion of the particles that are in contact with the fingertip, thus increasing the stimulus when they are released. Two theories can explain this phenomenon. In [14], the tangential force derives from the contact conditions between a vibrating plate and an elastic medium. In [9], the authors introduce the lateral viscous forces that are generated by a travelling wave, a phenomenon which is compatible with the existence of the squeeze-film. For both theories, the lateral force is a complex function of the normal and tangential displacements of the particles.

In a travelling wave, the ratio Λ between the normal and lateral vibration amplitude is fixed by the geometry of the vibrating plate. To cope for this issue,

[2] presents a prototype that can produce a longitudinal and a bending mode which resonant frequencies are equal. However, the actuator for the longitudinal mode is bulky, while the vibration amplitude of each mode are not controlled. With 2MoTact, the two components U_B and U_L can be independently set, offering the possibility to optimize the tuning of Λ in order to improve the lateral contact forces produced to enhance the click sensation. A study is then conducted in order to determine which ratio is optimal in terms of psychophysical threshold. To compare the thresholds between each other, the stimulation level is defined as $U = \sqrt{U_L^2 + U_B^2}$, which combines the vibration speed for the longitudinal and the bending mode. We include in the study the case of a pure bending mode displacement ($U_L = 0$, $\Lambda = 0$).

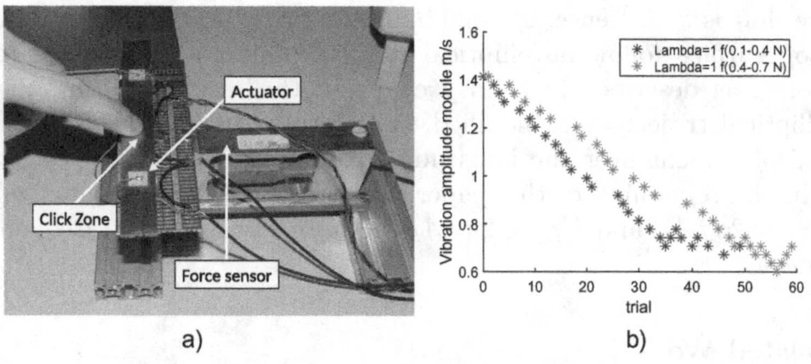

Fig. 2. The Experimental setup a) and typical example of 2 interlaced psychophysical staircases, which targets the 50% perceptual threshold by a one-up one-down test. At each trial, the implemented staircase is chosen at random with a probability of 1/2.

2.3 Design of the Plate

Our prototype consists of a $148 \times 18 \times 2$ mm^3 aluminium plate, actuated by nine piezoelectric ceramic actuators (5 mm × 7 mm × 0.5 mm from Noliac, Denmark) as presented Fig. 3c). The plate was designed such that both modes (longitudinal and bending) can be excited at the same frequency ($f = 34$ kHz). To energize the longitudinal mode, eight actuators placed on the upper and lower side of the plate are used. For the bending mode, only one ceramic placed in the middle of the plate on the lower side is used. Two other ceramic plates (14 mm × 2 mm × 0.3 mm, Noliac Denmark) are added as sensors. Two voltage amplifiers (WMA-300 from Falco, The Netherlands) can apply a voltage up to 300 V peak to peak to the actuators' terminals.

The deformation shape for each mode is presented in Fig. 3. Due to the deformation shape for each mode, the ratio Λ is not constant over the top surface, and its value is specified in the middle of the plate. The speed and phase of the vibration of each mode are controlled independently, using the vector control method [4,5]. From the vector control method we can define:

$$U_L = \sqrt{U_{Ld}^2 + U_{Lq}^2} \qquad U_B = \sqrt{U_{Bd}^2 + U_{Bq}^2} \tag{2}$$

For instance, an elliptical motion of the particle with the same vibration speed along x and y of 0.3 m/s is simply obtained by setting the references $U_{Ld} = U_{Bq} = 0.3$ and $U_{Lq} = U_{Bd} = 0$. The closed loop control is embedded into a DSP (STM32F405 from ST Microelectronics). The response time of the vibration speed for both modes is set to about 2 ms as presented in Fig. 3b).

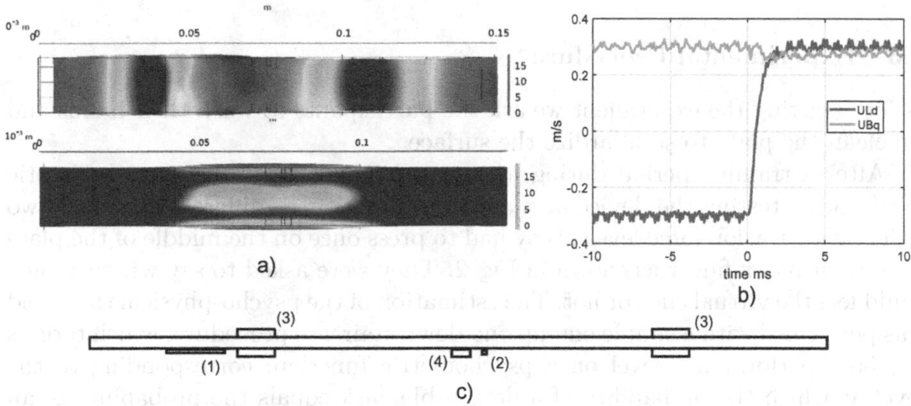

Fig. 3. Measurements on the device; a) deformation mode shape for the longitudinal mode (up) and the bending mode and b) transitory response of the vibration when Λ is switched from -1 to 1 at $U_B = U_L = 0.3$ m/s. c) Side view of the aluminium plate. (1) longitudinal sensor, (2) bending sensor, (3) longitudinal actuator (4) bending actuator

3 Materials and Methods

3.1 Participants

Data were collected from seven healthy volunteers aged between 18 and 40 (3 females). Participants were wearing noise-cancelling headphones in order to prevent noise disturbance. All participants gave written informed consent. The investigation conformed to the principles of the Declaration of Helsinki and experiments were performed in accordance with relevant guidelines and regulations.

3.2 Experimental Set-Up

The plate is mounted on a force sensor that measures the normal force at which a participant presses on the plate as shown in Fig. 2a). The keyclick rendering was performed as in [6]; two normal force thresholds were defined f_1 and $f_2 = f_1 + 0.3N$. When the user reaches f_1, we turn on the two modes, with specific vibration speeds U_L and U_B leading to a value for Λ. When f_2 is reached, the direction of the elliptical motion is reversed simply by inverting U_B. At the end, when the user releases his finger from the surface, we turn off the device. The force is sampled at 10 kHz by the DSP, and a Laptop PC is used to send the threshold value f_1 and the amplitude set points U_B and U_L to the DSP through USB.

3.3 Experimental Procedure

Before starting the experiment we ask the participants to wash their hands and we clean the plate to standardize the surface.

After a training period during when participants could discover the haptic feedback by testing the device at maximum intensity, at different Λ and at two different activation force levels, they had to press once on the middle of the plate with their index finger as shown in Fig. 2. They were asked to say whether they could feel the virtual click or not. The estimation of the psycho-physical threshold was performed with a simple one-up one-down staircase procedure, which targets the 50% performance level on a psychometric function, corresponding to the level at which the probability of a detectable click equals the probability of an undetectable one [8].

As in [6], two force levels $f_1 = [0.1N, 0.4N]$ are interleaved in order to avoid the prediction of the next stimulation intensity. Five different values for $\Lambda = [0.5\ 2/3\ 1\ 1.5\ 2\ \infty]$ were tested. The order of the conditions was pseudo-randomized across participants to avoid learning curve effects. The experiment ended when 6 turnovers or 60 trials were achieved.

4 Results

For all force criteria ($f_1 = [0.1N, 0.4N]$) and vibration ratio $\Lambda = [0\ 0.5\ 2/3\ 1\ 1.5\ 2\ \infty]$ we have computed the median level thresholds which are presented Fig. 4 and in Table 1.

The result shows very different threshold levels, depending on the value of Λ: for $\Lambda \in [1/2\ 2/3\ 1]$, the threshold is close to 0.7 m/s, and doesn't change much with the force level. For $\Lambda \in [1.5\ 2\ \infty]$, the threshold level is higher (around 0.9) and decreases when the force level f_1 increases. Overall, the averaged threshold level is minimal for $\Lambda = 1/2$ if $f_1 = 0.1N$ and for $\Lambda = 1$ when $f_1 = 0.4N$.

Table 1. Experimental results: 50% module threshold for different Λ and different activation pressure levels

		Λ					
f_1		0.5	2/3	1	1.5	2	∞
0.1N	Threshold (m.s-1)	0.73	0.72	0.70	0.98	0.95	0.97
	IQR	0.62–0.77	0.62–0.80	0.65–0.73	0.91–1.06	0.62–0.97	0.94–0.98
	var	0.017	0.027	0.003	0.020	0.033	0.00092
0.4N	Threshold (m.s-1)	0.70	0.75	0.76	0.91	0.87	0.97
	IQR	0.56–0.84	0.64–0.95	0.68–0.77	0.87–1.01	0.68–1.02	0.94–1
	var	0.024	0.023	0.002	0.010	0.019	0.00071

Interestingly, we also compare in Fig. 4c) and d) the variance for each condition. A F-Test ($\alpha = 0.05$, $F_{crit} = 4.28$) have been performed on the variance of threshold level for $\Lambda = 1$ against the others. The results are: for $f_1 = 0.1N$ $\{\Lambda = 0.5$, F $= 4.68$; $\Lambda = 2/3$, F $= 7.34$; $\Lambda = 1.5$, F $= 5.5$; $\Lambda = 2$, F $= 8.83\}$ and for $f_1 = 0.4N$ $\{\Lambda = 0.5$, F $= 9.73$; $\Lambda = 2/3$, F $= 9.42$; $\Lambda = 1.5$, F $= 4.08$; $\Lambda = 2$, F $= 7.69\}$. We observe that for both force conditions, the condition $\Lambda = 1$ produces less variance than other values.

5 Discussion

Our study shows that the elliptical motion of particles can indeed decrease the amount of vibration amplitude that is needed to give the illusion of a button click. Therefore, we can suggest that lateral displacement helps to increase the internal lateral stresses that are first stored when participants touch the plate, before they are released by inverting the direction of the particles' motion. Moreover, we have not seen many differences between the values for $\Lambda \leq 1$ on the detection threshold, and this threshold doesn't change with the force level; the condition $\Lambda = 0$ has been tested but the averaged detection threshold has been found to be higher than the capability of the device. Therefore, we hypothesize that the value of Λ does not change the lateral stresses when in the range $0.5 \leq \Lambda \leq 2/3$. We also show that the variance is minimal for $\Lambda = 1$. Therefore, this gives rise to the optimal value at which the plate should operate. However, for now we do not give an explanation for this specific behaviour; a modelling that takes into account the intermittent contact could be used for that aim [13].

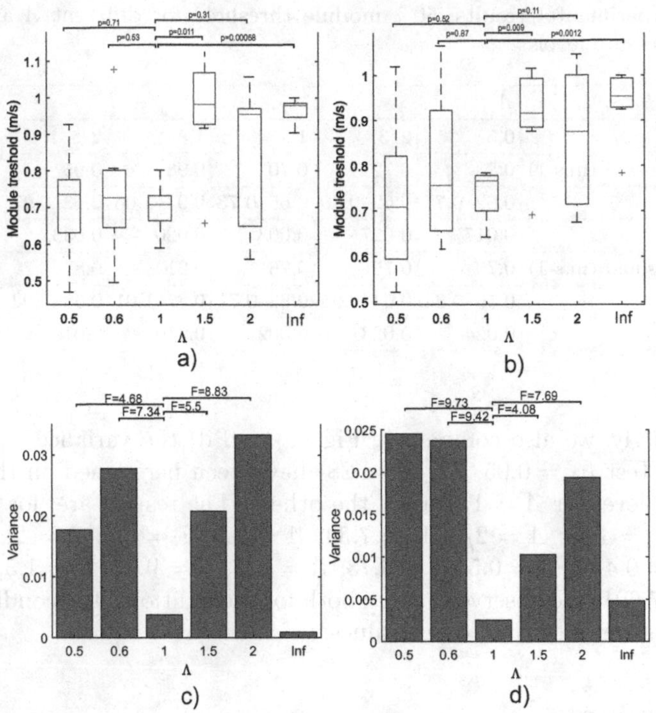

Fig. 4. Experimental results: 50% module threshold, compute for different Λ and different activation pressure levels : a) $f_1 = 0.4N$, b) $f_1 = 0.1$. The error bars, the whisker boxes and the horizontal bars show respectively the min. and max. values, their interquartile range and the median value and corresponding variance c) and d) respectively.

6 Conclusion

In this paper we used two modes simultaneously to produce the simulation of a click on a static finger. To create a simulation with the lowest detection threshold as well as the lowest variance between participants, an equal amount of normal and lateral displacement should be set. This finding will be useful to design and control new tactile interfaces.

Acknowledgement. This work is supported by IRCICA (Research Institute on software and hardware devices for information and Advanced communication, USR CNRS 3380).

References

1. Banter, B.: Touch screens and touch surfaces are enriched by haptic force-feedback. Inf. Display **26**(3), 26–30 (2010)
2. Dai, X., Colgate, J.E., Peshkin, M.A.: Lateralpad: a surface-haptic device that produces lateral forces on a bare finger. In: 2012 IEEE Haptics Symposium (HAPTICS), pp. 7–4, March 2012
3. Emgin, S.E., Aghakhani, A., Sezgin, T.M., Basdogan, C.: Haptable: An interactive tabletop providing online haptic feedback for touch gestures. IEEE Trans. Vis. Comput. Graph. **25**(9), 2749–2762 (2019)
4. Ghenna, S., Vezzoli, E., Giraud-Audine, C., Giraud, F., Amberg, M., Lemaire-Semail, B.: Enhancing variable friction tactile display using an ultrasonic travelling wave. IEEE Trans. Haptics **10**(2), 296–301 (2017)
5. Giraud, F., Giraud-Audine, C.: Piezoelectric Actuators: Vector Control Method. Butterworth-Heinemann, Oxford (2019)
6. Gueorguiev, D., Kaci, A., Amberg, M., Giraud, F., Lemaire-Semail, B.: Travelling ultrasonic wave enhances keyclick sensation. In: Prattichizzo, D., Shinoda, H., Tan, H.Z., Ruffaldi, E., Frisoli, A. (eds.) EuroHaptics 2018. LNCS, vol. 10894, pp. 302–312. Springer, Cham (2018). https://doi.org/10.1007/978-3-319-93399-3_27
7. Hudin, C., Panëels, S.: Localisation of vibrotactile stimuli with spatio-temporal inverse filtering. In: Prattichizzo, D., Shinoda, H., Tan, H.Z., Ruffaldi, E., Frisoli, A. (eds.) EuroHaptics 2018. LNCS, vol. 10894, pp. 338–350. Springer, Cham (2018). https://doi.org/10.1007/978-3-319-93399-3_30
8. Lekk, M.R.: Adaptive procedures in psychophysical research. Percept. Psychophys. **63**, 1279–1292 (2001). https://doi.org/10.3758/BF03194543
9. Minikes, A., Bucher, I.: Noncontacting lateral transportation using gas squeeze film generated by flexural traveling waves–numerical analysis. J. Acoust. Soci. Am. **113**(5), 2464–2473 (2003)
10. Monnoyer, J., Diaz, E., Bourdin, C., Wiertlewski, M.: Ultrasonic friction modulation while pressing induces a tactile feedback. In: Bello, F., Kajimoto, H., Visell, Y. (eds.) EuroHaptics 2016. LNCS, vol. 9774, pp. 171–179. Springer, Cham (2016). https://doi.org/10.1007/978-3-319-42321-0_16
11. Park, G., Choi, S., Hwang, K., Kim, S., Sa, J., Joung, M.: Tactile effect design and evaluation for virtual buttons on a mobile device touchscreen. In: Proceedings of the 13th MobileHCI conference, pp. 11–20 (2011)
12. Saleem, M.K., Yilmaz, C., Basdogan, C.: Psychophysical evaluation of change in friction on an ultrasonically-actuated touchscreen. IEEE Trans. Haptics **11**(4), 599–610 (2018)
13. Torres Guzman, D.A., Lemaire-Semail, B., Kaci, A., Giraud, F., Amberg, M.: Comparison between normal and lateral vibration on surface haptic devices. In: 2019 IEEE World Haptics Conference (WHC), pp. 199–204, July 2019
14. Wallaschek, J.: Contact mechanics of piezoelectric ultrasonic motors. Smart Mater. Struct. **7**(3), 369–381 (1998)
15. Xu, H., Klatzky, R.L., Peshkin, M.A., Colgate, J.E.: Localized rendering of button click sensation via active lateral force feedback. In: 2019 IEEE World Haptics Conference (WHC), pp. 509–514, July 2019

A Proposal and Investigation of Displaying Method by Passive Touch with Electrostatic Tactile Display

Hirobumi Tomita[1(✉)], Satoshi Saga[2], Shin Takahashi[1], and Hiroyuki Kajimoto[3]

[1] University of Tsukuba, 1-1-1 Tennodai, Tsukuba, Ibaraki, Japan
tomita@iplab.cs.tsukuba.ac.jp, shin@cs.tsukuba.ac.jp
[2] Kumamoto University, 2-39-1, Kurokami, Chuo-ku, Kumamoto, Japan
saga@saga-lab.org
[3] The University of Electro-Communications, 1-5-1, Chofugaoka, Chofu, Tokyo, Japan
kajimoto@kaji-lab.jp

Abstract. In the field of tactile displays, electrostatic force displays have been developed for presenting tactile stimulation on a screen. However, with the conventional electrostatic force displays, the user cannot feel the stimulation without rubbing on the display. In this paper, to solve this problem, we propose a new method to present a tactile sensation without moving a finger by creating a small space between an electrode and an insulating surface. Moreover, we evaluate the perceived threshold of the proposed display through the experiment.

Keywords: Electrostatic tactile display · Passive touch · Conductive thread

1 Introduction

In the field of tactile displays, electrostatic force displays have been developed for presenting tactile stimulation on a screen. The electrostatic tactile display consists of a high-voltage generator, an electrode, and an insulator. In this method, the user touches the insulating film on the electrode. If high voltage is applied to the electrode, he/she feels the tactile stimulation by rubbing on the film. In the 1950s, Mallinckrodt et al. discovered that a phenomenon such as vibration-like friction was generated through an insulating film by electrostatic force [5]. Related surveys, dealing with parameters like the effect of input frequencies, waveforms, or amplitude modulations were conducted by many researchers [1,3,4,7,9].

However, in this method, the tactile stimulation is not presented when the user stops moving the finger, and there is little research on displaying the method by passive touch with an electrostatic tactile display. Pyo et al. developed the

I. Nisky et al. (Eds.): EuroHaptics 2020, LNCS 12272, pp. 353–361, 2020.
https://doi.org/10.1007/978-3-030-58147-3_39

tactile display, which generates electrovibration and mechanical vibration by using the electrostatic parallel plate actuator [6]. This device enables passive touch with electrostatic force, however, it requires a voltage higher than 1 kV.

In this paper, we propose a new method to present tactile sensation without moving a finger; we also implement and evaluate a device applying the proposed method. If passive tactile presentation becomes possible by this technique, the range of applications of tactile presentation using electrostatic force will be widely expanded.

2 Implementation of Device for Passive Touch with Electrostatic Tactile Display

The electrostatic tactile display consists of a high-voltage generator, an electrode, and an insulator. In the conventional method, an insulating film covers a flat electrode shown in Fig. 1 (a), and a user moves his/her finger on the insulator. When a high voltage is applied to the electrode, the dielectric polarization is generated in the finger. In this state, the electrode applies an attractive static force to the finger. When the user slides his/her finger on the display, s/he feels a tactile stimulation. However, when the user does not slide his/her finger on the display, s/he cannot perceive the tactile sensation.

Here, we proposed a method that employs a parallel-plate electrostatic actuator mechanism [2,6] for displaying tactile stimulation to a stationary finger without significantly changing the configuration of the electrostatic tactile display. Figure 1 (b) shows the schematic sketch of the parallel-plate electrostatic actuator to generate mechanical vibration. The stationary electrode and non-stationary electrode are set in parallel. When high-voltage is applied to the stationary electrode, the attractive electrostatic force occurs on the non-stationary electrode. We considered that a passive tactile presentation was possible by adopting Pyo et al.'s method [6]. The finger is considered as the non-stationary electrode. We created a small space by using cylindrical electrodes, which have a small diameter, such as a conductive thread or wire, as shown in Fig. 1 (c, d). In this paper, we used 120 μm diameter stainless steel fiber as a conductive thread. These threads were arranged lengthwise without any gaps between them on PET resin (Fig. 2 (b)). The width of these threads arranged is 25 mm to allow enough space for one finger. All of these threads are conductive. They are connected to the high voltage generator. Then, we covered them with a polyvinylidene chloride insulation film. This insulating film is inexpensive and easy to replace. The thickness of this insulating film is 11 μm.

We considered that the slight vibration of the insulating film and finger surface became possible by applying periodic high-voltage from the high-voltage generator to the cylindrical electrode, and the vibration could be perceived by the finger. The high-voltage generator shown in Fig. 2 (c) is developed by Kajimoto laboratory (The University of Electro-Communications, Tokyo). The device includes an mbed LPC1768 microcontroller, which controls the output voltage at a maximum of 600 V by modifying the firmware.

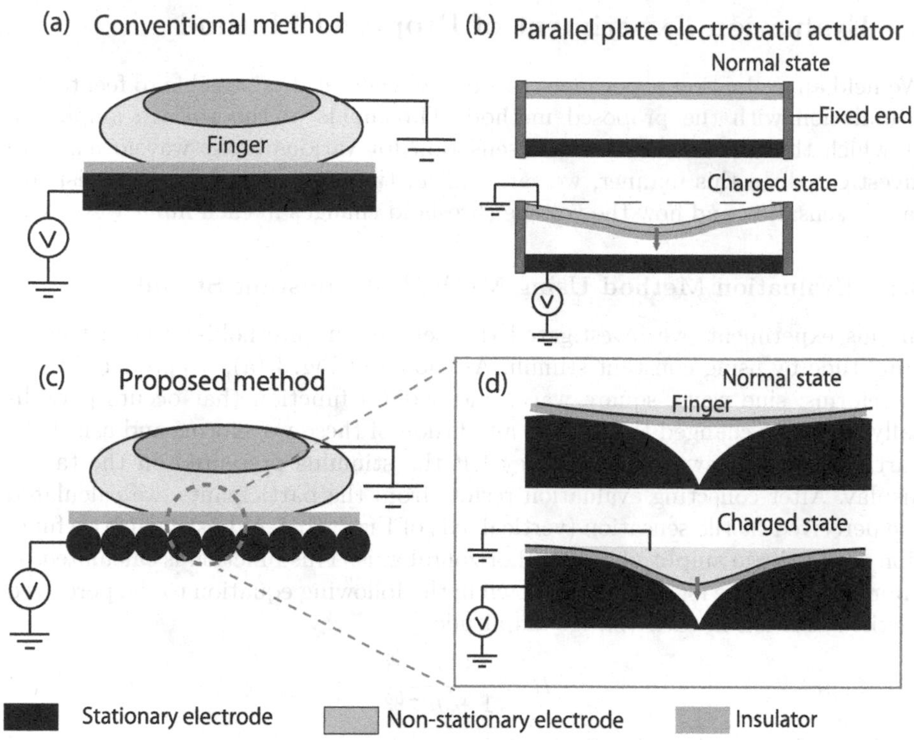

Fig. 1. (a) Conventional electrostatic tactile display: An insulating film is placed on a flat electrode. (b) Schematic sketch of the parallel plate electrostatic actuator [2,6]. (c) Proposed method: A soft insulating film is placed on a nonplanar electrode, such as a conductive thread or wire. (d) Enlarged view of finger surface with the proposed method.

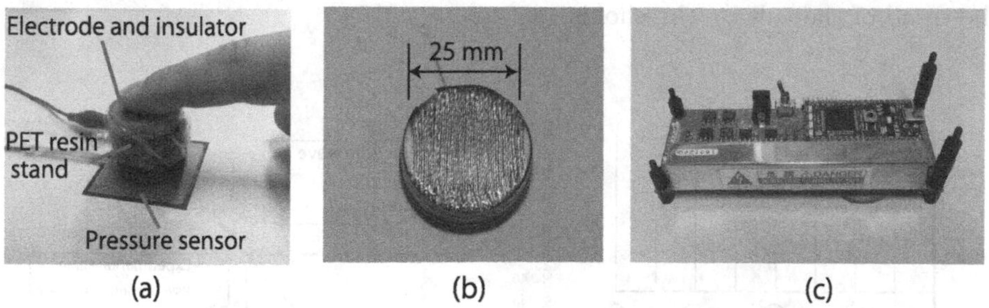

Fig. 2. (a) Overview of our display device, (b) a conductive thread arranged lengthwise and (c) the high-voltage generator.

3 Evaluation Experiment of Proposed Method

We held an evaluation experiment to explore whether it is possible to feel tactile stimulation with the proposed method. Thresholds of the voltage amplitude at which the user feels the tactile sensation for various input waveforms were investigated. In this manner, we can confirm the input waveform that displays tactile sensation and how the voltage threshold changes in each input waveform.

3.1 Evaluation Method Using Method of Constant Stimuli

In this experiment, we investigated the perception thresholds of each voltage amplitude by using constant stimuli. As shown in Fig. 3 (a), we prepared three waveforms: sine wave, square wave, and a delta function that occurs periodically. Then we changed the voltage amplitude of these waveforms and asked the participants to answer whether they felt the stimulus presented on the tactile display. After collecting evaluation results from the participants, we calculated the perceived tactile sensation (vertical axis of Fig. 3 (b)) and psychometric function with voltage amplitude on the horizontal axis. This function is calculated by approximating the logistic curve shown in the following equation to the perceived tactile sensation in each voltage amplitude.

$$R = \frac{1}{1 + ae^{-bv}} \tag{1}$$

R shows the perceived tactile sensation, v shows a voltage amplitude, a and b show constant parameters. By using this function, we estimate the threshold voltage at which the user feels the tactile sensation. In this experiment, a voltage where the psychometric function went across 50% of the perception rate was set as a threshold. An example of a psychometric function obtained from the averaged results of all participants was shown in Fig. 3 (b). By calculating the thresholds of every waveform and frequency of the input waves, we can estimate the trend of the voltage thresholds.

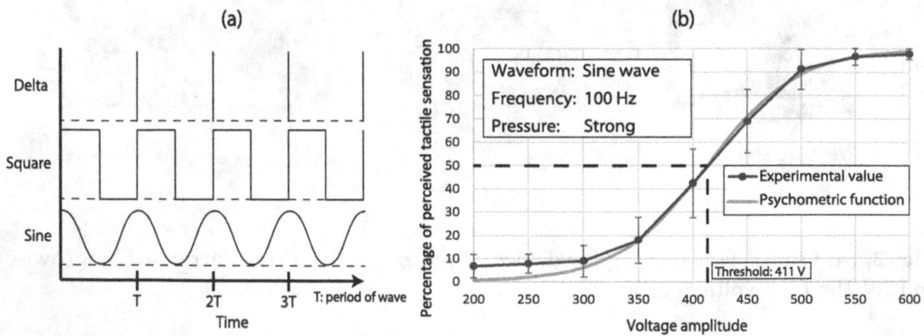

Fig. 3. (a) Three prepared waveforms and (b) an example of a psychometric function.

3.2 Outline of Evaluation Experiment

We conducted an evaluation experiment of constant stimuli with 9 university students (2 female), 22 to 25 years old.

We prepared the display device, a personal computer, a polyethylene terephthalate (PET) resin stand, and a pressure sensor (FSR406) for the experiment. The display device is put on the PET resin stand to isolate it from the floor and to make it easier to place the finger. The pressure sensor is installed under the PET resin stand (Fig. 2 (a)) to measure the pressing force of the finger. We fed back the value of the pressure sensor to the participants via an LCD monitor and instructed them to keep it constant.

We prepared three waveforms and 10 types of dominant frequencies from 10 Hz to 630 Hz of each waveform. Furthermore, we prepared 9 voltage amplitudes from 200 V to 600 V for calculating the voltage threshold, and each condition was displayed ten times. These conditions were selected in random order. In addition, we investigated whether the voltage threshold changes by the difference of the pressing force of the finger. The participants were requested to perform the experiments under the conditions of a weaker pressing force (between 35 gf and 44 gf) and a stronger one (between 75 gf and 84 gf). From these conditions, participants conducted 5400 trials evaluation (3 waveforms × 10 frequencies × 9 amplitudes × 2 pressure × 10 trials for each condition).

First of all, we explained the outline of the experiment and obtained written informed consent for participation in the study (based on ethical guidelines of University of Tsukuba) from all participants. For masking the external sound, white noise was applied to participants using a headphone during the experiment. We instructed them to place their right index finger on the display device and to keep the pressure condition during the experiment. After a waveform was inputted to the display device, participants were instructed to select either "feel something" or "feel nothing" by using the installed keyboard. After participants finished the evaluation with all the waveforms, we instructed them to keep the other pressure condition and repeat the evaluation. 5-minute breaks were taken three times during the experiment. After the experiment, we collected the participants' answers and calculated a voltage threshold for each input waveform.

3.3 Results and Discussion

The results of the evaluation experiment are shown in Fig. 4 and Fig. 5. The horizontal axis of these graphs shows the frequency of the input waveform. The blue line indicates the condition of a weaker pressure, and the orange line indicates the condition of a stronger pressure in these graphs. Figure 4 shows the number of participants who perceived some tactile sensation for each input waveform. Figure 5 shows the result of the voltage thresholds in each input waveform. These thresholds are derived from the intersection of the 50% line and the logistic curve. From Fig. 4 and Fig. 5, we observed that most of the participants could perceive the displayed tactile sensation in all conditions of square waves and limited frequency conditions of sine waves or delta functions. From the result, the

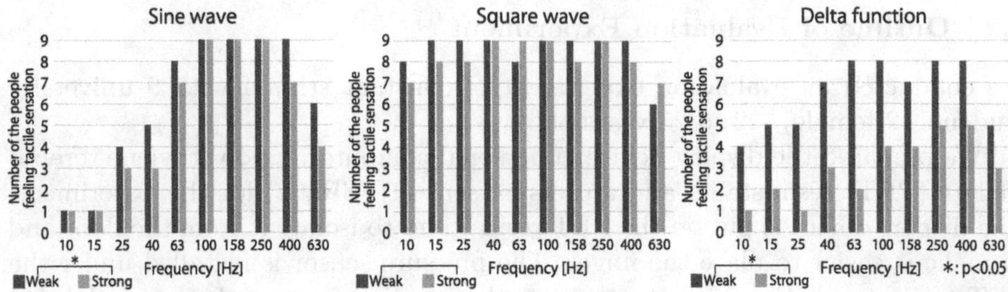

Fig. 4. Results of the number of participants who feel tactile sensation for each input waveform. (Color figure online)

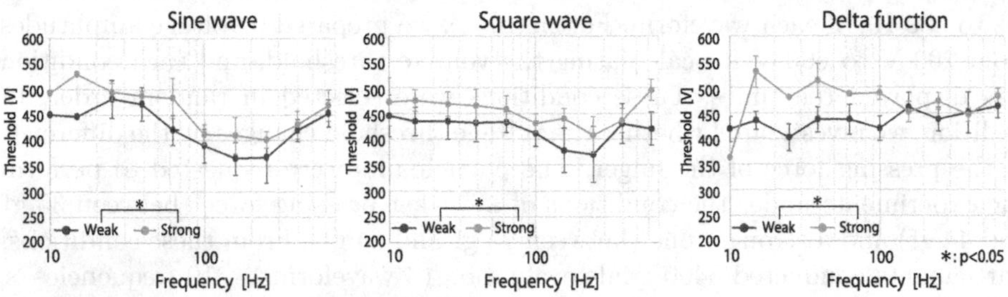

Fig. 5. Results of the threshold of voltage amplitude for each input waveform.

availability of the tactile display without moving a finger is confirmed by using our proposed method. Furthermore, the voltage thresholds are observed to be less than 500 V in most waveforms. In addition, we found the following two characteristics; the evaluation results differed according to the pressing force, and the outline of the graph differed according to the waveform.

Regarding the voltage thresholds, the results of 350 V to 500 V are very high compared to general tactile display devices. However, it is possible to reduce the voltage thresholds by making the insulation film thinner, and the advantage of this method is that there is almost no current flow and power at the display side. Then, it is possible to install many tactile displays or large area tactile displays. Therefore, even if a high voltage is required, we consider that this method of tactile presentation is effective for some specified applications.

Regarding the pressing force, we confirmed that the evaluation result differs depending on the pressing force. In Fig. 4, there is a significant difference in the number of participants who perceived the tactile sensation between the two pressing conditions in the sine waves and delta function. From Fig. 4, we observe that the number of participants who perceived the tactile sensation is larger under the weak pressure condition than that of under the strong pressure condition in most sine waves and delta functions. Most participants perceived the tactile sensation under both conditions in the square waves. In Fig. 5, there was a significant difference between the two pressure conditions at all waveforms. Furthermore, the averaged voltage threshold appeared to be smaller under the weak pressure condition than that under the strong pressure condition in all of the square waves, some sine waves and some delta function. We considered that the factor of this result is based on the ratio between the pressing force of the finger and a just noticeable difference. We focused on Weber's law and showed it in the following equation.

$$\frac{\Delta R}{R} = \text{Const} \tag{2}$$

where ΔR is a just noticeable difference in the tactile sensation, and R is a pressing force in this evaluation experiment. From Eq. (2), the more pressing force, the more a difference of perceived amount is needed. Therefore, we considered that the threshold of input voltage amplitude had to be high for decreasing a just noticeable difference when the pressing force was strong.

Regarding the trend difference of graphs between waveforms (Fig. 4), when square waves were displayed, most of the participants could perceive the displayed tactile sensation. However, the participants could not perceive the tactile sensation in some frequencies of sine waves and delta functions. As seen in Fig. 5, the graphs in the case of sine and square waves are downward convex with the minimum value around 200 Hz. The graph in the case of the delta function is gentle and upward right. The reason for the shapes is considered to be related to the frequency components of each input waveform and the frequency responses of human mechanoreceptors. Vardar et al. investigated how displayed waveforms affect haptic perceptions of vibration under electrostatic tactile display [8]. Then they observed that the participants were more sensitive to square wave stimuli than sine-wave stimuli for a dominant frequency lower than 60 Hz. Furthermore, they discussed that a low dominant frequency square wave still contains high-frequency components that stimulate the Pacinian channel. Our experimental results were similar to the results of theirs; voltage thresholds are higher in the case of sine wave stimuli than in the case of square one under 40 Hz frequencies. In the case of the delta function, these waves have many frequency components, however, they are smaller in amplitude than the other two waveforms. This could be because the waveforms did not have sufficient amplitude to stimulate mechanoreceptors at low dominant frequencies.

4 Conclusions

In this paper, we proposed a method that can display tactile stimulation without moving a finger with an electrostatic tactile display, and we conducted an evaluation experiment. The proposed method is based on the vibration of a thin insulating film, which is realized by creating a small space between the electrode and the insulator. We use a conductive thread as an electrode and place a plastic film on the conductive thread. We explored whether the user can feel tactile stimulations under several conditions. We collected the evaluation results of the participants under the conditions of waveforms, frequencies and pressing forces, and calculated the voltage threshold. From the results, the possibility of a tactile display without moving a finger is confirmed by using our proposed method.

Besides, we also revealed the following fact: the evaluation results differed according to the pressing force and the outline of the graph differed according to the waveform. Regarding the pressing force of the finger, we considered that the ratio between the pressing force and the suction force induce the difference of perceptibility. Regarding the difference in the shapes of the graphs, we considered that the frequency components of the input waveform and the frequency responses of mechanoreceptors is the reason.

In the near future, we plan to clarify the modeling of tactile perception using our proposed method and the difference in the width of tactile expression compared with the conventional method. Further, we will perform more precise experiments to observe the physical phenomenon between the surface of the finger and the display.

References

1. Bau, O., Poupyrev, I., Israr, A., Harrison, C.: Teslatouch: electrovibration for touch surfaces. In: Proceedings of the 23Nd Annual ACM Symposium on User Interface Software and Technology, UIST 2010, pp. 283–292. ACM, New York (2010). https://doi.org/10.1145/1866029.1866074
2. Burugupally, S.P., Perera, W.R.: Dynamics of a parallel-plate electrostatic actuator in viscous dielectric media. Sens. Actuators A: Phys. **295**, 366–373 (2019)
3. Jiao, J., et al.: Detection and discrimination thresholds for haptic gratings on electrostatic tactile displays. IEEE Trans. Haptics **12**(1), 34–42 (2018)
4. Kang, J., Kim, H., Choi, S., Kim, K.D., Ryu, J.: Investigation on low voltage operation of electrovibration display. IEEE Trans. Haptics **10**(3), 371–381 (2016)
5. Mallinckrodt, E., Hughes, A., Sleator Jr., W.: Perception by the skin of electrically induced vibrations. Science (1953)
6. Pyo, D., Ryu, S., Kim, S.-C., Kwon, D.-S.: A new surface display for 3D haptic rendering. In: Auvray, M., Duriez, C. (eds.) EUROHAPTICS 2014. LNCS, vol. 8618, pp. 487–495. Springer, Heidelberg (2014). https://doi.org/10.1007/978-3-662-44193-0_61
7. Strong, R.M., Troxel, D.E.: An electrotactile display. IEEE Trans. Man-Mach. Syst. **11**(1), 72–79 (1970)
8. Vardar, Y., Güçlü, B., Basdogan, C.: Effect of waveform on tactile perception by electrovibration displayed on touch screens. IEEE Trans. Haptics **10**(4), 488–499 (2017)
9. Vezzoli, E., Amberg, M., Giraud, F., Lemaire-Semail, B.: Electrovibration modeling analysis. In: Auvray, M., Duriez, C. (eds.) EUROHAPTICS 2014. LNCS, vol. 8619, pp. 369–376. Springer, Heidelberg (2014). https://doi.org/10.1007/978-3-662-44196-1_45

Sensing Ultrasonic Mid-Air Haptics with a Biomimetic Tactile Fingertip

Noor Alakhawand[1,2(✉)], William Frier[3], Kipp McAdam Freud[1,2], Orestis Georgiou[3], and Nathan F. Lepora[1,2]

[1] Department of Engineering Mathematics, University of Bristol, Bristol, UK
noor.alakhawand@bristol.ac.uk
[2] Bristol Robotics Laboratory, Bristol, UK
[3] Ultraleap Ltd., Bristol, UK

Abstract. Ultrasonic phased arrays are used to generate mid-air haptic feedback, allowing users to feel sensations in mid-air. In this work, we present a method for testing mid-air haptics with a biomimetic tactile sensor that is inspired by the human fingertip. Our experiments with point, line, and circular test stimuli provide insights on how the acoustic radiation pressure produced by the ultrasonic array deforms the skin-like material of the sensor. This allows us to produce detailed visualizations of the sensations in two-dimensional and three-dimensional space. This approach provides a detailed quantification of mid-air haptic stimuli of use as an investigative tool for improving the performance of haptic displays and for understanding the transduction of mid-air haptics by the human sense of touch.

Keywords: Tactile sensors · Biomimetic · Mid-air haptics

1 Introduction

Ultrasonic phased arrays can generate haptic sensations in mid-air. They focus acoustic radiation pressure in space, which deflects the skin to induce tactile sensation [1]. To evaluate whether the array is producing the desired haptic sensations, we need to understand how focal points of pressure interact with compliant skin to cause it to deform. In this paper, we propose a method for sensing mid-air haptics with a biomimetic tactile fingertip inspired by the human sense of touch. Using the data obtained from the sensor, we are able to visualize the different patterns produced by the haptic array.

Efforts to measure the haptic output from a phased ultrasonic array range from quantitative to qualitative. Quantitative methods include microphones to measure the sound pressure level of the generated focal points [1,8], directly measuring the ultrasonic output of the system without considering its interaction with other material. On the other hand, to quantitatively consider the interaction of the sensations with skin-like materials, Laser Doppler Vibrometry (LDV), a tool commonly used for non-contact vibration measurement, can give insight on

I. Nisky et al. (Eds.): EuroHaptics 2020, LNCS 12272, pp. 362–370, 2020.
https://doi.org/10.1007/978-3-030-58147-3_40

how the haptic stimuli would interact with human skin at high frequencies [2]. Alternatively, qualitative methods include pulsed schlieren imaging, which was used to visualize the pressure field produced by a focal point as it interacts with external materials [5]. Additionally, by projecting the focal points onto the surface of an oil bath, it can be used to visualize the patterns generated by the haptic array [6]. New research has used a microphone-based tactile sensor array to evaluate the vibrations of its surface due to ultrasonic haptic sensations [9], highlighting the potential of tactile sensors for testing the output of a haptic system.

In this work, we propose a method to sense and evaluate mid-air haptics using the TacTip, a biomimetic tactile fingertip. The TacTip is biologically inspired by glabrous (hairless) human skin, which has an intricate morphology of layers, microstructures, and sensory receptors that contribute to its functions [3,10]. We present a method for analyzing mid-air haptic sensations with a tactile sensor, allowing us to quantitatively test ultrasonic arrays with a method inspired by the human sense of touch.

2 Experimental Setup and Method

This work aims to develop a method for testing mid-air haptics with a biomimetic tactile sensor. We carried out experiments with the tactile fingertip mounted on a robot arm and an ultrasonic array (Fig. 1).

Fig. 1. The TacTip, a biomimetic tactile sensor (left): the flat-tipped model used in this study; the skin of the TacTip with 127 inner nodular pins (middle); and the experimental setup with the tactile sensor mounted on a robot arm to collect data over the ultrasonic phased array (right).

2.1 Biomimetic Tactile Sensor

The TacTip (Fig. 1, left panel) is a biomimetic tactile sensor developed at the Bristol Robotics Laboratory [3,10], based on the structure of glabrous skin. The human fingertip has *dermal papillae* where the dermis interdigitates with *intermediate ridges* in the epidermis. These ridges and papillae focus strain from the skin surface down to mechanoreceptors within the dermis. The TacTip mimics this structure with an outer rubber-like skin which connects to inner nodular pins (Fig. 1, middle panel). As the soft sensor interacts with objects, its skin deforms and the nodular pins transmit surface strain into inner mechanical movements, similar to human skin. An internal camera tracks the movement of its artificial papillae, making it possible to detect the shear deformation of the skin. The sensor has been used in many tasks in robot touch such as object exploration and slip detection [10]. The TacTip is manufactured using dual-material 3D printing, which prints both the sensor's plastic base and the soft rubber-like material for the skin. This allows for low-cost and rapid prototyping of different designs as well as its integration with robotic grippers and hands. Additionally, the design of the TacTip is modular, allowing for different tips to be used, such as varying the shape or texture of the skin or varying the layout of the nodular pins [10]. The tip of the sensor can be filled with gel to affect its compliance or be left unfilled. Since this is the first time the TacTip has been used to detect small forces on the order of millinewtons, we needed a more compliant tip; after testing tips with these variations, we found the flat-tipped TacTip without gel (Fig. 1, left panel) to be more sensitive, and thus suitable for this work.

2.2 Ultrasound Phased Array

To generate the mid-air haptic stimuli for the experiments, we used the Ultrahaptics Evaluation Kit (UHEV1) from Ultraleap. The array has a 16 by 16 grid of ultrasonic transducers which operate at 40 kHz to generate focal points in mid-air, with an update rate of 16 kHz. The device is accompanied by software which allows us to modulate these focal points so that they can be felt by users [1] and to generate various shapes and textures [6].

2.3 Experiment

We used a 6-DOF robotic arm (ABB IRB120) to move the tactile sensor over the haptic array. The robot arm moved the sensor in 10 mm increments over an 80 mm by 80 mm grid at a height of 200 mm above the haptic array. At each position, 30 frames were captured from the camera to image the TacTip's inner nodular pins at 30 fps. This was done for a focal point generated by the array, as well as two shapes (a line and a circle). The shapes were generated by the array using Amplitude Modulation (AM) and Spatiotemporal Modulation (STM), to see whether the sensor distinguishes between these two standard modulation techniques. AM generates focal points in the path of the desired pattern and modulates their intensity over time, while STM generates one focal point and moves it rapidly along the path.

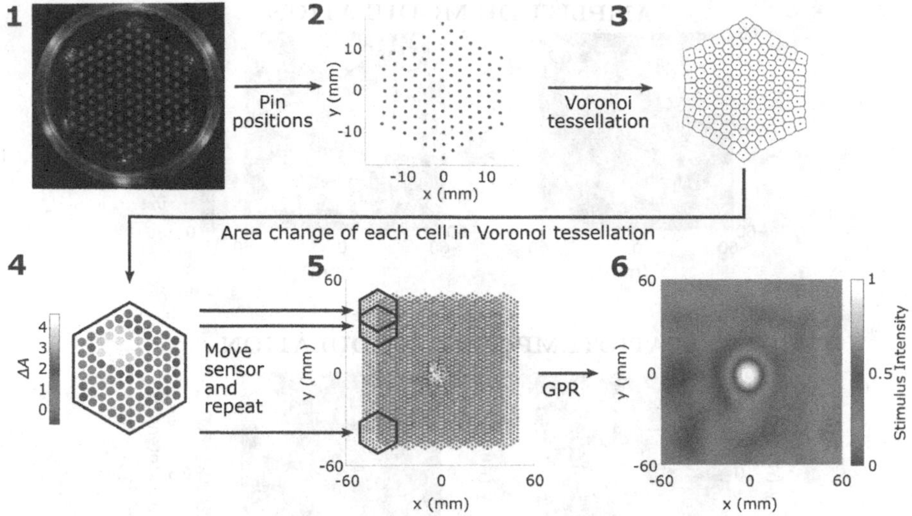

Fig. 2. Analysis method. We capture an image from the tactile sensor as it interacts with the mid-air haptic stimulus (1) and extract each pin position (2). Voronoi tessellation is generated with pin positions as the center point for each cell (3); the change of area of each cell compared to an unstimulated sensor, ΔA, is used as a measure of the stimulus intensity (4). This is repeated for readings over a grid (5). Gaussian Process Regression combines the data sets to produce detailed visualizations (6).

2.4 Analysis

In this work, we developed an analysis method to sense mid-air haptics with a biomimetic tactile fingertip (Fig. 2). The images captured from the tactile sensor as it interacts with the mid-air haptic stimulus were processed to find the positions of the nodular pins at each time step. Then we used the pin positions to generate a bounded Voronoi tessellation, shown by Cramphorn et al. to transduce a third dimension to the sensor data [4]. Voronoi tessellation partitions a plane based on the distance between points on that plane; each point along an edge is equidistant from two points, and each vertex is equidistant from at least three points. The areas of the cells give us information for tactile perception; increasing areas indicate a compression of the skin. Thus, the areas of each cell in the Voronoi tessellation were compared with a data set in which the sensor was not stimulated, and the difference between the two areas, ΔA, was used as a measure of the intensity of the stimulus as felt by the sensor. This was done for every time step, and then averaged over the 30 frames of data. The process was repeated for readings in a grid over the haptic display to populate a two-dimensional plane. Then we trained a Gaussian process regression (GPR) model for the measured intensity, represented by ΔA (using the MATLAB function *fitrgp* with the default squared exponential covariance function). The output values were then scaled between zero and one, to represent the relative intensity of the stimulus as felt by the tactile sensor.

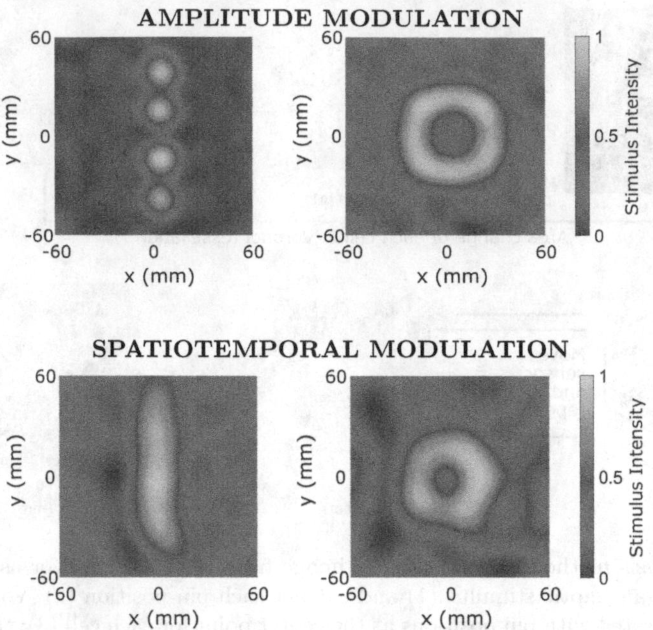

Fig. 3. The mid-air haptic shapes, line and circle, as felt by the biomimetic tactile sensor. The Amplitude Modulation (AM) cases use four focal points modulated with a 200 Hz sine wave. The Spatiotemporal Modulation (STM) cases use one focal point moving at a repeat frequency of 100 Hz along the path of the shape.

3 Results

In this work, we used a biomimetic tactile fingertip to sense mid-air haptics, to develop a method for testing the output of a haptic display. We measured the response of the tactile sensor to a focal point of pressure generated by an ultrasonic haptic array as well as two haptic shapes, each generated by Amplitude Modulation (AM) and Spatiotemporal Modulation (STM).

3.1 Sensing Mid-Air Haptics on a Two-Dimensional Grid

The experiments showed that the biomimetic tactile fingertip used in this study, the TacTip, is able to sense the mid-air haptic stimuli produced by the ultrasonic phased array using our developed method (Fig. 2). The focal points of pressure generated by the ultrasound caused the skin-like surface to deform, expanding the areas of the cells in the Voronoi tessellation, allowing us to identify the location and intensity of contact. Voronoi tessellation was a valuable tool in transducing a third dimension in the data, which allowed us to visualize the sensations produced by the ultrasonic array.

Our analysis methods enables us to produce detailed visualizations of the mid-air haptic sensations (Fig. 3), allowing us to distinguish between different patterns produced by the ultrasonic array. Additionally, the variation in the strength of the focal point can be clearly seen. The visual representation of the focal point shows that it creates a localized region of increased displacement (Fig. 2, lower right panel). The point is much stronger in the center, and then decreases in intensity as you move radially outwards. This is similar for the other shapes; the center path of the shape has increased intensity, which decreases as you move away (Fig. 3).

The visualizations produced by our method allow us to compare the shapes generated by the ultrasonic array using different modulation techniques. We see that the tactile sensor is able to distinguish the four focal points that make up the amplitude modulated line (Fig. 3, top left panel). A user of the ultrasonic array would not distinguish the points as the distance between them is small [1], and so it creates the illusion of a continuous line. The sensor can discriminate between the points because our analysis method is measuring the deformation of the tactile sensor's surface, which would correspond to the deformation of the user's skin rather than their perception of the sensation. On the other hand, the spatiotemporally modulated line (Fig. 3, lower left panel) is felt as a continuous line by the sensor. The focal point used to generate the line is moved along its path at very small increments. The distance between the points in this case is too small to be distinguished by the sensor, making the output more similar to how a user would sense the stimulus.

3.2 Sensing a Focal Point in Three-Dimensional Space

In the previous section, we presented our results when sensing various shapes over a two-dimensional grid. In this section, we extend our method to sense mid-air haptic stimuli in three dimensions. When a person interacts with the haptic array to sense the shapes it generates, they naturally move their hand around the display surface, which includes moving their hands up and down as they process the sensations they feel. Thus, understanding how the generated sensations vary with height is important to determine whether the desired effect is being produced and to check that there are no undesired artefacts. We repeated the data collection process described earlier for the point stimulus at different heights over the array, at 10 mm increments. This results in a three-dimensional grid on which we applied our presented analysis method.

This experiment allows us to see how the shape is sensed over a three-dimensional surface, looking at how the shape varies by height. The focal point is generated by the array at a specific height in space; however, there are still sensations at other points due to the interaction of the ultrasonic waves [8]. The point stimulus is sensed by the tactile fingertip as an elongated spheroid, with a localized region of increased intensity (Fig. 4). It appears the lower the intensity of the stimulus felt, the more elongated it is. As the sensor moves away from the center height, the stimulus becomes fainter.

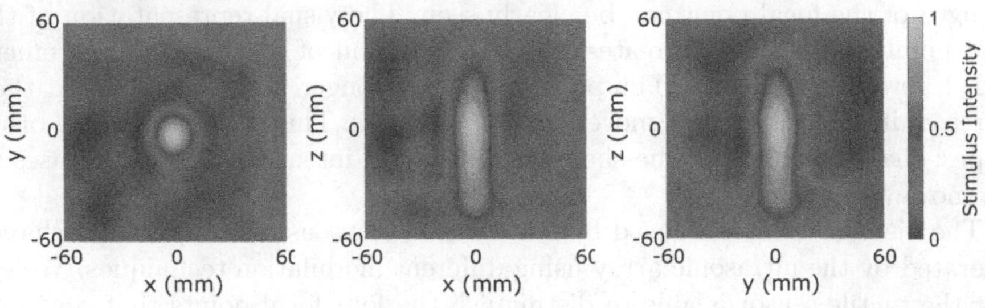

Fig. 4. The three-dimensional view of a focal point as felt by the biomimetic tactile sensor presented as cross sections at $z = 0$ (left), which corresponds to 200 mm above the array; $y = 0$ (middle); and $x = 0$ (right).

4 Discussion and Conclusion

Tactile sensors can further our understanding of the human sense of touch. Our experiments have shown that we can use a biomimetic tactile fingertip to sense the mid-air haptic stimuli produced by an ultrasonic phased array, providing insights on the deformation of the skin-like material of the TacTip due to ultrasonic mid-air haptic sensations. This allows us to produce detailed visualizations of the sensations produced by the device. Using our analysis methods, we were able to see the difference between shapes that are amplitude modulated versus spatiotemporally modulated by the ultrasonic array. Additionally, we were able to sense and visualize a focal point in three-dimensional space, providing insights on how the focal point varies by height.

The visualizations of the stimuli produced in our study are similar to those in other works which use alternative methods. For example, Laser Doppler Vibrometry was used to measure the deformation of skin-like material due to ultrasonic mid-air haptics [2]; it measured the high frequency vibrations (50 Hz and above) of the skin surface, and the root mean square (RMS) of the deformation was used to visualize the sensations. While we do not measure the high frequency vibrations, we get similar results. This could indicate that the data we collect is similar to the RMS of the skin deformation. Additionally, our three-dimensional measurements of the focal point look very similar to simulations of the same stimulus [8]. The elongated spheroid felt by the sensor looks like the higher values of acoustic field pressure in the simulation, suggesting that the tactile fingertip is able to sense the ultrasound when it crosses a threshold pressure.

This work has provided insights into the measurement of haptic stimuli, but it has areas for improvement. At this point, we have measured the intensity of the stimulus without relating it to a specific physical value. Further work is planned to determine the relationship of the measured stimulus intensity to the skin deformation, which would allow us to compare our results with other quantitative experiments. Additionally, we do not measure the skin deformation at high frequencies. While the results we get are similar to those which use vibrometry, studying the vibrations of the artificial skin could determine whether the sensor does behave similarly to human skin. One approach is to modify the sensor with a higher frame rate camera which could allow us to see the high-frequency deformations of the skin; another approach would be to add another high-frequency tactile sensing modality to the TacTip [7].

Our work has shown promising results for sensing mid-air haptics with a biomimetic tactile fingertip. The developed approach could be used as an investigative tool for evaluating and improving the capabilities of haptic displays. The insights gained from this work could also be used to investigate human perception. In the future, we could apply our methods to intelligent exploration of the haptic stimuli. This could allow us to develop an autonomous robotic system that is able to feel and interact with the sensations similar to how a person would explore mid-air haptic stimuli.

Acknowledgements. We thank A. Stinchcombe, K. Aquilina, and J. Lloyd for their help. N. Alakhawand was supported by an EPSRC CASE award with Ultraleap. N. Lepora was supported by an award from the Leverhulme Trust on 'A biomimetic forebrain for robot touch' (RL-2016-39).

References

1. Carter, T., Seah, S.A., Long, B., Drinkwater, B., Subramanian, S.: UltraHaptics: multi-point mid-air haptic feedback for touch surfaces. In: Proceedings of the 26th Annual ACM Symposium on User Interface Software and Technology, pp. 505–514 (2013)
2. Chilles, J., Frier, W., Abdouni, A., Giordano, M., Georgiou, O.: Laser doppler vibrometry and FEM simulations of ultrasonic mid-air haptics. In: 2019 IEEE World Haptics Conference, pp. 259–264 (2019)
3. Chorley, C., Melhuish, C., Pipe, T., Rossiter, J.: Development of a tactile sensor based on biologically inspired edge encoding. In: International Conference on Advanced Robotics (2009)
4. Cramphorn, L., Lloyd, J., Lepora, N.: Voronoi features for tactile sensing: direct inference of pressure, shear, and contact locations. In: 2018 IEEE International Conference on Robotics and Automation, pp. 2752–2757 (2018)
5. Iodice, M., Frier, W., Wilcox, J., Long, B., Georgiou, O.: Pulsed schlieren imaging of ultrasonic haptics and levitation using phased arrays. In: 25th International Congress on Sound and Vibration 2018, ICSV 2018 Hiroshima Call, pp. 1736–1743 (2018)
6. Long, B., Seah, S.A., Carter, T., Subramanian, S.: Rendering volumetric haptic shapes in mid-air using ultrasound. ACM Trans. Graph. **33**(6), 1–10 (2014)
7. Pestell, N., Lloyd, J., Rossiter, J., Lepora, N.: Dual-modal tactile perception and exploration. IEEE Robot. Autom. Lett. **3**(2), 1033–1040 (2018)
8. Price, A., Long, B.: Fibonacci spiral arranged ultrasound phased array for mid-air haptics. In: 2018 IEEE International Ultrasonics Symposium, pp. 1–4 (2018)
9. Sakiyama, E., Matsumoto, D., Fujiwara, M., Makino, Y., Shinoda, H.: Evaluation of multi-point dynamic pressure reproduction using microphone-based tactile sensor array. In: 2019 IEEE International Symposium on Haptic, Audio and Visual Environments Games (2019)
10. Ward-Cherrier, B., Pestell, N., Cramphorn, L., Winstone, B., Giannaccini, M.E., Rossiter, J., Lepora, N.: The TacTip family: soft optical tactile sensors with 3D-printed biomimetic morphologies. Soft Robot. **5**(2), 216–227 (2018)

BY

Soft-Wearable Device for the Estimation of Shoulder Orientation and Gesture

Aldo F. Contreras-González(✉), José Luis Samper-Escudero, David Pont-Esteban, Francisco Javier Sáez-Sáez, Miguel Ángel Sánchez-Urán, and Manuel Ferre

Centre for Automation and Robotics (CAR) UPM-CSIC, Universidad Politécnica de Madrid, 28006 Madrid, Spain
af.contreras@alumnos.upm.es
https://www.car.upm-csic.es/?portfolio=exoflex-flexible-exoskeleton

Abstract. This study presents the development of a wearable device that merges capacitive soft-flexion and surface electromyography (sEMG) sensors for the estimation of shoulder orientation and movement, evaluating five natural movement gestures of the human arm. The use of Time Series Networks (TSN) to estimate the arm orientation, and a pattern recognition method for the estimation of the classification of the gesture are proposed. It is demonstrated that it is possible to know the orientation of the shoulder, and that the algorithm is capable of recognising the five gestures proposed with two different configurations. The study is performed on people who reported healthy upper limbs.

Keywords: Soft robotics · Wearable sensors · UpperLimb · sEMG.

1 Introduction

There have been many attempts to identify the movement of the human body in a virtual way by monitoring the behaviour of the extremities for haptic interfaces [7], teleoperation tasks [1] and assistive and rehabilitation devices [4,16]. Robots increase the number of repetitions performed in a rehabilitation session, thus improving patient morale and motivation [24]. In recent years, rehabilitation devices use sEMG as main source of feedback [15] for control [8].

Several sEMG techniques are used for the identification and classification of movements [3], some of the most relevant being the Detrended Fluctuation Analysis (DFA) for the identification of low-level muscle activation [19], the sEMG signal decomposition into Motor Unit Action Potential Trains (MUAPTs) [18], the Tunable-Q factor Wavelet Transform (TQWT) based algorithm proposed for the classification of physical actions [2], and Convolutional Neural Network (CNN), recently confirmed as as a powerful tool for the classification of operator movements [25]. These methods, combined with appropriate signal filtering techniques [5], are useful for estimating the movement of the human body.

I. Nisky et al. (Eds.): EuroHaptics 2020, LNCS 12272, pp. 371–379, 2020.
https://doi.org/10.1007/978-3-030-58147-3_41

Data fusion [13] using sEMG sensors is widely used in rehabilitation. Movement recognition algorithms generally combine sEMG signals with the Inertial Measurement Unit (IMU) [11], or with force sensors [10]. There are particular cases where flexion sensors [22] are used to avoid the accumulated error on the measurement.

This paper focuses on the development of a soft-compressive jacket with a network of soft-flexion sensors, merged with sEMG sensors attached to it for movement detection of the upper limbs. This device allows the user to quickly start estimating shoulder orientation without the need for prior calibration.

2 Materials and Methods

Using a configuration of seven one-axis sensors, as in previous work [21], it is possible to obtain 95% of the variance of the principal components for the shoulder gestures. The configuration proposed in this paper places only an array of four flexion sensors in the intermediate positions due to the fact that they provide flexion measurements in two axes.

The array of four flexion sensors Sx (being 'x' the sensor number) was placed over a compression jacket (see Fig. 1). The capacitive flexion sensors are the *Two Axis Sensor of Bendlabs* [12] and its operation is explained and well detailed in [20]. The sensors have been attached to a compression jacket by sewing two small rigid pieces which hold and guide the sensor in the arm movement direction and to neglect properties such as wrinkles and stretching. The first support (FxA) (see Fig. 1b) holds the sensor in a fixed position while the second one (FxB) allows it to slide inside it and guides it over the arm (see Fig. 1c).

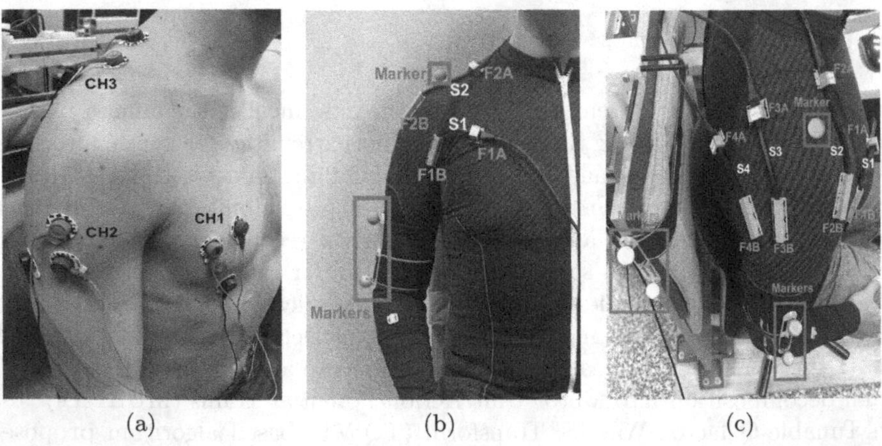

Fig. 1. Soft Sensor Device. sEMG location (1a): Trapezius Descendens (CH3), Deltoideus Medius (CH2) and Pectoralis major (CH1). Markers location: one over the shoulder Acromion bone, two on the arm (1b) and two vertically over the base (1c).

The sensor arrangement allows shoulder movement to be measured in a six-degree-of-freedom (DoF) work-space where arm rotation around its longitudinal axis is not included, this measurement is converted into two angle XY and YZ given by the conversion of the position of the ground truth. sEMG sensors [23], are allocated following the recommendations of Surface Electromyography for the Non-Invasive Assessment of Muscles (SENIAM) [9]. The electrodes are placed on the user (as shown on Fig. 1); then, the user puts on the compression jacket over the electrodes (not shown on Fig. 1b nor Fig. 1c). The design of this device allows the deformation and stretching of the fabric to be disregarded due to the small rigid pieces, in addition to not limiting the user's mobility on daily tasks.

The gestures performed were simplified to cover the natural range of arm movement [14] for daily tasks, and were assigned a number for further identification: **1**. Abduction/Adduction of the shoulder until the arm reaches 120° inclination; **2**. Flexion/Extension of the shoulder from 0° to 120°; **3**. Horizontal adduction/displacement of the arm at 90° flexion, hand crosses sagittal plane till arm reaches a 30° displacement; **4**. Closing/Swing drill movement of the arm inwards from 0° to 120°; and, **5**. Opening/Swing drill movement, starting with a flexion of 120° to 0°. The method developed in this study was evaluated in four healthy subjects; tests were spread over three different days to avoid exhaustion of the muscles. Each subject performed a total of five repetitions of each of the five gestures, continuously and without interruptions. Participants' ages ranged from 24 to 30 years old. All the gestures made by the subjects were performed in a chair facing a screen.

2.1 Data Acquisition

To start data collection, the sEMG sensors and the flex sensor compression jacket are placed on the subject's right arm. Then, OptiTrack [17] markers are located as shown in Fig. 1, in order to obtain the real pose of the subject's arm.

Both flexion and sEMG sensors are connected to a custom acquisition board based on the LAUNCHXL-F28379D development board. On the one hand, the sEMG sensors provide an amplified, rectified and integrated analogical signal

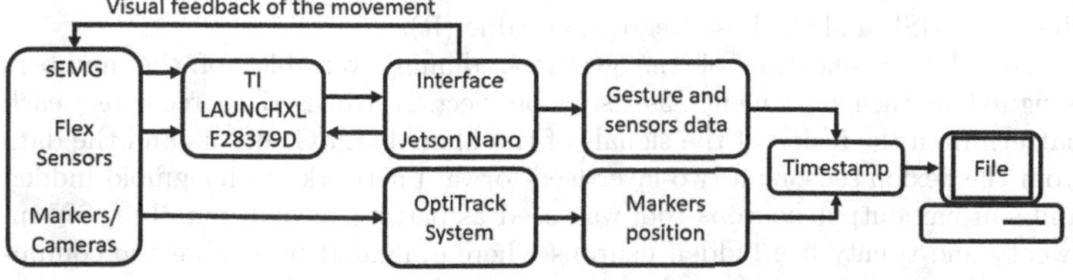

Fig. 2. The EMG signals and angles of the user's movements (first box on the left) were acquired using visual feedback generated by the interface on the Jetson Nano, which also stores this data. The OptiTrack system stores the position of the markers. In the end, both files are merged into one.

(AKA the EMG's envelope), which is obtained by the micro-controller at a rate of 1kHz. On the other hand, the flexion sensors communicate with the micro-controller via I2C protocol at a frequency of 200 Hz.

A graphical user interface has been developed to guide the speed and kind of movement of the participants while performing the gestures and to log all obtained data. This software has been implemented on an NVIDIA Jetson Nano. This device communicates with the acquisition board via SPI at 500 Hz and stores the data contained in every received message along with the timestamp and the gesture that is being performed in a plain text document. Simultaneously to the start of the sensors' data acquisition, Optitrack data acquiring is initiated at 240fps. In the end, a file with the positions of the markers belonging to the OptiTrack system is exported. The interaction of all elements is shown in Fig. 2.

2.2 Data Processing

The sessions for each subject are condensed into a single file. Given that the Optitrack system captures are made at 240 fps, an interpolation is performed to reach a frequency of 500 Hz in the data. The interpolation method consists on taking the Optitrack file which is the shortest and matching it with the number of samples with the Jetson Nano file by adding with a quadratic splines method the missing data. The signal from the sEMG sensors is filtered offline. To Smooth this data a Savitzky-Golay smoothing local regression using weighted linear least squares and a 2nd degree polynomial filter is used with an span of 0.7% of the total number of data points. The angle between the markers of the Optitrack system is obtained by calculating the angle generated between the line generated from the arm markers on the shoulder Acromion bone, and the vertical from the markers of the backrest of the rehabilitation system.

A Time series Network [6] with Levenberg-Marquardt algorithm is used to calculate the orientation of the arm using the angles given by the flexion sensors as input data, and the OptiTrack markers reference as target. With the use of Matlab's Machine Learning and Deep Learning Toolbox, it was possible to estimate that the best parameters for this task were 10 hidden neurons and consecutive samples. For the training of the neural network, 70% of the data was used for training, 15% for validation and 15% for testing, in order to find the lower MSE and the best Regression value (R).

For the classification of the gestures, dummy variables of the numbers assigned to each movement (as listed on Sect. 2) are used to recognise each pattern from the fusion of the signals of the filtered sEMG sensors and the data from the flexion sensors; a two-layer feed-forward network, with sigmoid hidden and softmax output neurons tool was used as pattern recognition. Ten, fifteen, twenty and twenty five hidden neurons where evaluated by testing the computation time and the number of iterations in order to get the best cross-entropy value; fifteen hidden neurons where the most appropriate for this task. The networks were trained with scaled conjugate gradient back-propagation (trainscg) using again, 70% of data for training, 15% for validation and 15% for testing.

3 Results and Discussion

For the estimation of the orientation, two Time series Network configurations were designed, one for dummy variables and one for the gestures as numbered in Sect. 2, resulting in a MSE of $1.49E-05$ with a Regression value of $9.99E-01$ and a MSE of $1.50E-04$ and R of $9.99E-01$, respectively. It can be concluded that the selection of either of the two target variables does not have a significant influence on the results, given that both have a regression value (R) of 0.99%, and the difference on the MSE is minimum.

In order to evaluate the trained networks for both orientation and gesture, a new single session is performed by one of the original subjects. It is observed that the proposed device is valid to find the orientation of an arm when the network is calibrated with a sub-millimeter system. This can be noted in the box on Fig. 3, corresponding to one portion of the whole closing drill movement; the data comes from flexion sensors only and is processed offline with the trained network and later compared with the ground truth data of that new session. It can be seen that the estimation is close to the calibration system, with an MSE of $1.32E-05$.

For gesture classification, the condensed data from the flex sensors is taken along with the sEMG data in order to train a pattern recognition neural network. To verify that the fusion of the data is feasible, three different networks are trained, one network only with the EMG data, another only with the flex sensors, and the third with the two of them. The resulting performance values are displayed in Table 1.

In order to compare the models, F-score is used. Given by: $Fscore = (2*Recall*Precision)/(Recall+Precision)$, where $Recall = TP/(TP+\sum FN)$ (being TP the true positive value and FN the false negative values); and

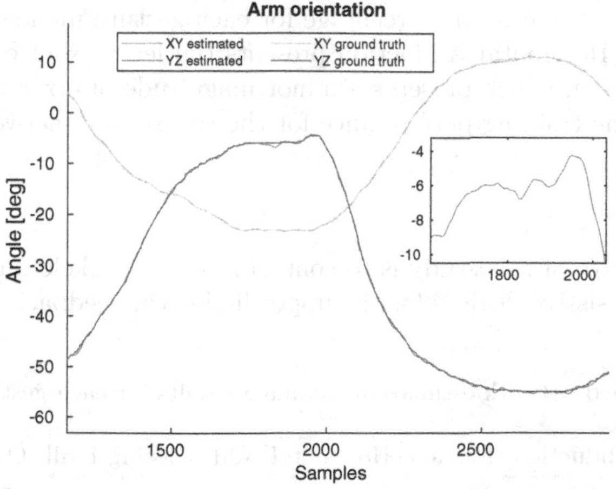

Fig. 3. Signal of the angle generated by the arm with the data of the ground truth together with the signal estimated by the network.

Table 1. Performance in percentage of the classification for sEMG (50.6%) and Soft-Flexion sensors (89.8%) and combined sEMG with Flexion (95.4%).

		sEMG		Flexion			sEMG + Flexion		
#	Gesture	Recall	Precision	Recall	Precision	**F-score**	Recall	Precision	**F-score**
1	Abduction	40.4	49.5	91.3	91.1	**91.2**	96.1	98.0	**97.5**
2	Flexion	66.8	56.8	90.5	87.1	**88.7**	97.2	94.3	**95.7**
3	Horizontal add	54.4	41.4	89.5	90.6	**90.0**	93.1	93.1	**93.1**
4	Closing drill	41.6	53.7	85.8	87.5	**86.6**	93.7	95.5	**94.6**
5	Opening drill	43.8	52.0	92.0	92.8	**92.4**	95.7	95.9	**95.8**

$Precision = TP/(TP + \sum FP)$, (being FP the false positive values for each of the Confusion Matrices).

The network using only the sEMGs shows poor results for this application, whilst the Flexion network and the combination of sEMG and Flexion sensors both have promising results. It could be said that flexion sensors are sufficient for the classification of movements for a certain type of application that does not require great sensitivity, while the fusion of both sensors denotes a great performance with minimum error. The overall performance of the network with the fusion of the two type of sensors is 95.4%; the best results were obtained by the Abduction gesture with 98% of precision, which could be a result of it being the only gesture generated in a different space and different muscle activation with respect to the other four gestures. On the other hand, Horizontal Adduction (93.1%) shares estimations with the Closing Drill and Opening Drill gestures; this can be given to the fact that they share movement space on certain spots. Using the data acquired for the new session and tested offline, it can be noted that the gestures, that coincide in the Flexion movement space such as the drilling gestures, cause an error in the estimation of the pattern. Table 2 depicts the response to the estimation in percentage for each gesture made during the new data collection. Horizontal Adduction presents the least exact estimation, contrary to Abduction, which presents a minor magnitude of error which coincides in a way with the training performance for the two sensors network.

3.1 Discussion

Since the objective of this study is to control flexible exoskeletons used in rehabilitation and assistive devices for the upper limbs, the feedback of the shoulder

Table 2. Trained network estimation: Ranking results for each gesture performed

	Abduction	Flexion	Horizontal Add	Closing Drill	Opening Drill
Recall	98.0	94.3	93.1	95.5	95.9
Precision	97.1	97.2	93.1	93.7	95.7

position and the gesture performed are extremely important for Control. This device, in its prototype mode, was created in a single compression shirt size, always being able to adapt in different sizes. This document does not present a study of the comparison of undefined gestures. It is estimated that the developed algorithm could be functional for new gestures as long as the data is processed and is not within the range of movement of the other gestures.

4 Conclusions

In this document, the signals of three electromyography sensors and an array of four flexion sensors are used to compose a flexible device to estimate five predefined gestures and the orientation of the shoulder. Two different algorithms are used to perform each characteristic, one for the identification of patterns to estimate the gesture being performed, and a recurrent neural network to estimate the orientation of the arm. The results show that the device consisting of an array of four flexion sensors is capable of estimating the gestures with a performance of 89.8%, with results showing improvement by adding the sEMG signal to the algorithm with a performance of 95.4%, there being an area of improvement in this last characteristic, such as filtering the sEMG signal online. Depending on the desired performance for the application, different arrangements can be used.

References

1. Artemiadis, P.K., Kyriakopoulos, K.J.: EMG-based teleoperation of a robot arm in planar catching movements using ARMAX model and trajectory monitoring techniques. In: Proceedings 2006 IEEE International Conference on Robotics and Automation. ICRA 2006, pp. 3244–3249. IEEE (2006)
2. Chada, S., Taran, S., Bajaj, V.: An efficient approach for physical actions classification using surface EMG signals. Health Inf. Sci. Syst. **8**(1), 3 (2020)
3. Chowdhury, R.H., Reaz, M.B., Ali, M.A.B.M., Bakar, A.A., Chellappan, K., Chang, T.G.: Surface electromyography signal processing and classification techniques. Sensors **13**(9), 12431–12466 (2013)
4. Cogollor, J.M., et al.: Handmade task tracking applied to cognitive rehabilitation. Sensors **12**(10), 14214–14231 (2012)
5. De Luca, C.J., Gilmore, L.D., Kuznetsov, M., Roy, S.H.: Filtering the surface EMG signal: movement artifact and baseline noise contamination. J. Biomech. **43**(8), 1573–1579 (2010)
6. Faust, O., Hagiwara, Y., Hong, T.J., Lih, O.S., Acharya, U.R.: Deep learning for healthcare applications based on physiological signals: a review. Comput. Methods Programs Biomed. **161**, 1–13 (2018)
7. Frisoli, A., Rocchi, F., Marcheschi, S., Dettori, A., Salsedo, F., Bergamasco, M.: A new force-feedback arm exoskeleton for haptic interaction in virtual environments. In: First Joint Eurohaptics Conference and Symposium on Haptic Interfaces for Virtual Environment and Teleoperator Systems. World Haptics Conference, pp. 195–201. IEEE (2005)
8. Gunasekara, J., Gopura, R., Jayawardane, T., Lalitharathne, S.: Control methodologies for upper limb exoskeleton robots. In: 2012 IEEE/SICE International Symposium on System Integration (SII), pp. 19–24. IEEE (2012)

9. Hermens, H.J., et al.: European recommendations for surface electromyography. Roessingh Res. Dev. **8**(2), 13–54 (1999)
10. Jimenez-Fabian, R., Verlinden, O.: Review of control algorithms for robotic ankle systems in lower-limb orthoses, prostheses, and exoskeletons. Med. Eng. Phys. **34**(4), 397–408 (2012)
11. Krasoulis, A., Vijayakumar, S., Nazarpour, K.: Multi-grip classification-based prosthesis control with two EMG-IMU sensor. IEEE Trans. Neural Syst. Rehabil. Eng. (2020)
12. Labs, B.: Bend labs. Internet draft (2018). https://www.bendlabs.com/products/2-axis-soft-flex-sensor/
13. López, N.M., di Sciascio, F., Soria, C.M., Valentinuzzi, M.E.: Robust EMG sensing system based on data fusion for myoelectric control of a robotic arm. Biomed. Eng. online **8**(1), 5 (2009)
14. Magermans, D., Chadwick, E., Veeger, H., Van Der Helm, F.: Requirements for upper extremity motions during activities of daily living. Clin. Biomech. **20**(6), 591–599 (2005)
15. McCabe, J.P., Henniger, D., Perkins, J., Skelly, M., Tatsuoka, C., Pundik, S.: Feasibility and clinical experience of implementing a myoelectric upper limb orthosis in the rehabilitation of chronic stroke patients: a clinical case series report. PloS One **14**(4) (2019)
16. Monroy, M., Ferre, M., Barrio, J., Eslava, V., Galiana, I.: Sensorized thimble for haptics applications. In: 2009 IEEE International Conference on Mechatronics, pp. 1–6. IEEE (2009)
17. NaturalPoint, I.: Optitrack. Internet draft (2019). https://optitrack.com
18. Nawab, S.H., Chang, S.S., De Luca, C.J.: High-yield decomposition of surface EMG signals. Clin. Neurophysiol. **121**(10), 1602–1615 (2010)
19. Phinyomark, A., Phukpattaranont, P., Limsakul, C.: Fractal analysis features for weak and single-channel upper-limb EMG signals. Expert Syst. Appl. **39**(12), 11156–11163 (2012)
20. Reese, S.P.: Angular displacement sensor of compliant material (Jan 27 2015), uS Patent 8,941,392
21. Samper-Escudero, J.L., Contreras-González, A.F., Ferre, M., Sánchez-Urán, M.A., Pont-Esteban, D.: Efficient multiaxial shoulder-motion tracking based on flexible resistive sensors applied to exosuits. Soft Robot. (2020)
22. Sankaran, S.: Robotic arm for the easy mobility of amputees. Int. J. Innov. Technol. Exploring Eng. 9 (2020). https://doi.org/10.35940/ijitee.B1151.1292S219
23. Technologies, A.: Myoware. Internet draft (2016). https://cdn.sparkfun.com/assets/a/3/a/f/a/AT-04-001.pdf
24. Washabaugh, E.P., Treadway, E., Gillespie, R.B., Remy, C.D., Krishnan, C.: Self-powered robots to reduce motor slacking during upper-extremity rehabilitation: a proof of concept study. Restorative Neurol. Neurosci. **36**(6), 693–708 (2018)
25. Yamanoi, Y., Ogiri, Y., Kato, R.: Emg-based posture classification using a convolutional neural network for a myoelectric hand. Biomed. Sig. Process. Control **55**, 101574 (2020)

Wearable Vibrotactile Interface Using Phantom Tactile Sensation for Human-Robot Interaction

Julian Seiler[1], Niklas Schäfer[1(✉)], Bastian Latsch[1], Romol Chadda[1], Markus Hessinger[1], Philipp Beckerle[2,3], and Mario Kupnik[1]

[1] Technische Universität Darmstadt, Measurement and Sensor Technology, Merckstraße 25, 64283 Darmstadt, Germany
schaefer@must.tu-darmstadt.de

[2] Technische Universität Dortmund, Robotics Research Institute, Otto-Hahn-Straße 8, 44227 Dortmund, Germany

[3] Technische Universität Darmstadt, Institute for Mechatronic Systems, Otto-Berndt-Straße 2, 64287 Darmstadt, Germany

Abstract. We present a wearable vibrotactile feedback device consisting of four linear resonant actuators (LRAs) that are able to generate virtual stimuli, known as phantom tactile sensation, for human-robot interaction. Using an energy model, we can control the location and intensity of the virtual stimuli independently. The device consists of mostly 3D-printed rigid and flexible components and uses commercially available haptic drivers for actuation. The actuators have a rated frequency of 175 Hz which is close to the highest skin sensitivity regarding vibrations (150 to 300 Hz). Our experiment was conducted with a prototype consisting of two bracelets applied to the forearm and upper arm of six participants. Eight possible circumferential angles were stimulated, of which four originated from real actuators and four were generated by virtual stimuli. The responses given by the participants showed a nearly linear relationship within $\pm 10^\circ$ for the responded angle against the presented stimulus angle. These results show that phantom tactile sensation allows for an increase of spatial resolution to design vibrotactile interfaces for human-robot interaction with fewer actuators.

Keywords: Vibrotactile feedback · Phantom tactile sensation · Human-robot interface

1 Introduction

Human-robot interaction (HRI) becomes more and more common due to progress in fields like robotics and artificial intelligence but also psychology. In particular, when a human and a robot are working on a common task, a bidirectional transfer of information is required [1,2].

Usually, the visual channel is already in use. Therefore, vibrotactile feedback can be helpful when added to provide additional information [3], especially

I. Nisky et al. (Eds.): EuroHaptics 2020, LNCS 12272, pp. 380–388, 2020.
https://doi.org/10.1007/978-3-030-58147-3_42

when multiple tasks are being performed and the workload is high [4]. Vibrotactile feedback devices can be used to present physical information, e.g., contact location, as well as abstract information, e.g., direction. They have been investigated in various applications such as robotic teleoperation [5], spatial awareness in virtual reality [6], navigation [7], and motion guidance [8].

Most collaborative tasks in the context of HRI, e.g., handling a tool or carrying an object, are performed by the human using hands and arms. Thus, vibrotactile feedback to the human arm offers a possibility to provide intuitively understandable information. In particular, circumferential feedback, i.e., feedback at different locations around the arm, enables providing information from different directions.

There are many vibrotactile feedback devices in research [9], including some for feedback around the arm [10–12]. Most of them use eccentric rotating mass motors (ERMs) due to the simplicity of control, small form factor, and low cost. However, the inherent coupling of amplitude and frequency of an ERM can be a limitation because the perceived intensity of a vibrotactile stimulus depends not only on its amplitude but also on its frequency [13].

A commercially available vibrotactile feedback device using ERMs is VibroTac (SENSODRIVE, Weßling, Germany). It was originally developed at the German Aerospace Center for the application on the human arm [12]. Due to the ergonomic design, it can be worn on a wide range of arm circumferences. It provides vibrotactile stimuli at six circumferential locations.

The spatial resolution of a vibrotactile feedback device can be increased by utilizing tactile illusions [14]. In [15] an illusion known as phantom tactile sensation was used to induce vibrotactile cues at any circumferential location around the wrist with six ERMs. Phantom tactile sensations can be classified regarding the perceived stimulus either being stationary or moving across the skin [16].

In this paper, we present a wearable vibrotactile feedback device for the human arm consisting of two bracelets. We investigate the feasibility of generating vibrotactile cues at eight circumferential locations around the arm with only four actuators by using stationary phantom tactile sensation. The application of linear resonant actuators (LRAs) ensures a constant frequency for all vibration amplitudes. The modular feedback device is evaluated experimentally on the forearm and the upper arm.

2 Fundamentals

The occurrence of phantom tactile sensations in haptics, not to be confused with phantom limb illusions, was first described by von Békésy in 1957 [17]. It terms the phenomenon that two simultaneous vibrotactile stimuli produced by two closely spaced actuators are perceived as one single vibration in between (Fig. 1a). This effect is based on sensory funneling [18]. Location and intensity of the phantom tactile sensation can be controlled by the intensities of the actuators. The location results from the actuators' relative magnitudes, whereas the intensity can be controlled by the actuators' absolute magnitudes. Two

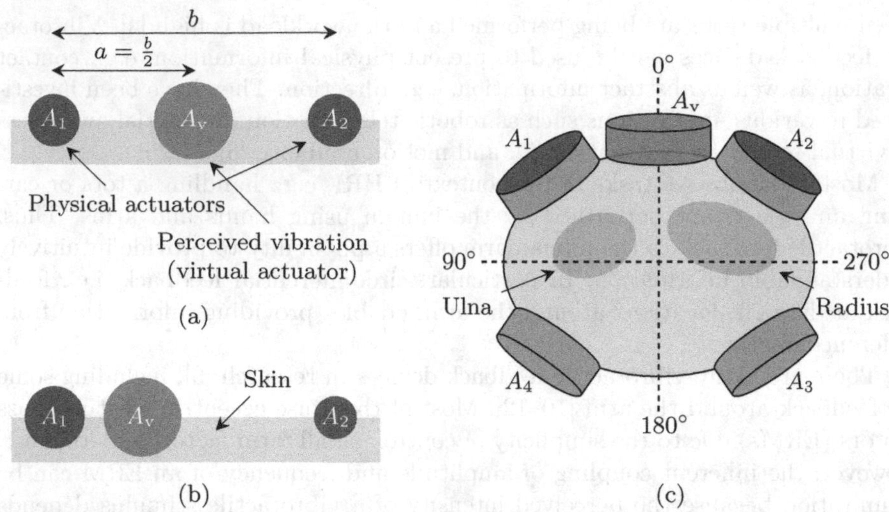

Fig. 1. Basic principle of phantom tactile sensation and placement of the actuators. The phantom tactile sensation is based on sensory funneling. Two simultaneous vibrotactile stimuli produced by two closely spaced actuators are perceived as one single vibration in between (a). Adjusting the actuators' vibration intensities allows shifting the perception closer to the actuator with higher magnitude (b). In order to provide directional cues, four LRAs are arranged equidistant around the left human forearm (c). A phantom tactile sensation is created by activating two adjacent actuators. Location and intensity of the virtual actuator are controlled based on an energy model [19].

vibrations with equal intensities result in a centered phantom tactile sensation. Adjusting the magnitudes equally leads to a sensation with the same location but different intensity. If the intensities are different, the phantom tactile sensation is located closer to the actuator with the higher magnitude (Fig. 1b).

The energy model proposed in [19] allows controlling the relative location $\beta = \frac{a}{b}$ and the intensity A_v of a virtual actuator induced by phantom tactile sensation independently. The required intensities of the two physical actuators are

$$A_1 = \sqrt{1-\beta} \cdot A_v, \quad A_2 = \sqrt{\beta} \cdot A_v. \tag{1}$$

This energy model is based on two assumptions. First, the vibration frequencies of both actuators are equal. Second, the skin sensitivity thresholds at the locations of the physical actuators are identical.

3 Design and Construction

In order to provide directional vibrotactile feedback, four LRAs (G1036002D, Jinlong Machinery & Electronics, Wenzhou, China) are arranged equidistant around the arm (Fig. 1c). Unlike ERMs, the vibration amplitude of an LRA can be adjusted without changing the vibration frequency. This satisfies the energy model's first assumption of equal vibration frequencies. With 175 Hz, the rated frequency of the LRAs is close to the highest sensitivity of the Pacinian corpuscles, which is usually found between 150 and 300 Hz [20, 21].

The feedback device is designed to be wearable as a bracelet on the arm. It consists of multiple 3D-printed segments, which are mounted on an elastic cord (Fig. 2). There are two types of segments. The actuator segments consist of a rigid basis (polylactic acid) and a comparatively flexible mounting (thermoplastic polyurethane, shore hardness 98 A) for an LRA which aims at reduction of vibration propagation into the mechanical structure. The intermediate segments carry the electronics as well as the control unit and are used for cable routing. The alternating arrangement of the segments creates a zigzag pattern, which increases the overall elasticity and an equidistant actuator arrangement [12].

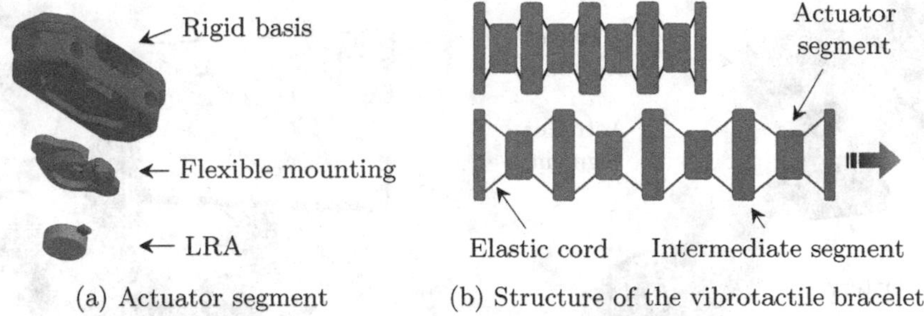

(a) Actuator segment (b) Structure of the vibrotactile bracelet

Fig. 2. Each actuator segment consists of a rigid basis, a flexible mounting, and an LRA (a). Actuator segments and intermediate segments are arranged alternating on an elastic cord (b). The zigzag pattern of the elastic cord ensures equal distances between the actuators in relaxed and stretched state [12].

The LRAs require a sinusoidal driving voltage at their resonance frequency. We use commercially available haptic drivers (DRV2605L, Texas Instruments, Dallas, Texas, U.S.) in combination with an ESP32-WROOM-32 module (Espressif Systems, Shanghai, China) controlling the amplitudes. The control unit receives commands containing direction and magnitude from a PC via Bluetooth. The desired vibrotactile cues between two physical actuators are generated by a control algorithm that applies the energy model (Eq. 1).

4 System Evaluation

The goal of our evaluation experiment was to investigate the perceived direction of vibrotactile cues generated by the developed prototype (Fig. 3a). The locations of the vibrotactile cues resulted either from one real actuator at its own location or from two actuators by inducing phantom tactile sensation in between. Both, forearm and upper arm, were stimulated with vibrotactile cues. In our first test, six participants (1 female, 5 male, 22.7 ± 1.6 years) gave prior informed consent and took part in the experiment. They were informed that their vibrotactile perception is investigated in the experiment but no information about phantom tactile sensation and placement of the actuators was given.

After measuring the circumferences of the forearm (25.8 ± 3.3 cm) and the upper arm (30.8 ± 2.3 cm), the participants were requested to wear the system on the left arm. The vibrotactile bracelets were placed in the middle of the respective arm segment, ensuring an equidistant arrangement of actuators. The origin of the reference frame for each arm segment was defined to be collocated with the driver electronics. The bracelet with the control unit and the battery was always worn on the upper arm. The elastic cord was adjusted to the individual arm circumference of each participant in order to ensure that the vibrotactile bracelets could be worn comfortably.

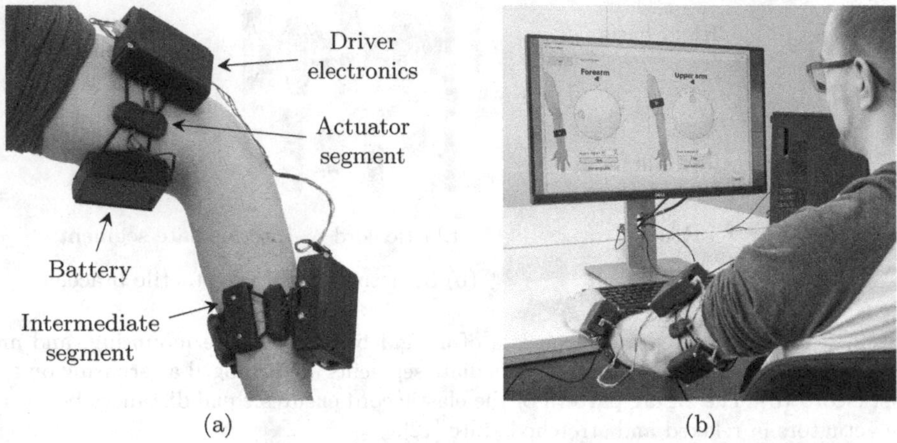

Fig. 3. Prototype of the vibrotactile feedback device consisting of two bracelets worn on the forearm and the upper arm (a). The bracelet on the upper arm contains the control unit (hidden behind the upper arm), the driver electronics, and the power supply. Due to the modular system architecture, up to eight devices with up to eight actuators each can be connected. During the evaluation experiment, the subjects were asked to adjust the rotary knob to the perceived stimulus angle (b).

4.1 Experimental Procedure

The participants were seated in front of a computer screen with a graphical user interface (GUI) consisting of two rotary knobs for an intuitive selection of the vibrotactile cue direction at the forearm and the upper arm, respectively (Fig. 3b). Each arm segment was stimulated with 32 vibrotactile cues. These vibrotactile cues pointed in one of eight possible directions [0°, 45°, 90°, 135°,

180°, 225°, 270°, and 315° (Fig. 1c)], which were tested four times each in random order. Each vibrotactile cue lasted until a response was given by the participant. The LRAs were driven by a sinusoidal voltage with a frequency of 175 Hz. When stimulating at the location of a real actuator, the rated amplitude of 2.0 V_{RMS} was used. In the case of phantom tactile sensation, the amplitudes of the two actuators in use were set such that the intensity of the resulting virtual actuator corresponded to the intensity of one real actuator at rated amplitude (Eq. 1). The participant indicated the direction perceived after each stimulus using the rotary knobs in the GUI (resolution of 1°). In addition to the perceived direction, the time for locating the vibrotactile cues was measured as well. After finishing one arm segment, a short break was taken and the participant was requested to rate the difficulty of locating the vibrotactile cues on a scale from 1 to 10 (1 meaning very easy and 10 meaning very hard). The experiment lasted approximately 25 min for each participant.

4.2 Result and Analysis

The results of the averaged perceived directions of the vibrotactile cues over the stimulus directions show a nearly linear relationship for both arm segments (Fig. 4). For all but two stimulus angles, the mean response deviates less than ±10° from the real value. It should be noted though that visual and auditory modalities were not controlled, which may have affected the results.

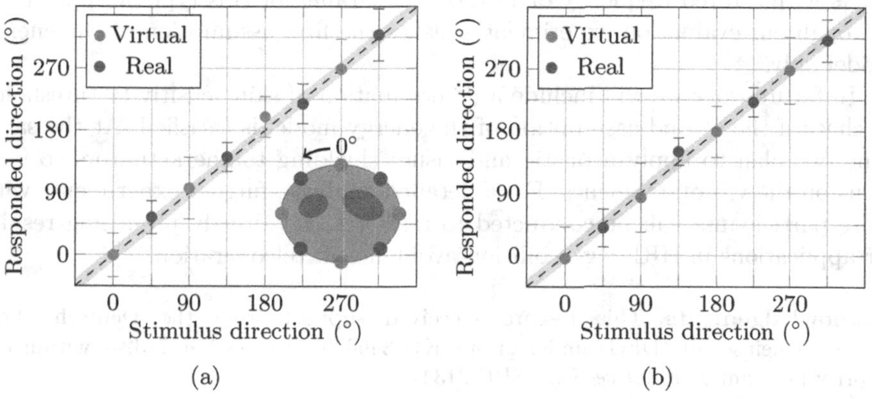

Fig. 4. Mean values of the responded angles against the presented stimulus angles for the forearm (a) and the upper arm (b). The standard deviations, achieved with the virtual actuators, are higher compared to those observed with the real actuators. For all but two angles, the mean response deviates less than ±10° (gray band) from the real value. The experiment was always conducted on the left arm (Fig. 1c).

It is noticeable that the perceived vibrotactile cues of the real actuators (forearm $R^2 = 91.81\,\%$, upper arm $R^2 = 93.45\,\%$) deviate less than those induced by virtual actuators (forearm $R^2 = 87.20\,\%$, upper arm $R^2 = 83.72\,\%$) with respect to the ideal linear response. Comparing the results of both arm segments shows that the scattering of the perceived virtually-generated stimuli on the upper arm is stronger than on the forearm. Furthermore, a Wilcoxon signed-rank test ($W = 0$, $p = 0.0156$) indicated that the difficulty of locating the vibrotactile cues was rated significantly higher for the upper arm (mean 4.67) in contrast to the forearm (mean 3.17). One possible reason for the latter two observations is the lower innervation density of the mechanoreceptors on the upper arm [22].

In addition to the quantitative results, some of the participants expressed their subjective opinions. They reported that it was difficult to assign the vibrotactile cues using the rotary knobs in the GUI, in particular for the upper arm. This impression is consistent with the higher rating of difficulty for the upper arm. Furthermore, the participants estimated their accuracy of locating the vibrotactile cues between 20 to 30°.

5 Conclusion and Outlook

The wearable feedback device developed is capable of generating vibrotactile cues across all circumferential locations around the human arm. This is achieved with only four actuators. In our evaluation experiment the spatial resolution of the vibrotactile feedback can be doubled from four to eight by inducing phantom tactile sensations midway between two actuators. Therefore, we conclude that LRAs with a rated frequency of 175 Hz are suitable for this type of application although our evaluation experiment satisfies the first assumption of the energy model only.

In future work we will include a determination of skin sensitivity thresholds to check if the second assumption of the energy model is satisfied. At the same time, we plan to apply acoustic and visual shielding to the actuators to gain focus on the vibrotactile cues. For a detailed analysis, further experiments with more participants will be conducted to reinforce the already promising results for applications in HRI, e.g., collision avoidance in teleoperation.

Acknowledgements. This research received support from the Deutsche Forschungsgemeinschaft (DFG) under grants KU 3498/3-1 and KA 417/32-1 within the priority program *The Active Self* (SPP 2134).

References

1. Beckerle, P., et al.: A human-robot interaction perspective on assistive and rehabilitation robotics. Front. Neurorobot. **11**, 24 (2017)
2. Ajoudani, A., et al.: Progress and prospects of the human-robot collaboration. Auton. Robot. **42**, 957–975 (2018)
3. Elliott, L.R., Coovert, M.D., Redden, E.S.: Overview of meta-analyses investigating vibrotactile versus visual display options. In: Jacko, J.A. (ed.) HCI 2009. LNCS, vol. 5611, pp. 435–443. Springer, Heidelberg (2009)

4. Burke, J.L., et al.: Comparing the effects of visual-auditory and visual-tactile feedback on user performance: a meta-analysis. In: Proceedings of the 8th International Conference on Multimodal Interfaces - ICMI 2006, Banff, Alberta, Canada, p. 108. ACM Press (2006)
5. Bimbo, J., et al.: Teleoperation in cluttered environments using wearable haptic feedback. In: 2017 IEEE/RSJ International Conference on Intelligent Robots and Systems (IROS), Vancouver, BC, pp. 3401–3408. IEEE, September 2017
6. Louison, C., Ferlay, F., Mestre, D.R.: Spatialized vibrotactile feedback improves goal-directed movements in cluttered virtual environments. Int. J. Hum.-Comput. Interact. **34**, 1015–1031 (2018)
7. Erp, J.B.F.V., et al.: Waypoint navigation with a vibrotactile waist belt. ACM Trans. Appl. Percept. **2**, 106–117 (2005)
8. Bark, K., et al.: Effects of vibrotactile feedback on human learning of arm motions. IEEE Trans. Neural Syst. Rehabil. Eng. **23**, 51–63 (2015)
9. Pacchierotti, C., Sinclair, S., Solazzi, M., Frisoli, A., Hayward, V., Prattichizzo, D.: Wearable Haptic Systems for the Fingertip and the Hand: taxonomy, review, and perspectives. IEEE Trans. Haptics **10**, 580–600 (2017)
10. Pezent, E., et al.: Tasbi: multisensory squeeze and vibrotactile wrist haptics for augmented and virtual reality. In: 2019 IEEE World Haptics Conference (WHC), Tokyo, Japan, pp. 1–6. IEEE, July 2019
11. Tsetserukou, D., Tachi, S.: Efficient object exploration and object presentation in TeleTA, Teleoperation system with Tactile feedback. In: World Haptics 2009 - Third Joint EuroHaptics conference and Symposium on Haptic Interfaces for Virtual Environment and Teleoperator Systems, Salt Lake City, UT, USA, pp. 97–102. IEEE (2009)
12. Schaetzle, S., et al.: VibroTac: an ergonomic and versatile usable vibrotactile feedback device. In: 19th International Symposium in Robot and Human Interactive Communication, Viareggio, Italy, pp. 670–675. IEEE, September 2010
13. Verrillo, R.T., Fraioli, A.J., Smith, R.L.: Sensation magnitude of vibrotactile stimuli. Percept. Psychophysics **6**, 366–372 (1969)
14. Barghout, A., Cha, J., El Saddik, A., Kammerl, J., Steinbach, E.: Spatial resolution of vibrotactile perception on the human forearm when exploiting funneling illusion. In: 2009 IEEE International Workshop on Haptic Audio visual Environments and Games, Lecco, Italy, pp. 19–23. IEEE, November 2009
15. Salazar Luces, J.V., Okabe, K., Murao, Y., Hirata, Y.: A phantom-sensation based paradigm for continuous vibrotactile wrist guidance in two-dimensional space. IEEE Robot. Autom. Lett. **3**, 163–170 (2018)
16. Park, G., Choi, S.: Tactile information transmission by 2D stationary phantom sensations. In: Proceedings of the 2018 CHI Conference on Human Factors in Computing Systems - CHI 2018, Montreal QC, Canada, pp. 1–12. ACM Press (2018)
17. Békésy, G.V.: Sensations on the skin similar to directional hearing, beats, and harmonics of the ear. J. Acoust. Soc. Am. **29**, 489–501 (1957)
18. Békésy, G.V.: Funneling in the nervous system and its role in loudness and sensation intensity on the skin. J. Acoust. Soc. Am. **30**, 399–412 (1958)
19. Israr, A., Poupyrev, I.: Tactile brush: drawing on skin with a tactile grid display. In: Proceedings of the 2011 Annual Conference on Human Factors in Computing Systems - CHI 2011, Vancouver, BC, Canada, p. 2019. ACM Press (2011)
20. Choi, S., Kuchenbecker, K.J.: Vibrotactile display: Perception, technology, and applications. Proc. IEEE **101**, 2093–2104 (2013)

21. Bolanowski, S.J., Gescheider, G.A., Verrillo, R.T.: Hairy skin: psychophysical channels and their physiological substrates. Somatosens. Mot. Res. **11**, 279–290 (1994)
22. Békésy, G.V., Wever, E.G.: Experiments in Hearing. McGraw-Hill, New York City (1960)

A Parallel Elastic Haptic Thimble for Wide Bandwidth Cutaneous Feedback

Daniele Leonardis(✉), Massimiliano Gabardi, Massimiliano Solazzi, and Antonio Frisoli

Percro Laboratory, Scuola Superiore Sant'Anna of Pisa, Pisa, Italy
d.leonardis@santannapisa.it
https://www.santannapisa.it/it

Abstract. Design of wearable fingertip haptic devices is often a compromise between conflicting features: lightness and compactness, against rich and neat haptic feedback. On one side direct drive actuators (i.e. voice coils) provide a clean haptic feedback with high dynamics, with limited maximum output forces. On the other side mechanical transmissions with reduction can increase output force of micro sized motors, at the cost of slower and often noisy output signals. In this work we present a compact fingertip haptic device based on a parallel elastic mechanism: it merges the output of two differently designed actuators in a single, wide bandwidth haptic feedback. Each actuator is designed with a different role: one for rendering fast, high frequency force components, the other for rendering constant to low frequency components. In the work we present design and implementation of the device, followed by experimental characterization of its performance in terms of frequency response and rendering capabilities.

Keywords: Cutaneous feedback · Haptics · Bandwidth · Fingertip · Wearable · Parallel elastic

1 Introduction

In recent years, rendering of the sense ot touch in teleoperated or virtual reality has become a rich field of research, especially concerning highly wearable haptic devices. In particular the scientific literature shows the development of different devices able to provide the user with a specific cutaneous feedback, such as thermal [5], vibratory [14], contact orientation [2], contact force [7], or a combination of the mentioned feedback [4,16]. In [12], a complete review of portable and wearable haptic devices for the fingertips can be found.

Rendering the correct physical interaction is a challenging objective [1], and concerning portable and wearable haptic devices, practical requirements such as wearability and portability of the devices determine limits to features of the rendered feedback. Limitations can be, for instance, in terms of bandwidth and

I. Nisky et al. (Eds.): EuroHaptics 2020, LNCS 12272, pp. 389–397, 2020.
https://doi.org/10.1007/978-3-030-58147-3_43

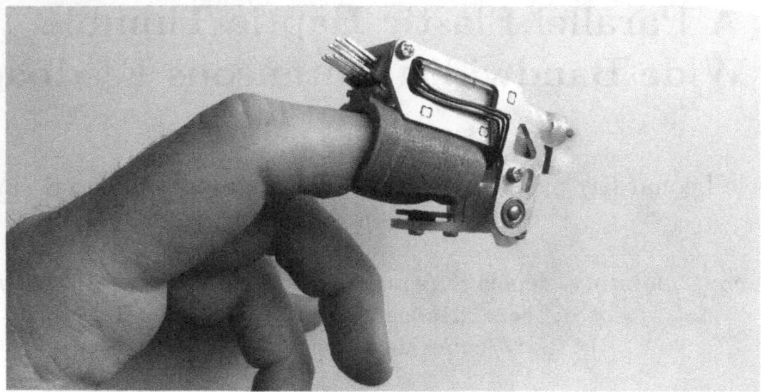

Fig. 1. The presented Parallel Elastic Thimble, featuring two actuators coupled by an elastic element to render from static to high frequency cutaneous feedback

force amplitude for force rendering devices, range of motion for shape rendering thimbles or heat flux for thermal devices. The choice of the actuation system is a trade-off between feedback performance (max. force, bandwidth, max. stroke, noise) and device requirements (mass, dimensions, wearability). In fact, the actuation system is usually the heaviest part of a portable fingertip haptic interface. Typical electromagnetic actuators (DC motors or voice coils) used in haptic devices allow for high quality haptic rendering, ranging from constant to high frequency force components, with the drawback of a limited maximum output force. Small actuators provided with mechanical reduction can be used to amplify the output force, yet at the cost of reduced output bandwidth and degraded quality of the haptic feedback in terms of noise and backlash. A different solution to increase the feedback bandwidth while keeping a reasonable constant force is obtained by coupling in a serial manner a macro-actuator (or a small reduced actuator) featuring low bandwidth and high output force, with a micro-actuator, with wide bandwidth but low output force. Such a solution has been explored in grounded haptic devices in a series configuration, i.e. a very compact "micro" actuator is placed at the end-effector of a desktop or grounded haptic device [9,11,15] to enhance its dynamic response. Concept of parallel micro-macro configuration via elastic parallel transmission was proposed in [10] and then developed for robotic arm manipulators in [13] and [17]. To the knowledge of authors, micro-macro solutions have never been applied to a fully wearable and compact fingertip haptic display. In this paper, the proposed novel fingertip device features a parallel elastic actuation system, obtained by combining the output of two micro DC motors of the same size: one with high mechanical reduction and considerable continuous force, the second with low reduction and high dynamics of the output force (Fig. 1).

2 The Parallel Elastic Thimble

The fingertip device proposed in this work is conceived to render 1 dof cutaneous feedback at the fingerpad, featuring rendering of the no-contact to contact transition and modulation of the contact force. In order to efficiently render from static to high-frequency force components, a parallel elastic configuration implementing two parallel micro-sized motors has been experimented. Overview of the design and placement of the different components is shown in Fig. 2. Actuators have been placed on the finger dorsum, and the whole device design has been studied in order to minimize lateral interference with other fingers and interference with the hand workspace. The device is 16 mm wide and weights 21 g. Importantly, the slim design at the sides of the device allow for easy switching between thimbles of different sizes. Different thimble sizes were fabricated by 3D printing in TPU (Thermoplastic Polyurethane) soft polymer, resulting in a compliant and precise fit of the device to the specific finger shape, not requiring additional fastening elements (velcro or clips). A slider mechanism allows easy and rapid switch between different thimbles (Fig. 3).

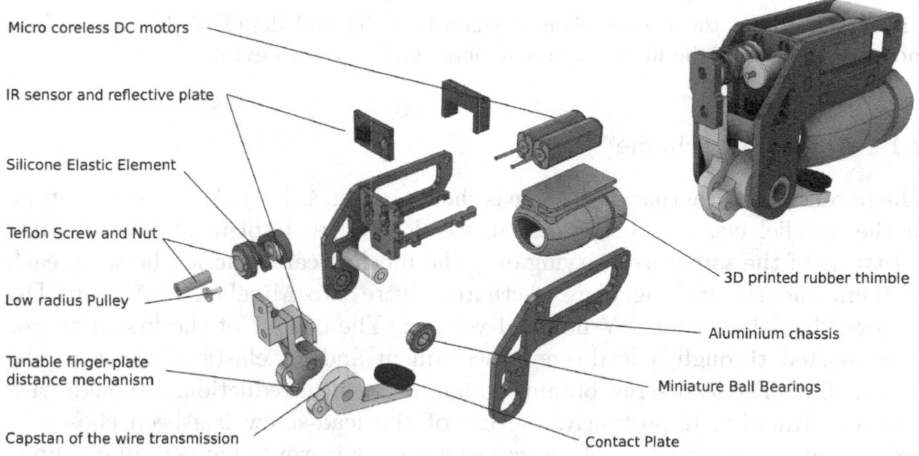

Fig. 2. Mechanical components of the Parallel Elastic Haptic Thimble

The moving plate in contact with the fingerpad has been implemented through a 1 dof link with a revolute joint. A slot mechanism with a fastening screw has been designed in order to tune distance of the plate with respect of the finger surface, thus minimizing the required displacement of the plate. Considering that a displacement of few millimeters of the contact plate is a sufficient range in cutaneous haptic devices [6,8], the consequent angular displacement, due to the revolute joint, can be negligible (18 mm radius). Also, the revolute joint with miniaturized ball bearings allow to minimize friction with respect to a linear slider mechanism.

Fig. 3. The slim lateral profile of the device (left) allows easy switch of rubber thimbles with different finger sizes through a slider mechanism (right)

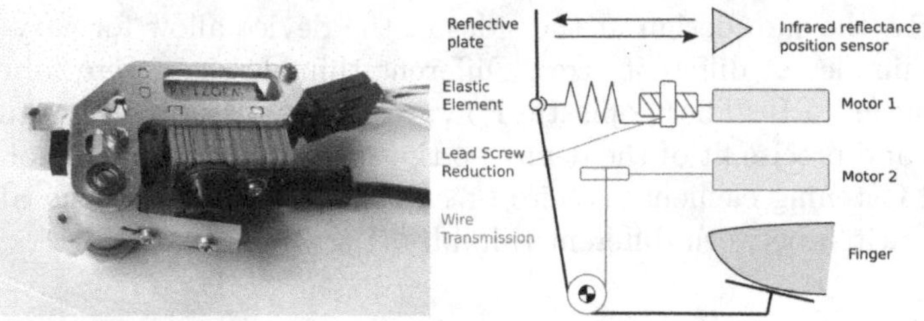

Fig. 4. Scheme of the transmission mechanism (left) and detail of the force sensor mounted in place of the fingertip for characterization of the device

2.1 Actuation Scheme

The parallel elastic actuation scheme is shown in Fig. 4 (left). As a first prototype of the parallel elastic concept design, we decided to implement two identical actuators of the same size, varying only the mechanical reduction between each of them and the moving plate. Actuators were two Minebea K30 micro DC motors, diameter 8 mm, 5 V nominal voltage. The output of the first actuator is connected through a lead-screw mechanism and an elastic element to the moving link. The lead-screw obtains a high mechanical reduction, although with no-backdrivability. Importantly, the use of the lead-screw has been chosen in order to avoid introduction of sources of noise, as it would happen, in example, for a more conventional gear reduction. The second actuator is coupled to the moving link by means of a one branch wire transmission: a capstan (radius 8 mm at the moving link is connected by the actuation wire to a pulley (radius 1 mm) at the output shaft of the motor. It results in a low reduction, highly reversible mechanical transmission. The elastic element has been fabricated from a silicon tube: after preliminary experiments it was preferred with respect to a steel spring due to the inherent presence of a damping factor. A position sensor has been embedded into the device, measuring displacement of the moving link with the contact plate. We used a reflectance infrared sensor, due to its very compact size, to the measurement range particularly suitable for the device and to the sensitivity of the sensor to small displacement (measured noise of 0.01 mm in the middle point of the measuring range).

3 Experimental Characterization

The experimental activity was conducted to characterize the novel (for the size of a fingertip mechanism) compact parallel elastic structure in terms of frequency response and output forces and displacement. A holder for a compact force sensor (Optoforce 10N with resolution of 1 mN) was fabricated to be mounted in place of the rubber thimble (Fig. 4 left). A microcontroller board (Teensy 3.6) was used to implement the low level control of the device, to acquire the analog infrared position sensor, and to drive motors through a dual H-bridge (Texas Instruments DRV8835). Sample time of the low-level control loop was 1 KHz. Communication with a host PC was implemented through a Wiz5500 Ethernet module and UDP communication. A Matlab Simulink Desktop-Real Time model, executed on the host PC, implemented the high level control interface and data recording.

The first experimental activity consisted in measuring the force output of the two actuators. The contact plate was positioned at the contact threshold with the force sensor. Then, a slow voltage reference ramp was commanded to each motor separately for ten repetitions. The obtained current intensity to force characteristics are shown in Fig. 5 (left). The graph highlights the different mechanical reduction of the two identical actuators: the first obtains a higher output force, presenting non-linearity due to friction of the lead-screw mechanism. The second actuator shows a lower output force and a more linear characteristic. A position control loop was then tuned for the first actuator, in order to control displacement of the moving plate. Step response of the position control loop is shown in Fig. 5 (middle). Bandwidth of the two actuators was then measured. The first actuator was controlled in closed loop with a chirp reference position signal, ranging from 0.5 to 40 Hz. The obtained frequency response shows a limit of the first actuator bandwidth at 15 Hz (Fig. 5, right). The second actuator was commanded in open loop with a chirp voltage reference ranging from 5 to 250 Hz. The second actuator response shows a cutoff frequency of 120 Hz, which is about

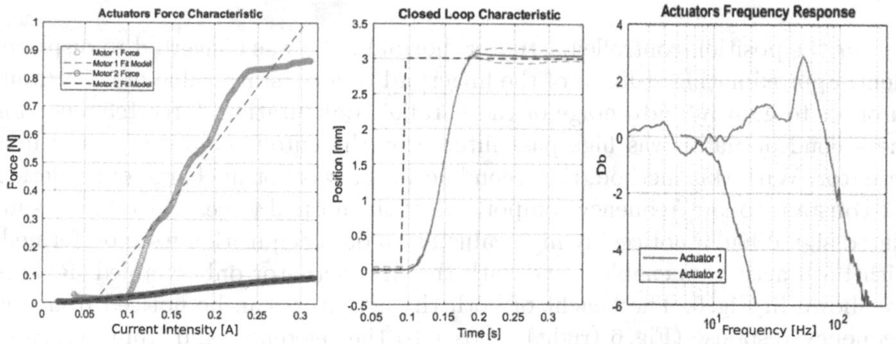

Fig. 5. Force characterization of the two actuators (left), closed loop step response of the first actuator (middle) and frequency response of the two actuators (right)

one order of magnitude greater than the first actuator. Also, a peak appears at 55 Hz, possibly due to a resonant frequency introduced by the elastic element of the system.

3.1 Experimental Evaluation of Sample Texture Rendering

Overall device performance was finally evaluated using a pre-recorded signal involving contact with a texturized surface. The device was evaluated by wearing it onto the experimenter's index finger. Displacement measurements of the contact plate were recorded through the embedded infrared position sensor. The texture pattern was taken from the "Sandpaper 100" object of the texture library of the Penn Haptic Texture Toolkit [3].

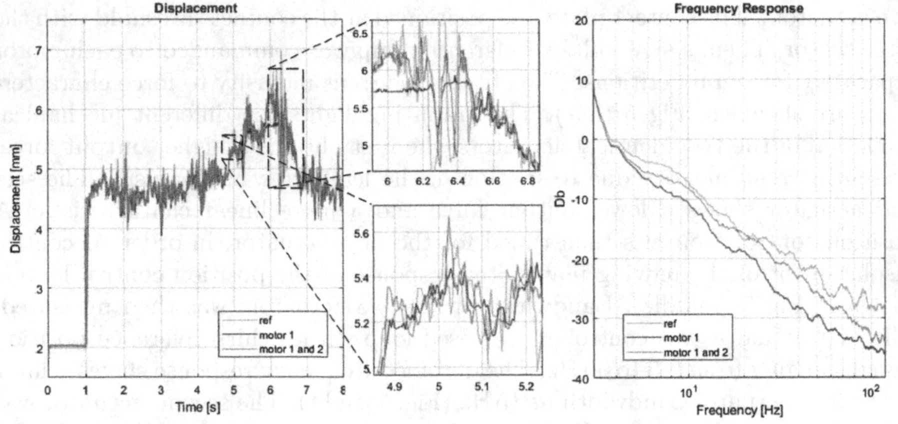

Fig. 6. Bench test of the device rendering a sample texture (sandpaper) with contact transition and non-zero constant component of the normal force

For the position-controlled actuator, normal force was converted to displacement, approximating stiffness of the fingerpad to a constant value of 0.5 N/mm. In order to gain full advantage of the parallel configuration, force reference for the second actuator was high-pass filtered at the cutoff frequency of the first actuator. With this method, the second actuator was not in charge of rendering the constant to low frequency components of the normal force, which in the simulated signal had a noticeably high value. Two data acquisition were performed: with both actuators enabled, and with the first actuator only enabled. Results are shown in Fig. 6. The benefit of both the actuators can be noticed from the frequency response (Fig. 6 (right)), closer to the reference, and from details of Fig. 6 (middle). The output of second actuator produces more crisp and reactive dynamics of the plate, whereas the first actuator alone, especially at the higher indentation levels (top detail), tends to a more flat response.

4 Conclusions

The novel design of a wearable fingertip device implementing a parallel elastic mechanism was proposed. The idea was originated from the contrasting requirements of fingertip haptic devices, involving compactness and wearability of the device and rendering of relatively high constant forces together with wide bandwidth tactile cues. The parallel structure allows to optimize each actuator for different purposes: the first, with high mechanical reduction, for rendering static to low frequency cues, which in typical applications can be noticeably high (i.e. when grasping a virtual object or exploring a surface). The second actuator, with low reduction and high transparency, was designed to render high frequency tactile cues, which typically have a reduced amplitude with respect to the static and slow force components.

The obtained prototype included two miniaturized motors with noiseless mechanical reduction (a lead-screw and a capstan wire transmission) and an elastic element to couple the two actuators. Mechanical design of the prototype was focused on enhancing wearability by minimizing mass and dimensions (16 mm total width, 21 g mass), by optimizing arrangement of actuators, and by implementing a user's tailored and switchable soft thimble design.

Force characterization and frequency response confirmed the desired different behavior of the two actuators (same motors with different reduction) in complementary frequency ranges. A resonant peak was noticeable, and further investigation is required in order to obtain a more flat frequency response. A deeper study of the mechanical model of the system can guide the choice of the elastic element stiffness, with the aim of optimizing interaction between the two actuators.

The final evaluation with the sample texture evidenced how the reference signal can be conveniently split between the two actuators. Although more investigation is required to obtain proper optimization of the developed device, the proposed method can result in more compact wearable devices with better energy efficiency and better capabilities, in terms of quality of the output signal and hi-fidelity rendering.

References

1. Caldwell, D.G., Tsagarakis, N., Wardle, A.: Mechano thermo and proprioceptor feedback for integrated haptic feedback. In: 1997 Proceedings of the IEEE International Conference on Robotics and Automation, vol. 3, pp. 2491–2496. IEEE (1997)
2. Chinello, F., Malvezzi, M., Pacchierotti, C., Prattichizzo, D.: Design and development of a 3RRS wearable fingertip cutaneous device. In: 2015 IEEE International Conference on Advanced Intelligent Mechatronics (AIM), pp. 293–298. IEEE (2015)
3. Culbertson, H., Lopez Delgado, J.J., Kuchenbecker, K.J.: The Penn haptic texture toolkit for modeling, rendering, and evaluating haptic virtual textures (2014)

4. Gabardi, M., Leonardis, D., Solazzi, M., Frisoli, A.: Development of a miniaturized thermal module designed for integration in a wearable haptic device. In: 2018 IEEE Haptics Symposium (HAPTICS), pp. 100–105. IEEE (2018)
5. Gallo, S., Rognini, G., Santos-Carreras, L., Vouga, T., Blanke, O., Bleuler, H.: Encoded and crossmodal thermal stimulation through a fingertip-sized haptic display. Front. Robot. AI **2**, 25 (2015)
6. Gleeson, B.T., Horschel, S.K., Provancher, W.R.: Design of a fingertip-mounted tactile display with tangential skin displacement feedback. IEEE Trans. Haptics **3**(4), 297–301 (2010)
7. Leonardis, D., Solazzi, M., Bortone, I., Frisoli, A.: A wearable fingertip haptic device with 3 DoF asymmetric 3-RSR kinematics. In: 2015 IEEE World Haptics Conference (WHC), pp. 388–393. IEEE (2015)
8. Leonardis, D., Solazzi, M., Bortone, I., Frisoli, A.: A 3-RSR haptic wearable device for rendering fingertip contact forces. IEEE Trans. Haptics **10**(3), 305–316 (2016)
9. Lu, T., Pacoret, C., Hériban, D., Mohand-Ousaid, A., Regnier, S., Hayward, V.: Kilohertz bandwidth, dual-stage haptic device lets you touch brownian motion. IEEE Trans. Haptics **10**(3), 382–390 (2016)
10. Morrell, J.B., Salisbury, J.K.: Parallel-coupled micro-macro actuators. Int. J. Robot. Res. **17**(7), 773–791 (1998)
11. Pacchierotti, C., Prattichizzo, D., Kuchenbecker, K.J.: Cutaneous feedback of fingertip deformation and vibration for palpation in robotic surgery. IEEE Trans. Biomed. Eng. **63**(2), 278–287 (2015)
12. Pacchierotti, C., Sinclair, S., Solazzi, M., Frisoli, A., Hayward, V., Prattichizzo, D.: Wearable haptic systems for the fingertip and the hand: taxonomy, review, and perspectives. IEEE Trans. Haptics **10**(4), 580–600 (2017)
13. Shin, D., Sardellitti, I., Khatib, O.: A hybrid actuation approach for human-friendly robot design. In: 2008 IEEE International Conference on Robotics and Automation, pp. 1747–1752. IEEE (2008)
14. Solazzi, M., Frisoli, A., Bergamasco, M.: Design of a novel finger haptic interface for contact and orientation display. In: 2010 IEEE Haptics Symposium, pp. 129–132. IEEE (2010)
15. Wall, S.A., Harwin, W.: A high bandwidth interface for haptic human computer interaction. Mechatronics **11**(4), 371–387 (2001)
16. Wang, D., Ohnishi, K., Xu, W.: Multimodal haptic display for virtual reality: a survey. IEEE Trans. Ind. Electron. **67**(1), 610–623 (2019)
17. Zinn, M., Khatib, O., Roth, B., Salisbury, J.K.: Large workspace haptic devices-a new actuation approach. In: 2008 Symposium on Haptic Interfaces for Virtual Environment and Teleoperator Systems, pp. 185–192. IEEE (2008)

Instrumenting Hand-Held Surgical Drills with a Pneumatic Sensing Cover for Haptic Feedback

Chiara Gaudeni[1(✉)], Tommaso Lisini Baldi[1], Gabriele M. Achilli[2], Marco Mandalà[3], and Domenico Prattichizzo[1,4]

[1] Department of Information Engineering and Mathematics, University of Siena, Siena, Italy
{gaudeni,lisini,prattichizzo}@diism.unisi.it

[2] Department of Engineering, University of Perugia, Perugia, Italy

[3] Department of Medicine, Surgery and Neuroscience, University of Siena, Siena, Italy

[4] Department of Advanced Robotics, Istituto Italiano di Tecnologia, Genoa, Italy

Abstract. Despite the recent achievements in the development of open surgery tools, preserving the haptic capabilities during drilling tasks is still an open issue. In this paper, we propose a novel tool for hand-held drills composed of a cover for force sensing and a haptic display for force feedback. A pneumatic device has been developed to estimate the contact force occurring during the interaction between drill bit and bones. A performance comparison with a precise commercial force sensor proved the reliability of the measurements. A haptic ring is in charge of providing cutaneous sensations helping the surgeon in performing the task. The effectiveness of our method has been confirmed by experimental results and supported by statistical analysis.

1 Introduction

Technological advancements in surgical tools have expanded the field of robot-assisted surgery to newer specialties. Even if the achievements in the last years have been impressive, current robotic surgical systems are still limited by the lack of haptic feedback. It has been proved that restoring the haptic capability in robotic surgery contributes to improve accuracy and safety in performing complex and delicate surgical tasks [9]. On the other hand, also open surgery may suffer from a reduction of tactile perception. In fact, even if in these procedures surgeons directly interact with the patient's body, some surgical tools, *e.g.* drills, may limit the haptic perception. As a matter of fact, a common issue in surgical drilling is that vibrations generated by the tool affect the perception of the surgeon, reducing, for instance, the capability in discerning different tissues and detecting the break-through force [5].

In this paper, we focus on otologic procedures, where a precise control of the surgical drill is required because the critical anatomy within the middle ear,

I. Nisky et al. (Eds.): EuroHaptics 2020, LNCS 12272, pp. 398–406, 2020.
https://doi.org/10.1007/978-3-030-58147-3_44

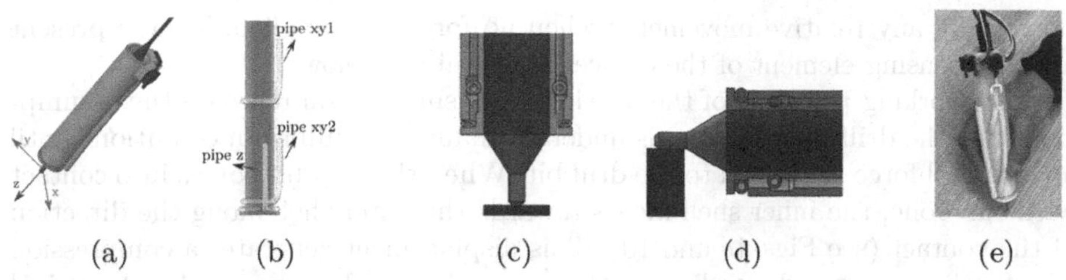

Fig. 1. The developed sensing system: (a) CAD model; (b) attachment of pipes to the inner shell; (c)(d) details of the sensing mechanism measuring perpendicular and tangential forces, respectively; (e) a user holding a surgical drill enriched with the sensing cover. Outer soft silicone pipes are covered by rigid housings to prevent the surgeon from touching them and affecting the measurements.

inner ear, and skull base can be accessed by drilling within the temporal bone for operations which demand high precision and accuracy [8]. Several researchers pointed out benefits of restoring haptic feedback in hand-held drilling procedures. In [1], force sensing has been elected as the appropriate way to obtain controlled penetration in the patient's body and automatic discrimination among layers of different tissues. Hessinger *et al.* integrated a thrust force sensor into the drill to enable high accuracy during pedicle screw positioning [3]. In the aforementioned works, the force measure is obtained integrating sensors into the tool mechanism. In this paper, we propose a pneumatic method to measure the contact force between the drill bit and the bone without modifying the internal structure of the tool. The aim is to create an instrumented cover that can be easily customized and adapted to the off-the-shelf hand-held drills. Moreover, the proposed system is capable of rendering the force feedback to the user by means of a haptic ring. To the best of our knowledge, this represents the first attempt to assist a surgeon with haptic feedback in open surgery without modifying the existing equipment.

2 Design of the Pneumatic Force Sensor

Working Principle

Measuring contact forces between the drill bit and a surface without modifying existing tools encounters several non-trivial challenges. Because of the drilling task, the sensing system has to be placed far from the contact point. Common precise and accurate 3-axis force sensors are bulky and not suitable for small devices. Thus, we developed a pneumatic system capable of estimating forces using pipes and air pressure sensors. The great advantage of using a pneumatic system is that it is lightweight, tiny, and measurement information is transferred by means of a gas to the sensors, which can be located out of the operational workspace. We exploit a sensing structure consisting of two concentric cylindrical shells separated by a gap, as shown in Fig. 1a. The inner shell (Fig. 1b) is rigidly attached to the body of the drill, while the outer shell is the one held by the surgeon. Soft silicone pipes are placed between the two shells to fill the gap, preventing any relative movements when no forces are applied. They represent also the sensing element of the device, as explained below.

The working principle of the developed sensing system relies on the assumption that the drill-hand system is under mechanical equilibrium conditions until an external force is applied to the drill bit. When the drill bit comes into contact with the bone, the inner shell moves towards the outer shell along the direction of the contact (see Figs. 1c and 1d). This displacement generates a compression of soft silicone pipes depending on the external force. In the considered surgical scenarios, torque components are treated as negligible, because the drill bit does not enter deeply into the bone generating significant values of torque. It is worth noting that the grasp squeezing forces applied by the surgeon do not cause any structural deformation of the pipes thanks to the high stiffness of the external shell. Then, the only deformations are due to the forces applied to the drill bit. To estimate the entity of the compression, the increase of pipes internal pressure is measured by means of air pressure sensors placed outside the outer shell. Pipes are not additionally inflated: at the steady state they are at the equilibrium with the external air pressure.

Hardware Implementation
As depicted in Fig. 1e, we present a proof of concept in which the pneumatic force sensor is composed of two 3D-printed parts made of ABSPlus (Stratasys Inc., USA), soft silicone pipes (ID 2.5 mm, OD 3.5 mm), and two 2 kPa differential pressure sensors (MPXV7002DP, NXP Semiconductors, NL) measuring forces along the z-axis and on the xy-plane, respectively (as in the reference system of Fig. 1a). The total weight of the cover is 51 g. Each pipe is ring-shaped and firmly attached to the inner shell as shown in Fig. 1b, with one end leak-proof sealed. The opposite side of the pipe conveys the pneumatic information to the pressure sensors. Forces on the xy-plane are measured by a sensor connected to *pipes xy*. Two pipes spaced along the length of the drill are required to prevent possible relative movements between the two shells when no forces are applied. It is important to notice that each pipe goes out of the outer shell as soon as the loop close, through a small hole. In this way, the measurement resolution is the same in all the directions of the xy-plane. Both the shells contain grooves for enclosing the pipes so as to reduce the width of the device. In this way, the gap between the shells corresponds to the internal diameter of the pipes. In the outer shell, the cavity is vertically extended so as not to detect pressure variations when the only force involved is along the z-axis. Taking advantage of the flange on the lower part of the shell, it is possible to measure with a second pressure sensor the forces exerted on the z-axis by means of *pipe z* (see Fig. 1b). The effect of gravitational force has been considered negligible with respect to the forces at work, due to the lightweight of the drill. Moreover, we supposed that in a such accurate task the surgeon compensates the weight by his hands and the contact force is equivalent to the force exerted by the surgeon. Vibrations generated by the drill are filtered by means of a hardware R-C filter with an experimentally selected cut-off frequency of 144.68 Hz. To further isolate sensors from vibrations, two tiny sponge layers are placed under the sensor housings.

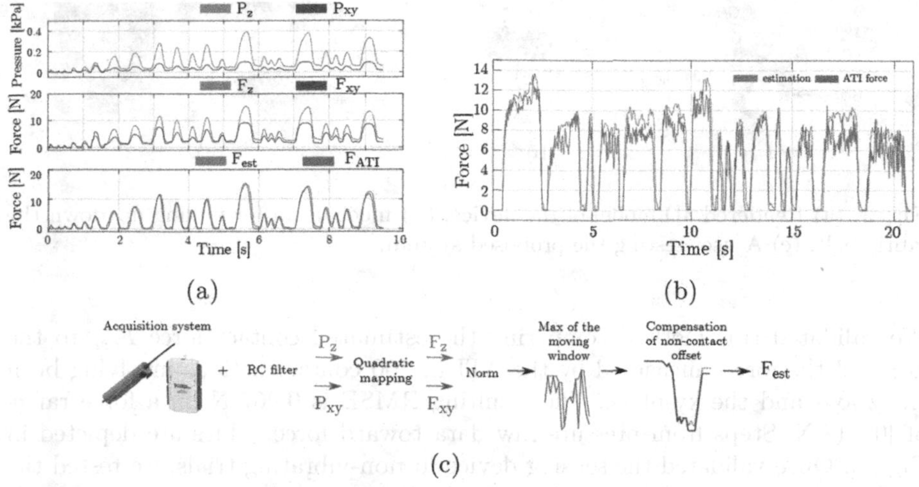

Fig. 2. Force estimation. In (a) raw pressure values, the correspondent forces, and the comparison between the final force estimation and the ground-truth value are reported. Values obtained in a representative trial, affected by vibrational noise, are in (b). In (c), the steps of the force estimation are detailed.

Analog data from the sensors are acquired using a NI USB-6218 DAQ. Finally, a software algorithm, described below, processes the signal.

Software Implementation
A calibration procedure is required to correctly transform data from pressure sensors into forces. The initial calibration and the subsequent validation phase are performed with the drill switched off. A high precision ATI Gamma F/T sensor (ATI Industrial Automation, USA) is used to identify the force-pressure relation. A separate procedure is required for calibrating the two sensors. For what concerns the z-axis calibration, the drill was vertically pushed toward the ATI sensor for a total of 50 contact actions, so that the generated force deforms only *pipe z*, as depicted in Fig. 1c. In accordance with [2], data gathered from the pressure sensor and the ATI were quadratically interpolated using a Matlab© algorithm. The same procedure was repeated for the xy-plane, keeping the drill horizontally (Fig. 1d). For the tool exploited in this work, the two found relations are: $F_z = 5034{\cdot}P_z - 2280{\cdot}P_z^2$ and $F_{xy} = 9542{\cdot}P_{xy} - 1017{\cdot}P_{xy}^2$, being P_{xy} and P_z the pressure values. As a final step, the norm of F_z and F_{xy} is computed. Indeed, from the user point of view, there is no need to distinguish the three components of the force: the surgeon just needs to have a feedback on the total force exerted, which corresponds to the norm of F_z and F_{xy}. As noticeable in Fig. 2a, this value is affected by the relatively slow dynamic of the pneumatic system. After breaking contact, the pressure of the pipes does not immediately reach the initial zero-value. A rapid decrease of the pressure is followed by a slow downwelling. Thus, we introduced a compensation algorithm which brings to zero the *"non-contact offset"*, identified as a flat trend after a significant negative slope.

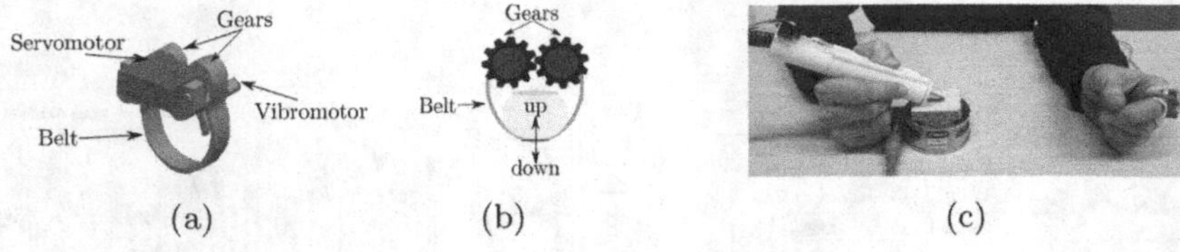

Fig. 3. (a) Rendered 3D model of the device; (b) mechanism for pulling up/down the fabric belt. (c) A user testing the proposed system.

We validated this method comparing the estimated contact force F_{est} to the norm of the forces measured by the ATI in 200 contact actions involving both the z-axis and the xy-plane. The resulting RMSE is 0.967 N, in a force range of $[0-18]$N. Steps from pressure raw data toward force values are depicted in Fig. 2a. Once validated the sensing device in non-vibrating trials, we tested the proposed system switching on the drill. Adding vibrations introduces a significant modification in the force profile, as depicted in Fig. 2b. To compensate this negative effect, a software filter has been implemented. The filtered force value corresponds to the maximum value in a moving window of 33 ms. In this way, the downward peaks are ignored guaranteeing a safer overestimation. This implies that it is not possible to compare the filtered force estimation with the measurements of the ATI in drilling tasks. The duration of the moving window was selected to obtain the best compromise between filter performance and response delay. All the steps of the force estimation are reported in Fig. 2c.

3 Force Feedback

Contextually with the force sensing system, we developed a haptic ring capable of generating cutaneous and vibrotactile force feedback. To have a lightweight device with a limited encumbrance, we employed a single servo-motor (HS-35HD Ultra Nano, HITEC Inc., USA) controlling a flexible belt for generating cutaneous stimuli and an eccentric-mass motor (EMM) to generate vibrotactile stimuli [7]. The device is controlled by the same DAQ board used for sensors data acquisition through an ad-hoc library. The servo motor generates the rotation of a master gear that moves a slave gear. Such mechanism results in opposite spinning directions of the gears, that translate the belt along the vertical axis. The workings are depicted in Figs. 3a and 3b. The maximum range of the belt motion in the vertical direction is 23 mm and it depends on the external diameter of the gears (11 mm), the length of the belt (95 mm), and the maximum rotation range of the servo motor (120°). We selected these values considering that also fingertips bigger than the average should fit. The maximum exploitable displacement range for force generation is 6 mm, so that the device can apply a maximum force of 3 N considering a stiffness of 0.5 N/m as elastic behavior of the finger pulp. Interested readers are referred to [6] and [4] for further details on the force feedback generation. A manual calibration is performed for each participant to adjust the initial position of the belt. The vibrational motor is placed horizontally alongside the device. It generates vibrations (1 g at 3.6 V) to notify the force threshold over-reaching.

4 Experimental Validation

The experimental evaluation was carried out with a twofold aim: *i*) demonstrating the effectiveness of the haptic feedback in the aforementioned surgery and *ii*) identifying the best feedback approach. Ten users (6 males, age 23–56, all right handed) took part in the experiment. One was a surgeon with many years of experience, three were medical students with 5 years of experience, while the remaining six were medical students with lower/no experience in performing open surgical procedures.

The experiment aimed at simulating a cochlear implant surgery. Participants were asked to completely remove a blue colored rectangle (0.6 cm × 2.0 cm) from a piece of plywood using the instrumented drill (rotating at 15.000 rpm). Users wore the haptic ring on the left hand (see Fig. 3c), where a clear perception of the haptic feedback is allowed by the absence of vibrations. Participants were told that the task was considered successfully accomplished when the drilling force was maintained in a specific range, *i.e.* [0–7.5]N, without overreaching the limit time of 13 s. The time limit has been introduced to prevent subjects from being excessively slow in order to completely remove the blue color using low forces. Three feedback conditions were evaluated: *i)* no feedback *(N)*; *ii)* vibratory *(V)* alert in case of exceeding the force threshold; *iii)* vibratory alert and cutaneous feedback proportional to the exerted force *(C)*. A proportional scale factor was used to map the measured maximum force into the admissible range of the ring. Each user performed a set of three trials per each feedback condition (pseudo-randomly selected), resulting in a total of 9 trials. Time to complete the task and impulse (the integral of the force out of the boundaries over the time interval) were considered as metrics for evaluating the task performance. A familiarization period of 2 min was provided to acquaint participants with the system. In this phase, users tested the overreaching of the force limit with and without haptics. In the first case, the exerted force was displayed by the haptic ring, in the latter by a graphical indicator on a LCD screen.

4.1 Results and Discussion

Data collected in the experimental phase were analyzed by means of statistical tests. For each participant completion time and impulse were computed (see Figs. 4b and 4c). All the participants were able to completely remove the blue color in the time limit of 13 s both with and without the haptic feedback. The average completion time among all the trials is 10.85 ± 2.54 s, 10.23 ± 1.72 s, 9.32 ± 1.16 s for *N*, *V*, and *C* feedback conditions, respectively. Statistical analysis revealed that there is no statistically significant difference in the completion time of the task using different feedback. As shown also in Fig. 4a-upper, the

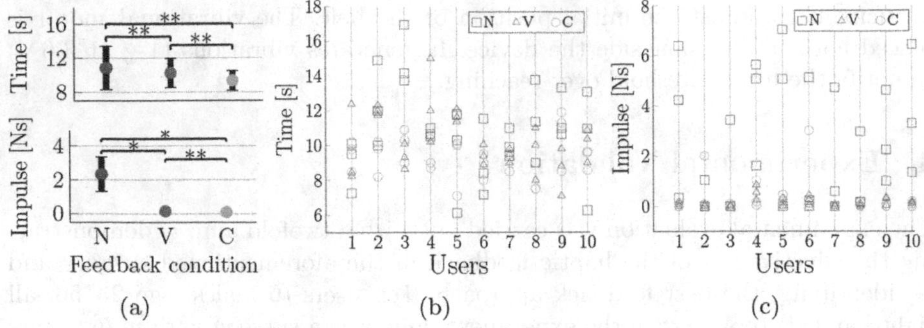

Fig. 4. In (a) mean and 95% CI for all the feedback conditions are reported for time (upper panel) and impulse of the force over the threshold (lower panel). The *p*-values are reported on top of the error bars, $**$ and $*$ indicate $p > 0.05$ and $p < 0.0005$, respectively. In (b) and (c) users' results for each trial are shown.

task execution is not slowed down by the increasing number of stimuli to be focused on. Concerning the impulse of the force over the threshold, participant exceeded the limit with 2.33 ± 1.25 Ns, 0.15 ± 0.16 Ns, 0.10 ± 0.13 Ns testing the setup with *N*, *V*, and *C* feedback, respectively. Moreover, a one-way repeated measures ANOVA was conducted to determine whether there were statistically significant differences in impulse over the different feedback. There were no outliers. Data were transformed using the squareroot transformation and passed the ShapiroWilk normality test ($p > 0.05$). The assumption of sphericity was violated, as assessed by Mauchly's test ($\chi^2(2) = 7.87$, $p < 0.05$). Therefore, a Greenhouse-Geisser correction was applied. The results of the test (reported in Fig. 4a-lower) assessed that the feedback modality elicited statistically significant changes in over-applied forces ($p < 0.0005$). Post hoc analysis with Bonferroni adjustment revealed that the reduction of impulse was statistically significant. More in detail, the test revealed that there is statistically significant difference in performing the task with or without haptic feedback. In case of feedback, the impulse error had a almost complete reduction, which implies a more controlled penetration in the plywood. For what concerns the difference between the two haptic feedback, the difference is lower, but not statistically significant. Supported by the outcomes of the statistical analysis, we can affirm that haptic feedback can enhance the safety in surgical hand-held drilling tasks, maintaining the drilling force in a specific range. In addition, participants to the experimental campaign reported positive qualitative feedback on the haptic-assisted experience and on the positioning of the ring in the contralateral side. They argued that the cover did not interfere with the task and it would be useful to introduce the device in real surgical procedures, after appropriate refinements.

5 Conclusion and Future Work

In this work, we presented a novel approach to measure the force exerted on bones during drilling tasks in open surgery. A pneumatic sensing cover for drills and a haptic ring to reproduce such forces were developed. We tested our sensing device in a comparison with a high-resolution/accuracy commercial force sensor, demonstrating the robustness of our approach. To properly reconstruct the force profile, we implemented both a hardware and a software filters. The advantage of our sensing system is that it can be easily adapted to any surgical drills, changing only few design parameters in the CAD model (e.g. introducing some shims to modify only the internal profile of the inner shell). The resolution and the range of our sensor are customizable: they can be modified changing the silicone pipes and using pressure sensors with different resolution.

We evaluated the effectiveness of our haptic-assisted hand-held drill with long experience and novice surgeons. Forces and vibrations were exploited to help the users in evaluating the real exerted forces. We compared the performance of the participants with and without haptic feedback, proving that haptic enhancement outperformed the haptic-free technique.

The presented results pave the way for numerous interesting research directions that will be the subject for future works. Different feedback policies and locations will be tested in a future experimental campaign. Additional metrics, such as tissue discrimination, will be considered to evaluate device and feedback. Learning curves in performing the task with and without feedback will be evaluated in a more careful future study, involving a larger sample. In further developments, the cover can be instrumented with additional sensors (e.g. an accelerometer) to measure the inclination of the drill and improve the calibration procedure. Finally, ergonomic studies will be taken into consideration.

References

1. Allotta, B., Giacalone, G., Rinaldi, L.: A hand-held drilling tool for orthopedic surgery. IEEE/ASME Trans. Mech. **2**(4), 218–229 (1997)
2. Gaudeni, C., Meli, L., Prattichizzo, D.: A novel pneumatic force sensor for robot-assisted surgery. In: Prattichizzo, D., Shinoda, H., Tan, H.Z., Ruffaldi, E., Frisoli, A. (eds.) EuroHaptics 2018. LNCS, vol. 10894, pp. 587–599. Springer, Cham (2018)
3. Hessinger, M., Hielscher, J., Pott, P.P., Werthschützky, R.: Handheld surgical drill with integrated thrust force recognition. In: Proceedings of IEEE International Conference on E-Health and Bioengineering, pp. 1–4 (2013)
4. Lisini Baldi, T., Scheggi, S., Meli, L., Mohammadi, M., Prattichizzo, D.: GESTO: a glove for enhanced sensing and touching based on inertial and magnetic sensors for hand tracking and cutaneous feedback. IEEE Trans. Human-Mach. Syst. **47**(6), 1066–1076 (2017)
5. Louredo, M., Diaz, I., Gil, J.J.: DRIBON: a mechatronic bone drilling tool. Mechatronics **22**(8), 1060–1066 (2012)
6. Park, K.H., Kim, B.H., Hirai, S.: Development of a soft-fingertip and its modeling based on force distribution. In: Proceedings of the IEEE International Conference on Robotics and Automation, vol. 3, pp. 3169–3174 (2003)

7. Precision Microdrives: Model No. 304–002 4mm Vibration Motor - 8mm Type Datasheet. https://www.precisionmicrodrives.com/product/datasheet/304-002-4mm-vibration-motor-8mm-type-datasheet.pdf
8. Sang, H., Monfaredi, R., Wilson, E., Fooladi, H., Preciado, D., Cleary, K.: A new surgical drill instrument with force sensing and force feedback forrobotically assisted otologic surgery. J. Med. Dev. **11**(3) (2017)
9. Wagner, C.R., Stylopoulos, N., Howe, R.D.: The role of force feedback in surgery: analysis of blunt dissection. In: Proceedings of the IEEE Haptics Symposium, pp. 68–74 (2002)

Rendering Ultrasound Pressure Distribution on Hand Surface in Real-Time

Atsushi Matsubayashi(✉), Yasutoshi Makino, and Hiroyuki Shinoda

The University of Tokyo, Tokyo, Japan
Matsubayashi@hapis.k.u-tokyo.ac.jp

Abstract. In this paper, we propose a method for rendering the pressure distribution on the skin surface of a hand in real-time using an ultrasonic phased array. Our method generates a polygon mesh model representing the hand shape by fitting a rigid template to the point cloud captured by depth sensors. Obtaining the entire hand shape as a mesh model enables to solve scattering problem to generate a precise distribution on the hand surface. Therefor, for example, our method can control the width of the distribution on the fingertip according to the size of the contact area with the virtual object. We have experimentally verified that considering the scattering on the mesh model contributes to accurate pressure pattern reproduction.

Keywords: Mid-air haptics · Scattering problem · Airborne ultrasound

1 Introduction

Ultrasound haptics is a technology to generate a tactile sensation on a human skin by creating a point with high sound pressure using an ultrasound phased array. Since a tactile presentation by airborne ultrasound was first demonstrated in 2008 [9], many attempts have been made to create haptic images in the air using this technology. In order to create a pressure distribution with the desired shape in the air, many methods solve the inverse problem using the relationship between the sound pressure of control points and the complex amplitudes of the transducers [4,6,10]. Since the intensity of a stimulus felt on the skin depends only on the amplitude of the sound pressure, the phase of the target sound pressure distribution can be set arbitrarily in the inverse problem. Properly setting this phase by solving eigenproblem [10] or phase retrieval problem [6] widens the range of reproducible haptic image. Some method, furthermore, present a stronger stimulus by tracking the hand and generating a pressure distribution only on the contact area of the hand touching the image, allowing the user to identify the shape of the haptic image more clearly [10,11]. However, since these above methods do not consider scattering on the hand surface, the pressure

I. Nisky et al. (Eds.): EuroHaptics 2020, LNCS 12272, pp. 407–415, 2020.
https://doi.org/10.1007/978-3-030-58147-3_45

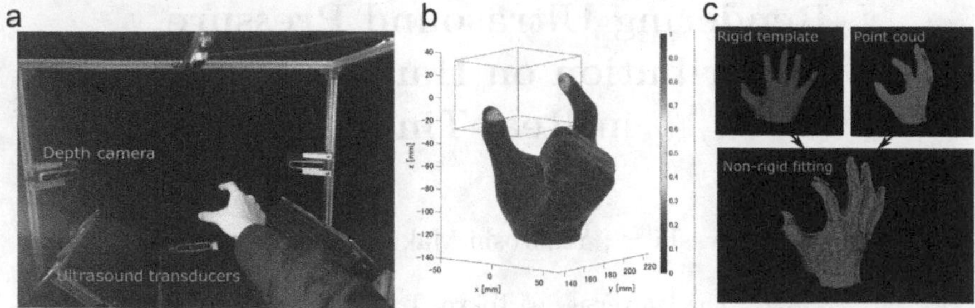

Fig. 1. a, b) The phase of the ultrasound transducers is determined according to the hand mesh model generated from the depth information acquired by depth cameras, and the desired pressure distribution is generated at the fingertip. c) By fitting the template rigid model to the point cloud, a non-rigid mesh model is dynamically generated in real-time.

distribution actually generated on the skin surface differs from target pressure distribution to be reproduced. Solving the scattering problem will enable more accurate reconstruction of the pressure pattern. Inoue et al. have proposed a method for generating a stronger focal point by considering the scattering on a polygon mesh model of a finger [7]. They demonstrated that a stronger ultrasound focus could be created by solving a scattering problem using a static mesh model, but it has not been possible to control the pressure distribution on the mesh model dynamically generated in real-time.

In this paper, we propose a method to render the pressure distribution in real-time on the polygon mesh model deforming according to the hand shape. In this method, the shape of the hand placed above a transducer array is aquired with multiple cameras as shown in Fig. 1 (a), and the hand polygon mesh model deforms non-rigidly to fit this shape. The scattering model formulates the relationship between a sound pressure pattern on the mesh model and phases of the ultrasound transducers in the style of the boundary element method. Based on this relationship, our method optimize both phases of transducers and phases of target pressure distribution to generate the desired pattern at any position on the hand surface. For example, as shown in Fig. 1 (b) our method can generate a distribution according to the size of the contacting region with a virtual object.

2 Method

2.1 Generation of Mesh Model

In order to dynamically generate a polygon mesh model of the hand, we used a mesh reconstruction technique similar to that proposed by Zollhöfer et al. [12]. Mesh reconstruction process consists of two phases. First, a rigid template is created by scanning the hand with a fixed form using multiple depth cameras. Then, as shown in Fig. 1 (c), a non-rigid mesh model is generated by fitting the rigid template to the point cloud obtained from the depth cameras. We used the fitting method proposed by Dou et al. [5]. which has the advantage of being able to generate a mesh model that closely matches the skin surface of the actual hand compared to skeletal hand tracking methods used, for instance, in Leap Motion [2]. This is an important property for controlling the pressure distribution on the hand surface.

2.2 Scattering Model of Hand Surface

The relationship between the sound pressure on the faces of the mesh model and the phases of the ultrasound transducers is formulated in style of the boundary element method, similar to Inoue's adaptive focusing method [7]. The sound pressure $p(\boldsymbol{r}) \in \mathbb{C}$ scattered on the smooth surface Ω of a sound-hard rigid body is given by the following boundary integral Eq. [3].

$$\frac{1}{2}p(\boldsymbol{r}) = p_{inc}(\boldsymbol{r}) - \int_{\Omega} p(\boldsymbol{r}) \frac{\partial g(\boldsymbol{r}, \boldsymbol{s})}{\partial \boldsymbol{n}} dS, \tag{1}$$

where g is the Helmholtz green function, and p_{inc} is the incident wave from the transducers. In our method, this is simplified by a spherical wave with directivity D_n as

$$p_{inc}(\boldsymbol{r}) = \sum_{n} D_n(\boldsymbol{r}) \frac{e^{-jk\|\boldsymbol{r}-\boldsymbol{x}_n\|}}{\|\boldsymbol{r} - \boldsymbol{x}_n\|} a_n e^{\phi_n}, \tag{2}$$

where $a_n \in \mathbb{R}, \phi_n \in \mathbb{R}$ and $\boldsymbol{x}_n \in \mathbb{R}^3$ are the amplitude, phase and position of a transducer $n \in \{1, \cdots N\}$ respectively. To reduce computational cost of the optimization, we set the amplitude constant, i.e. $a_n = a$.

When the boundary surface is represented by a polygon mesh and the sound pressure and the gradient of the Green's function are approximated to be constant on each face of the mesh, the boundary integral equation (1) is discretized as follows:

$$B \left(p_1, \cdots, p_M\right)^{\mathrm{T}} = G \left(e^{\phi_1}, \cdots, e^{\phi_N}\right)^{\mathrm{T}}, \tag{3}$$

where

$$B = \begin{pmatrix} \frac{\partial g}{\partial n}(\boldsymbol{y}_1, \boldsymbol{y}_1)A_1 + \frac{1}{2} & \cdots & \frac{\partial g}{\partial n}(\boldsymbol{y}_1, \boldsymbol{y}_M)A_M \\ \vdots & \ddots & \vdots \\ \frac{\partial g}{\partial n}(\boldsymbol{y}_M, \boldsymbol{y}_1)A_1 & \cdots & \frac{\partial g}{\partial n}(\boldsymbol{y}_M, \boldsymbol{y}_M)A_M + \frac{1}{2} \end{pmatrix}, \tag{4}$$

$$G = \begin{pmatrix} D_1(\boldsymbol{y}_1) \frac{ae^{-jk\|\boldsymbol{y}_1 - \boldsymbol{x}_1\|}}{\|\boldsymbol{y}_1 - \boldsymbol{x}_1\|} & \cdots & D_N(\boldsymbol{y}_1) \frac{ae^{-jk\|\boldsymbol{y}_1 - \boldsymbol{x}_N\|}}{\|\boldsymbol{y}_1 - \boldsymbol{x}_N\|} \\ \vdots & \ddots & \vdots \\ D_1(\boldsymbol{y}_M) \frac{ae^{-jk\|\boldsymbol{y}_M - \boldsymbol{x}_1\|}}{\|\boldsymbol{y}_M - \boldsymbol{x}_1\|} & \cdots & D_N(\boldsymbol{y}_M) \frac{ae^{-jk\|\boldsymbol{y}_M - \boldsymbol{x}_N\|}}{\|\boldsymbol{y}_M - \boldsymbol{x}_N\|} \end{pmatrix}, \tag{5}$$

and $p_m \in \mathbb{C}, \boldsymbol{y}_m \in \mathbb{R}^3$ and $A_m \in \mathbb{R}$ is the sound pressure, position, and area of a face $m \in \{1, \cdots, M\}$ of the mesh model respectively.

2.3 Optimizing Phases of Ultrasound Transducers

Given the target sound pressure amplitude $\boldsymbol{p}' = (p'_1, \cdots, p'_M)^{\mathrm{T}} \in \mathbb{R}^M$, we want to determine the phases of the transducers $\boldsymbol{\phi} = (\phi_1, \cdots, \phi_N)^{\mathrm{T}} \in \mathbb{R}^N$ and phases of the sound pressure $\boldsymbol{\theta} = (\theta_1, \cdots, \theta_M)^{\mathrm{T}} \in \mathbb{R}^M$ that minimize the least square error $\| \left(p'_1 e^{\theta_1}, \cdots, p'_M e^{\theta_M}\right)^{\mathrm{T}} - B^{-1} G \left(e^{\phi_1}, \cdots, e^{\phi_N}\right)^{\mathrm{T}} \|_2^2$. However, calculating the inverse of B is very computationally expensive, so we solve the following optimization problem instead (Fig. 2).

$$\min_{\boldsymbol{\phi}, \boldsymbol{\theta}} \|B \left(p'_1 e^{\theta_1}, \cdots, p'_M e^{\theta_M}\right)^{\mathrm{T}} - G \left(e^{\phi_1}, \cdots, e^{\phi_N}\right)^{\mathrm{T}} \|_2^2. \tag{6}$$

We solve this problem iteratively using the Levenberg-Marquardt method. At each iteration, parameters $\boldsymbol{t} = (\phi_1, \cdots, \phi_N, \theta_1, \cdots, \theta_M)^{\mathrm{T}}$ is updated as follows:

$$\boldsymbol{t} \leftarrow \boldsymbol{t} - (J^{\mathrm{T}} J + \lambda I)^{-1} J^{\mathrm{T}} \boldsymbol{f}, \tag{7}$$

where the residual vector $\boldsymbol{f} = B \left(p'_1 e^{\theta_1}, \cdots, p'_M e^{\theta_M}\right)^{\mathrm{T}} - G \left(e^{\phi_1}, \cdots, e^{\phi_N}\right)^{\mathrm{T}}$, and J is the Jacobian of $\boldsymbol{f}$. If the number of parameters $M + N$ is large, the time taken for an iteration will be very long, but when pressure is generated only in a local part such as a fingertip, excluding zero pressure faces can save computation time.

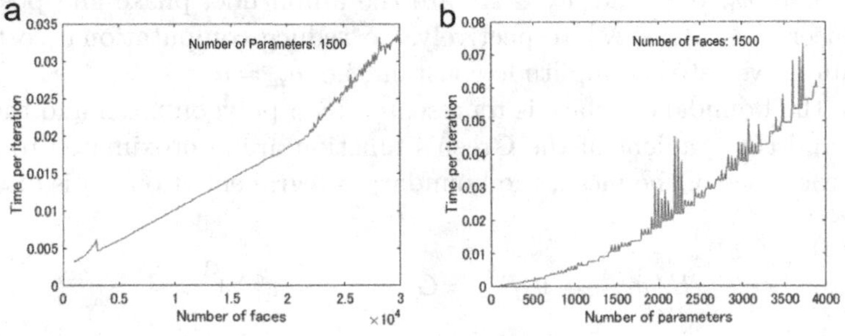

Fig. 2. a) The time per iteration against the number of faces. b) the time per iteration against the number of parameters

3 Implementation

The above algorithms were implemented with CUDA on two GeForce RTX 2080 Ti GPUs. One is used for mesh generation algorithm and the other is used for phase optimization. We measured the time taken for one iteration of the phase optimization in this environment. Figure 3 shows the time per iteration against the number of faces and parameters. The number of parameters is the sum of the number of transducers and the number of faces with non-zero pressure. In consideration of this result, possible resolution of the pressure distribution and

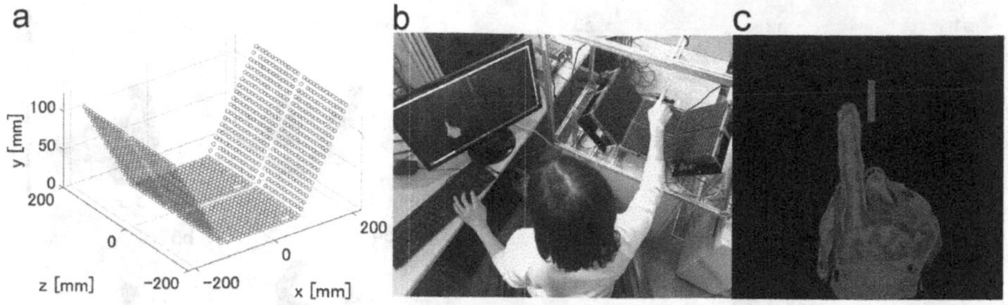

Fig. 3. a) The coordinate system and the arrangement of the ultrasound transducers in the experimental setup. b,c) The participants sat in front of the system and touched the box checking the position of the hand and box displayed on a LCD.

the limitations of the devices used, we set the number of faces to about 10,000 and the number of transducers to 1496. Therefore the time per iteration is about 10 ms, so we set the update frequency of the phase to 20 Hz with five iteration.

We constructed an experimental setup as shown in Fig. 1 (a). We installed Intel RealSense Depth Camera D415 [1] to measure the hand. The resolution of the depth image captured by each camera is 640 × 360, and the refresh rate as well as the update frequency of the mesh generation is 30 Hz. The architecture of the ultrasound transducer array unit is that proposed by Inoue et al [8]. The resonant frequency of the transducer is 40 kHz, and 200 Hz amplitude modulation is applied to make the tactile stimulus easier to perceive. Figure 3 (a) shows the coordinate system and arrangement of the ultrasound transducers in our setup.

4 Numerical Analysis

We performed numerical simulation to verify how close the distribution could be to the target in the experimental setup.

Figure 4 shows the simulation results of our method. In the target distribution (a1-a3), a constant pressure is applied to faces inside a box-shaped region. The width of the box is 5 mm, 8 mm and 11 mm in a1, a2 and a3 respectively. It can be seen that the distribution generated by our method (b1-b3) is close to the target and changes according to the width of the box. Figure 4 (c1-c3) shows the simulation result of the phase optimization performed without consideration of the scattering, which means replacing matrix B in our algorithm with the identity matrix. The result suggest an appropriate distribution cannot be generated without considering scattering. Our method can be applied to the case of touching with multiple fingers as well. The Fig. 4 (a4, b4 and c4) shows the simulation result of the case where a virtual box is grasped.

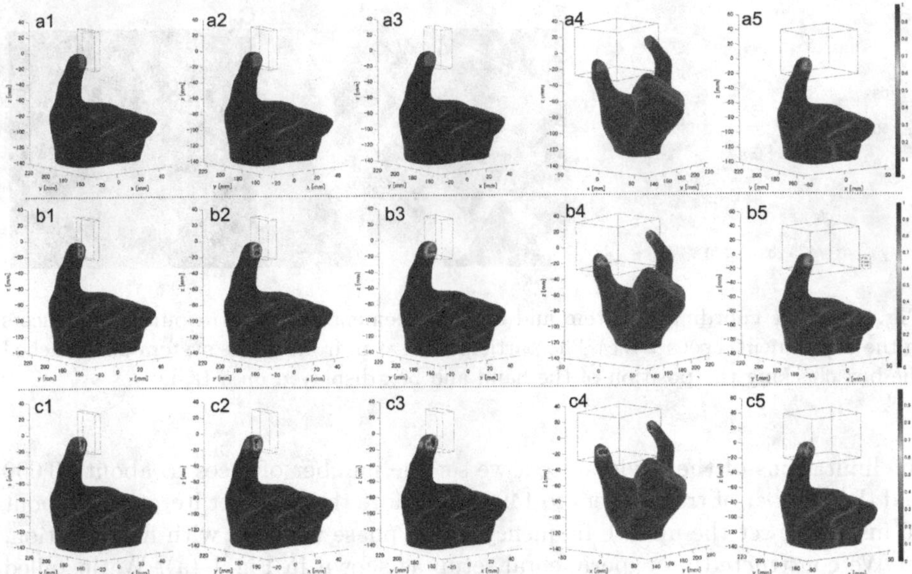

Fig. 4. Simulation results. a1-a5) Target pressure distribution. b1-b5) Distribution reproduced by our method. c1–c5) Distribution reproduced without scattering model.

When actually touching an object, the pressure on the contact area is not uniform, but greater toward the center. Figure 4 (a5, b5 and c5) shows the simulation result of a simplified model that a strong pressure is presented proportional to the penetration distance into the box. Although the shape is somewhat deformed, target distribution is reproduced. In these simulations, 30 iterations were performed in the phase optimization with the initial value as the zero vector. However, it has been empirically known that by setting the phases of the previous frame to the initial value, convergence can be sufficiently achieved in 5 to 10 iterations.

5 User Study

To verify if our method can generate a discernible difference in the pressure distribution, we conducted a user study. In this study, participants performed tasks of touching and identifying three types of distribution. We compared the accuracy of identification between the two methods. One is our method using the scattering model, and the other is the method without considering scattering.

Procedure. The participants sat in front of the system and placed his hand above the transducer array. For the stability of mesh generation, form of the hand was limited to only the index finger up throughout the experiment as shown in Fig. 3 (b). Then, the participant's hand was scanned and a rigid template was

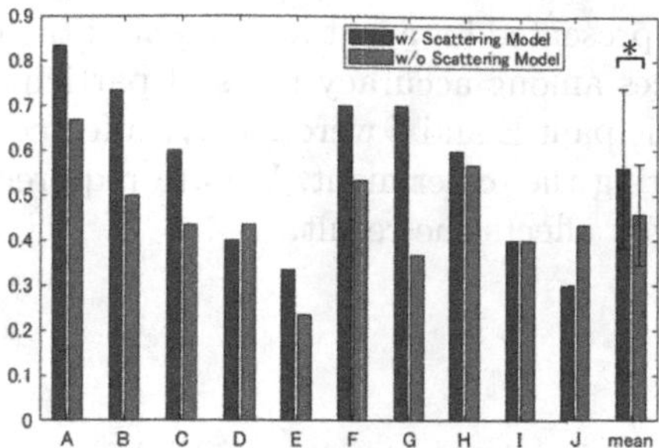

Fig. 5. Accuracy rates of the participants and the corresponding mean value.

created, which takes about 20 sec. After confirming that the mesh generation was working properly, the participants were asked to experience three different widths of pressure pattern for 15 sec each. As in the simulation, a uniform pressure is applied to the part that enters a box-shaped region. See the Fig. 4 (b1-b3) for the size of the region. The participants were not allowed to move their fingers horizontally to feel the width, but only to move vertically checking their hand and the box-shaped region displayed on a LCD as shown in the Fig. 3 (c). Then, the participants were asked to repeatedly perform the tasks to identify the width. After 15 s of touching, the tactile presentation was stopped, and participants answered one of three widths. Participants performed three task as a practice and then performed 10 tasks for each width (total 30 tasks). In either case, no answer was taught to the participants. The order of the tasks is randomized. The above process was done separately under two methods to avoid confusion between methods. Five of the ten participants performed the experiment with scattering model and the other five performed the experiment without scattering model.

Result and Discussion. Ten participants (eight males and two females), aged between 23 to 24, took part in the experiment. Figure 5 shows the accuracy rate of the participants. The participant A to E performed the experiment with scattering model first and the participant F to J performed the experiment without scattering model first. The mean value of the accuracy rates among participants was 0.56 with scattering model and 0.47 without scattering model, and the Wilcoxon signed-rank test yielded a significant difference ($p < 0.05$) between the two methods. The mean accuracy indicates that the differences in the distributions generated by our method are discernable to some extent, but not perfect. One of the reasons for this may be that an accurate mesh model could not be generated due to the error of the depth camera. However, the difference in accuracy between the two methods suggests that it is effective to consider scattering even in such a case. Also, in this experiment, we did not give instructions on the appropriate speed of touching. Since there is a delay between capturing the hand and presenting the tactile sensation, an appropriate distribution cannot be presented for a fast movement of the finger. This may have led to a large differences among accuracy rates of participants. In particular, it is considered that participant E and J were greatly affected by the delay because their fingers shook during the experiment. We are required to verify how much delay there is and how it affects the result.

6 Conclusion

In this paper, we proposed and examined a method for rendering an ultrasound pressure distribution by solving the scattering problem. Although in the experiment, the pressure presentation was limited to the fingertip, our method can produce a pressure pattern on the entire hand surface, so there is still room for verification as to what kind of and how high the pressure distribution can be generated. At present, there is limitations on the temporal and spatial resolutions of the distribution due to computational cost, sensing accuracy, and transducer's resonant frequency. However, these problems will be solved with the advancement of the devices. One of our future work is the complete reproduction of the pressure distribution when touching a soft object using the presented approach.

References

1. Intel corporation (2018). https://www.intel.com/
2. Ultraleap ltd. (2020). https://www.ultraleap.com/
3. Bai, M.R., Ih, J.G., Benesty, J.: Acoustic Array Systems: Theory,implementation, and Application. Wiley, Hoboken (2013)
4. Carter, T., Seah, S.A., Long, B., Drinkwater, B., Subramanian, S.: UltraHaptics: multi-point mid-air haptic feedback for touch surfaces. In: Proceedings of the 26th Annual ACM Symposium on User Interface Software and Technology, pp. 505–514. ACM (2013)
5. Dou, M., et al.: Fusion4D: real-time performance capture of challenging scenes. ACM Trans. Graph. (TOG) **35**(4), 1–13 (2016)
6. Inoue, S., Makino, Y., Shinoda, H.: Active touch perception produced by airborne ultrasonic haptic hologram. In: Proceedings of the World Haptics Conference, pp. 362–367. IEEE (2015)
7. Inoue, S., Makino, Y., Shinoda, H.: Mid-air ultrasonic pressure control on skin by adaptive focusing. In: Bello, F., Kajimoto, H., Visell, Y. (eds.) EuroHaptics 2016. LNCS, vol. 9774, pp. 68–77. Springer, Cham (2016). https://doi.org/10.1007/978-3-319-42321-0_7
8. Inoue, S., Makino, Y., Shinoda, H.: Scalable architecture for airborne ultrasound tactile display. In: Hasegawa, S., Konyo, M., Kyung, K.-U., Nojima, T., Kajimoto, H. (eds.) AsiaHaptics 2016. LNEE, vol. 432, pp. 99–103. Springer, Singapore (2018). https://doi.org/10.1007/978-981-10-4157-0_17
9. Iwamoto, T., Tatezono, M., Shinoda, H.: Non-contact method for producing tactile sensation using airborne ultrasound. In: Ferre, M. (ed.) EuroHaptics 2008. LNCS, vol. 5024, pp. 504–513. Springer, Heidelberg (2008). https://doi.org/10.1007/978-3-540-69057-3_64

10. Long, B., Seah, S.A., Carter, T., Subramanian, S.: Rendering volumetric haptic shapes in mid-air using ultrasound. ACM Trans. Graph. **33**(6), 181:1–181:10 (2014)
11. Matsubayashi, A., Oikawa, H., Mizutani, S., Makino, Y., Shinoda, H.: Display of haptic shape using ultrasound pressure distribution forming cross-sectional shape. In: 2019 IEEE World Haptics Conference (WHC), pp. 419–424. IEEE (2019)
12. Zollhöfer, M., et al.: Real-time non-rigid reconstruction using an RGB-D camera. ACM Trans. Graph. (ToG) **33**(4), 1–12 (2014)

Energy Analysis of Lateral vs. Normal Vibration Modes for Ultrasonic Surface Haptic Devices

Diana Angelica Torres Guzman(✉), Betty Lemaire-Semail, Frederic Giraud, Christophe Giraud-Audine, and Michel Amberg

Univ. Lille, Arts et Metiers Institute of Technology, Centrale Lille, Yncrea Hauts de-France, L2EP – ULR 2697, 59000 Lille, France
diana.torres-guzman@univ-lille.fr

Abstract. In this paper, we propose a new device in order to produce normal and lateral ultrasonic vibrations in a plate, using an array of piezoelectric ceramics. This setup serves to continue the comparative analysis between the two vibration modes for tactile feedback rendering, by including an energetic characterization. With the help of a tribological analysis, this study will help to examine the energy performance of each vibration mode in terms of active power consumption against friction contrast (which is linked to perception). Using a simplified second order plate model, the energetic results are analyzed. The results show a better energy efficiency for the lateral vibration for low exploration speeds. The tribological analysis helps as well to evaluate the effect of frequency increase in terms of friction reduction vs. vibration amplitude for both vibration modes.

Keywords: Surface haptics · Ultrasonic · Out of plane vibrations · Lateral vibration · Friction reduction · Power measurement

1 Introduction

Haptic devices for texture discrimination utilize techniques to achieve friction modulation, since differences in friction may create a perception of texture [1]. In order to achieve this, ultrasonic vibration may be used to produce friction reduction on a surface, thus creating a sensation of 'smoothness' [2] (active lubrication). The amount of friction reduction is dependent on the vibration amplitude and frequency [3], and properties of the probing object [4, 5].

Generally, ultrasonic tactile feedback surfaces use out of plane vibration. However, lateral modes for friction reduction may also be interesting for several purposes, such as mechanical integration and noise reduction, and may be used additionally or in place of normal modes. The phenomenon of friction reduction with ultrasonic vibration has been thoroughly explored for 'out-of-plane' [6–9], or for combined vibration modes [10]. But the interaction mechanism through which friction is reduced with purely lateral vibration is less explored.

In [6], a simplified finger model, is used to explore the effects of lateral ultrasonic vibration on the grip function. In [7] this model is used to analyze the comparison

I. Nisky et al. (Eds.): EuroHaptics 2020, LNCS 12272, pp. 416–424, 2020.
https://doi.org/10.1007/978-3-030-58147-3_46

between lateral and normal vibration modes, by measuring the perception of friction contrast for each mode at a set of vibration amplitudes, for a group of participants. With the test conditions explained in [7], the psychophysical measures indicate that the large majority of subjects are more sensitive to normal rather than lateral vibration modulation for a given value of wave amplitude, up to about 3 $\mu m_{p\text{-}p}$, at frequencies around 30 kHz. The measurements have also shown that the finger exploration speed affects significantly the result with lateral vibration, requiring larger amplitudes at higher speeds to produce the same perception.

The energetic performance could also be a determinant factor in the mode choice for haptic devices. Indeed, if the haptic device is to be used with a battery, for example, a better energetic performance may help increasing operational time or reducing battery capacity specifications.

It is therefore interesting to evaluate the energetic requirement for a given friction contrast ($\Delta\mu/\mu$: where $\Delta\mu$ is equal to the friction coefficient without vibration (μ) minus the friction coefficient with vibration), for a given subject, since this value is related to perception [8]. In order to do so, a new device has been conceived and built. This device allows exciting either a normal or a lateral mode (at similar frequencies), with a same set of piezo-ceramics. Two experiments are then designed, for both vibration modes. In the first one, the active power required to reach a given vibration amplitude is measured, with and without load (finger). The second experiment consists of a tribology analysis, which serves to link the vibration amplitude with the friction modulation. Additionally, the tribology analysis will help studying the effect of increasing the frequency, in comparison with the results presented in [7], and knowing whether the exploration speed continues to influence the result.

Section 2 explains the conception and setup of the ultrasonic device. The lateral and normal modes created in this device are characterized and evaluated in terms of energy vs. vibration amplitude in Sect. 3, while Sect. 4.

2 Device Producing Lateral and Normal Modes on a Plate

A new ultrasonic device is designed to produce a pure lateral mode and a normal mode vibration on the same structure, with the same motion sources. The resonance frequencies of both modes must be close to each other without causing interference, and they must be higher than 30 kHz, which is the frequency already tested in [7], in order to explore the effect of increasing frequency on lateral haptic devices. For this reason, the design is made at about 60 kHz.

The conceived structure consists of an aluminum plate with twelve piezoelectric ceramics glued to the center of the plate on both sides (see Fig. 1), similar to the setup proposed in [11]. Ten of these ceramics serve as actuators, and two are sensors. A damp-proof polymeric material is glued to the top face of the aluminum beam on both sides. This design allows an exploration area through a length of about 3 cm between two nodes. The dimensions of the resonator are 128 mm × 30 mm × 1.94 mm, and the ceramics are 5 mm × 9 mm × 0.3 mm. The resonator is attached to an immobile section through a series of isthmus situated approximately at the vibrational nodes of both modes.

An optimization algorithm is performed in order to estimate the dimensions which may minimize the difference between the resonance frequencies of the two modes, without having them interfere with each other, as they are meant to be comparable but excited independently. It is important to mention that the device is not created to be optimal in terms of energy consumption, as in [11, 12], but built in such a way as to render the two modes comparable. Once these dimensions are estimated, a modal analysis is performed using finite element analysis. The results of the analysis will help calculating in precision the geometry of the device and the placement of the piezo-ceramics.

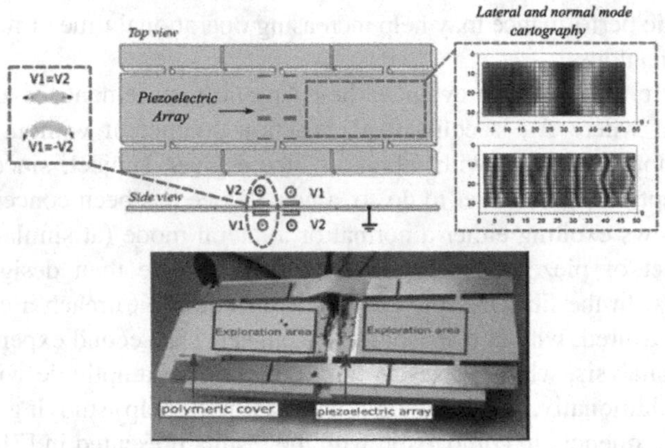

Fig. 1. Plate design and setup to perform mode comparison on the same surface haptic device. At top right, the cartography of lateral (up) and normal (down) modes on one side. The cartography confirms that the modes are almost completely pure. The scanned area does not include the portion with the ceramics, which is why nodal lines look asymmetrical.

The 'top' and 'bottom' side ceramics are connected to different voltage sources (V1 and V2 on Fig. 1), with the common ground connected to the conductive plate. All ceramics deform in a d_{31} mode (stretch-compress), creating a surface tension on the aluminum surface. When V1 = V2, a symmetrical stretch-compress deformation is produced simultaneously on both sides of the aluminum plate, thus inducing the lateral mode. When V1 = −V2, one surface will be stretched, while the other compressed, 'bending' the material, thus inducing the normal mode. This method allows creating relatively pure lateral and normal modes.

3 Energy Analysis

3.1 Dynamic Model of the Plate at no-Load Condition

In order to explain the differences we observe with the two vibration modes, let us come back to the mechanical behavior of the plate. As it has been explained in [13], the dynamics of a device at resonance can be simplified as a second order model (1).

$$M_{L/N}\ddot{w} + D_{L/N}\dot{w} + K_{L/N}w = N_{L/N}V \tag{1}$$

The index L/N indicates that the equation is accurate for both lateral and normal modes. Assuming that a single mode is being excited at its resonance frequency, this equation serves to represent the modal parameters of mass $M_{L/N}$, dampening $D_{L/N}$, and elasticity $K_{L/N}$. The state variables w, $\dot{w}$ and $\ddot{w}$, represent the instantaneous displacement, speed and acceleration, respectively. Since the motion source is a piezoelectric ceramic, the electrical part of the equation, representing the motion force from the piezoelectric transducer, may be written as $N_{L/N}V$, with $N_{L/N}$ representing the electro-mechanical transformation factor, and V value of the input voltage. The parameters of the plate are identified experimentally, as described in [14], with 40 V pk-to-pk applied to the lateral mode, and 12 V pk to pk to the normal mode. Their values are listed in Table 1. The question of the voltage difference will be addressed in Sect. 5.

Table 1. Modal parameters identified for the lateral and normal modes induced on the device

Parameter	Mode	Symbol	Value
Electro-mechanical transformation factor	Lateral	N_L	0.0773 N/V
Modal Mass	Lateral	M_L	15.4 g
Modal Dampening	Lateral	D_L	27 N s/m
Modal Elasticity	Lateral	K_L	2017 MPa
Electro-mechanical transformation factor	Normal	N_N	0.3 N/V
Modal Mass	Normal	M_N	13.8 g
Modal Dampening	Normal	D_N	22.1 N s/m
Modal Elasticity	Normal	K_N	1678 MPa

3.2 Active Power Measurement

In order to measure the active power consumption, a Fluke Norma 4000 power analyzer is connected to the motor ceramics. A frequency sweep is made at about ±1 kHz around the resonance at three different voltage amplitudes. The voltages are set for each mode to produce a vibration amplitude at resonance of 0.8 μm_{p-p}, 0.6 μm_{p-p} and 0.3 μm_{p-p}. The vibration amplitude values are recorded together with the total active power measurement for each frequency. The measurements are performed three times: first at no load, then with a static finger pressing over the surface at a normal force of 0.5 N, and finally with a finger pressing at 1 N. The results are shown in Fig. 2.

The normal and lateral modes show a similar behavior with the power evolving proportionally to the square of the amplitude at no load. For these conditions, the lateral mode requires marginally less power for a given wave amplitude. It is also possible to observe that the presence of the finger produces an attenuation of the amplitude, and a slight shift of the resonance frequency. These phenomena impact more significantly the normal mode, with an attenuation of over 54%, against 23% for the lateral mode. In Fig. 2, it can be perceived that with a load, the power required to achieve a given amplitude is increased. An interpolation of the evolution of the power vs. amplitude at resonance can be made for each curve, with a quadratic fit of the measured data. This result (see Fig. 3) provides a relation of the wave amplitude versus active power for each studied case.

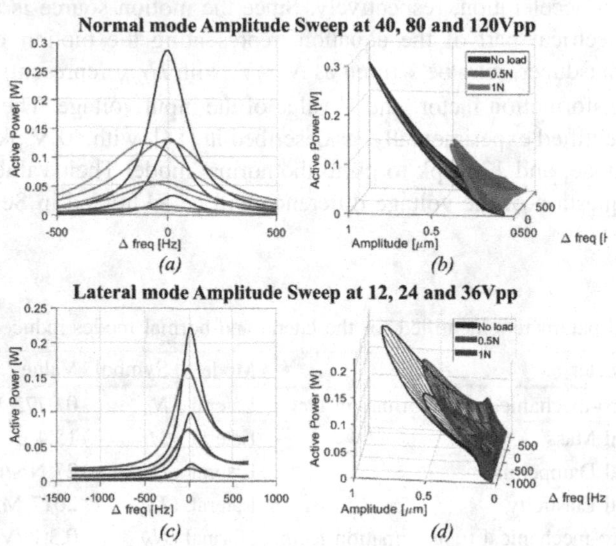

Fig. 2. Active Power for a frequency sweep for normal and lateral modes, for three voltage supplies and three finger pressures. (a) and (c) Amplitude vs. frequency shift. The graph shows the attenuation due to a finger pressing on the surface. (b) and (d) Power vs. Amplitude.

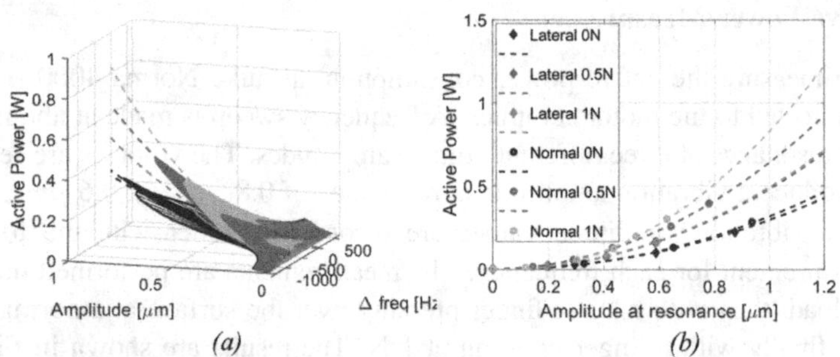

Fig. 3. Active power vs. amplitude, relation extrapolation. (a) Evolution of power vs. amplitude at resonance from the sweep data. (b) Power vs. Amplitude relation vs. the points for the measured data at resonance. The quadratic fit of the lateral power measurements at loads of 0.5 N and 1 N are superposed.

4 Tribology Analysis

4.1 Frequency and Speed Effects for a Hard Probe and for a Finger

In order to compare the friction reduction for both modes at different frequencies, the relative friction coefficient $\mu' = \mu_k/\mu_0$ (with μ_k a measurement at a given amplitude, and μ_0 the measurement without vibration) is deduced thanks to a tribometer, either using an artificial finger (a probe already used in [7]), or with a real finger.

The measurements are performed for different finger or probe speeds, and for the two vibration modes. These results will allow determining the relation of active power vs. friction reduction for each mode. The mechanism through which friction is reduced with purely lateral modes is explained in [7], based on the model proposed in [6] (Fig. 4).

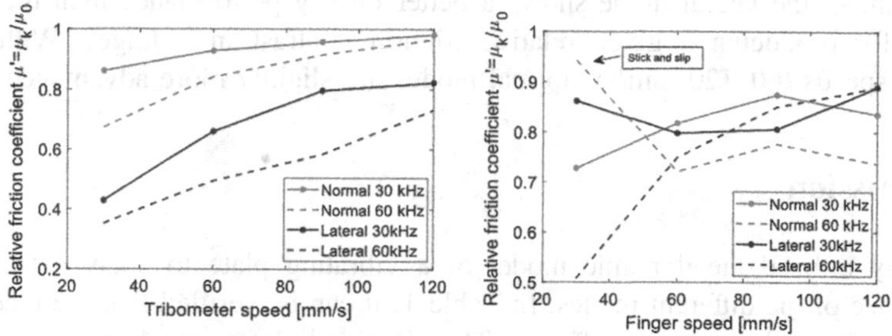

Fig. 4. Tribology measurements at 30 kHz vs. 60 kHz, at a vibration amplitude of 1.2 μm_{p-p}, and a pressing force of 0.5 N. Left: Tribometer measurements. Right: mean value of the measurements on a moving finger. Stick and slip was felt by the finger for normal vibration with a speed of 30 mm/s, which may explain the relatively high measurement. Moreover, active exploration with a moving finger may produce inaccuracies in the friction measurement of up to ±0.2

The tribology results are compared with results gotten with the devices previously used in [7]. These devices worked at a resonance frequency about 30 kHz. At the same exploration speed, the same vibration amplitude ranges, the same probe and the same finger, this comparison allows an analysis of the vibration frequency influence (30 kHz versus 60 kHz). Indeed, the measurements taken with the tribometer at 1.2 $\mu m_{p\text{-}p}$ wave amplitude, show that the increase in frequency improves the lubrication for both modes.

4.2 Active Power vs. Friction Contrast

As the amplitude of vibration increases, so does the friction contrast of the surface with and without vibration. It is possible to use the measured friction data at different amplitudes, and combine it to the amplitude vs. power relation found in Sect. 3.2 (Fig. 3), in order to estimate the actual energy required to produce a given friction contrast on a plate when using either a normal or a lateral mode (see Fig. 5).

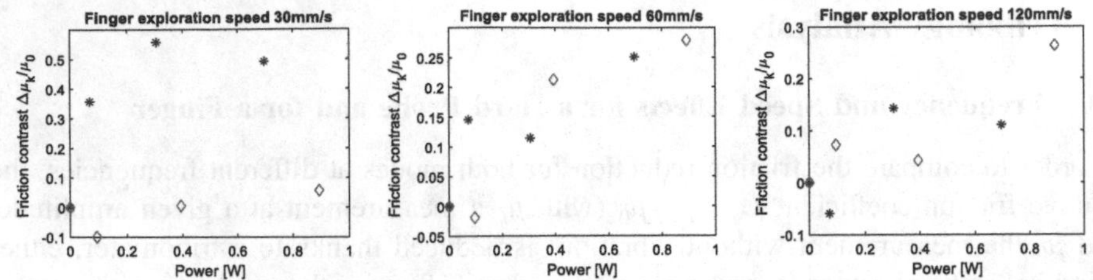

Fig. 5. Friction contrast vs. active power for 30 mm/s, 60 mm/s and 120 mm/s exploration speed. Blue: lateral vibration. Red: Normal vibration. Negative values may be a product of measurement imprecisions, or because of the presence of sick and slip which may increase the friction. (Color figure online)

The results show that, for a frequency about 60 kHz, for slow exploration speeds (30–60 mm/s), the lateral mode shows a better energy performance than the normal devices, for producing a given relative friction contrast in a finger. With higher scanning speeds (60–120 mm/s), normal modes are slightly more advantageous.

5 Discussion

This study utilized the dynamic model of a vibrating plate to analyze the energy performance of the different modes. In Table 1, it can be verified that the mechanical parameters for each mode are different. The identified electro-mechanical transformation factor indicates that the piezoelectric array produces a force about 5.5 larger when 'bending' the plate (to produce the normal mode), than when 'stretching' it (to produce the lateral mode), since the ceramics are better coupled with this mode. This can be explained by the difference of the wavelength of each mode. This produces a more important deformation of the ceramic in the normal mode than in the lateral mode. It can be seen as well that the damping factor (which is related to active power consumption) of the normal mode is higher than the one for the lateral mode, hence the active power consumption is higher as well.

Thanks to the tribology experiment, it was possible to observe a difference in friction reduction at different finger speeds for the lateral mode at 60 kHz as it is observed for 30 kHz in [6]. It is also confirmed that the frequency increase improves the active lubrication results for both modes. This phenomenon affects more the lateral vibration. For this reason, the relation of amplitude vs. sensation found in [6] may no longer be factual at 60 kHz vibration.

6 Conclusions

In this article, a device was created in order to perform an energetic comparison between two vibration modes at 60 kHz vibration frequency. The active power requirements show that lateral modes are generally advantageous in terms of the energy requirement to reach a given vibration amplitude, especially in the presence of a load.

The presence of the finger affects more significantly one mode than the other. This can be explained by the nature of the contact, a subject which may be explored in further studies.

When comparing the relation of power against friction contrast for both modes, it is the exploration speed which influences most the results. Indeed, at exploration speeds of 30–60 mm/s, the lateral vibration mode appears to require less active power than the normal mode to reach the same friction contrast. It is comparable or slightly worse to the normal mode for higher exploring speeds of 60–120 mm/s.

As a follow-up to this study, a psychophysical test will be performed in order to relate this study with perception for a set of different participants. This will also help evaluate the variability of friction contrast in terms of power from one subject to another.

Acknowledgements. This work is supported by IRCICA (Research Institute on software and hardware devices for information and Advanced communication, USR CNRS 3380).

References

1. Adams, M.J., et al.: Finger pad friction and its role in grip and touch. J. R. Soc. Interface **10** (80), 20120467 (2012)
2. Biet, M., Giraud, F., Lemaire-Semail, B.: Squeeze film effect for the design of an ultrasonic tactile plate. IEEE Trans. Ultrason. Ferroelectr. Freq. Control **54**(12), 2678–2688 (2007)
3. Sednaoui, T., Vezzoli, E., Dzidek, B., Lemaire-Semail, B., Chappaz, C., Adams, M.: Friction reduction through ultrasonic vibration part 2: experimental evaluation of intermittent contact and squeeze film levitation. IEEE Trans. Haptics **10**(2), 208–216 (2017)
4. Friesen, R.F., Wiertlewski, M., Colgate, J.E.: The role of damping in ultrasonic friction reduction. In: IEEE Haptics Symposium 2016, Philadelphia, pp. 167–172 (2016)
5. Janko, M., Wiertlewski, M., Visell, Y.: Contact geometry and mechanics predict friction forces during tactile surface exploration. Sci. Rep. **8**, 4868 (2018)
6. Vezzoli, E., Dzidek, B., Sednaoui, T., Giraud, F., Adams, M., Lemaire-Semail, B.: Role of fingerprint mechanics and non-Coulombic friction in ultrasonic devices. In: IEEE World Haptics Conference (WHC) 2016, Evanston, IL, pp. 43–48 (2015)
7. Torres Guzman, D.A., Lemaire-Semail, B., Kaci, A., Giraud, F., Amberg, M.: Comparison between normal and lateral vibration on surface haptic devices. In: IEEE World Haptics Conference (WHC) 2019, Tokyo, pp. 199–204 (2019)
8. Messaoud, W.B., Bueno, M.-A., Lemaire-Semail, B.: Relation between human perceived friction and finger friction characteristics. Tribol. Int. **98**, 261–269 (2016)
9. Winfield, L., Glassmire, J., Colgate, J.E., Peshkin, M.: T-PaD: tactile pattern display through variable friction reduction. In: Second Joint EuroHaptics Conference and Symposium on Haptic Interfaces for Virtual Environment and Teleoperator Systems (WHC 2007), Tsukuba, Japan, pp. 421–426 (2007)
10. Dai, X., Colgate, J.E., Peshkin, M.A.: LateralPaD: a surface-haptic device that produces lateral forces on a bare finger. In: IEEE Haptics Symposium (HAPTICS), pp. 7–14 (2012)
11. Wiertlewski, M., Colgate, J.E.: Power optimization of ultrasonic friction-modulation tactile interfaces. IEEE Trans. Haptics **8**(1), 43–53 (2015)

12. Yang, Y., Lemaire-Semail, B., Giraud, F., Amberg, M., Zhang, Y., Giraud-Audine, C.: Power analysis for the design of a large area ultrasonic tactile touch panel. Eur. Phys. J. Appl. Phys. **72**(1), 11101 (2015)
13. Ghenna, S., Giraud, F., Giraud-Audine, C., Amberg, M.: Vector control of piezoelectric transducers and ultrasonic actuators. IEEE Trans. Ind. Electron. **65**(6), 4880–4888 (2018)
14. Giraud, F., Giraud-Audine, C.: Piezoelectric Actuators: Vector Control Method, 1st edn. Elsevier, Cambridge (2019)

Midair Tactile Reproduction of Real Objects

Emiri Sakiyama[1(✉)], Atsushi Matsubayashi[1], Daichi Matsumoto[2], Masahiro Fujiwara[2], Yasutoshi Makino[2], and Hiroyuki Shinoda[2]

[1] The University of Tokyo, 7-3-1 Hongo, Bunkyo-ku, Tokyo, Japan
sakiyama@hapis.k.u-tokyo.ac.jp
[2] The University of Tokyo, 5-1-5 Kashiwanoha, Kashiwa-shi, Chiba-ken, Japan

Abstract. Midair tactile display using ultrasound radiation pressure is suitable for tactile reproduction because of its reproducibility and controllability. This paper is the first report that compares the tactile feelings of real objects with those associated with artificial stimuli reproduced through a sequence of sensing, processing, and reproduction. We previously proposed the concept of a midair tactile reproduction system and examined the basic properties of the sensing part, but had not achieved the tactile display to the human skin. In this paper, we report a practical method for the pressure reproduction from the measured data and examine the accuracy of the reproduced stimulation. The psychophysical experiments evaluate the fidelity of the tactile reproduction for some objects such as brushes, sponges, and towels.

Keywords: Midair haptics · Tactile reproduction · Tactile sensor

1 Introduction

An airborne ultrasound tactile display (AUTD) [1,2] can present tactile sensations with high spatio-temporal controllability in a non-contact manner. This advantage is suitable for tactile reproduction of real objects. However, the fidelity of tactile reproduction performance has not been studied in detail since the development of the device in 2008 [1]. In this study, we achieve a midair tactile reproduction system which consists of the AUTD and a tactile array sensor.

The concept behind the system was first introduced in our previous work-in-progress paper [3]. Further, we examined the design of the sensor and the reproduction scheme in [4]; however, the reproduced pressure distribution was strongly distorted owing to problems in the reproduction algorithm.

This paper is the first report that compares the tactile feelings of real objects to those associated with artificial stimuli reproduced through a sequence of sensing, processing, and reproduction. As a key part of the system, we report a practical method of the pressure reproduction. Conventional methods [4,5] have

Supported in part by JSPS Kakenhi 16H06303 and JST CREST JPMJCR18A2.

I. Nisky et al. (Eds.): EuroHaptics 2020, LNCS 12272, pp. 425–433, 2020.
https://doi.org/10.1007/978-3-030-58147-3_47

no constraints on the upper limit of the output amplitude, and sometimes, it is impossible for devices to output the optimization results. In this study, we acquire a feasible solution by executing the Levenberg–Marquardt algorithm (LMA) [6,7] with the constraint that the amplitudes of the transducers must all be equal. After confirming the basic physical performance of pressure reproduction, the fidelity of the reproduced sensation was evaluated in comparison with that of real objects by psychophysical experiments.

2 Tactile Presentation by Ultrasound Phased Array

2.1 Acoustic Radiation Pressure

Midair tactile stimulation using ultrasound is based on the acoustic radiation pressure. This pressure is a non-linear acoustic phenomenon which generates DC positive pressure on the boundary between the two types of media with different acoustic impedances. The acoustic radiation pressure P for plane ultrasound wave is described as $P = \alpha \frac{p_0^2}{\rho c^2} (> 0)$, where ρ is the density of the medium on the incident side; c is the speed of sound in the medium; p_0 is the RMS sound pressure of the ultrasound; and α is a constant that is dependent on the power reflection coefficient R such that $\alpha \equiv 1 + R$, and $1 \leq \alpha \leq 2$. In this study, we consider acoustic radiation pressure on solid surfaces in standard air; hence, $\rho = 1.25\ \mathrm{kg/m^3}$, $c = 340\ \mathrm{m/s}$, and $\alpha = 2$.

2.2 Sound Field Generated by Phased Array

In this study, a desired pressure distribution is reproduced by the radiation pressure. We control the radiation pressure field by controlling the sound field.

Assuming a steady sinusoidal wave, a complex sound amplitude vector $\boldsymbol{p}$ generated by a phased array is expressed as follows [5]:

$$\boldsymbol{p} = \boldsymbol{G}\boldsymbol{q}, \tag{1}$$

$$\boldsymbol{p} = [p_1, p_2, \ldots, p_M]^\top, \ p_m = p(\boldsymbol{r}_m) = p_{\mathrm{amp},m} e^{j\psi_m}, \tag{2}$$

$$G_{mn} = C \frac{D(\theta_{mn})}{|\boldsymbol{r}_m - \boldsymbol{r}_n|} e^{-\beta|\boldsymbol{r}_m - \boldsymbol{r}_n|} e^{jk|\boldsymbol{r}_m - \boldsymbol{r}_n|}, \tag{3}$$

$$\boldsymbol{q} = [q_1, q_2, \ldots, q_N]^\top, \ q_n = A_n e^{j\phi_n}. \tag{4}$$

Here, $p_m (m = 1, \ldots, M)$ is a complex sound amplitude at the m-th control point $\boldsymbol{r}_m$, and $q_n (n = 1, \ldots, N)$ is a complex amplitude of the surface vibration velocity of the n-th transducer at $\boldsymbol{r}_n$. $\boldsymbol{G}$ is the transfer matrix, where C is a constant, $D(\theta)$ is the directivity function of a transducer, θ_{mn} is the angle between the transducer normal and $\boldsymbol{r}_m - \boldsymbol{r}_n$, β is the attenuation coefficient and k is the wave number.

2.3 Controlling Pressure Field by Phased Array

In the case of tactile presentation, the driving signal of the phased array $\boldsymbol{q}$ is decided based on the sound pressure amplitude distribution $\boldsymbol{p}_{\text{amp}}$ converted from the desired instantaneous pressure distribution $\boldsymbol{P}$. The proposed method solves the following problem; find $\boldsymbol{q}$ s.t. $\boldsymbol{p}_{\text{amp}} = |\boldsymbol{G}\boldsymbol{q}|$ at each time. The tactile presentation is performed by outputting the time series of $\boldsymbol{q}$ in a quasi-stationary manner. The phases of the complex sound distribution $\boldsymbol{p}$ are unconstrained. A more efficient output can essentially be achieved by modifying these phases [8].

3 Tactile Sensing

In order to measure a slight contact force of around 0.01 N/cm^2($\simeq$the maximum presentation force per 1 AUTD) at sufficient sampling rate, we adopted the new tactile sensor that we had designed previously [4] using high-sensitivity microphones for use in this study. The sensor is composed of a 4×4 channels microphone array (Fig. 1a). The contact pressure of the sensor surface causes pressure change in the sensor cavity and the microphone detects the change (Fig. 1b). Using this sensor, we can acquire a pressure distribution of 4×4 channels at 11 mm intervals by a sampling frequency of 1 kHz. The frequency characteristics are compensated for components with 20 – 470 Hz in this study. The components under 20 Hz and over 470 Hz are removed because of the low sensitivity of the microphones.

In order to achieve high fidelity of tactile sensation by this system, the vibrations produced on the tactile sensor must be similar to the actual vibrations on the human skin. Therefore, the mechanical characteristics of the sensor surface are required to be similar to those of the human skin. HITOHADA® gel sheet (hardness: Askar C0, thickness: 1 mm) of EXSEAL Co., Ltd., Japan is used as the contact surface. This is a super-soft urethane sheet comparable to human skin. Moreover, human fingerprints affect the generation and detection of slip. In this study, we compared the reproduction fidelity of tactile sensation between two conditions of the contact surface shape: non-treated condition and concentric circle groove pattern imitating fingerprints, as shown in Fig. 2.

4 Tactile Reproduction

4.1 Data Preprocessing

Before the optimization of the driving signal of AUTD, four preprocesses are performed on the measured data. First, half-wave rectification of the measured data is performed because the radiation pressure is always positive. Though this process generates unnecessary harmonics, we adopt it for the easiness and clearness of the process. Second, the pressure distribution data P is converted to sound pressure amplitude distribution p_{amp} by $p_{\text{amp}} = \sqrt{2}p_0 = c\sqrt{\frac{2\rho P}{\alpha}}$.

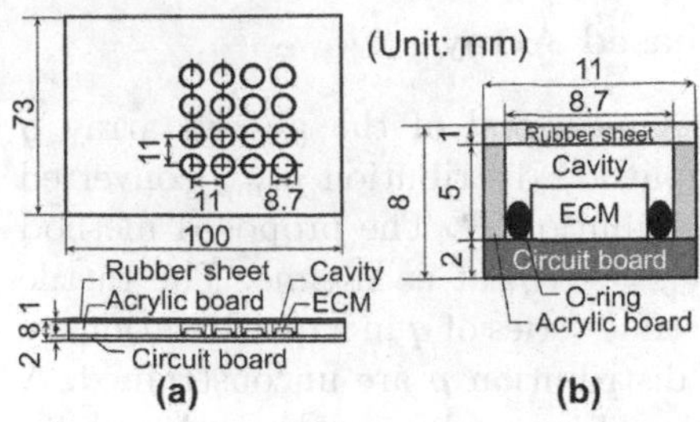

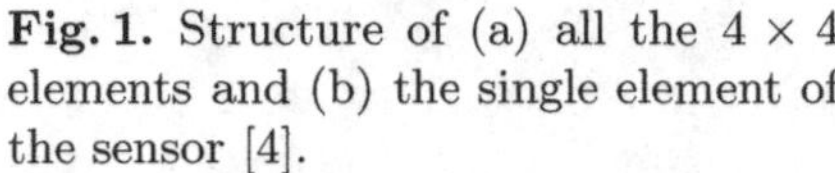

Fig. 1. Structure of (a) all the 4 × 4 elements and (b) the single element of the sensor [4].

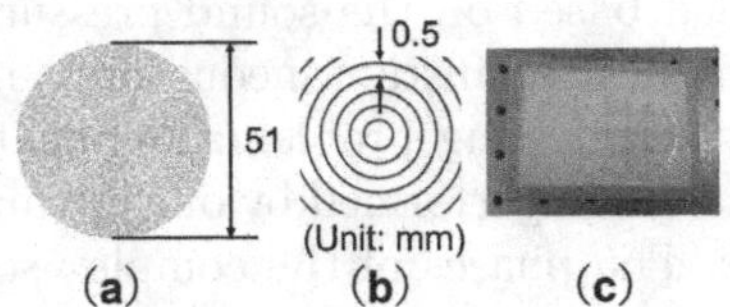

Fig. 2. The concentric circle pattern of the sensor surface. (a) The whole pattern. (b) The enlarged view. (c) The appearance of the sensor.

In order to limit the effects of pressures outside the region of interest, 20 control points are added around the region at 11-mm intervals ($4 \times 4 \rightarrow 6 \times 6$ points), and the sound pressure amplitudes are set to zero at the peripheral points. Subsequently, the 36-point data with 11 mm intervals are converted into 256-point data with 11/3 mm intervals by linear interpolation of the original data on the assumption that the pattern is smooth.

4.2 Optimization of Driving Signal of AUTD

In this study, the LMA [6,7] was used to optimize the driving signal of AUTD. LMA is an iterative algorithm for solving non-linear least squares problems. In general, to solve a problem $\min_{\boldsymbol{\theta}} \|\boldsymbol{f}(\boldsymbol{\theta})\|_2^2$ by LMA, the parameter $\boldsymbol{\theta}$ is updated to $\boldsymbol{\theta}^{k+1} = \boldsymbol{\theta}^k - [(\boldsymbol{J}^k)^\top \boldsymbol{J}^k + \lambda^k \boldsymbol{I}]^{-1} \boldsymbol{f}(\boldsymbol{\theta}^k)$, where λ is the damping parameter and $\boldsymbol{J}$ is the Jacobian matrix of $\boldsymbol{f}(\boldsymbol{\theta})$.

We solved the problem using the LMA with the constraint that the amplitudes of the transducers $A_1, \ldots, A_N$ must all be equal. In this case, the optimization problem is described as follows:

$$\min_{\boldsymbol{\theta}} \left\| \begin{matrix} \mathrm{Re}(\boldsymbol{Gq} - \mathrm{diag}(\boldsymbol{p}_{\mathrm{amp}})\boldsymbol{u}) \\ \mathrm{Im}(\boldsymbol{Gq} - \mathrm{diag}(\boldsymbol{p}_{\mathrm{amp}})\boldsymbol{u}) \end{matrix} \right\|_2^2 \tag{5}$$

$$\boldsymbol{q} = A \times \left[e^{j\theta_1}, e^{j\theta_2}, ..., e^{j\theta_N}\right]^\top, \boldsymbol{u} = \left[e^{j\theta_{N+1}}, e^{j\theta_{N+2}}, ..., e^{j\theta_{N+M}}\right]^\top \tag{6}$$

where Re and Im are the real and imaginary parts of the complex vector, respectively. The results of the LMA depend on the initial value of $\boldsymbol{\theta}_0$ because the algorithm determines only a local minimum. In this study, we use a zero vector or a final value of a previous time frame as the initial value.

5 Experiments

5.1 Numerical Simulation

First, the reproduction performance of arbitrary static pressure distribution was evaluated via numerical simulation. We generated binary distributions on the

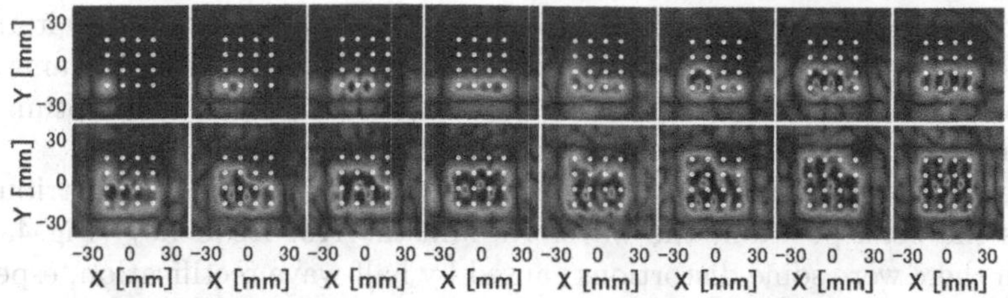

Fig. 3. An example of the simulated optimization results: The white circle indicates the position of each channel. They were optimized with 996 transducers (4 AUTDs), 256 control points at 50 iterations.

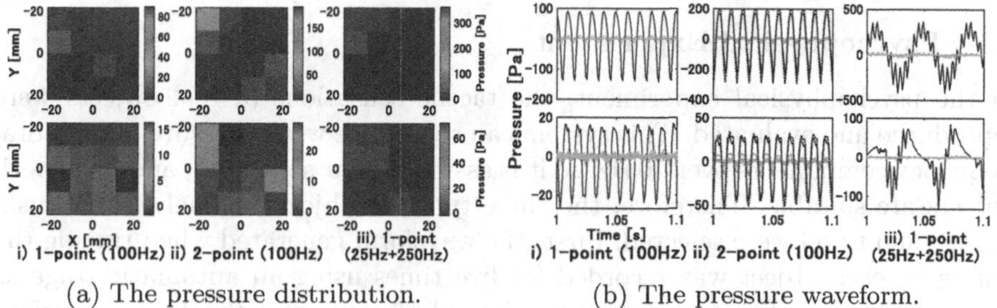

(a) The pressure distribution. (b) The pressure waveform.

Fig. 4. The results of the reproduction of AUTD stimuli recorded by the sensor. **Top:** Original data. **Bottom:** Reproduction data. In (a), the top-left corner: 0 ch, bottom-right corner: 15 ch. In (b), Black solid line: 0 ch, black dotted line: 15 ch, and gray line: other channels.

control points such that the sound pressure amplitude of each channel was 0 or 1. The number of distributions was $2^{16}-1$ in total for 16 control points, with the exception of a distribution in which all the amplitudes were zero. Subsequently, we reproduced these distributions based on the point source model using MATLAB software. The examples of the simulated optimization results are shown in Fig. 3. The average error $\frac{1}{M}\sqrt{\|\boldsymbol{Gq}-\boldsymbol{p}\|_2^2/\|\boldsymbol{p}\|_2^2}$ was approximately $5-15\%$.

5.2 Reproduction of Dynamic Radiation Pressure Distribution

Due to half-wave rectification and optimization errors in the reproduction algorithm, the waveform may change to some extent in the actual reproduction. Therefore, in the preliminary experiment, the dynamic pressure distribution generated by AUTD was reproduced by the system and again recorded by the sensor. Thus, the degree of distortion was examined and the distortion was corrected.

The presented stimuli were i) 1-point amplitude modulation (AM) stimulus to the 0 channel position of the sensor, ii) 2-point AM stimulus to the 0 and 15 channel positions, and iii) 1-point AM containing multiple frequency components to the 0 channel position. The AM frequency of i) and ii) was 100 Hz. As for iii), the AM 25 Hz waveform and the AM 250 Hz waveforms were added together, where the amplitude ratio was 3:2. Except for the first-time frame, 20 iterations of the LMA were performed using a final value of a previous time frame as an initial value each time. The other settings of equipment layout and optimization were the same as those of the numerical simulation.

As a result, the system reproduced some features of the original distributions, such as the peak position, the waveform and the AM frequency (Fig. 4a, 4b). Though there were some distortions caused by half-wave rectification, especially in iii), we left these distortions as they were. The amplification factor between the AUTD-driving voltage and the observed sensor output voltage is determined experimentally. We determined it so that the focused beam to the sensor just below the AUTD center produces an expected output amplitude.

5.3 Psychophysical Experiment

In the psychophysical experiment, the tactile sensations of real objects were reproduced and evaluated. This system can reproduce weak pressure of temporal frequency components over 20 Hz, so it is assumed that soft, light, and unsmooth objects are suitable. Therefore, the three types of objects, namely, soft brush, sponge and towel were selected. First, the waveform generated when tracing the sensor by each object was recorded for five times using an automatic stage as shown in Fig. 5a. The speed is approximately 5 cm/s and the contact surface is of circular shape, approximately 2 cm in diameter. This measurement was performed in each condition of the sensor surface with and without concentric circle grooving treatment. The reproduction stimulation was then generated on a palm according to the proposed method. Simultaneously, amplitude correction was carried out based on the result of the preliminary experiment.

The experiment was carried out in two stages. First, the ability of the real object to discriminate tactile sensation was confirmed. The tactile sensation of the real object was randomly presented 12 times to the blindfolded participants by a human hand. They answered which object was presented each time. Next, the reproduction stimulation was evaluated. In practice, each type of reproduced stimulation was once presented without the prior information of what object was reproduced. Subsequently, the remaining three types of stimulation were randomly presented four times. The number of trials was minimized considering the fatigue of the participants. For each stimulus, participants were asked to answer two types of questions, i.e., A) how much they were similar to the tactile sensation of the brush/sponge/towel (1: totally different $\sim$ 7: very similar), and B) which object they felt it was closest to. This experiment was carried out twice by changing the surface treatment conditions. The participants were 10 people in their 20's, two of whom were females, and all stimuli were presented on the palm of the dominant hand (one male was left-handed). While presenting the reproduction stimulus, the participants were listening to white noise by a headphone, and were allowed to touch the real objects freely.

As shown in Fig. 6a, the real objects could be discriminated with a correct answer rate of over 95%. In contrast, the correct answer rate of the reproduction stimulus was lower(Fig. 6b, 6c). In Fig. 7, the F-measure was calculated from

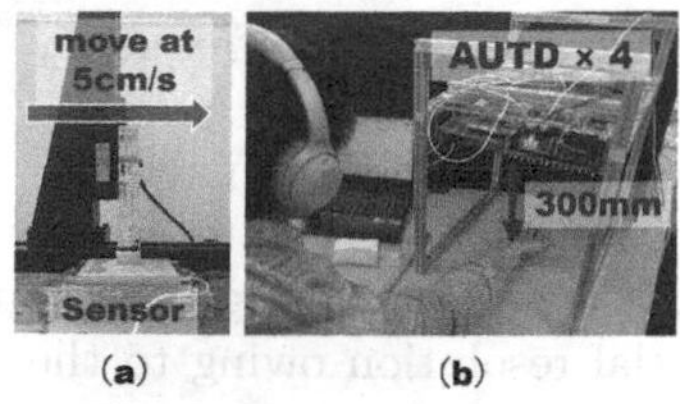

Fig. 5. (a) The measurement of the tactile sensations of the real objects. (b) The presentation of the reproduction stimulation.

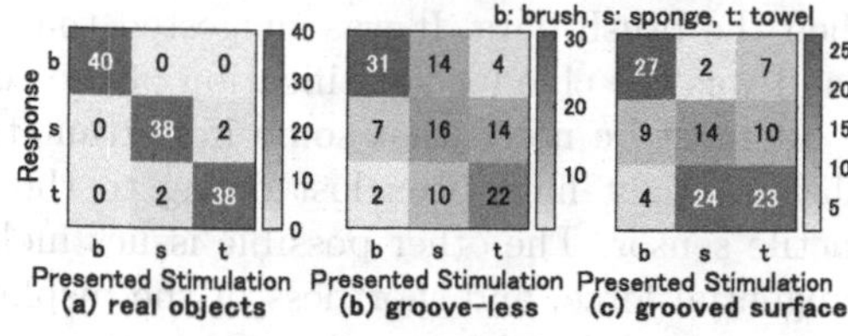

Fig. 6. The responses to the question "which object they felt it was closest to." The number indicates the number of responses.(b)/(c) is the result of the groove-less/grooved sensor surface condition.

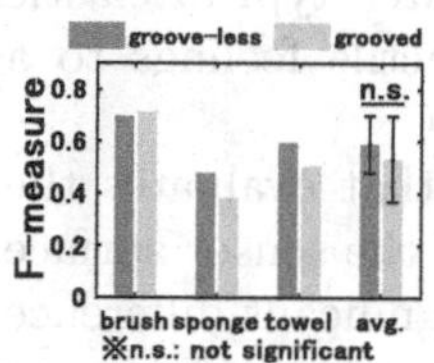

Fig. 7. The F-measure calculated from the responses to the question B.

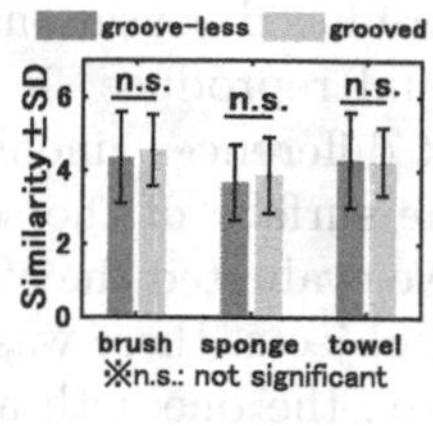

Fig. 8. The similarity to the tactile sensation of real objects that the participants answered.

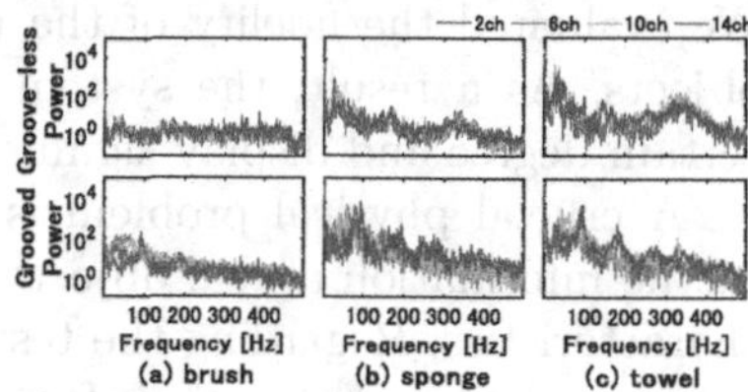

Fig. 9. The power spectrum of the original tactile sensation data at the four major channels that the objects passed through (2, 6, 10, 14 ch). These are all after calibration and half-wave rectification.

the answer to question B) shown in Fig. 6b, 6c. As the result indicates, the F-measure exceeds 0.33, which is the chance rate. Moreover, a paired t-test did not show a significant difference between non-treated condition and concentric circle treated condition($t(2) = 1.60, p = .25$). In addition, a three-way repeated measures ANOVA was carried out on the answers to question A). As shown in Fig. 8, there is also no significant difference in the subjective similarity evaluation by the surface treatment condition.

Based on these results, it was confirmed that three types of tactile sensation could be discriminated to some extent by this system without prior information. However, there were no significant differences of subjective evaluation and discrimination results between two surface conditions. Therefore, it is assumed that the reproduced stimulation contains some important tactile features, regardless of the surface treatment, for identification of the three types of objects. As shown in Fig. 9, the intensity increased by the surface groove, especially in the components under 300 Hz. In both surface conditions, the intensity of high-frequency band over 300 Hz was largest in the towel data, followed by the sponge data, and then the brush data. It was suggested that the intensity of the high-frequency band was the clue for identification of the three types of objects.

It must be noted that some important tactile features in the original stimulation might have been lost owing to the spatial resolution of 11 mm of the tactile sensor. The other possible issues include the acoustic streaming, lack of tangential force, and shortness of the displayed spatial resolution owing to the limitation of the ultrasound wavelength.

6 Conclusion

In this study, we proposed an accurate tactile reproduction system by introducing the LMA for determining the driving signals of transducers on the AUTD. We evaluated the fidelity of the reproduced sensations of three typical elastic objects. As a result, the system could reproduce similar tactile feelings to a certain degree and display significant differences among them.

A crucial physical problem is the surface of the sensor that evaluates the tactile information of real objects. We evaluated the effect of the sensor surface characteristics. Regarding the tested objects, there was no significant difference between the two types of surfaces, i.e., the one with a concentric circle groove pattern and the other with no groove pattern.

The most demanding future work is to improve the spatial resolution of the sensor. The system can be used as a useful tool to clarify the human tactile sensation and as a practical system that records and reproduces the tactile feelings associated with various products.

References

1. Iwamoto, T., Tatezono, M., Shinoda, H.: Non-contact method for producing tactile sensation using airborne ultrasound. In: Ferre, M. (ed.) EuroHaptics 2008. LNCS, vol. 5024, pp. 504–513. Springer, Heidelberg (2008). https://doi.org/10.1007/978-3-540-69057-3_64
2. Hoshi, T., Takahashi, M., Iwamoto, T., Shinoda, H.: Noncontact tactile display based on radiation pressure of airborne ultrasound. IEEE Trans. Haptics **3**(3), 155–165 (2010)
3. Sakiyama, E., Matsumoto, D., Fujiwara, M., Makino, Y., Shinoda, H.: Midair tactile reproduction of real objects using microphone-based tactile sensor array. In: IEEE World Haptics Conference, WPI.35(Work-in-Progress Papers), 9–12 July, Tokyo, Japan (2019)
4. Sakiyama, E., Matsumoto, D., Fujiwara, M., Makino, Y., Shinoda, H.: Evaluation of multi-point dynamic pressure reproduction using microphone-based tactile sensor array. In: IEEE International Symposium on Haptic Audio-Visual Environments and Games, Sunway, Malaysia, 3–4 October 2019
5. Inoue, S., Makino, Y., Shinoda, H.: Active touch perception produced by airborne ultrasonic haptic hologram. In: 2015 IEEE World Haptics Conference, pp. 362–367 (2015)
6. Levenberg, K.: A method for the solution of certain non-linear problems in least squares. Q. J. Appl. Math. II (2), 164–168 (1944)

7. Marquardt, D.W.: An algorithm for least-squares estimation of non-linear parameters. J. Soc. Ind. Appl. Math. **11**(2), 431–441 (1963)
8. Long, B., Seah, S.A., Carter, T., Subramanian, S.: Rendering volumetric haptic shapes in mid-air using ultrasound. ACM Trans. Graph. **33**(6), Article 181 (2014)

LinkRing: A Wearable Haptic Display for Delivering Multi-contact and Multi-modal Stimuli at the Finger Pads

Aysien Ivanov(✉), Daria Trinitatova(✉), and Dzmitry Tsetserukou(✉)

Skolkovo Institute of Science and Technology (Skoltech), Moscow 121205, Russia
{aysien.ivanov,daria.trinitatova,d.tsetserukou}@skoltech.ru

Abstract. LinkRing is a novel wearable tactile display for providing multi-contact and multi-modal stimuli at the finger. The system of two five-bar linkage mechanisms is designed to operate with two independent contact points, which combined can provide such stimulation as shear force and twist stimuli, slippage, and pressure. The proposed display has a lightweight and easy to wear structure. Two experiments were carried out in order to determine the sensitivity of the finger surface, the first one aimed to determine the location of the contact points, and the other for discrimination the slippage with varying rates. The results of the experiments showed a high level of pattern recognition.

Keywords: Haptic display · Wearable tactile devices · Multi-contact stimuli · Five-bar linkage

1 Introduction

Recent developments in computer graphics and head-mounted displays contributed to the appearance of different applications and games that implement VR technologies. Nevertheless, there is still a lack of providing haptic feedback, which is a crucial component to accomplish the full user immersion in a virtual environment.

Tactile information obtained from the fingers is one of the most important tools for interacting with the environment as fingertips have a rich set of mechanoreceptors. Thereby, the finger area plays an essential role in haptic perception. Nowadays, there are many research projects aimed at the design and application of fingertip haptic devices. One of the common methods for providing cutaneous feedback to the fingertips is by a moving platform shifting on the finger pad [2,4]. Several works are focused on delivering mechanical stimulations and can reliably generate normal forces to the fingertip [8], as well as shear deformation of the skin [9]. The most general method to simulate the object texture is using vibration [7,12], and the friction sensation is usually generated by electrostatic force [5]. A number of studies investigate the perception of object

A. Ivanov and D. Trinitatova—Both authors contributed equally to the paper.

I. Nisky et al. (Eds.): EuroHaptics 2020, LNCS 12272, pp. 434–441, 2020.
https://doi.org/10.1007/978-3-030-58147-3_48

softness [3,15]. Villa Salazar et al. [11] examined the impact of combining simple passive tangible VR objects and wearable haptic display on haptic sensation by modifying the stiffness, friction, and shape perception of tangible objects in VR. Some research works explore multimodal tactile stimulation. In [13], Yem et al. presented FinGAR haptic device, which combines electrical and mechanical stimulation for generating skin deformation, high/low-frequency vibration, and pressure. Similarly, the work [14] presents the fingertip haptic interface for providing electrical, thermal, and vibrotactile stimulation. Still, there is a need for a device that allows users to interact with virtual objects in a more realistic way. The interaction can be improved by the physical perception of the objects and their dimensions. Our approach suggests the use of a wearable haptic device, which can provide a multi-contact interaction on the finger pad that a person can experience when interacting with a real object.

In this paper, we propose LinkRing, a novel wearable haptic device with 4-DoF, which provides multi-contact and multi-modal cutaneous feedback on the finger pad (Fig. 1(a)). The haptic display is designed as the system of two five-bar linkage mechanisms which operate with two independent contact points. The proposed device can deliver a wide range of tactile sensations, such as contact, pressure, twist stimuli, and slippage.

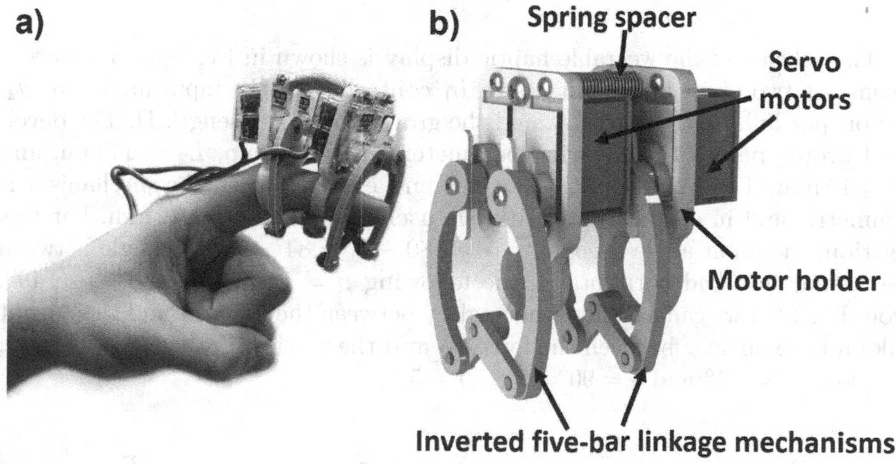

Fig. 1. a) A wearable haptic display LinkRing. b) A CAD model of wearable tactile display LinkRing.

2 Design of Haptic Display LinkRing

LinkRing is designed to deliver multi-modal and multi-contact stimuli at the finger pads. The proposed device is based on the planar parallel mechanism, which was previously applied in several works [1,6,10]. Two parallel inverted five-bar mechanisms with 2-DoF each deliver tactile feedback in two independent points on the finger pads. Each of them can generate normal and shear forces at the contact point. And in combination, they can simulate different sizes of grasped objects, the feeling of spinning objects, and the sense of sliding a finger on the surface to the left, to the right, and around the axis perpendicular to the finger pad plane. The prototype consists of 3D-printed parts, namely, motor holders fastened to the finger and links made of flexible PLA and PLA, respectively, and metal spring spacers with diameter 6 mm connecting the system of inverted five-bar linkages (Fig. 1(b)). The distance between the end-effectors is 26 mm, and it can vary by choosing the different lengths of the spacers. The fastening for the finger was designed as a snap ring to be suitable for different sizes. The specification of the device is shown in Table 1.

Table 1. Technical specification of LinkRing.

Motors	Hitec HS-40
Material of the motor holders	Soft PLA
Material of the links	PLA
Weight [g]	33
Link length L1, L2 [mm]	35, 17
Max. normal force at each contact point [N]	1.5 ± 0.15

The scheme of the wearable haptic display is shown in Fig. 2(a). The mechanism has two input links with length L_1 controlled by the input angles θ_1, θ_2, two output links with length L_2, and the ground link with length D. The developed prototype has the following parameters: $L_1 = 35$ mm, $L_2 = 17$ mm, and $D = 15$ mm. The initial position for the end effector, when the mechanism is symmetric and in contact state with the user's finger, is $H = 22$ mm. For this position, the input angle is equal to $\alpha = 180 - \theta_1 = 84°$, and the angle between the output link and horizontal is the following $\beta = arcsin(\frac{L_1 sin\alpha - H}{L_2}) = 49°$. From Fig. 2(b) we can calculate the angle γ between the force F_1 and the output link and the angle φ between the force F_2 and the vertical in the following way: $\varphi = 90° - \beta = 41°$ and $\gamma = 90° - \alpha + \beta = 55°$.

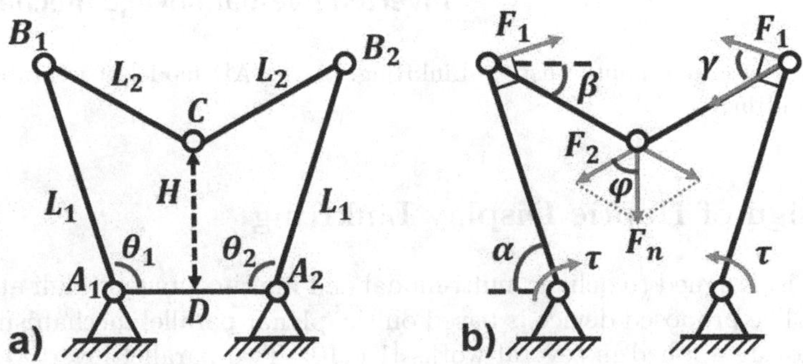

Fig. 2. a) Device kinematic scheme. b) Static force diagram.

The device is actuated by Hitec HS-40 servo motors with a low mass and rather high output torque (Weight is 4.8 g, dimensions are 20 × 8.6 × 17 mm, maximum torque is 0.6 kg-cm). The maximum force is achieved with the symmetric case when the resulting force applied to the finger is a normal force (directed vertically). In this case: $F_1 = \tau/L_1$; $F_2 = F_1 \cdot cos(\gamma)$; $F_n = 2 \cdot F_2 \cdot cos(\varphi)$. The maximum value of the normal force is $F_n = 1.46$ N. To verify this result, we experimentally measured the force with a calibrated force sensing resistor (FSR 400). The experiment showed that the maximum generated force is $F = 1.5\,\mathrm{N} \pm 0.15\,\mathrm{N}$.

3 User Study

We evaluated the performance of the developed haptic display in two user studies. Firstly, we estimated the distinction in the perception of different static contact patterns simulated on the user's finger. Secondly, we tested the ability of users to distinguish the sliding speed of end effectors at the contact area. For the two experiments, the user was asked to sit in front of a desk, and to wear the LinkRing display on the left index finger. To increase the purity of the experiment, we asked the users to use headphones and fenced off the hand with a device from the user by a barrier (Fig. 3 (a)).

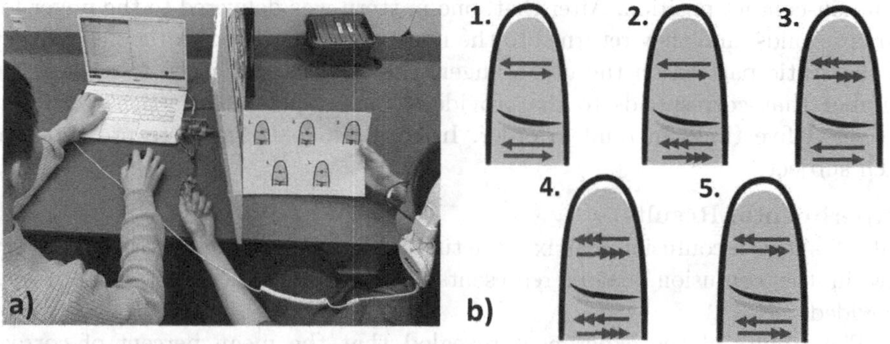

Fig. 3. a) Overview of the experiment. b) Slippage patterns for the experiment. Single arrows indicate the slow speed in the contact point, double arrows mean middle rate, and triple arrows represent fast speed. The direction of the arrow indicates the side of the contact point motion.

3.1 Static Pattern Experiment

The purpose of the experiment was to study the recognition of various patterns on the user's finger differing in the location of the contact points. Nine patterns with different positions of static points were designed (Fig. 4). In total, ten subjects took part in the experiment, two women and eight men, from 21 to 30 years old.

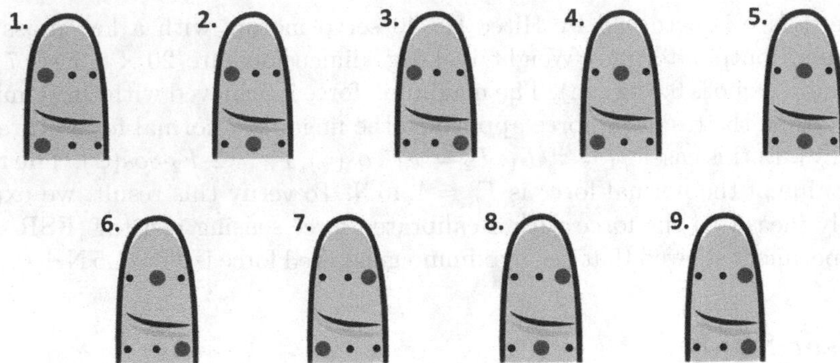

Fig. 4. Patterns for the experiment with static points. Blue dots represent pressed point, and back dots show the possible contact locations. (Color figure online)

Before the experiment, the calibration was conducted for each participant to provide all the contact points on the finger. The calibration was based on two parameters: the thickness and width of the finger. During the training session, each pattern was delivered two times to the user.

During the experiment, the participant was provided with a visual guide of the designed patterns. For each trial, end effectors reached a specified location in a non-contact position. After that, one pattern was delivered to the finger for three seconds, and they returned to the non-contact position. After the delivery of the static pattern on the user's finger, the subject was asked to specify the number that corresponds to the provided contact pattern. Each pattern was delivered five times in random order. In total, 45 patterns were presented to each subject.

Experimental Results
Table 2 shows a confusion matrix for actual and perceived static patterns. Every row in the confusion matrix represents all 50 times a contact pattern was provided.

The results of the experiment revealed that the mean percent of correct answers for each pattern averaged over all the participants ranged from 68% to 100%. The mean percentage of the correct answers is 90%. The most recognizable contact positions were 1st, 4th, and 5th, with a recognition rate of 98%, 98%, and 100%. And the least distinct pattern was 9th with a rate of 68%. It can be observed that a high number of participants confused pattern 9 with 8, which are very close.

In order to understand if there is a real difference between pattern perception, the experimental results were analyzed using one-factor ANOVA without replication with a chosen significance level of $p < 0.05$. According to the test findings, there is a statistically significant difference in the recognition rates for the different contact patterns ($F(8, 81) = 2.43,\ p = 0.02 < 0.05$). It was significantly more difficult for participants to recognize pattern $9th$ than $5th(F(1, 18) = 16,\ p = 8.4 \cdot 10^{-4} < 0.05)$, $1st$ and $4th(F(1, 18) = 13.2,\ p = 1.88 \cdot 10^{-3} < 0.05)$, and $2nd$, $6th$ and $7th(F(1, 18) = 6.23,\ p = 0.022 < 0.05)$.

Table 2. Confusion matrix for actual and perceived static patterns across all the subjects

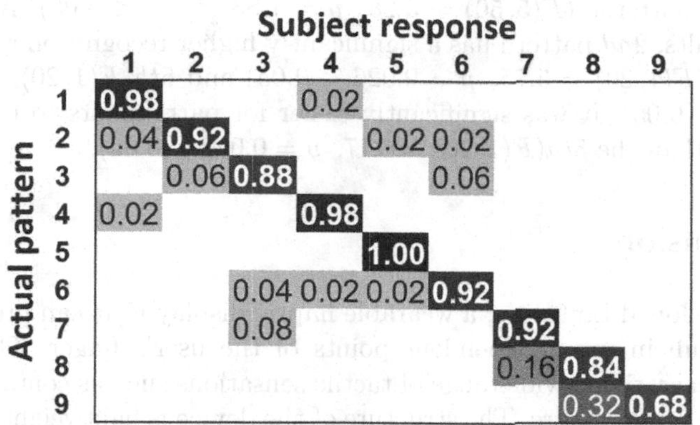

Actual pattern \ Subject response	1	2	3	4	5	6	7	8	9
1	0.98			0.02					
2	0.04	0.92			0.02	0.02			
3		0.06	0.88			0.06			
4	0.02			0.98					
5					1.00				
6			0.04	0.02	0.02	0.92			
7			0.08				0.92		
8							0.16	0.84	
9								0.32	0.68

3.2 Experiment of the Recognition of Moving Patterns

The objective of the experiment was to study the user recognition of end effector slippage with various speeds which were delivered to the finger at the same time by both contact points. Eleven participants volunteered into the experiments, two females and nine males, from 21 to 31 years old.

Three different speeds for slippage on the finger were used: slow (43 mm/s), middle (60 mm/s), and fast (86 mm/s). The five slippage patterns were designed to study the sensing of the end effectors velocity. Second and third patterns transmit different speeds on contact points, and other patterns deliver equal rates (Fig. 3 (b)). Each pattern was presented five times in random order, thus, 25 patters were provided to each subject.

Experimental Results

The results of the experiment are summarized in a confusion matrix (Table 3). The mean percentage of the correct answers is 81%. The most distinctive speed patterns are 2nd and 3rd, with a recognition rate of 94% and 89%. They have represented patterns with different velocities of the end effector in two contact points. And the most confusing pattern is 5th with a recognition rate of 65%.

Table 3. Confusion matrix for actual and perceived slippage patterns across all the subjects

Actual pattern \ Subject response	1	2	3	4	5
1	0.76	0.02	0.02		0.2
2		0.94	0.02	0.04	
3			0.89	0.02	0.09
4		0.02	0.05	0.82	0.11
5	0.25		0.04	0.06	0.65

Using the one-factor ANOVA without replications, with a chosen significance level of $p < 0.05$, we found a statistically significant difference among the different dynamic patterns ($F(5, 50) = 3.28,\ p = 1.83 \cdot 10^{-2} < 0.05$). According to ANOVA results, $2nd$ pattern has a significantly higher recognition rate than the patterns $1st(F(1, 20) = 5.75,\ p = 0.026 < 0.05)$ and $5th(F(1, 20) = 11.4,\ p = 2.97 \cdot 10^{-3} < 0.05)$. It was significantly easier for participants to recognize the $3rd$ pattern than the $5th(F(1, 20) = 6.17,\ p = 0.022 < 0.05)$.

4 Conclusion

We have developed LinkRing, a wearable haptic display that can provide multi-contact stimuli in two independent points of the user's finger. The device is capable of generating a wide range of tactile sensations such as contact, slippage, twist stimuli, and pressure. The structure of the device is lightweight and easy to wear. The user study revealed high recognition rates in discrimination of static and dynamic patterns delivered to the finger pads. The obtained results allow us to determine the most suitable patterns for further presenting the static and moving object for the finger perception with the proposed display.

The future work will be aimed at expending multi-modal stimuli by adding vibration motors to the end effectors as well as improving the design of the device by reducing its dimensions and increasing its ergonomics. Various virtual applications are going to be developed to study virtual immersion quality and fidelity of multi-modal tactile stimuli. The developed haptic display can potentially bring a highly immersive VR experience in the guiding blind navigation systems, teleoperation, and medical VR simulators.

References

1. Cabrera, M.A., Tsetserukou, D.: LinkGlide: a wearable haptic display with inverted five-bar linkages for delivering multi-contact and multi-modal tactile stimuli. In: Kajimoto, H., Lee, D., Kim, S.-Y., Konyo, M., Kyung, K.-U. (eds.) AsiaHaptics 2018. LNEE, vol. 535, pp. 149–154. Springer, Singapore (2019). https://doi.org/10.1007/978-981-13-3194-7_33
2. Chinello, F., Malvezzi, M., Pacchierotti, C., Prattichizzo, D.: Design and development of a 3RRS wearable fingertip cutaneous device. In: 2015 IEEE International Conference on Advanced Intelligent Mechatronics (AIM), pp. 293–298. IEEE (2015)
3. Frediani, G., Mazzei, D., De Rossi, D.E., Carpi, F.: Wearable wireless tactile display for virtual interactions with soft bodies. Front. Bioeng. Biotechnol. **2**, 31 (2014)
4. Gabardi, M., Solazzi, M., Leonardis, D., Frisoli, A.: A new wearable fingertip haptic interface for the rendering of virtual shapes and surface features. In: 2016 IEEE Haptics Symposium (HAPTICS), pp. 140–146. IEEE (2016)
5. Meyer, D.J., Wiertlewski, M., Peshkin, M.A., Colgate, J.E.: Dynamics of ultrasonic and electrostatic friction modulation for rendering texture on haptic surfaces. In: 2014 IEEE Haptics Symposium (HAPTICS), pp. 63–67, February 2014. https://doi.org/10.1109/HAPTICS.2014.6775434

6. Moriyama, T.K., Nishi, A., Sakuragi, R., Nakamura, T., Kajimoto, H.: Development of a wearable haptic device that presents haptics sensation of the finger pad to the forearm. In: 2018 IEEE Haptics Symposium (HAPTICS), pp. 180–185. IEEE (2018)
7. Romano, J.M., Yoshioka, T., Kuchenbecker, K.J.: Automatic filter design for synthesis of haptic textures from recorded acceleration data. In: 2010 IEEE International Conference on Robotics and Automation, pp. 1815–1821, May 2010. https://doi.org/10.1109/ROBOT.2010.5509853
8. Scheggi, S., Meli, L., Pacchierotti, C., Prattichizzo, D.: Touch the virtual reality: using the leap motion controller for hand tracking and wearable tactile devices for immersive haptic rendering. In: ACM SIGGRAPH 2015 Posters. Association for Computing Machinery (2015)
9. Schorr, S.B., Okamura, A.M.: Three-dimensional skin deformation as force substitution: wearable device design and performance during haptic exploration of virtual environments. IEEE Trans. Haptics **10**(3), 418–430 (2017)
10. Tsetserukou, D., Hosokawa, S., Terashima, K.: LinkTouch: a wearable haptic device with five-bar linkage mechanism for presentation of two-DOF force feedback at the fingerpad. In: 2014 IEEE Haptics Symposium (HAPTICS), pp. 307–312, February 2014. https://doi.org/10.1109/HAPTICS.2014.6775473
11. Villa Salazar, D.S., Pacchierotti, C., De Tinguy De La Girouliere, X., Maciel, A., Marchal, M.: Altering the stiffness, friction, and shape perception of tangible objects in virtual reality using wearable haptics. IEEE Trans. Haptics **13**(1), 167–174 (2020). https://doi.org/10.1109/TOH.2020.2967389
12. Yatani, K., Truong, K.N.: SemFeel: a user interface with semantic tactile feedback for mobile touch-screen devices. In: Proceedings of the 22nd Annual ACM Symposium on User Interface Software and Technology, UIST 2009, pp. 111–120 (2009). https://doi.org/10.1145/1622176.1622198
13. Yem, V., Kajimoto, H.: Wearable tactile device using mechanical and electrical stimulation for fingertip interaction with virtual world. In: Proceedings - IEEE Virtual Reality, pp. 99–104 (2017). https://doi.org/10.1109/VR.2017.7892236
14. Yamamoto, S., Mori, H. (eds.): HCII 2019. LNCS, vol. 11570. Springer, Cham (2019). https://doi.org/10.1007/978-3-030-22649-7
15. Yem, V., Vu, K., Kon, Y., Kajimoto, H.: Softness-hardness and stickiness feedback using electrical stimulation while touching a virtual object. In: 25th IEEE Conference on Virtual Reality and 3D User Interfaces, VR 2018 - Proceedings (2018). https://doi.org/10.1109/VR.2018.8446516

ElectroAR: Distributed Electro-Tactile Stimulation for Tactile Transfer

Jonathan Tirado[1(✉)], Vladislav Panov[1(✉)], Vibol Yem[2(✉)], Dzmitry Tsetserukou[1(✉)], and Hiroyuki Kajimoto[3(✉)]

[1] Skolkovo Institute of Science and Technology (Skoltech), Moscow 121205, Russia
{jonathan.tirado,vladislav.panov,D.tsetserukou}@skoltech.ru

[2] Tokyo Metropolitan University, Tokyo, Japan
yemvibol@tmu.ac.jp

[3] The University of Electro-Communications, 1-5-1 Chofugaoka, Chofu, Tokyo 182-8585, Japan
kajimoto@kaji-lab.jp

Abstract. We present ElectroAR, a visual and tactile sharing system for hand skills training. This system comprises a head-mounted display (HMD), two cameras, a tactile sensing glove, and an electro-tactile stimulation glove. The trainee wears the tactile sensing glove that gets pressure data from touching different objects. His/her movements are recorded by two cameras, which are located in front and top side of the workspace. In the remote site, the trainer wears the electro-tactile stimulation glove. This glove transforms the remotely collected pressure data to electro-tactile stimuli. Additionally, the trainer wears an HMD to see and guide the movements of the trainee. The key part of this project is to combine distributed tactile sensor and electro-tactile display to let the trainer understand what the trainee is doing. Results show our system supports a higher user's recognition performance.

Keywords: Tactile display · Tactile sensor · Tactile transmission · Virtual reality

1 Introduction

There are several tasks that incorporate hand-skill training, such as surgery, palpation, handwriting, etc. We are developing an environment where a skilled person (trainer), who actually works at a different place, can collaborate with a non-skilled person (trainee) in high precision activities. The trainer needs to feel as if he/she exists at the place and work there. The trainee can improve his/her performance with the trainer's help. This can be regarded as one type of telexistence [1], in which remote robot is replaced by trainee.

We especially focus on the situation that incorporates finger contact. This requires a tactile sensor on the trainee's side and tactile display on the trainer's side. The trainee handles real objects, and the tactile sensor-display pair enables

I. Nisky et al. (Eds.): EuroHaptics 2020, LNCS 12272, pp. 442–450, 2020.
https://doi.org/10.1007/978-3-030-58147-3_49

the trainer to feel the same tactile experience as the trainee; thus, he/she can command or show what the trainee should do next.

For tactile sensors, a wide variety of these pads have been developed in the past for robotics and medical applications, using resistive, capacitive, piezoelectric, or optical elements. These pads have often been placed in gloves to monitor hand manipulation. While some of them are bulky and inevitably deteriorate the human haptic sense, recently, several researches are focused on reducing this problem by using thinner and more flexible force-sensing pads [3]. In this study, we use a similar tactile sensor array with high spatio-temporal resolution.

For tactile display, there were also several researches on wearable tactile displays [4]. They are simple, yet cannot present distributed tactile information that our sensor can detect. As we believe that distributed tactile information is important, especially when we recognize shapes, we need some way to present distributed tactile information to fingertips. There were also several works on pin-array type tactile display [2, 5]. We employ electro-tactile display [6, 7], since it is durable, light-weight, and easy to be made small and extends to several fingers.

This paper is an initial report of our system, especially focuses on how well the shape information can be transmitted through our system.

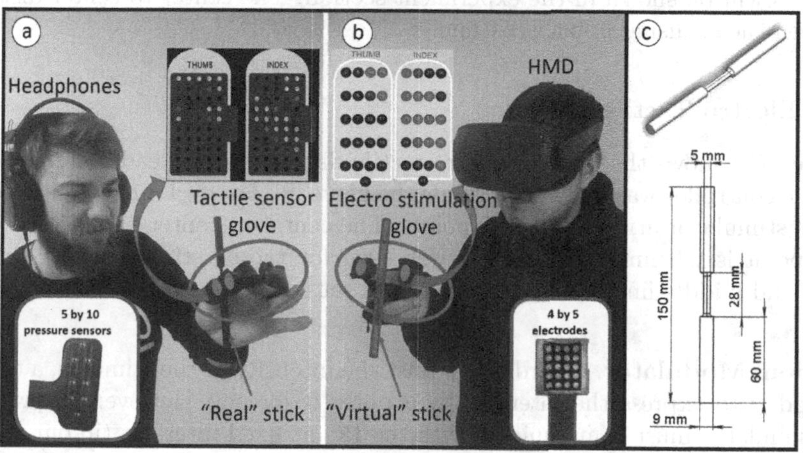

Fig. 1. ElectroAR. (a) Follower side. (b) Leader side. (c) Cylindrical stick with regular prismatic shape

2 System Overview

As shown in Fig. 1, the system consists of three main components. On the follower (trainee's) side, the user wears a tactile sensing glove. The glove gets the pressure data from touching objects. The data of pressure sensors were spatially filtered by using Eq. (1),

$$p'_{i,j} = \frac{p_{i,j} + p_{i+1,j} + p_{i,j+1} + p_{i+1,j+1}}{4} \tag{1}$$

where p is a pressure value and p' is a filtered pressure value, i and j are order number on the axis of width and height of the sensor array [9].

The leader's glove transforms the filtered pressure data to electro-tactile stimuli at fingertips. They are linked not only with haptic feedback but also with visual and audio feedback. Visual feedback gives for the leader side full information of the movement on the follower side, and audio feedback provides for the follower side commands from the leader.

2.1 Tactile Sensor Glove

We are using a glove that contains three tactile sensor arrays [9]. These sensor arrays are located on the three fingers of the right hand (thumb, index and middle). Figure 1 (a) shows the internal distribution of the pressure sensors in the array of 5 by 10 for each finger. The force range of sensing element was not accurately measured, but it can discriminate edge shapes by natural pressing force, as will be shown in the experiment section. The center-to-center distance between each sensing point is 2.0 mm.

2.2 Electro-Tactile Glove

Figure 1 (b) shows the glove of electro-tactile display for the leader user [9]. The module controller was embedded inside the glove [8]. For each finger, the electro-tactile stimulator array has 4 by 5 points. The center-to-center distance between each point is 2.0 mm. This module was used for tactile stimulation of thumb, index and middle finger. The pulse width is set to 100 *us*.

Random Modulator. In order to adjust the intensity of the stimulus, a typical method is to express the intensity by a pulse frequency. However, in practice, the stimulator must communicate with the PC at fixed intervals (in our case at 120 Hz). Therefore, although it is relatively easy to set the pulse frequency to, for example, 30 Hz, 60 Hz, or 120 Hz, it is a little difficult to perform electrical stimulation of an arbitrary frequency.

Here, we propose a method to change the probability of stimulation as a substitute for setting pulse frequency. For each time interval (in our case 1/120 *second*), the system gives the probability of stimulating each electrode. The higher the probability, the higher the average stimulus frequency. The algorithm is expressed as follows.

$$\textbf{if}\ \ \textit{rand}\,() \leq p\ \ \textbf{then}\ \ \textit{stimulate}\,() \tag{2}$$

Where *rand* () is a uniformly distributed random variable from 0 to 1. If it is less than or equal to a value p, the electrode is stimulated. Otherwise, it is not stimulated. The probability that the electrode is stimulated is hence p. This calculation is performed for the electrode every cycle, resulting in an average stimulation cycle of $120 * p$ Hz.

The value p represents the probability of stimulation, and a function representing the relationship between p and the subjective stimulus intensity S is required. In general, higher stimulus frequency gives stronger subjective stimulus, so this function is considered to be a monotonically increasing function.

$$S = F(p) \tag{3}$$

If F is obtained, the inverse function can be used to determine how the stimulus probability p should be set for the intensity S to be expressed as follows.

$$p = F^{-1}(s) \tag{4}$$

2.3 View Sharing System

Ideally, the view sharing system should be bi-directional. However, as the scope of this paper is to examine the ability of our tactile sensor-display pair, we used a simplified visual system only for the trainer.

As shown in Fig. 1 (b), the trainer wears an HMD. At the remote side, two cameras are installed for having full view information for the trainer, both from the top and from the side. This information is presented in virtual screens which are located in front and the horizontal view. Although the view is not three-dimensional, it can provide sufficient information of the trainee's hand movement, and the trainer can mimic the movement while perceiving the tactile sensation by the electro-tactile display glove.

3 Experiment

3.1 Preliminary Experiment : Random Modulator's Function

The proposed random modulation method needs a function F, which can represent the relationship between strength perception and the probability of stimulating each electrode. This preliminary experiment has the objective of collect data for fitting function F. In the whole experiment, the base stimulation frequency was 120 Hz. For example, if the probability is 1, the stimulation is done at 120 *pps* (pulses per second).

Experimental Method. The strength of stimulation was evaluated by the magnitude estimation method. First, the user's right index fingertip was put on the electrodes' array, and exposed to a pulsatory stimulation, provided by electrodes. The user was asked to find a comfortable and recognizable level (absolute stimulation level), which was set as 100.

In the second part, we prepared six probability levels: 0.1, 0.2, 0.4, 0.6, 0.8, 1.0. There were five trials for each level, 30 trials in total in random order. Each trial was composed of an initial one-second impulse with the 100 intensity level, followed by a one-second randomly modulated stimulation with assigned probability. After each trial, the user must determine how lower or higher was the second stimulus presented. We recruited seven participants, five males and two females aged 21–27; all right-handed and all without previous training.

Result. The result in Fig. 2 (a) shows a sigmoid function tendency. Thus, the data were fitted using Matlab, as shown in Fig. 2 (b).

Once we know the function, we calculated the inverse function that determines the stimulation probability from desired strength, which is the function F^{-1}, described in the Eq. (5), where a, b and k are coefficients of the sigmoid function, p is the probability of the electrode being stimulated and S is the subjective stimulus intensity.

$$p = \frac{a - log(\frac{k}{S} - 1)}{b} \tag{5}$$

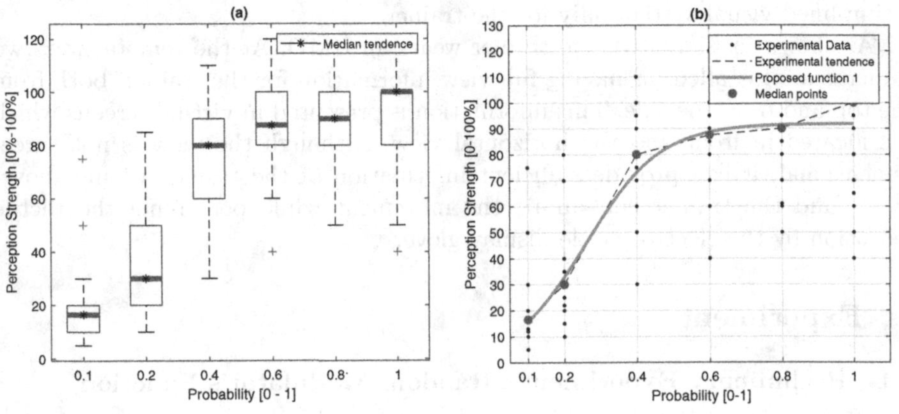

Fig. 2. Random Modulation. (a) Experimental results. The quantitative relation between cumulative probability distribution and the strength perception percentage estimated for the volunteers. (b) Sigmoid function regression. Experimental data were fitted to sigmoid function by logistic regression

3.2 Experiment 1: Static Shapes Recognition

The following two experiments try to validate that our system is capable of transmitting tactile information necessary for tactile skill transfer. In many haptic related tasks, we typically use a pen-type device that we pinch by our index finger and thumb. These can be a scalpel, a driver, a tweezer, or a pencil. In such situations, we identify the orientation of the device with tactile sense.

Our series of experiments try to reproduce part of these situations. Experiment 1 was carried out to assess the electro-tactile display's capacity for presenting bar-shape in different orientations.

Experimental Method. Four patterns, which are line with inclinations of 0, 45, 90 and 135° were presented on the right index finger. The experiment was divided into three steps. The first step was to identify a suitable stimulus level. The second step was the training phase, in which each pattern was presented twice to the volunteers.

After a two minutes break, the evaluation stage was performed. They were asked to try randomly chosen pattern and chose from the four candidates. The recognition time was also recorded. We recruited ten volunteers, nine males and one female, aged 21–27; all right-handed. There were seven trials per pattern, 28 in total.

Result. Figure 3 (a) shows a numerical comparison of the effective recognition level for each proposed pattern. The four patterns have a similar range of recognition, being the 90° pattern pointed the lowest (73% accuracy) and the 0° pattern the highest (87% accuracy). The result also indicates that the 90° pattern is often confused with the 135° (10% error), and in the same way the 45° is confused with 90° pattern (10% error).

Figure 3 (b) shows that the recognition time for the majority of the volunteers ranges between 4 and 10 s for all of the patterns. The median time is close to 6 s.

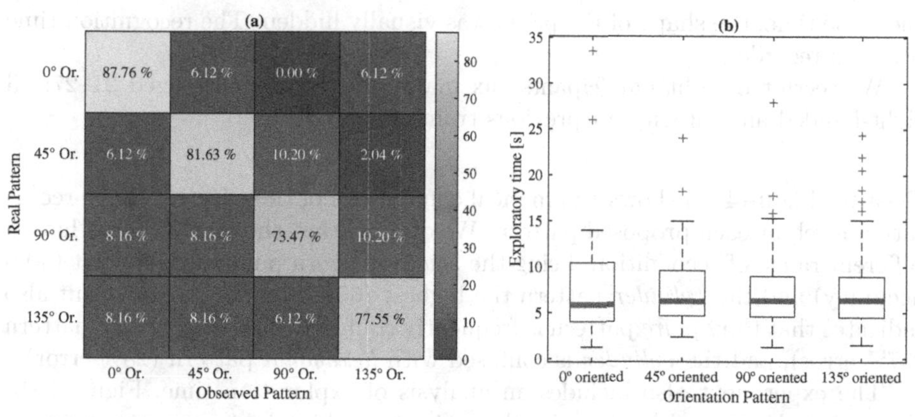

Fig. 3. Experiment 1. (a) Confusion Matrix Pattern recognition rate. (b) Exploratory time comparative evaluation

3.3 Experiment 2: Dynamic Pattern Perception

Experiment 2 was carried out to assess our system's capacity to convey dynamic tactile information. As mentioned before, we focused on the situation of handling a bar-shaped device. We confirmed if we can identify different "devices" that we handle with our index finger and thumb.

Data Set Acquisition. Four cylindrical sticks with regular prismatic shape in their middle section were designed for the experiment (Fig. 1 (c)). The total length of each stick is 150 mm, and 28 mm for their middle section. Every prism has a different cross-section: circle, triangle, square, and hexagon. The radius of the sticks was 9 mm and the circumradius of the prisms 5 mm. This special design visually covers the middle section for avoiding the possibility of answering only by observation. On this way, we provide only the motion of the hand as visual feedback.

Using the tactile sensing glove, one of the authors grasped the stick in a 90° orientation, and he slowly scrolled the bar back and force between two fingers, repeating for ten times. The pressure patterns were recorded, and the video was taken by two cameras that we described in the previous section.

Experimental Method. In the main experiment, the recorded videos were replayed so that the user can mimic the hand motion. Simultaneously, the tactile feedback was delivered to two fingertips (right index finger and thumb) using the recorded pressure patterns.

A set of twenty randomly ordered samples was presented, and the user must associate this visual and tactile sensation with one of the previously indicated shapes. Visual feedback was provided to show the motion of the hand, but at the same time, the shape of the prism was visually hidden. The recognition time was also recorded.

We recruited eight participants, six males and two females aged 21–27; all right-handed and all without previous training.

Result. Figure 4 (a) shows a numerical comparison of the effective shape recognition level for each proposed pattern. We observe that the four patterns have a different range of recognition, being the *square* pattern pointed the lowest (40% accuracy) and the *cylinder* pattern the highest (65% accuracy). The result also indicates that the *square* pattern is frequently confused with the *triangle* pattern (37% error), and the *cylinder* is confused with *hexagonal* pattern (25% error).

The experiment also includes an analysis of exploration time. Figure 4 (b) shows that the recognition time for the majority of the volunteers ranges between 8 and 18 s. The median time is close to 13 s also for all of the cases, except for the *triangle* pattern which median exploratory time is 16 s.

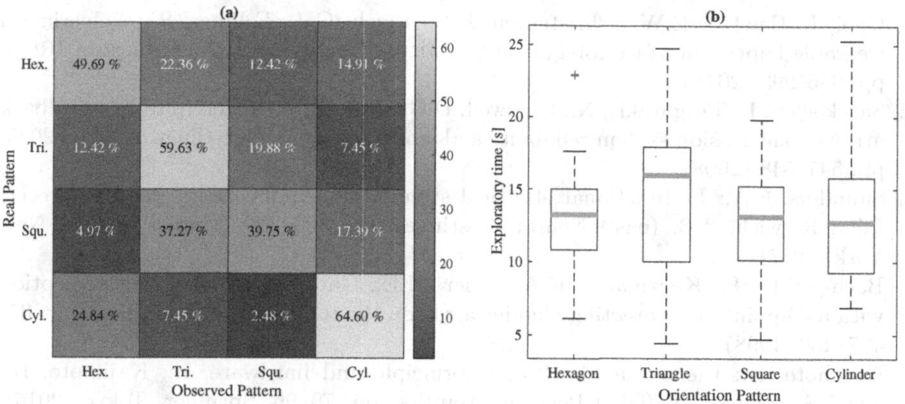

Fig. 4. Experiment 2. (a) Confusion Matrix about 3D Shape recognition rate. (b) Exploratory time comparative evaluation for 3D shape recognition

4 Conclusion

In this paper, we mainly developed a haptic feedback component of the virtual reality system for remote training. We implemented a simple tactile communication capable of transmitting shape sensations produced at the moment of manipulating 3D objects with two fingers: thumb and index fingers. The follower side comprises a tactile-sensor glove and the leader side comprises an electro-tactile display glove.

We tested our system with two experiments: static shape perception and dynamic pattern perception, both assuming the situation of grasping a bar-like object. The results confirmed our expectations, that this system has the ability to deliver information of 3D bar-like object.

There are several limitations to the current work. The visual part of the system is incomplete; the follower side should see the hand gesture of the trainer, and the leader side should see 3D visual information of the follower by the use of 3D display technologies. Tactile display and sensor are slightly small, and it must be enlarged to cover the whole fingertips. Roughness and temperature sensations must be considered for providing material sense. All these will be handled in our future work.

References

1. Tachi, S.: Tele-existence - toward virtual existence in real and/or virtual worlds. In: Proceedings of ICAT 1991, pp. 85–94 (1991)
2. Kim, S.-C., et al.: Small and lightweight tactile display (SaLT) and its application. In: Proceedings WorldHaptics, pp. 69–74 (2009)
3. Beebe, D., Denton, D., Radwin, R., Webster, J.: A silicon-based tactile sensor for finger-mounted applications. IEEE Trans. Biomed. Eng. **45**, 151–159 (1998)

4. Choi, I., Hawkes, E.W., Christensen, D.L., Ploch, C.J., Follmer, S.: Wolverine: a wearable haptic interface for grasping in virtual reality. In. Proceedings of IROS, pp. 986–993 (2016)
5. Sarakoglou, I., Tsagarakis, N., Caldwell, D.G.: A portable fingertip tactile feedback array - transmission system reliability and modelling. In: Proceedings of WHC 2005, pp. 547–548 (2008)
6. Saunders, F.A.: In functional electrical stimulation: applications. In: Hambrecht, F.T., Reswick, J.B. (eds.) Neural Prostheses, pp. 303–309. Marcel Dekker, New York (1977)
7. Bach-y-Rita, P., Kaczmarek, K.A., Tyler, M.E., Garcia-Lara, J.: Form perception with a 49-point electrotactile stimulus array on the tongue. J. Rehab. Res. Dev. **35**, 427–430 (1998)
8. Kajimoto, H.: Electro-tactile display: principle and hardware. In: Kajimoto, H., Saga, S., Konyo, M. (eds.) Pervasive Haptics, pp. 79–96. Springer, Tokyo (2016). https://doi.org/10.1007/978-4-431-55772-2_5
9. Yem, V., Kajimoto, H., Sato, K., Yoshihara, H.: A system of tactile transmission on the fingertips with electrical-thermal and vibration stimulation. In: Yamamoto, S., Mori, H. (eds.) HCII 2019. LNCS, vol. 11570, pp. 101–113. Springer, Cham (2019). https://doi.org/10.1007/978-3-030-22649-7_9

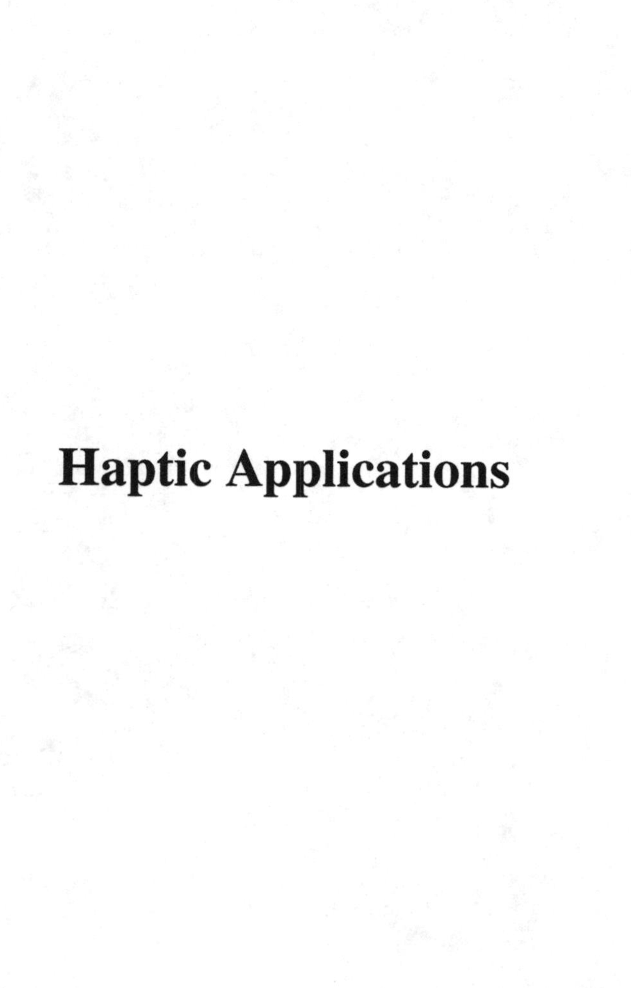

Haptic Applications

Identification Rate of Simple and Complex Tactile Alerts in MUM-T Setup

Dana Rosenblum(✉), Nuphar Katzman, and Tal Oron-Gilad

Human Factors and Ergonomics Laboratory, Department of Industrial Engineering and Management, Ben-Gurion University of the Negev, Beersheba, Israel
{danafa, nuphar}@post.bgu.ac.il, orontal@bgu.ac.il

Abstract. Vibro-tactile interfaces were proposed as an alternative to enhance human-machine communication in information-rich domains. The current study aims to examine the effectiveness of two levels of tactile alerts when combined with visual alerts, in MUM-T (Manned UnManned Teaming) setup. In MUM-T, aside from their primary mission, mounted operators are responsible for supportive unmanned systems and must attend to their health. On the simple level, the alert provides information about a threat or a failure in the supportive unmanned systems, while in the complex level, the alert includes more specific information about the source of failure, that may require more effort to interpret. The experiment simulates an operational mission in which participants ride an autonomous ground patrol vehicle while identifying threats and targets in the area and being supported by two unmanned systems. Response accuracy to alerts and threat identification rates were measured. Results indicate that tactile alerts given in addition to visual alerts in a visually loaded and auditory noisy scene, improve task performance. Moreover, the complex level of tactile alerts did not impair performance compared to the simple level of tactile alerts and led to higher rate of identification in specific cases. Nevertheless, relatively high rates of false alarms (FA) for threats were observed, especially when tactile alerts were present, which can be explained by the payment matrix (no penalty) or by the assumption that adding tactile alerts may lead participants to be more vigilant, which can lead to higher correct identifications, but also to higher FA rates.

Keywords: Tactile alert · Level of alert · Operational missions · MUM-T setup

1 Introduction

Military operational activities require obtaining large quantities of data from multiple sources in a short period of time while running multiple tasks simultaneously [1]. Operators are expected to detect changes quickly and respond systematically and fast [2]. Unmanned systems are potent force multipliers, in the MUM-T (Manned UnManned Teaming) operational concept, they are controlled from a moving ground or aerial platform, and are used for tasks that otherwise would have been taken over by other manned or remotely operated platforms. MUM-T can cause significant task load increase for its operating crew and key design elements are necessary for achieving the

I. Nisky et al. (Eds.): EuroHaptics 2020, LNCS 12272, pp. 453–461, 2020.
https://doi.org/10.1007/978-3-030-58147-3_50

workload reduction necessary to facilitate it [3]. We focus on the use of tactile alerts to improve operators' ability to detect threats and attend to failures. Tactile alerts may enhance attentional abilities by distributing information between several resources [4], as visual alerts alone are less suited for this level of complexity [5]. The tactile channel can be used to alleviate workload and attract operators' attention to mission related notifications at various levels and events [6].

Yet, like the visual and auditory channels, frequent use of tactile alerts may increase workload and impair performance [7]. Moreover, multiple alerts can lead to neglect, where the alerts are ignored [8], mainly when the task is unclear and the transfer of information to the operator is impaired [9]. Therefore, the tactile alerting system must be used properly. Tactile cues can be used as signals with lower information processing requirements [10]. According to Elliot et al. [11], when tactile cues are added to existing visual cues, performance improves. An important question is content related what kind of tactile messages should be transferred by a tactile display.

The current study was aimed to examine the level of tactile alert (simple/complex) that should be used in a MUM-T operational setup. For this, a burdening, visually and auditory loaded operational mission was examined. We defined two levels of tactile alerts - simple and complex: A simple alert provides information that is easy to interpret about the occurrence of a pre-determined event. It is defined as a binary happened/did-not-happen alert; A complex alert requires more effort to interpret, but includes more specific informative information. It provides the operator with at least one more layer of information, e.g., source, direction, distance, etc. [6]. Two types of tactile alerts were given; A "danger" alert which indicates upon the presence of hostile targets, and a "failure" alert which indicates upon a technical failure in one of the supportive unmanned systems. "Danger" alert was defined as more immediate and required faster responses.

It was hypothesized that *H:* Identification rate of alerts will be higher when tactile alerts are given in addition to visual ones, more so, for the tactile alerts provided in the "complex" level for both "danger" and "failure" alert types.

2 Method

The study aimed to examine the effectiveness of two levels of tactile alerts when combined with visual alerts. Each participant was placed in a simulated workstation of an autonomous ground patrol vehicle that navigated autonomously. While driven along the route, the operator was required to identify threats and targets in the mission area and manage two unmanned vehicles (UVs), an unmanned ground vehicle (UGV) and a drone, that support the mission in a MUM-T setup. Occasionally a failure occurred in one of the supporting vehicles and the operator had to attend to it while continuously seeking for threats and targets. Each participant executed three different scenarios, one scenario for each level of tactile alerts (none, simple, complex). Visual alerts were presented on a computer screen (#3 in Fig. 1b); Tactile alerts were transmitted through a wearable tactile display located on the soldiers' forearms (Fig. 1a).

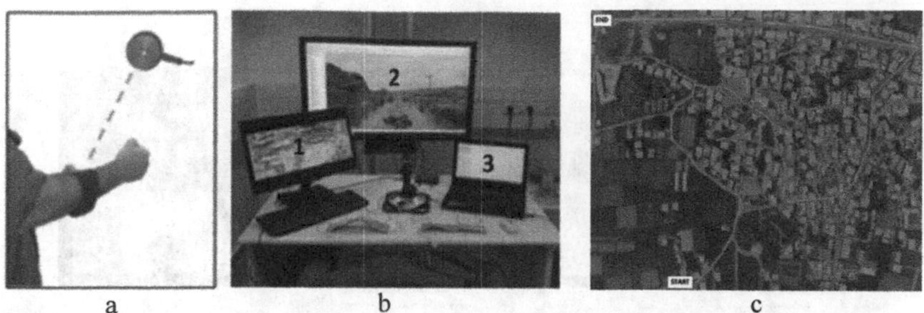

a b c

Fig. 1. a) Tactile system: two tactor forearm bracelets (one on each arm) and C2 tactor; b) Experimental setup. Screen #1 navigation map, screen #2 main navigation display, and screen #3 visual display in which participants receive information and react; c) Example of a navigation map (screen #1)

Three females and 23 males, students aged 24–30 (M = 27.1, SD = 1.6) all military reserve soldiers on active duty at least once in the year prior to the experiment participated in the study. Recruitment was via social media and compensation was 40 NIS for one hour. Nine participants' objective performance data was withdrawn from the study due to a technical problem in saving their data, leaving a total of 17 participants. These withdrawn participants did not change the population's parameters.

2.1 Apparatus

Experimental Environment. The study was conducted at the Human Performance Lab. The simulator consists of three visual displays in different sizes: screen #1 - 22′, screen #2 - 42′, screen #3 - 13.3′, running a simulation of the operational mission (see Fig. 1b). Participants sat about one meter away from the main screen (#2 in Fig. 1b). The laboratory is temperature and noise controlled.

Operational Tasks. To accomplish the mission three task types were required. Ongoing: the task was to detect targets and mark them by pressing a "target" button on visual display (#3). The target button was blinking in red three times when a target was identified by one of the supportive UVs (Fig. 2), 14–16 targets per session with a variation of 10–85 s between two consecutive targets. UV status: the participant was asked to identify failures in the UVs and report them, 11–13 failure alerts per session, with a variation of 5–40 s gap between two alerts. Gray and blue buttons were blinking when there was a failure in one of UVs (gray for drone and blue for UGV) and stopped blinking when the operator pressed the corresponding button on the visual display (#3; Fig. 2). A secondary task was to report upon identification of vehicles via the communications device: "car identified". The purpose of this task was to serve as a distracting element to the main mission tasks.

Headphones. (Samson SR850) were used to transmit typical military radio communication, presented at a range of 40 dB with short peaks of up to 80 dB, to add to the realism of the simulation and increase mental demand.

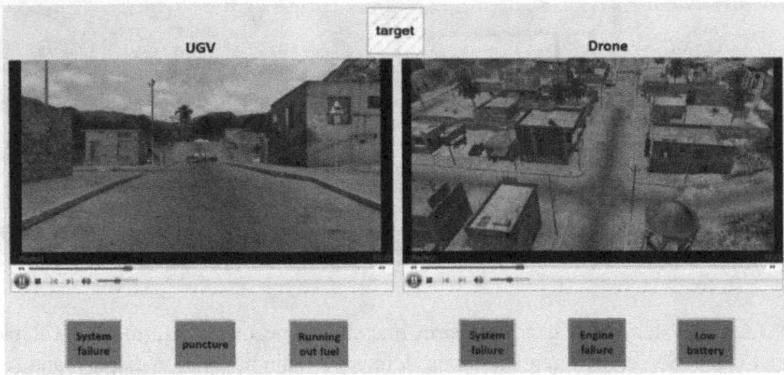

Fig. 2. The visual display where participants received the alerts and responeded to them (touchscreen, screen #3). On the top is the target indicator and on the bottom are the various failures that could have occurred for the UGV and drone.

Tactile Interface. Military applications strive for simplicity on one hand, and redundancy on the other hand for robustness, therefore a two-tactor system was chosen but with one signal [12]. The two C2 tactors (i.e., vibrating tactile actuators) powered by the Eval 2.0 controller (Engineering Acoustics Inc (EAI)) and were stitched to two elastic-fiber straps. The straps were worn around the forearms, one on each hand, over the participant's skin (see Fig. 1a). The tactile alerts included four different temporal patterns (on-off and duration modulation) through the tactile system (see *Tactile alerts' design*). The gain scale ranges from 1–255 units, as determined by the EAI apparatus.

Tactile Alerts' Design. Two types of tactile alerts were chosen: one for conveying "danger" and one for conveying "failure". The tactile signals for each alert type were chosen following a pre-experiment. For the pre-experiment five tactile cues were designed according to a list of criteria such as the alerts' overall length, gain and pulse rate (see Fig. 3a). The experimental part of the pre-experiment was divided into two. The first part included a navigation task, in which each participant was exposed to all five different cues once in random order. Time gap between two cues was at least two minutes. Participant had to choose which kind of meaning better describes the alert that has just been executed: "danger" or "failure". Participants responded while the navigation mission continued. In the second part, participants were able to feel all five cues for an unlimited period of time, by pressing on the alert buttons. They were asked to choose the most appropriate alert for "danger" and afterwards for "failure". From the pre-experiment, for the "danger" alert, all five alert types were chosen almost at the same rate in the first part, while in the second part alerts C, D and E were hardly chosen at all, therefore, it was decided not to use them for the context of danger. Among alerts A and B, it was decided to choose alert B for "danger" although the rates of the two alerts were close, with a small favor for B in both parts of the pre-experiment. For the "failure" alert, alerts A and B were less chosen by participants in the first part of the pre-experiment, and there was almost an equality between alerts C, D and E. To choose

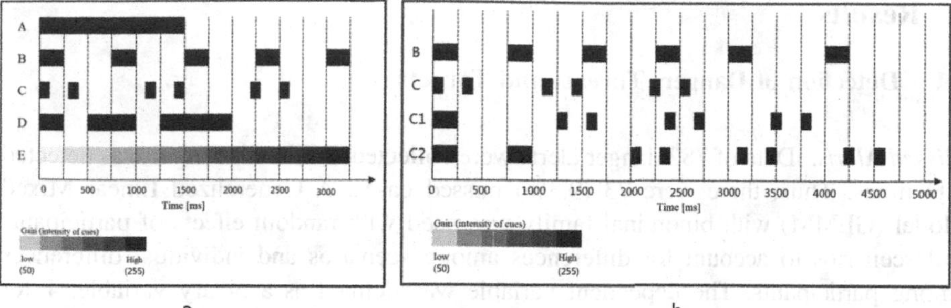

Fig. 3. a) Tactile cue patterns used in the pre-experiment. b) Tactile cue patterns selected to the experiment. In the simple condition: Alert B for "danger", alert C for "failure", in the complex condition: Alert B for "danger", alert C1 for failure in the UGV and alert C2 for failure in the drone.

the fitted alert for "failure", we used the second part of the pre-experiment, where there was an unequivocal majority for alert C. It was decided to choose alert C as "failure" alert.

Thus, alert B was chosen for "danger" and alert C for "failure" for the current experiment. The "simple" tactile alert condition consisted of alerts B for danger/threat and C for "failure" in any one of the UVs. The "complex" tactile alert condition consisted of the same alert B for danger/threat and two alerts for "failure" depending on the source of the failure: alert C1 for failure in the UGV and alert C2 for failure in the drone (see Fig. 3b).

2.2 Procedure

Participants were invited solely to the lab for one-hour long sessions. At the beginning of the session, each participant read the instructions and signed informed consent. Following a short verbal briefing, each participant got familiarized with the task (the way targets were marked and how to detect them, how to identify failures of the vehicles and how to report them). A separate briefing about the tactile and visual displays was given in order to familiarize participants with the alerts.

The experimental part of the study included a trial scenario (of two minutes) in which participants received both visual and tactile alerts ("simple" condition), and three operational scenarios, each participant performed all scenarios. Participants completed each scenario with one of the three modality conditions. The scenarios' order was fixed, but the modality conditions were balanced across participants. The length of each section was approximately nine minutes, with a break of a few minutes between sessions to prepare participants for the next session.

3 Results

3.1 Detection of Danger, Threats and Targets

Missed Alerts. Data of 782 danger alerts were collected. All in all, danger was detected 709 times. Thus, there were 73 (9.3%) missed cases. A Generalized Linear Mixed Model (GLMM) with binominal family was used with random effects of participants and scenarios to account for differences among scenarios and individual differences among participants. The dependent variable was defined as a binary variable: 1 for reacting to danger, and 0 for missing a danger, and the independent variable was the alert level (none, simple, complex). The final model includes a main effect of alert level (χ^2 (df = 2) = 7.037, $p < .01$) and two random effects, for subject and scenario. Post-Hoc (Tukey-HSD) analyses revealed a significant difference between the "none" condition and the "complex" level of tactile alert; 15% missed cases without tactile alerts out of all miss alerts, and 5% with "complex" level of tactile alerts. Results remain stable in perspective of missing dangers per person according to the different conditions as shown in Fig. 4b.

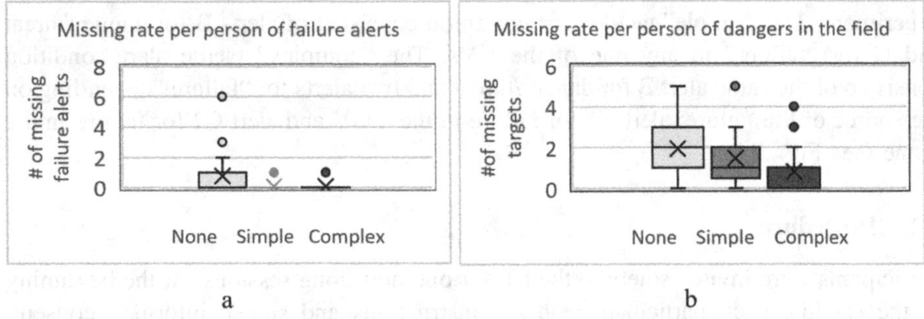

Fig. 4. a) Missing rate per person of failure alerts by condition. b) Missing rate per person of dangers in the field of operation by condition.

False Alarms. There was no penalty for FA. On average, there were 8.87 (SE = 1.54, ME = 7.50) incorrect responses (clicking on the "danger" button when no target was present) per participant, of those 2.25 occurred in the "none" condition, 3.12 occurred during the "simple" condition, and 3.5 in the "complex" condition.

3.2 Identification of Failure Alerts

Missed Alerts. Data from 672 failure alerts were collected. There were 13 (1.9%) missed notifications (no response), only 2 of which with a tactile alert. A Generalized Linear Mixed Model (GLMM) with binominal family was used with random effects of participants and scenarios to account for differences among scenarios and individual differences among participants. The dependent variable was defined as a binary variable: 1 for reaction for a failure alert, and 0 for missing a failure alert, and the independent variable was the alert level (none, simple, complex). The final model includes main effect of alert level (χ^2 (df = 2) = 5.797, $p < .01$) and two random effects, for subject and scenario. Post-Hoc (Tukey-HSD) analyses revealed that there was a significant difference between the "none" condition to the "simple" level of tactile alert; 5% missed cases without tactile alerts out of all miss alerts and 0.48% missed cases with "simple" level of tactile alerts out of all missed alerts, and also between the "none" condition to the "complex" level of tactile alert; 5% missed cases without tactile alerts out of all miss alerts, and 0.49% with "complex" level of tactile alerts. Results remain stable in perspective of missing failure alerts per person according to the different conditions as shown in Fig. 4a.

False Alarms. There were no incorrect responses (clicking on one of the "failure" buttons when no failure occurred).

4 Discussion

The aim of the experiment was to examine the effectiveness of two tactile alert levels in a multi-task operation with high workload. Previous studies have shown that tactile cues can capture attention, which may be helpful under high workload [13]. How to define the level of detail and complexity necessary of the tactile alerts is important for this operational research field. The results of the identification rate of both "danger" and "failure" alerts were significantly higher when tactile alerts were added than when only visual alerts were shown. Thus, hypothesis *H* was confirmed. Rates were 95.6% identification of tactile notifications compared to 91.1% for visual notifications only (without tactile alerts). Moreover, the "complex" level of tactile alerts provided higher rates of "danger" identification compared to the "none" condition, while the rate of "failure" identification for both the "simple" and the "complex" levels was significantly higher than for the "none". This indicates that having higher granularity of tactile alerts did not interrupt operators with performing their main task, and moreover, improved task performance. More so, "complex" tactile alerts can improve performance better than the "simple" ones when looking at the identification of "danger" alerts.

False alarm (FA) for a presence of a target was higher when tactile alerts were given. Incorrect responses (report about a target when there is no target in the field) in the military context is a severe mistake. One explanation for the high rate of FA is that due to the experimental design, operators pressed the "target" button right when they received a "danger" alert, even though this kind of alert is designed to warn the operator of a close danger/target observed by the supporting unmanned vehicles, and not an immediate one. Another explanation may lie on the "payment matrix" [14], in which when gain or loss for a specific behavior are not defined, people tend to pay less attention to false alarms. Accordingly, participants were not told that they will be fined for incorrect responses, which may have caused higher rates of FA. No fine for FA could make them prefer to react in more leniently in order to miss fewer targets. Further study should use more specific instructions regarding the fine for incorrect responses in order to avoid high rate of FA. Moreover, the data regarding the higher rate of FA in the tactile alert conditions can testify that operators did not distinguish between the alert types ("danger" and "failure") and pressed on the "target" button even when a "failure" alert was activated. However, it may also indicate that the additional tactile alerts condition led participants to be tenser and more vigilant, which led to higher correct identifications, but also to higher FA rates.

Alongside the supportive results to use the tactile alerts there are few limitations. First, in the "complex" level condition, the three tactile alerts shared a similar initiation pattern, which forces the operator to wait until the middle or the end of the alert before responding, or to use the visual display right at the beginning of the tactile alert without attending to it until the end. By that participants actually treat the "complex" level of tactile alerts as the "simple" level. Second, participant had limited time to learn and train on the tactile alert patterns, while the visual alerts are more common and straight forward. A longer learning session of the tactile patterns may lead to different results and should be taken into consideration in further researches. A follow-up study should also include a simpler level of only one simple tactile cue with no specific information regarding "danger" or "failure" (a basic level), that will be compared to the levels tested here. Possibly, the "complex" tactile alerts may create a burden on the operator and a basic level would be more effective in multi-task operation.

In conclusion, in the current study it was found that tactile alerts can be combined with visual alerts, in a multi-task operation in MUM-T setup without reducing performance. Moreover, tactile alerts in this kind of operation reduce the missing of "danger" alerts, which is important for operational setups where missing alerts can lead to severe consequences.

Acknowledgement. This work was supported by the US Army Research Laboratory through the GDLS subcontract no: GDLS PO 40253724 (B.G. Negev Technologies and Applications Ltd) under Prime Contract no W911MF-10-2-0016 (Robotics Consortium), Robotics CTA 2015-2020, T2C1S3C, Michael Barnes, Technical Monitor. The views expressed in this work are those of the authors and do not reflect an official Army policy. The work is unclassified, Approved for public release.

References

1. Katzman, N., Oron-Gilad, T.: Towards a taxonomy of vibro-tactile cues for operational missions. In: EuroHaptics, no. 2018 (2018)
2. Oron-Gilad, T., Redden, E.S., Minkov, Y.: Robotic displays for dismounted warfighters: a field study. J. Cogn. Eng. Decis. Making **5**(1), 29–54 (2011)
3. Strenzke, R., Schulte, A.: Human-automation cooperation issues in manned-unmanned teaming. In: 2012 AUVSI Unmanned Systems North America Conference, vol. 2, pp. 851–870 (2012)
4. Wickens, C.D.: Multiple resources and mental workload. Hum. Factors **50**(3), 449–455 (2008)
5. Sklar, A.E., Sarter, N.B.: Good vibrations: tactile feedback in support of attention allocation and human-automation coordination in event-driven domains. Hum. Factors J. Hum. Factors Ergon. Soc. **41**(4), 543–552 (1999)

6. Katzman, N., Oron-Gilad, T., Salzer, Y.: Tactile interfaces for dismounted soldiers: user-perceptions on content, context and loci. In: Proceedings of the Human Factors and Ergonomics Society Annual Meeting (2015)
7. Ferris, T., Stringfield, K., Sarter, N.: Tactile 'change blindness' in the detection of vibration intensity, pp. 1316–1320 (2010)
8. Marsh, E.B., Hillis, A.E.: Dissociation between egocentric and allocentric visuospatial and tactile neglect in acute stroke. Cortex **44**(9), 1215–1220 (2008)
9. Mortimer, B.J.P., Elliott, L.R.: Identifying errors in tactile displays and best practice usage guidelines. In: Chen, J. (ed.) Advances in Human Factors in Robots and Unmanned Systems. AISC, vol. 595, pp. 226–235. Springer, Cham (2017). https://doi.org/10.1007/978-3-319-60384-1_22
10. Baldwin, C.L., et al.: Multimodal cueing: the relative benefits of the auditory, visual, and tactile channels in complex environments. In: Proceedings of the Human Factors and Ergonomics Society Annual Meeting, pp. 1431–1435 (2012)
11. Elliott, L.R., Coovert, M.D., Prewett, M., Walvord, A.G., Saboe, K., Johnson, R.: A review and meta analysis of vibrotactile and visual information displays, September 2009
12. Elliott, L.R., Schmeisser, E.T., Redden, E.S.: Development of tactile and haptic systems for U.S. infantry navigation and communication. In: Smith, M.J., Salvendy, G. (eds.) Human Interface 2011. LNCS, vol. 6771, pp. 399–407. Springer, Heidelberg (2011). https://doi.org/10.1007/978-3-642-21793-7_45
13. Spence, C.: Crossmodal spatial attention. Ann. N. Y. Acad. Sci. **1191**, 182–200 (2010)
14. Zeeman, E.C.: Population dynamics from game theory. In: Nitecki, Z., Robinson, C. (eds.) Global Theory of Dynamical Systems. LNM, vol. 819, pp. 471–497. Springer, Heidelberg (1980). https://doi.org/10.1007/BFb0087009

Attention-Based Robot Learning of Haptic Interaction

Alexandra Moringen, Sascha Fleer(✉), Guillaume Walck, and Helge Ritter

Neuroinformatics Group, Bielefeld University, Bielefeld, Germany
{abarch,sfleer}@techfak.uni-bielefeld.de

Abstract. Haptic interaction involved in almost any physical interaction with the environment performed by humans is a highly sophisticated and to a large extent a computationally unmodelled process. Unlike humans, who seamlessly handle a complex mixture of haptic features and profit from their integration over space and time, even the most advanced robots are strongly constrained in performing contact-rich interaction tasks. In this work we approach the described problem by demonstrating the success of our online haptic interaction learning approach on an example task: haptic identification of four unknown objects. Building upon our previous work performed with a floating haptic sensor array, here we show functionality of our approach within a fully-fledged robot simulation. To this end, we utilize the haptic attention model (HAM), a meta-controller neural network architecture trained with reinforcement learning. HAM is able to learn to optimally parameterize a sequence of so-called haptic glances, primitive actions of haptic control derived from elementary human haptic interaction. By coupling a simulated KUKA robot arm with the haptic attention model, we pursue to mimic the functionality of a finger.

Our modeling strategy allowed us to arrive at a tactile reinforcement learning architecture and characterize some of its advantages. Owing to a rudimentary experimental setting and an easy acquisition of simulated data, we believe our approach to be particularly useful for both time-efficient robot training and a flexible algorithm prototyping.

Keywords: Haptic interaction in 3D · Reinforcement learning · Haptic attention · Robot control

1 Introduction

Most activities such as sports, high-precision dexterous handling of tools or playing musical instruments take place through *haptic interaction* with the 3D environment.

A. Moringen, S. Fleer and G. Walck—The authors contributed equally to this work.

Electronic supplementary material The online version of this chapter (https://doi.org/10.1007/978-3-030-58147-3_51) contains supplementary material, which is available to authorized users.

I. Nisky et al. (Eds.): EuroHaptics 2020, LNCS 12272, pp. 462–470, 2020.
https://doi.org/10.1007/978-3-030-58147-3_51

Under *haptic interaction* we understand a physical interaction with objects established by active touch. A number of examples demonstrate the importance of haptics in order to perform a dexterous task successfully. The most famous example is the experiment in which a study participant was asked to light up a match with anesthetized fingers [4] and encounters extreme difficulties doing so. Unlike humans, who, after years of developmental process [15], seamlessly handle a complex mixture of haptic features [9] and profit from their integration over space and time [6], even the most advanced robots are strongly constrained in performing contact-rich interaction tasks [2]. This is due to several reasons. In contrast to other fields such as computer vision, encompassing haptic interaction benchmark sets do not exist yet. Many questions about a general approach to haptic interaction modeling are still unanswered, e.g.: How to represent multimodal haptic characteristics of the explored object, such as rigidity, temperature, or texture? How to integrate over space and time, and how to organize the corresponding haptic memory? How to perform efficient control that – in turn – produces the most suitable data to ensure a successful task progress? How to represent a primitive haptic action? Overall, it remains an open question, how to enable robots to perform haptic interaction with the 3D environment, a skill that should finally allow them to e.g. achieve a human or even superhuman level of dexterity, compared to results achieved on the basis of computer vision alone [3]. In this work we propose to advance in this direction with our rudimentary tactile reinforcement learning infrastructure which we believe to be useful for a versatile development of the robot dexterous manipulation.

Framework. Because ideally all the above issues should be addressed within one framework, the contribution of this work is a systematic approach integrating four highly modular components: 1) primitive haptic actions, haptic glances (HGs) [1] 2) a haptic attention model (HAM), illustrated in Fig. 1, that performs an optimal sequence of primitive haptic actions given a high-level goal specification [1], 3) a modular world model MHSB [5,7,8] that can serve as a platform for a given task specification in a three-dimensional space, 4) a physics-driven simulation environment Gazebo incorporating all experimental components, the robot arm equipped with a tactile sensor array and the 3D objects (see Fig. 2). We show how the above framework enables us to perform haptic interaction learning in simulation by successfully solving an object classification task.

Contributions. The major contribution of the proposed work, compared to previous approaches to perform haptic interaction with robots (e.g. [11,12,14]), is the absolutely minimal amount of hard-coded inputs, hand-crafted preprocessing or other prior knowledge, e.g. human demonstrations, necessary for a successful task performance. The skill represented by a control policy is learned from scratch, and the only input consists of a specification of a primitive haptic action type, and a reward for a successful task execution. In our previous work [1], we have developed a learning architecture that is able to shape the exploration process of a floating tactile sensor through directing its attention on salient tactile features of objects. As a first step towards real robot integration, the present paper is now porting the designed learning architecture into a realistic robot simulation. In this work we also show that our model optimized with a limited cached data set, generalizes well to a performance with new data acquired online. Due to the above results, we believe that our work may be particularly suitable as a foundation for learning of more complex tasks, such as e.g. assembly or search. We encourage you to watch the video that is provided within the supplementary material for a quick overview of the presented paper.

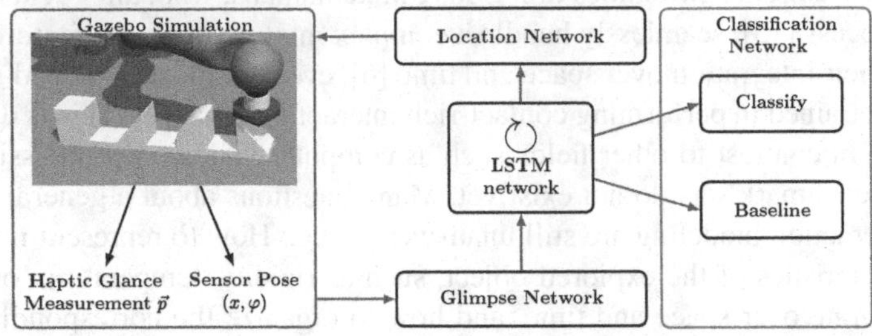

Fig. 1. Illustration of HAM. Generated tactile data *p* of the simulation is combined with the position and the orientation of the sensor within the *glimpse network*. The generated features are then processed through the *LSTM network* in order to either classify the given object, employing the *classification network* or to generate a new pose for the sensor by utilizing the *location network*. The pose is employed to perform a further haptic glance within the simulation.

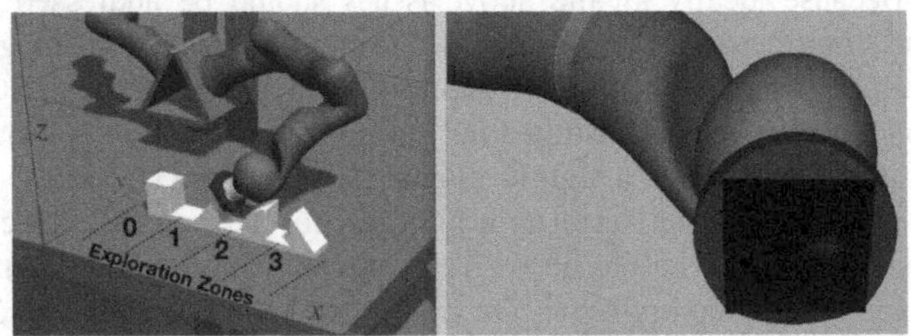

Fig. 2. Simulation of the KUKA LWR4 robot arm (7 degree of freedom) with the Myrmex sensor array attached to the end-effector. Left: the robot performs a haptic glance by establishing contact with the object of $\sim$10 cm^2 footprint placed in exploration zone 1. Right: corresponding visualization of the resulting tactile measurement. A simulated version of the sensor was created for Gazebo that mimics the contact distribution over the array through the use of a mixture of Gaussians around each contact point, weighted according to the local contact force. (Color figure online)

2 Robot Control and Data Acquisition

Even if the proposed method is generic and can be applied in many different scenarios, we chose to demonstrate the applicability of the concept with a setup designed to acquire tactile signals with a tactile sensor array mounted on a robot arm, while exploring a stimulus. Our inspiration is therefore the functionality of a finger performing haptic interaction with fingertip-sized objects, similar to [14].

Exploration Zones. In order to learn an exploration policy that is independent of the object's pose within the global coordinate system, we introduce *exploration zones* (see again Fig. 2). Each exploration zone is a pre-defined region ~20 cm wide in front of the robot, in which the ~10 cm wide objects are centred for exploration. The exploration zones define their own local reference frame, with normalized coordinates to cover the range $[-1, 1]$ around the origin. After specification of the exploration zone, two out of six pose parameters of the tactile sensor can be modified by the HAM: the position $x \in [-1, 1]$ along the x-axis within the coordinate frame of the corresponding exploration zone, and the orientation angle $\varphi \in [-0.3\pi, 0.3\pi]$ around the y-axis.

Robot Arm and Control. The robotic setup consist of a KUKA LRW4 robot arm with 7 degrees of freedom ensuring a range of motion similar to a human arm. Its end-effector is equipped with the tactile sensor Myrmex mounted on an ATI force-torque sensor as shown in Fig. 2. The purpose of the setup is to explore the stimuli with the sensing surface in a safe manner. Robotic interactions with the environment always require great care to avoid damage due to unintended high contact forces. Therefore, unplanned contacts are usually not desired. Since the exploration procedure is guided by the learning system, the various sensor poses executed on the robot are not known in advance. Moreover, for more realism, the shapes to explore are also unknown but contained in a ~10 cm^2 footprint, which forbids any planning for obstacle avoidance.

Hence, the robot arm should move and rely on events to react accordingly when touching the environment. Tactile events are not sufficient to stop motion, because the contact could also occur on non-sensorized surfaces of the robot. To complement the tactile events, two other events are taken into account. First the force-torque sensor mounted between the tactile sensor and the last joint of the arm triggers an event when a force threshold is reached. The force is induced by the contact between the environment and any part of the tactile sensor, even on the non-pressure reactive surfaces of the assembly (dark orange part in Fig. 2). The force threshold is higher than the minimum force needed to trigger the tactile array to ensure pressure data can be acquired before stopping motion due to contact forces. Secondly, an event is generated when other parts of the robot touch the environment. We rely on the joint-impedance control mode of the robot, that permits to select the stiffness of the arm when reaching a certain posture. The "softer" the stiffness is set, the larger can the deviation be relative to the desired posture. This allows to execute a motion that penetrates or collides with an obstacle (here the stimuli), but exerts small forces on contact without crushing the sensor or damaging the arm limbs. The deviation between the actual pose and the desired one can be monitored and an event is triggered if the deviation is too large, meaning the robot was stopped by an obstacle. To summarize, the motion is stopped either by a tactile event, which is a successful data acquisition, or by a too high force on contact between the end-effector and the environment, or by a too large deviation between the desired joint target and the actual in case of a contact with other robot body parts, both latter events being considered as a failed data acquisition. As a first step, the whole robotic system was recreated in simulation, using Gazebo and a simulated LWR robot controller providing impedance control in joint space. The real-time control loop consisting of a Cartesian controller and of a Cartesian trajectory controller (interpolating motions and monitoring deviation), is exactly the same for the real-world robot, and permits to validate the safety mechanism and the algorithms in the virtual environment first. Due to time constraints, the data used is currently from simulation only, which could be acquired rapidly in unattended mode.

Haptic Glances and Haptic Glance Controller. Haptic Glance Controller (HGC) is the interface between HAM and the robot simulation. The robot is controlled by the HGC via a state machine receiving the target pose for each individual HG from the HAM as depicted in Fig. 3. HGC requests new exploration poses which the state machine executes following a sequence of three states. In *To Pos* the sensor is moved to the pose (x, φ) above the objects, while z remains at the constant pre-defined level. Then, the *Go Down* state queries a slow downwards motion, while monitoring the high-force, deviation and tactile pressure events. On any of the events, the state-machine switches to the *Go Up* state, moving the sensor away from the object. In the case of a tactile event, the data is transmitted back to the HAM, completing one haptic glance.

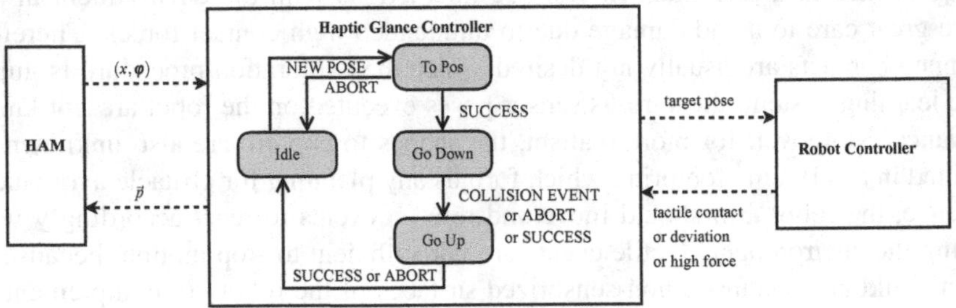

Fig. 3. Haptic Glance Controller receives the pose from the HAM and forwards it to the state machine employed for the robot control. During haptic interaction the robot switches between three main states, *To Pos* (move to a given pose), *Go Down* (establish contact by going down until collision is detected), and *Go Up* (go up after contact has been established).

A HG in this work is implemented as a movement downwards towards the object, while sustaining a given pose, until a contact is established. Each HG that is executed within the Gazebo simulation is represented by the pose (x, φ) of the tactile sensor within the associated exploration zone, but converted to a 6D coordinate for the robot end-effector to reach. Before the execution of a haptic glance the sensor is placed at the specified pose, with the height of the sensor above the given exploration zone being predefined manually. In order to establish the contact, the sensor is moved down along the z-axis, and a corresponding pressure vector p is recorded once any sensitive cell of the sensor reaches a pre-defined threshold.

3 Experiments and Results

To avoid repeated acquisitions of the same or very similar haptic glances in simulation, and to enable an efficient evaluation of the model hyperparameters, we create a dataset of pose-pressure tuples. To this end, we tessellate the whole location-orientation space that can be accessed by the sensor and generate a cache of haptic glances that are then stored in a dataset for learning. Hence, our experiment is split in two parts. As in our previous work, we train our model by using the dataset until a high accuracy is learned. We then show that our trained model is able to generalize beyond the recorded data, we utilize the best model for all trained number of glances and test their performance by classifying the four objects within the fully-fledged simulation online.

Dataset. The dataset is generated by recording tuples $d_o = (p, x, \varphi)$ of the normalized pressure data p, together with the corresponding normalized location x and orientation φ of the sensor. For the data to be independent of the object global pose, the location data $x \in [-1, 1]$ is given within the local coordinate frame of the exploration zone. After reaching the corresponding exploration zone with the robot, the recording of data points starts at $x = -1$ with the orientation $\varphi = -0.3\pi$. After covering 41 discrete orientations φ with a step size $\Delta_\varphi = \pi \cdot 0.05$, the location is incremented by $\Delta_x = 0.05$ and the recording of 41 orientations starts anew, until 41 locations are covered. Leading to 41×41 pre-recordings per object and to a full dataset of 6724 data points. During training, the model generates location-orientation pairs (x, φ) for which the corresponding pressure vector p is directly extracted from the dataset at data point d_o that best matches (x, φ), instead of re-measuring the pressure vector in simulation.

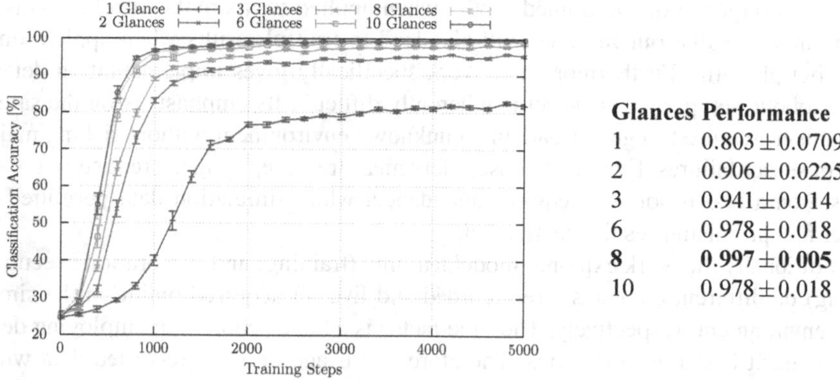

Glances	Performance
1	0.803 ± 0.0709
2	0.906 ± 0.0225
3	0.941 ± 0.014
6	0.978 ± 0.018
8	**0.997 ± 0.005**
10	0.978 ± 0.018

Fig. 4. Evaluation of the models learning performance when trained on the pre-recorded dataset. The speed and efficiency of the learning is depending on the number of executed glances per classification step. The table on the right lists the measured classification performance, recorded in simulation using the best model, pre-trained on the recorded dataset. The results are averaged over 4 trials.

Model Training. The pre-recorded dataset is used to train the designed model on a different number of glances for 5000 training steps. For evaluation, the training is stopped after a predefined step interval. The current policy is then evaluated on 100 test batches in which all of the four objects have to be identified an equal number of times[1]. Even for such a small dataset, the designed model is able to identify the different objects with a nearly perfect score of ≈100% for 10 glances (see Fig. 4).

Testing the Model on the Simulated Robot Arm Online. After successfully training a model that is able to classify the four objects with high accuracy while only using the limited data of the pre-recorded dataset, the learned model is now tested within simulation. For testing the quality, every object is presented 20 times for classification within 4 distinct trials. The results are listed in the table in Fig. 4. Even the use of one single haptic glance per object leads to an accuracy of more than 80%. Again, the classification performance is increasing when more glances are used. While adding a second glance increases the success rate about 10%, the third one only adds a gain of ≈4% and further increases with every additional glance added. Nevertheless, an accuracy of more than 99% can be reached for this simple task when 8 glances are used. For 10 glances, the accuracy is slightly dropping to about 98%.

4 Discussion

A physics-driven control of the robot arm employed in this work vs. the position-based control performed without gravity in our previous work, resulted in a more realistic and less noisy haptic data. As a result we could demonstrate a higher reliability and a faster convergence of the trained model when applied to a simulated robot. This is a good indicator that our research will also lead to fruitful results when applied on a real robot platform. Furthermore, this work specifically gives implementation details on the robotic setup, as this aspect is inherently difficult. Its emphasis is on the safety mechanisms required to gather data in an unknown environment without risking major robot or sensor failures. Even with those safety measures, preparing the reduced 41×41 data set on a real robot still requires attendance, while simulation data permitted to extract first promising results unattended.

Additionally the work explored model learning (training) and exploration execution (testing) on different data sets, pre-recorded and live set acquired online in the simulation environment, respectively. The time factor is a huge problem in employing deep reinforcement learning in robotics. Therefore, the usage of a pre-recorded data without new generation of data in each test iteration may be a promising methodology for a development of algorithms and useful for transition to real-world data sets. Importantly, the usefulness of the pre-recorded sets remains to be tested w.r.t. its advantages for the transition to the real-world performance.

Altogether, we believe that this work may serve as a foundation that brings the known framework of active vision and glances to a different modality with haptic glances. It integrates haptic glances in RL and performs learning of haptic interaction based on physics-driven robot arm control, leading to faster convergence and increased reliability of the resulting model. This opens different possible above-mentioned research directions.

[1] As a batch size of 64 is used for training, every object has to be identified 1600 times during the performance evaluation.

5 Conclusion and Future Work

This work presents an approach for teaching a simulated robot equipped with a tactile sensor, how to classify four objects from data gathered with haptic glances at one or more sensor poses. In order to answer the question how these poses should be selected in an optimal way, we adapted the *haptic attention model*. This model enables us to learn efficient haptic interaction by integrating over the time-series of acquired tactile sensor data while simultaneously improving the current policy. In order to enable fast hyperparameter optimization and avoid multiple calculations of the same data in simulation, we have pre-recorded a dataset of haptic glances (p, x, φ). First tested on this small set, our approach reaches nearly optimal classification performance. With the goal to evaluate the generalizability we then exploited the same model for performing the classification task within the simulation environment online. Despite a relatively small training set compared to the number of trainable variables, the network shows good generalization performance as demonstrated by the results achieved in the online simulation. This is in line with findings in literature, that state that a large overparameterization does not necessarily lead to overfitting [10].

Training of the model solely on a pre-recorded data set might not be enough for more complicated tasks on the one hand. On the other hand, a full training even within the online simulation is likely to be time consuming. Therefore further approaches need to be investigated. One possibility is to use a *transfer learning* approach [13] by first training the model on a pre-recorded dataset and then adding refinement to the learned policy by training the same model for a smaller number of training steps directly on the simulated robot setup. A next step would be to make the transition from the simulated robot to a real-world setup, using the proposed safety mechanisms, but performing unplanned poses still requires attendance. Hence, a reasonable intermediate step would be again to pre-record a dataset with predictable safe poses. Training with this real-world dataset should show how well the model can deal with the noise within the data that is inevitably present when working with a real robotic setup.

References

1. Fleer, S., Moringen, A., Klatzky, R.L., Ritter, H.: Learning efficient haptic shape exploration with a rigid tactile sensor array. PLOS ONE **15**(1), 1–22 (2020). https://doi.org/10.1371/journal.pone.0226880
2. Lee, M.A., et al.: Making sense of vision and touch: Self-supervised learning of multimodal representations for contact-rich tasks. arxiv (2019)
3. Levine, S., Finn, C., Darrell, T., Abbeel, P.: End-to-end training of deep visuomotor policies. J. Mach. Learn. Res. **17**(1), 1334–1373 (2016). http://dl.acm.org/citation.cfm?id=2946645.2946684
4. Match lighting experiment. https://www.youtube.com/watch?v=0LfJ3M3Kn80
5. Modular haptic stimulus board. https://www.youtube.com/watch?v=CftpCCrIAuw
6. Morash, V.S., Pensky, A.E.C., Miele, J.A.: Effects of using multiple hands and fingers on haptic performance. Perception **42**(7), 759–777 (2013)
7. Moringen, A., Haschke, R., Ritter, H.: Search procedures during haptic search in an unstructured 3D display. In: IEEE Haptics Symposium (2016)

8. Moringen, A., Aswolinskij, W., Buescher, G., Walck, G., Haschke, R., Ritter, H.: Modeling target-distractor discrimination for haptic search in a 3D environment (2018). bioRob
9. Panday, V., Tiest, W.M.B., Kappers, A.M.L.: Bimanual integration of position and curvature in haptic perception. IEEE Trans. Haptics **6**(3), 285–295 (2013). https://doi.org/10.1109/TOH.2013.8
10. Poggio, T.A., et al.: Theory of deep learning III: explaining the non-overfitting puzzle. CoRR abs/1801.00173 (2018). http://arxiv.org/abs/1801.00173
11. Shenoi, A.A., Bhattacharjee, T., Kemp, C.C.: A CRF that combines touch and vision for haptic mapping. In: 2016 IEEE/RSJ International Conference on Intelligent Robots and Systems (IROS), pp. 2255–2262, October 2016. https://doi.org/10.1109/IROS.2016.7759353
12. Sommer, N., Billard, A.: Multi-contact haptic exploration and grasping with tactile sensors. Robot. Auton. Syst. **85**, 48 – 61 (2016). https://doi.org/10.1016/j.robot.2016.08.007, http://www.sciencedirect.com/science/article/pii/S0921889016301610
13. Taylor, M.E., Stone, P.: An introduction to intertask transfer for reinforcement learning. AI Mag. **32**(1), 15 (2011). https://doi.org/10.1609/aimag.v32i1.2329, http://www.aaai.org/ojs/index.php/aimagazine/article/view/2329
14. Tian, S., et al.: Manipulation by feel: Touch-based control with deep predictive models. arxiv (2019)
15. Withagen, A., Kappers, A.M.L., Vervloed, M.P.J., Knoors, H., Verhoeven, L.: The use of exploratory procedures by blind and sighted adults and children. Attention, Percept. Psychophys. **75**(7), 1451–1464 (2013). https://doi.org/10.3758/s13414-013-0479-0

Motion Guidance Using Translational Force and Torque Feedback by Induced Pulling Illusion

Takeshi Tanabe[1(✉)], Hiroaki Yano[2], Hiroshi Endo[1], Shuichi Ino[1], and Hiroo Iwata[2]

[1] Human Informatics Research Institute, National Institute of Advanced Industrial Science and Technology (AIST), Central 6, 1-1-1 Higashi, Tsukuba, Ibaraki 305-8566, Japan
{t-tanabe,hiroshi-endou,s-ino}@aist.go.jp

[2] Faculty of Engineering, Information and Systems, University of Tsukuba, 1-1-1 Tennodai, Tsukuba, Ibaraki 305-8577, Japan
yano@iit.tsukuba.ac.jp, iwata@kz.tsukuba.ac.jp
https://unit.aist.go.jp/hiri/hi-fitness/index.html,
http://eva.vrlab.esys.tsukuba.ac.jp

Abstract. It is known that humans experience a kinesthetic illusion similar to a pulling sensation in a particular direction, when subjected to asymmetric vibrations. In our previous study, we developed a device that can apply a translational force and a torque to induce this illusion. The illusory translational force might induce a reaching motion of the upper limb, and the applied torque might induce a flexion–extension motion of the wrist. In the present study, we experimentally verified whether these motions can be induced. The results confirmed that the device could guide the upper limb with a success rate of 94.3% when switching between the application of the translational force and the torque. The results suggested that torque application could be a cue for the user to determine the movement direction intuitively.

Keywords: Illusory force perception · Asymmetric vibration · Non-grounded haptic interface · Motion guidance · Skill transfer

1 Introduction

Conventional methods of motion guidance include a verbal method, in which an expert verbally teaches a motion to a trainee, and a non-verbal method, in which the expert directly touches the trainee's body to induce a targeted motion. In particular, non-verbal methods are used in a wide range of fields, such as rehabilitation, craftsmanship, medical techniques, and sports. These instructions are called extrinsic feedback and are known to be effective in motor learning [1]. In recent years, motion guidance methods using haptic interfaces were proposed for high reproducibility and quantitative training [2,3]. These devices guide the

I. Nisky et al. (Eds.): EuroHaptics 2020, LNCS 12272, pp. 471–479, 2020.
https://doi.org/10.1007/978-3-030-58147-3_52

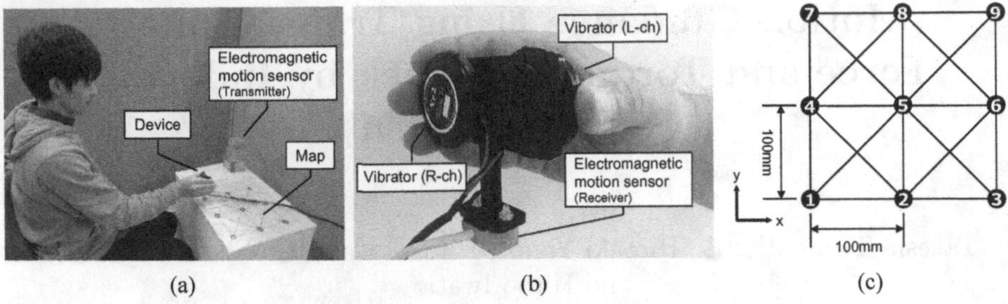

Fig. 1. Upper-limb guidance system: (a) overview, (b) enlarged view of the device, and (c) map of the navigation system.

user in performing the target motion by applying a force or torque to the user's body. However, these haptic interfaces need a large workspace because they must be grounded. Other studies [4,5] proposed a motion guidance method using vibrotactile cues, in which vibrators can be worn on the body because these actuators are smaller. This method has been applied to training for violin playing [5] and snowboarding [6]. Moreover, the application of a vibrotactile cue to an end effector, such as the hand [7] or wrist [4], is more effective than the application to a joint because the motion of the upper limb depends mainly on the hand position. However, the vibration stimuli have no directional cue such as a force or torque. The motions corresponding to vibrotactile cues, such as the movement of an arm in the direction in which vibration was perceived [4], must be mapped in advance. Such mapping of motions is difficult for trainees with a high degree of freedom of motion [8]. Therefore, to achieve efficient motion guidance, the actuator must be able to provide compelling directional information easily to the end effector.

In recent years, haptic interfaces that utilize a kinesthetic sensory illusion have been proposed. It is known that the sensory properties of humans are nonlinear. When strong and weak stimuli are applied sequentially, the user perceives the former but does not clearly perceive the latter. Based on this finding, Amemiya et al. proposed a method of applying vibrations with asymmetric acceleration to induce the perception of force toward a single direction in the user's hand [9]. Furthermore, we previously proposed a method of inducing pulling illusion by using a small voice-coil-type vibrator [10]. In addition, we developed a holdable device that presents an illusory translational force and torque by combining the two force vectors [11]. A new motion guidance method can be developed based on this pulling illusion for applications such as rehabilitation and skill transfer because compelling directional information can be easily provided to the fingertips by only vibration stimuli.

Reaching is a basic motion of the upper limb, in which the hand extends to a target position. Thus, the pulling illusion can be induced to guide the upper-limb motion of reaching. The application of a translational force might be effective for the guidance of reaching because the user's arm can be led by the translational force. However, when applying a translational force in one dimension, only the pushing and pulling motions of the upper limb can be induced. Reaching requires motion guidance in two or three dimensions, which can be achieved by the vector synthesis of translational force. On the other hand, translational force can be applied to guide only large movements of the user's arm to in the forward–backward, upward–downward, and leftward–rightward directions. To guide more complex movements, which are required for applications such as rehabilitation and skill transfer, it is necessary to combine the reaching motion with the motion of an end effector, such as the flexion–extension movement of the wrist. The application of a torque might be effective to induce a rotational motion of the wrist joint. In other words, if the reaching of the upper limb by a translational force and wrist flexion–extension movement by a torque can be combined, a wide range of motions from large to complex movements can be induced. In particular, torque application is important because it can provide a directional cue to perform reaching by changing the angle of the wrist. Therefore, it is hypothesized that torque application can provide intuitive feedback for the direction in which the upper limb should be moved.

In the present study, we experimentally verified whether the motion of upper limb can be induced based on the pulling illusion by using a combination of a translational force and torque. An upper-limb guidance system was developed, and a guidance experiment was performed for a reaching task in two dimensions as a basic study. The characteristics of guidance with and without torque application were investigated.

2 Method

2.1 Participant

Seven right-handed males aged 22–24 years participated in the experiment. The experimental procedure was in accordance with the Declaration of Helsinki. Informed consent was obtained from all participants.

2.2 Upper-Limb Guidance System

Figure 1 shows the upper-limb guidance system, which consists of a device that can apply a translational force and torque to induce the pulling illusion, an electromagnetic motion sensor (Polhemus Inc., 3SPACE FASTRACK), and a printed map. The device consists of two voice-coil-type vibrators (Acouve Lab Inc., Vibration Transducer Vp210). Two vibrators were placed in a side-by-side configuration. When a user holds these between the thumb and index finger, as shown in Fig. 1 (b), the force generated by the adjacent vibrator is primarily perceived by each digit. When the direction of the force exerted by each vibrator is controlled, a user can perceive both a translational force and torque. If forces are applied in the same direction on the participant's thumb and index finger, the participant perceives a translational force in this direction. In contrast, if forces in opposite directions are applied by the vibrator, the participant

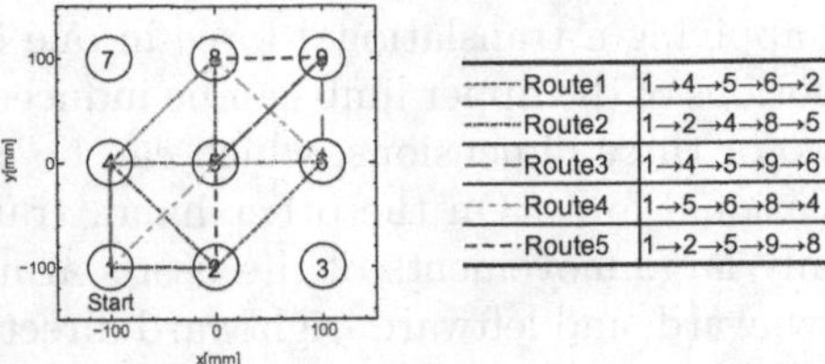

Fig. 2. Routes used in the experiment.

perceives a torque. In other words, this device can apply forces in four directions, in which a forward or backward translational force and a clockwise (CW) or counterclockwise (CCW) torque, by combination of the force vectors. These directions are controlled using an asymmetric-amplitude signal with a two-cycle sine wave that is inverted for a half-cycle [10]. Each signal was amplified with an amplifier circuit using a power amplifier IC (Texas Instruments Inc., LM386) with a maximum output voltage of ±4.5 V. The frequency of the input signal was 75 Hz. More details of this device can be found in Ref. [11].

The transmitter of the motion sensor and map were set on the table in front of the participants (Fig.1 (a)). The receiver of motion sensor is attached to the device (Fig. 1 (b)), and the position and posture of the device on the map were measured. The upper limb of the participant was induced by switching and presenting the translational force and the torque. The map and the device are not fixed, and participants can move the device freely on the map. Numbered circles on the map indicate nodes, and solid lines indicate routes (Fig. 1 (c)). The map has three nodes in the x direction and three nodes in the y direction. A total of nine nodes are arranged at 100 mm intervals. Each node and adjacent nodes in 45° steps were connected by routes. Routes in up to eight directions are connected at one node.

Next, the guidance algorithm is described. In this system, a translational force was applied if the posture of the device faced the direction of the next target node, and a torque was applied to face the target direction otherwise. The target angle θ_P between the position of the device $D(D_x, D_y)$ and the next target node $P(P_x, P_y)$ is

$$\theta_P = \tan^{-1} \frac{P_y - D_y}{P_x - D_x}. \tag{1}$$

The posture around the vertical axis of the device θ_D was measured, and a CW or CCW torque was applied when $\theta_D > \theta_P$ or $\theta_D < \theta_P$, respectively. Considering the range of motion of the wrist and the measurement accuracy, when the difference between the posture of the device and the target angle was less than ±15°, switching from torque to translational force was performed. When the target node was in the positive translational direction from the device, a forward translational force was applied, and when the target node was in the negative direction, a backward translational force was applied. When the device was within a certain distance from the center of the target node (set to

10 mm in this system), the target node was switched to the next node. In this algorithm, even if the position of the device deviates from the predetermined route, the torque and translation force are applied in accordance with the angle and distance to the target node. This system is intended for use with the right hand, and the posture of the device was set to 0° when facing the front (see Fig. 1 (b)), 45°, −45°, and −90° (CW is positive). The upper limb of the participants was guided in eight directions by applying a forward or backward translational force in the above four postures. The angle of the turning motion (difference between the device postures at the start and end of turning, hereinafter referred to as the rotation angle) was 0° at minimum and 135° at maximum.

The route was set to five types, each passing through five out of nine nodes with one stroke (Fig. 2). The start position was fixed at node 1. These routes were selected to make the number of occurrences of each rotation angle approximately equal.

2.3 Procedure

Under the hypothesis that the torque application can provide intuitive feedback for the direction in which the upper limb should be moved, the experiment was conducted with (w.) and without (w/o) torque application. Because the wrist posture could not be guided in the w/o torque condition, the participants performed active exploration for the next target posture by rotating the device. In the w. torque condition, the posture was guided by torque application as explained in Sect. 2.2. Under both conditions, a forward or backward translational force was presented when the device was guiding to the next node.

The experimental tasks are as follows. Participants moved their upper limbs as guided by the device to trace a solid line when traveling along a route and to perform rotation within a node. In total, 30 trials were performed for each participant, with 3 trials for each condition (2 levels) and each route (5 levels), and each condition and each route was randomized. To minimize fatigue during the experiment, the trials were divided into three blocks of 10 trials each, and the participants were given a two-minute break between blocks. In this experiment, audiovisual information was not blocked because actual motion guidance was assumed. Before the trial, the experimenter verbally guided the participant about the condition of torque application. A sound effect was output from the speaker only when the goal was achieved, indicating that the trial was completed. A 5-min practice period was set up so that participants could learn how to operate the device before the experiment.

3 Results

Figure 3 shows typical examples of the device's trajectory. The solid lines show the trajectories in which the device was moved by the participants, and the broken lines show the target routes. The task was determined as successful when the device passed the nodes in the set order.

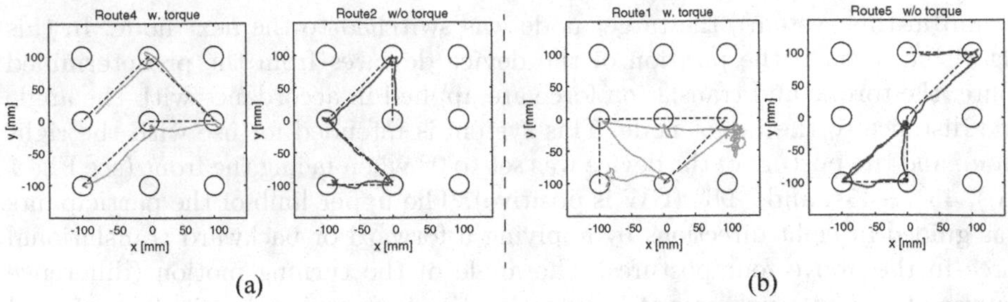

Fig. 3. Typical examples of the device's trajectory: (a) successful task; (b) failed task.

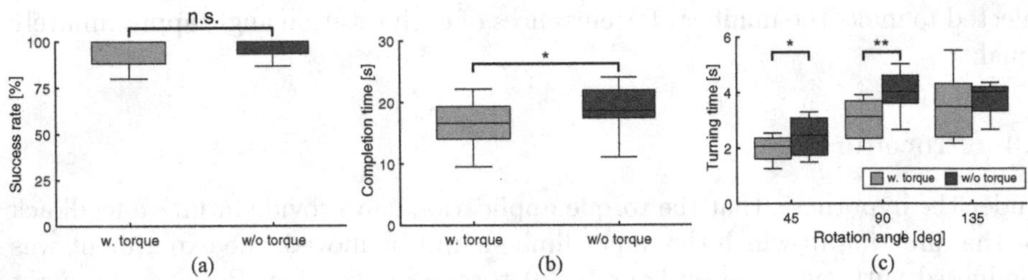

Fig. 4. Experimental results: (a) success rate of guiding along routes, (b) task completion time, and (c) turning time (n.s.: not significant, *: $p < 0.05$, **: $p < 0.01$).

Figure 4 (a) shows the success rate for guiding routes. The average success rate in the w. torque condition was 94.3%, and the rate in the w/o torque condition was 96.2%. In a paired t-test between the conditions, no significant difference was found [$t(6) = 1.00$, $p = 0.37$, $d = 0.28$].

The task completion time and turning time were analyzed for only successful trials. Figure 4 (b) shows the task completion time. The completion time indicates the time from the start signal provided by the experimenter to time at which the fifth node is reached. In a paired t-test between the conditions of torque application, a significant difference was found [$t(6) = 2.49$, $p < 0.05$, $d = 0.53$].

The turning time indicates the time from the moment when a node is reached to the time at which the turn at the node is completed. Figure 4 (c) shows the turning time at each rotation angle. Since the change in turning time was small with respect to the rotation direction, the rotation directions were merged for each rotation angle. Furthermore, the rotation angle was not considered a factor, because the turning time increased as the rotation angle increased. In a t-test between the conditions of torque application, significant differences were found for 45° [$t(6) = 2.67$, $p < 0.05$, $d = 0.68$] and 90° [$t(6) = 3.96$, $p < 0.01$, $d = 1.32$], whereas no significant difference was found for 135° [$t(6) = 0.57$, $p = 0.59$, $d = 0.24$].

4 Discussion

Because the success rates of guidance were high under both conditions, it was shown that the upper-limb motion could be guided by the pulling illusion. This result suggests that the user can be guided to perform large and complex movements by combining the reaching and wrist motions. However, although no significant difference was found between the conditions, the success rate was lower under the w. torque condition. This might have been caused by the adaptation to stimuli induced by the long-term application of asymmetric vibration. We confirmed that the subjective sensitivity of illusory force decreased with the continuous presentation of asymmetric vibration [12]. In the w. torque condition, the asymmetric vibration was always presented, whereas in the w/o torque condition, the stimulus during the turning motion was not presented. The sum of the vibration-stimulation times was larger in the w. torque condition, and the adaptation might be induced earlier. Therefore, an application method that considers adaptation is required.

The completion time and turning time were significantly shortened on applying the torque. Thus, it is considered that the participants intuitively determined the direction in which the wrist must be flexed with the torque application. On the other hand, although the difference in turning time between the torque conditions was remarkable when the turning angle was small, such as 45° or 90°, the effect of the torque application was smaller for a turning angle of 135° ($d = 0.24$). When the turning angle was 135°, either the CW rotation from the −45° posture or the CCW rotation from the 45° posture was performed. Thus, the participant could determine the next direction to rotate to without directional information when the turning angle was 135°. In other words, this result supports the hypothesis that the torque application is effective when it is necessary to determine the direction. Therefore, the illusion may be useful as a trigger of motion.

This method is expected to be effective in an environment where visual information can also be presented because the task is to guide the user along routes on a map. Moreover, it is difficult to induce high-speed motion by using this method because the user needs to move the upper limb by using a force or torque cue. Therefore, guidance for slow motions, such as rehabilitation using a pegboard, welding, and calligraphy, might be feasible. Particularly in the field of rehabilitation, it has been reported that reaching assisted by a robot contributes to the recovery of motor function [13]. Motor learning might be promoted by supporting reaching using our device. However, although this study showed that the pulling illusion could guide reaching and wrist motions, the effects on motor learning were not clarified. In the future, long-term verification of motor learning using our device is required.

5 Conclusion

In this study, we verified whether the motion of upper limb can be induced by using the translational force and torque presentation based on the pulling illusion. As a result, it was confirmed that this device could be guided with a success rate of 94.3% in the condition of switching the presentation of the translational force or the torque. It was also suggested that torque presentation could be a cue to intuitively determine the direction. In the future, we investigate the relationship the pulling illusion and motor learning, and develop the applications, such as rehabilitation or skill transfer.

Acknowledgments. This work was supported by JSPS KAKENHI Grant No. 19K24374.

References

1. Tani, H.: Are Instructions and Feedback by a Therapist Effective in the Motor Learning?. Rigakuryoho Kagaku, vol. 21, no. 1, pp. 67–73, 2006. (in Japanese)
2. Yem, V., Kuzuoka, H., Yamashita, N., Ohta, S., Takeuchi, Y.: Hand-skill learning using outer-covering haptic display. In: Auvray, M., Duriez, C. (eds.) EUROHAPTICS 2014. LNCS, vol. 8618, pp. 201–207. Springer, Heidelberg (2014). https://doi.org/10.1007/978-3-662-44193-0_26
3. Feygin, D., Keehner, M., Tendick, F.: Haptic guidance: experimental evaluation of a haptic training method for a perceptual motor skill. In: Proceedings of 10th Symposium on Haptic Interfaces for Virtual Environment and Teleoperator System (HAPTICS 2002), pp. 40–48 (2002)
4. Salazar, J., Okabe, K., Murao, Y., Hirata, Y.: A phantom-sensation based paradigm for continuous vibrotactile wrist guidance in two-dimensional space. IEEE Robot. Autom. Lett. **3**(1), 163–170 (2018)
5. van der Linden, J., Schoonderwaldt, E., Bird, J., Johnson, R.: MusicJacket-combining motion capture and vibrotactile feedback to teach violin bowing. IEEE Trans. Inst. Meas. **60**(1), 104–113 (2010)
6. Spelmezan, D., Schanowski, A., Borchers, J.: Haptic guidance: experimental evaluation of a haptic training method for a perceptual motor skill. In: Proceedings 4th International ICST Conference Body Area Networks, pp. 1–8 (2009)
7. Basu, S., Tsai, J., Majewicz, A.: Evaluation of tactile guidance cue mappings for emergency percutaneous needle insertion. In: Proceeding of IEEE Haptics Symposium 2016, pp. 106–112 (2016)
8. Bark, K., et al.: Effects of vibrotactile feedback on human learning of arm motions. IEEE Trans. Neural Syst. Rehabil. Eng. **23**(1), 52–63 (2015)
9. Amemiya, T., Ando, H., Maeda, T.: Lead-me interface for a pulling sensation from hand-held devices. ACM Trans. Appl. Percept. **5**(3), art. 15, 1–17 (2008)
10. Tanabe, T., Yano, H., Iwata, H.: Evaluation of the perceptual characteristics of a force induced by asymmetric vibrations. IEEE Trans. Haptics **11**(2), 220–231 (2018)
11. Tanabe, T., Yano, H., Iwata, H.: Proposal and implementation of non-grounded translational force and torque display using two vibration speakers. In: Hasegawa, S., Konyo, M., Kyung, K.-U., Nojima, T., Kajimoto, H. (eds.) AsiaHaptics 2016. LNEE, vol. 432, pp. 187–192. Springer, Singapore (2018). https://doi.org/10.1007/978-981-10-4157-0_32
12. Tanabe, T., Yano, H., Iwata, H.: Temporal characteristics of non-grounded translational force and torque display using asymmetric vibrations. In: Proceedings of IEEE World Haptics Conference, pp. 310–315 (2017)
13. Takahashi, K., et al.: Efficacy of upper extremity robotic therapy in subacute poststroke hemiplegia. Stroke **47**(5), 1385–1388 (2016)

BY

Perceptually Compressive Communication of Interactive Telehaptic Signal

Suhas Kakade(✉) and Subhasis Chaudhuri

Department of Electrical Engineering, Indian Institute of Technology Bombay, Mumbai, India
{suhaskakade,sc}@ee.iitb.ac.in

Abstract. During telehaptic applications over a shared communication medium, Weber's law of perception based adaptive sampling scheme can be applied to reduce the data rate without degrading the perceptual quality of the haptic signal. However, the perceptual threshold (JND) is often unknown for a user. An experimental design involving bidirectional communication of haptic data over the Internet between the operator and teleoperator is carried out in this paper, which provides a real-time estimate of the Weber threshold for the telehaptic backward channel force signal. Using the proposed data-driven experimental protocol, we are able to reduce the packet rate significantly. We provide a subjective evaluation of the proposed technique to substantiate its usefulness.

Keywords: Adaptive sampling · Telehaptics · Weber threshold

1 Introduction

In recent years, the transmission of haptic data has received the attention of many researchers working in the area of telehaptics. This bidirectional transmission creates a global control loop between the operator and the teleoperator over a communication channel. Contrary to video or audio signal, in order to make the global control loop stable and to maintain the *Quality of Service* (QoS), a maximum communication delay of 30 ms is allowed [8]. Since the haptic signal is typically sampled at 1 kHz, to limit the packet rate while maintaining the QoS, perceptually significant adaptive sampling schemes based on Weber's law of perception [9] or event detection [12] have been applied. Weber's law of perception depicts a logarithmic relation between perceptual stimuli and human perception. Weber threshold is given by $\rho = \frac{\Delta I}{I}$ where ρ represents Weber threshold, ΔI represents just noticeable difference (JND), i.e., the minimum change in signal magnitude required to produce a noticeable variation in signal perception and I is the reference signal magnitude.

In a perceptually significant adaptive sampling scheme, sampling of the kinesthetic force signal will be carried out at instances at which the relative difference exceeds the Weber threshold, i.e., $\frac{|F_i - F_{i-1}|}{|F_{i-1}|} \geq \rho$ where F_i represents force

I. Nisky et al. (Eds.): EuroHaptics 2020, LNCS 12272, pp. 480–488, 2020.
https://doi.org/10.1007/978-3-030-58147-3_53

sample at time t_i and F_{i-1} represents the last perceived force sample at time $t_{(i-1)}$. Then these samples are transmitted. This helps in the reduction of packet rate. Hence the study of estimation of Weber threshold is essential. However, the JND depends on perceptual capabilities of an individual as well as the task one is performing.

A lot of research work has been carried out on the estimation of JND and Weber threshold. Though [5] discusses the estimation of JND on off-line data and their application in adaptive sampling, there is little research on the estimation of JND from real-time data. For the multidimensional signal, [5,13] analyzed the variation in Weber threshold due to the change in force direction. Researchers have applied techniques like psychometric function [6,15] mostly on off-line data to determine the Weber threshold of force signal. The work in [9] demonstrated adaptive sampling for telehaptic signal but with a fixed set of Weber threshold values. In this paper, we try to adaptively determine the Weber threshold directly from on-line data acquired in the given environment in real-time and simultaneously use it for adaptive sampling of the data on the backward channel. Haptics over Internet Protocol (HoIP) has been proposed in [7] to carry the haptic signal over the internet. [8] uses HoIP for the propagation of the telehapatic signal and also performs Weber sampling of the velocity signal. The stability of the global control loop is achieved through appropriate approaches, as specified in [10,11,14]. In this work, we assume that our global control loop satisfies the stability criteria, and we did not observe any instability of the global control loop during experimentation.

In this paper, we apply Weber's law of perception based adaptive sampling scheme to the interactive telehaptic force signal under an appropriately simulated environment. The force signal is generated through signed distance field based haptic rendering of different watertight mesh models. This rendering is carried out at the teleoperator end, and the force signal is transmitted from the teleoperator to the operator on the backward channel. The force signal is adaptively sampled based on a real-time data-driven estimate of JND to reduce the packet rate over the Internet. In order to adaptively sample the haptic signal, we need to know the JND for the given task. We use an adaptively updatable probing threshold for sampling the signal on the backward channel while the user provides an intermittent task-nondisruptive feedback to the teleoperator through the forward channel. Our experiment showed that there is a substantial reduction in the haptic packet rate in-spite of not having any prior knowledge of the Weber threshold for the user. It is also well known that the user experience of haptic communication is dependent on network delay. We perform experiments to study this and show how it affects the users.

2 Proposed Method

2.1 Hardware Setup

The experiment is designed in two modes - 1) standalone and 2) networking. In standalone mode, only one haptic device is used with a computer. This mode is used only to study the effect of packetization delay and to serve as the reference (baseline experiment) during user studies. The theme of this paper is to carry out the experimentation in networking mode. Here two phantom omni haptic devices are used, one as an operator and the other as a teleoperator. These haptic devices are connected to two separate computers. These computers acting as end systems are connected through another node computer. A virtual network of these three computers is managed through Wide Area Network Emulator (WaNem) [3]. The unidirectional communication link capacity is set at 1000 Mbps (to avoid any network congestion), and the maximum packet rate is set to 1000 packets per second. The operator device is kept at a distance of 40 cm from the shoulder of the user. Arm and wrist are kept at fixed positions while perceiving the force. To analyze the effect of propagation delay P_d on the estimated Weber threshold, one-way P_d is emulated on the system and is varied between 0–15 ms, increasing in steps of 5 ms. The average packetization delay is observed to be 0.13 ms in the operator side and 0.45 ms in the teleoperator side, which is much lower than the introduced propagation delay P_d.

2.2 Signed Distance Field Based Haptic Rendering

To generate the force signal at the teleoperator, haptic rendering of watertight objects like bunny [1] and teddy [2] are carried out. The bunny model used in the experiment has 4976 faces and 2490 vertices, and it is voxelized into $135 \times 142 \times 90$ grid cells, whereas the teddy model has 3192 faces and 1598 vertices. The teddy model is voxelized into $117 \times 143 \times 97$ voxels. Given a haptic interaction point H_{IP} and the object, the shortest distance between the point and the bounding object surface is obtained. Depending upon the sign of the distance $\phi(H_{IP})$, we can determine whether the interaction point lies inside $(\phi(H_{IP}) > 0)$/outside $(\phi(H_{IP}) < 0)$/on $(\phi(H_{IP}) = 0)$ the surface of an object.

The signed distance field is created for an object as in [4] and stored on a three-dimensional grid. During rendering, the H_{IP} position is first obtained, and depending upon the sign of distance $\phi(H_{IP})$, the presence of H_{IP} inside/outside the object is detected. If H_{IP} lies inside the object, force $F(H_{IP})$ is generated from the gradient using a suitable scaling constant k as follows, $F(H_{IP}) = k\phi(H_{IP})\nabla\phi(H_{IP}) \|\nabla\phi(H_{IP})\|_2^{-1}$.

2.3 Proposed Sampling Scheme

In this section, we discuss the proposed framework for adaptive sampling of force signal along with simultaneous Weber threshold estimation for an individual user.

As shown in Fig. 1, at the teleoperator, haptic and visual renderings of a solid object are carried out to simulate the teleoperation. The operator holds the device stylus and remotely explores the object at the teleoperator's side through the means of video and haptic information transmitted from teleoperator to operator. This interaction is captured and transmitted in the form of position and velocity on the forward channel to the teleoperator. The end-effector running at the teleoperator end generates a force signal which is then adaptively sampled,

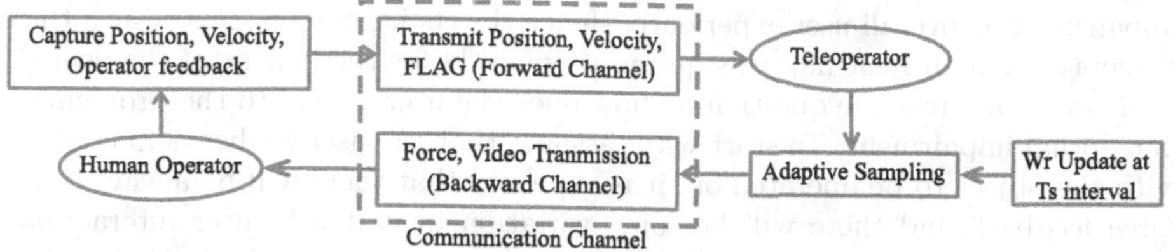

Fig. 1. Proposed scheme for real-time interaction driven adaptive sampling.

and if its relative magnitude is greater than the Weber threshold ρ at that instant, it is transmitted back to the operator on the backward channel. However ρ is not known apriori and needs to be estimated from the interaction itself.

We start with a very small value of ρ (let us call it W_r) at the beginning and use it to adaptively sample the data. Since the user experience at such a small value of W_r will not impede the perception of the user, after Ts interval W_r is incremented by 5% of its previous value, and the adaptive sampler is changed accordingly. This continues until the operator's experience is impaired. At this point, the operator sends a feedback signal (setting the FLAG bit TRUE in Fig. 1) through the forward channel to the teleoperator end. At the teleoperator end, when the FLAG is found to be TRUE, it is understood (see comments later) that the impairment in perception is due to loss of perceptually important samples and hence the threshold value W_r is decremented agressively by 20% of its current value while passing the signal through the adaptive Weber sampler before sending it through the backward channel after packetization. At the operator end, if the user experience is not compromised then the threshold W_r is again raised slowly by 5% else it is further decremented by 20%. The process continues until the teleoperation ends.

A few comments are in order at this point. We implicitly assume that (a) there is no communication delay during teleoperation as it introduces a hysteresis behaviour between exerted and perceived forces, (b) the control loop under which the teleoperation is performed does not suffer from transparency related impairments and (c) any perceptual impairment during teleoperation is due to a higher choice of the Weber threshold W_r when the loss in information received at the operator end becomes perceptually significant.

It is well accepted in the literature that the Weber threshold ρ lies in between $0.04 \leq \rho \leq 0.20$. Hence instead of starting the teleoperator with a value of W_r to be zero, we set $W_{r_{min}} = 0.04$. As the W_r value is slowly incremented, we saturate $W_{r_{max}}$ at 0.20, if required. The choice of an incremental step size of 5% was adhoc, but was found to be a good compromise between being aggressive or sluggish in approaching $W_r \rightarrow \rho$ (true value of the threshold) when the user continues to enjoy the interaction perceptually. A higher choice of increment requires the user to provide feedback more frequently, thus distracting the user away from the assigned task. Similarly, the choice of the decremental step size to be 20% is motivated by the fact that the system should quickly come out of the impaired deadzone of sampling to reduce the scope of continuation of perceptual impairment on overall user experience. Hence the choice of step sizes is such that perceptual impairment happens quite infrequently for the benefit of the user.

Now we address the question on how does the user attend to the problem of perceptual impairment. To start with, assume that the user is always in contact with the object to be operated on, it is expected that there will be always some force feedback and there will be some variation in it due to user interaction. When W_r is small, the user will feel such changes more often and when W_r becomes very high (say, close to 0.20), the user will not be experiencing much changes while still manipulating the object, thus impairing the experience. An immediate reduction in the W_r tries to make the changes in force feedback a lot more perceivable. In the event the end-effector is not in contact with the user, any choice of value for W_r is good as there will be no force feedback and W_r will rapidly approach the value $W_{r_{max}}$.

We now discuss how the FLAG is set to TRUE during the communication over the forward channel. The default value of FLAG is FALSE when W_r is continuously incremented by 5% at the backward channel. We attach a sound detector (a microphone that detects any audio utterance by the user, up to an appropriate choice of threshold to take care of ambient noise) to the user. Whenever the user experience worsens, the user is required to make some utterances which when detected, sets the FLAG to TRUE momentarily before being reset to FALSE. Thus the variable FLAG works on a monostable mode. When the received packet header at the teleoperator end has a FLAG value TRUE, the threshold W_r is decremented by 20%. It may be mentioned that we use HoIP (Haptics over Internet Protocol), as proposed in [7], which has the provision of padding such flags in the packet header for bilateral communication. It may also be noted that one may instead use a call button feedback or a vision based experience impairment detection technique through user specified gesticulation. However, we have found that the audio based technique is less distractive and has comparatively a lower latency without impairing the dexterity of the operator.

The Weber threshold at a given instant depends on the transmission of the FLAG sequence through the forward channel, which in turn, depends on the operator feedback. As is common in any telehaptic communication, we use a sampling rate of 1 kHz, generating up to a maximum of 1000 packets per second when no Weber threshold based adaptive sampler is used. A user can provide her feedback as per her experience on interaction while the Weber threshold W_r is updated every Ts. We select $T = 2$ s or $T = 3$ s, and within this interval, the value of W_r is held constant.

3 Experimental Results

Data Collection: The data is collected voluntarily from 16 male users following the standard ethical clearance of the university and the process of acquiring the data through informed consent of the users. The users, all right handed, were in the age-group of 22–34 years, out of which 10 users were unaware of the working of any haptic device before starting the experiment. The users did not have any history of neurophysiological illness. At the beginning of the experiment, every

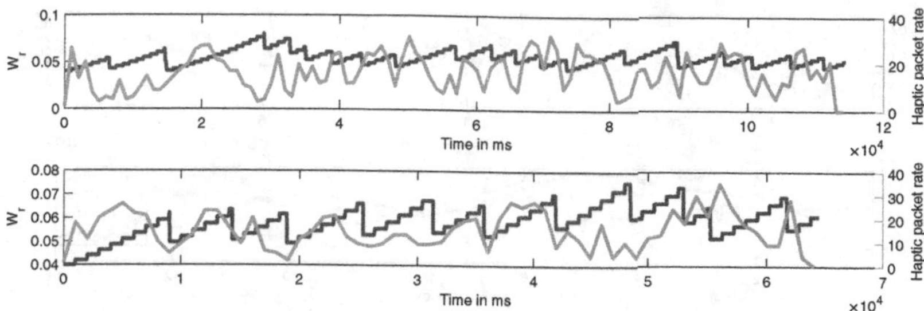

Fig. 2. Computed $W_r(t)$ and haptic packet rate for a given user for the object (top) bunny and (bottom) teddy with $T = 2\,\text{s}$ and $P_d = 0\,\text{ms}$.

user was introduced to the device, and an appropriate training was given to get them familiar with the device to perceive the kinesthetic force signals properly. The experiment was also explained to them in detail.

The experimental protocol was as follows. 3D point cloud data of some solid objects are rendered for haptic interaction using a standard distance field based method. The user is requested to move the interaction stylus in the YZ plane only (Y being the vertical direction), so that the rendered force is along the Z (away from the body) direction only. This allows us to avoid user sensitivity to directional change in force, if any [5]. The rendered force is maintained within the range [0, 3N]. At the emulator level, the propagation delay P_d is set at one of the following four values - 0, 5, 10 and 15 ms (see Sect. 2.1). While sending the force on the backward channel, the Weber threshold based sampling was done. The user orally gives the feedback whenever his interaction is perceived to be poor, as explained earlier. The Weber threshold W_r is updated at 2 s and 3 s interval (T) in two different experiments.

The total time required for the experiment is approximately 1 h per user per rendered object. To avoid physical or device fatigue, users were asked to perform the experiment in multiple sittings with each sitting not exceeding 20 min. During experimentation, users were asked to listen to music using headphone to avoid distractions in the laboratory, if any. Users were also asked to evaluate the experiment on a 5 point scale, with 5 being excellent and 1 being very poor, for different values of P_d and T based on the perceptual parameters: smoothness and experience [8]. The users were asked to rate the experiment relative to the baseline experiment (Sect. 2.1) on the standalone system with zero propagation and packetization delay and with no Weber sampling.

Results: For a given user, the computed W_r as a function of time is plotted for two different objects (bunny and teddy) in Fig. 2 for a propagation delay $P_d = 0\,\text{ms}$ and backward channel update time $T = 2\,\text{s}$. We observe that in both cases, the estimated Weber threshold ρ is quite reasonable (about 7%) and follows a similarly repetitive pattern. The saving in packet rate is quite substantial (about 20 packets per second are being transmitted) to warrant the use of the proposed method. The corresponding SNR of the received force signal is above 30 dB, possibly good enough for most haptic interactions.

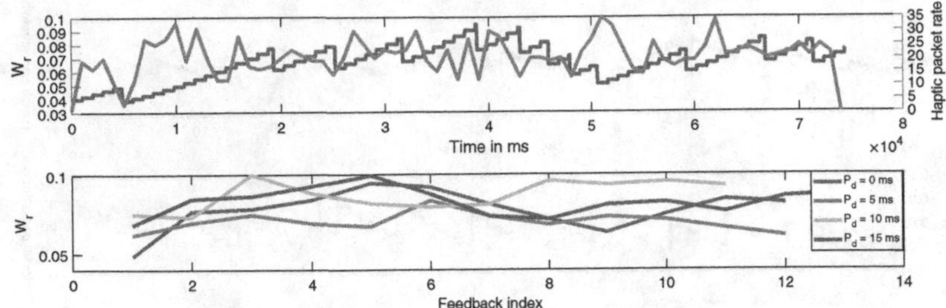

Fig. 3. (top) Computed $W_r(t)$ and haptic packet rate for the object teddy with $T = 3$ s and $P_d = 0$ ms. (bottom) Effect of varying propagation delay P_d on estimated Weber threshold $W_r(t)$ for the object teddy with $T = 3$ s for the same subject.

Table 1. Mean opinion score (MoS) and standard deviation (SD) on user study.

Parameters	$T = 2$ s								$T = 3$ s							
	$Pd = 0$ ms		$Pd = 5$ ms		$Pd = 10$ ms		$Pd = 15$ ms		$Pd = 0$ ms		$Pd = 5$ ms		$Pd = 10$ ms		$Pd = 15$ ms	
	MoS	SD	MoS	SD	MoS	SD	MoS	SD	MoS	SD	MoS	SD	MoS	SD	MoS	SD
Smoothness	4.75	0.45	4.06	0.77	3.00	0.63	2.19	1.05	4.56	0.51	4.00	0.52	3.06	0.68	2.19	0.91
Experience	4.75	0.45	4.38	0.62	3.31	0.60	2.56	0.89	4.75	0.45	4.25	0.45	3.31	0.60	2.48	0.96

As we slow down the backward channel update time T to 3 s, we notice from Fig. 3 (top) that the nature of the corresponding plot of $W_r(t)$ remains more or less the same. However, the updating being slow, as expected, there is a marginal increase in packet rate. Ideally the values of Weber threshold (ρ) will be slightly less than the $W_r(t)$ value at which the user provides a negative feedback. In Fig. 3 (bottom), we plot only the local peaks of $W_r(t)$ function against feedback index (at which the user responded). We repeat the experiment for the same user and for the same object for different values of P_d to study the effect of propagation delay. It may be observed that even though the haptic interaction becomes poorer as P_d increases (see Table 1), the computed value of ρ remains fairly unchanged, substantiating the appropriateness of the proposed method.

We present the result of subjective analysis of user experience for all 16 users for various values of propagation delay P_d and update interval T on two specific perceptual parameters, namely interaction smoothness and user experience, in Table 1. It can be seen that the user experience gets quite impaired if the propagation delay increases to 10 ms and beyond, suggesting that the proposed technique should work under the upcoming 5G communication, but not on the current 4G technology. We also note that both the suggested update intervals $T = 2$ s or $T = 3$ s are acceptable to the user.

4 Conclusions

In this paper, we demonstrated an experimental protocol on how to reduce the interactive haptic signal packet rate over the Internet between the operator and teleoperator when the Weber threshold is not known. However, it is observed that as the propagation delay increases, the immersiveness of the virtual system decreases. However, within the specified QoS of 5 ms delay of 5G technology, the proposed method will work well. Hence the method has the potential for application in telehaptic application over a short-haul communication. However, the stability of the proposed telehaptic loop under the quantization error and delay jitter requires further investigation. One also needs to study the suitability of the proposed method for two handed interactions when the Weber threshold may be different for each hand.

References

1. https://cs.uwaterloo.ca/~c2batty/bunny_watertight.obj
2. https://groups.csail.mit.edu/graphics/classes/6.837/F03/models/teddy.obj
3. WaNem. http://wanem.sourceforge.net/
4. Bridson, R.: Fluid Simulation for Computer Graphics. AK Peters/CRC Press (2015)
5. Chaudhuri, S., Bhardwaj, A.: Kinesthetic Perception: A Machine Learning Approach. Springer, Heidelberg (2017). https://doi.org/10.1007/978-981-10-6692-4
6. García-Pérez, M.A., Alcalá-Quintana, R., Woods, R.L., Peli, E.: Psychometric functions for detection and discrimination with and without flankers. Attention Percept. Psychophys. **73**(3), 829–853 (2011)
7. Gokhale, V., Dabeer, O., Chaudhuri, S.: HoIP: haptics over internet protocol. In: IEEE Haptic Audio Visual Environments and Games (HAVE), pp. 45–50. IEEE (2013)
8. Gokhale, V., Nair, J., Chaudhuri, S., Kakade, S.: Network-aware adaptive sampling for low bitrate telehaptic communication. In: Prattichizzo, D., Shinoda, H., Tan, H.Z., Ruffaldi, E., Frisoli, A. (eds.) EuroHaptics 2018. LNCS, vol. 10894, pp. 660–672. Springer, Cham (2018). https://doi.org/10.1007/978-3-319-93399-3_56
9. Hinterseer, P., Hirche, S., Chaudhuri, S., Steinbach, E., Buss, M.: Perception-based data reduction and transmission of haptic data in telepresence and teleaction systems. IEEE Trans. Sig. Process. **56**(2), 588–597 (2008)
10. Hirche, S., Buss, M.: Transparent data reduction in networked telepresence and teleaction systems. Part II: time-delayed communication. Presence: Teleoper. Virt. Environ. **16**(5), 532–542 (2007)
11. Jafari, A., Nabeel, M., Singh, H., Ryu, J.: Stable and transparent teleoperation over communication time-delay: observer-based input-to-state stable approach. In: IEEE Haptics Symposium, pp. 235–240 (2016)
12. Nadjarbashi, O.F., Najdovski, Z., Nahavandi, S.: Event-driven data transmission in variable-delay network. In: IEEE International Conference on Systems, Man, and Cybernetics, pp. 1681–1686 (2017)
13. Pongrac, H., et al.: Limitations of human 3D force discrimination. Human-Centered Robotics Systems (2006)
14. Uddin, R., Ryu, J.: Predictive control approaches for bilateral teleoperation. Ann. Rev. Control **42**, 82–99 (2016)
15. Wichmann, F.A., Hill, N.J.: The psychometric function: I. fitting, sampling, and goodness of fit. Percept. Psychophys. **63**(8), 1293–1313 (2001)

Sound-Image Icon with Aerial Haptic Feedback

Seunggoo Rim, Shun Suzuki(✉), Yutaro Toide(✉), Masahiro Fujiwara(✉), Yasutoshi Makino(✉), and Hiroyuki Shinoda(✉)

The University of Tokyo,
5-1-5 Kashiwanoha, Kashiwa-shi, Chiba-ken 277-8561, Japan
{rim,suzuki,toide}@hapis.k.u-tokyo.ac.jp,
Masahiro_Fujiwara@ipc.i.u-tokyo.ac.jp,
yasutoshi_makino@k.u-tokyo.ac.jp,
hiroyuki_shinoda@k.u-tokyo.ac.jp

Abstract. In this study, we attempt to define a novel invisible mid-air three-dimensional (3D) object, which informs users of its existence and location via sound and haptic feedback. The correlation between the senses of hearing and touch creates the feeling of touching a sound source, which the user recognizes as a virtual object: sound-image icon. The sound from the icon instantaneously notifies the user about its position without requiring vision. In addition, aerial tactile sensation enables users to freely interact with and manipulate these icons with no need to wear any devices. Therefore, this approach exhibits enormous potential in various situations, such as surgical operations, works in factories, driving cars, and button/switch operations in daily life. In this study, we prototyped the sound-image icon and experimentally examined their feasibility. We confirmed that users could estimate the location of the icons and measured the time required to access these icons. The results indicate that the sound-image icon is feasible as a novel 3D interface.

Keywords: Mid-air haptics · Haptic display · Sound-source localization

1 Introduction

In this study, we propose a sound-image icon, which is an invisible 3D object that integrates sound-source localization and mid-air tactile sensation. This sound and haptic feedback creates a virtual object with no visual appearance, without requiring the user to wear any devices.

A typical method of reproducing tactile sensation in mid-air [2,3,6,8] is installing an airborne ultrasound tactile display (AUTD) that presents the moving stimulus on a skin by remotely producing the radiation pressure of focused aerial ultrasound. Such mid-air haptic feedback has been integrated with visual floating images in previously conducted studies [9,11]. From the viewpoint of interface design, vision is the most efficient channel to transmit the spatial

I. Nisky et al. (Eds.): EuroHaptics 2020, LNCS 12272, pp. 489–496, 2020.
https://doi.org/10.1007/978-3-030-58147-3_54

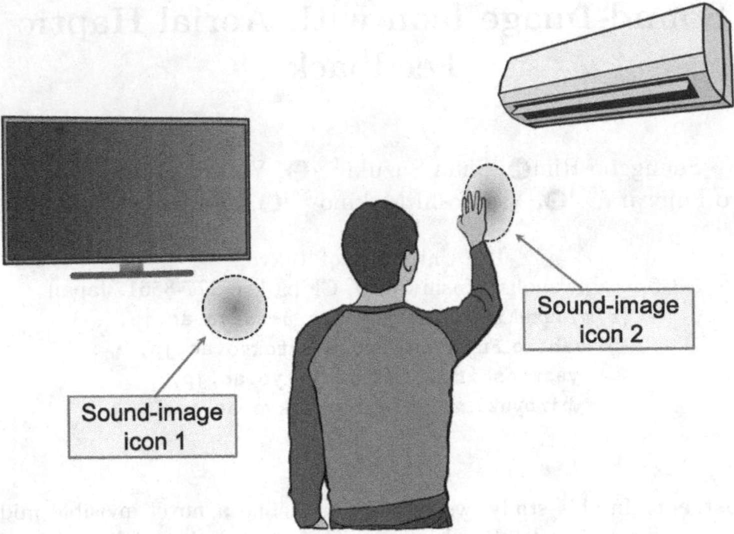

Fig. 1. Concept of a sound-image icon. The person selects the sound-image icon of the air conditioner to control the temperature. The sound from the icon informs the user about the position of itself. The icon provides tactile feedback without any visual image.

arrangement of an object to a user, while haptics facilitates the transmission of the will of a user to a computer system. Therefore, a 3D visual interface with mid-air haptics is a reasonable integration as an efficient interface. However, the eyes are sometimes occupied by a specific task such as in-car driving or a surgical operation. In addition, glassless 3D vision is still immature, where it is difficult to secure a wide view angle, while a head-mounted display sometimes causes fatigue and VR sickness. Instead of vision, the use of sound is another option to display the object position, as humans can instantaneously identify the direction of the sound when they are in an environment where sound can be clearly heard [1]. In addition, the sound can transmit words and tones that express various attributes.

Many studies have been performed to create virtual sound sources at specific locations under various conditions and environments. Recent studies have focused on virtual sound source positioning for acoustic navigation in unknown spaces [14] or vector base amplitude panning for creating 2D or 3D sound fields without considering the placements of any number of loudspeakers [13]. Although the aforementioned technologies form a wide research domain, we could not find studies that integrated virtual sound sources and mid-air haptics technologies. It would be intriguing to investigate whether auditory and haptic perceptions effectively complement each other.

The concept of a sound-image icon is depicted in Fig. 1. The user hears a binaural sound and identifies the direction of the sound-source. We refer to this sound source as a "sound image." The sound image can represent various functions, and users can recognize the role of the object by its sound. For example, an icon that represents an air-conditioner generates a sound to explain it in words, while an audio-volume icon produces a pleasant musical sound. The users utilize this sound as a clue to reach for the icons. Using tactile cues, they can recognize the exact positions of the icons and then perform fine tasks; for example, a user who accepts the objects of the air-conditioner can control the temperature by operating the sound-image icon. The haptic feedback is critical not only to improve the operability during the control but also to reliably guide and hold the user's hand to the starting point of the operation.

In this study, we prototype sound-image icons and experimentally verify their feasibility. We aim to realize a system where a user can select the desired icon among multiple icons and operate it. In this study, for the first step, we examine whether users can find an icon and measure the time required to access it. The combination of auditory and tactile sensations enables users to accurately and effectively locate the icons.

2 Proposed Method

A sound-image icon is realized using a sound source with the sense of touch provided via acoustic radiation pressure. In this section, we describe the method used to create the sound-image icon.

2.1 Producing a Sound Image

The user specifies the direction of the sound source using a binaural sound. We plan to provide binaural sounds using ultrasound beams to reach the ears. However, in this experimental system, the binaural sound was provided to the users by an in-ear binaural headset (CS-10EM, Roland). The binaural sound was recorded using the microphone in the headset that was fixed to the ears of one of the authors, keeping the sound source at the icon position. By reproducing the recorded sound in both the ears, the listener perceived the same 3D sound image as the real sound, under the assumption that the head related transfer functions are common [10].

2.2 Aerial Haptic Feedback

The aerial tactile sensation is presented at the icon location by using AUTD. The users actively search for the ultrasound focus, where the acoustic radiation pressure produces a tactile sensation of the virtual icon.

AUTD is a phased array that generates an ultrasonic focal point at an arbitrary position in the air [5,6]. The acoustic radiation pressure is proportional to the sound energy density on the skin surface [3]. Though an AUTD can produce various pressure patterns by controlling the amplitude and phase of each transducer at the frame-rate of 1 kHz [4], a single focus is created in this prototype. The users can perceive a certain stimulus around the focal point, where the tactile feel becomes vivid when the ultrasound amplitude is modulated in the amplitude or the focus position is laterally vibrated on the skin [15].

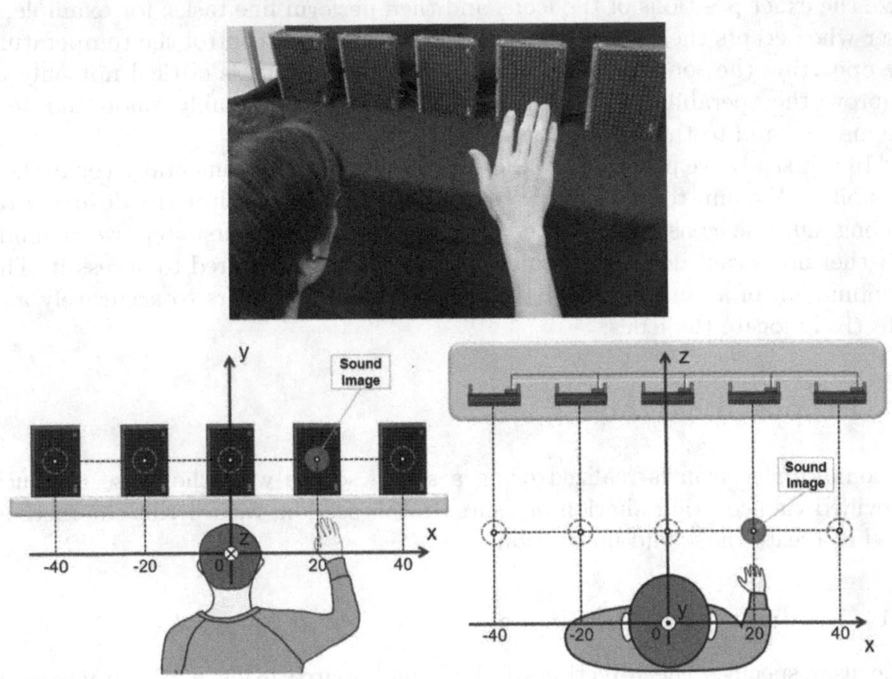

Fig. 2. Experimental setup. Five AUTDs were deployed 30 cm behind the five spots where the sound-image icon must be placed. The units of the numbers are in cm.

3 Experiment

We experimentally verified that humans could haptically identify the location of a particular icon by following the perception of sound. We measured the accuracy and time for haptic identification.

3.1 Procedure

The experimental setup is depicted in Fig. 2. In this experiment, we displayed five icons and examined whether the participants could identify them, and then we measured the time required for the identification. The icons were placed at $(-40, 20, 20)$, $(-20, 20, 20)$, $(0, 20, 20)$, $(20, 20, 20)$, and $(40, 20, 20)$ cm, where the origin was the center of the head, z-axis was parallel to the front direction, y-axis was parallel to the vertical direction, and x-axis is set as forming a right-handed system. The five AUTDs were placed 30 cm behind the sound image. A single corresponding AUTD emitted a focused ultrasound with 200 Hz sinusoidal amplitude modulation. The unit of the AUTD is an ultrasound phased array (SSC-HCT1, Shinko Shoji Co., Ltd.) with 14×18 elements. The maximum force displayed by a single unit is 10 mN.

As sound sources, we recorded the solos of the following five kinds of musical instruments: drum, bass, acoustic guitar, piano, and flute. We defined the drum sound as the target sound, as it was the easiest one to locate the position. We instructed the participants that the drum was the sound of the target sound-image icon to locate. Several sound sources (five at maximum), which included the target sound, were randomly selected and played at random positions in the five locations. As tactile feedback, a single ultrasound focus of the target icon was created at the target position.

The finding-target experiment was conducted considering the following two conditions: auditory-only and auditory-with-tactile. In the auditory-only condition, only the binaural sound was presented, while in the auditory-with-tactile condition, both sound and tactile stimulation were simultaneously presented.

We asked the participants to estimate the location of the target sound-image icon as soon as possible after they recognized the start cue, i.e., the moment the audio was played. In addition, we instructed them to close their eyes during fumble to prevent the visual effect. After determining the position of the icon, the participants indicated the position number among the five options from 1 to 5 with the keyboard. We applied white noise to eliminate the effects of AUTD driving noise.

The participants in this experiment were twelve men in their twenties who had no problems with hearing or health.

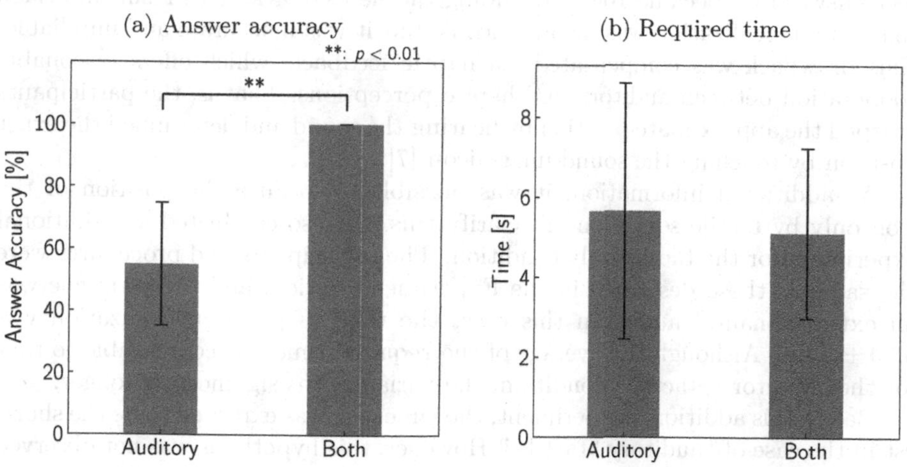

Fig. 3. Average of participants' (a) answer accuracy and (b) required time.

3.2 Results and Evaluations

The results are depicted in Figs. 3 (a) and (b). According to Fig. 3 (a), the answer accuracy was 55% when only the sound was informed of the location to the participants. Despite the high error rate of the "auditory-only condition," the accuracy rate of the "auditory+tactile condition" was almost 100%. Using the *t*-test, we examined whether the correct answer rate could be significantly improved by adding a tactile sensation to the sound cue. As a result of the test, the *p*-value between "auditory" and "auditory+tactile" was smaller than 0.01. This result means that the participants could exactly pinpoint the location of the sound image when they were able to search using tactile sensations.

In addition, Fig. 3 (b) shows that the average required times of each case were almost the same. Accordingly, the average time for "auditory" was 5.67 s, and that for "auditory+tactile" was 5.12 s. The standard deviation for "auditory+tactile" was less than that for "auditory." The standard deviation for "auditory+tactile" was approximately 2.12 s, and that for "auditory" was 3.20 s. This indicates that presenting both the stimuli reduced their standard deviations. For clarification, we used Levene's test for the standard deviations and Welch's *t*-test for the averages. As a result of Levene's test, the *p*-value between "auditory" and "auditory+tactile" was smaller than 0.01. According to the test, a significant difference was observed between the variances of required time for the two conditions. Additionally, the significance between the average amounts was not noticed, as the *p*-value from the Welch's *t*-test was 0.121.

4 Discussion

As depicted in Fig. 3 (a), in the case of sound alone, the exact position of the icon could not be estimated, and mistakes occurred. On the other hand, the correct answer rate became 100% by adding tactile feedback. This result indicates that sound localization was instantaneous, but it was inaccurate and unreliable. This drawback was compensated via haptic feedback, which offers reasonable cooperation between auditory and haptic perceptions. That is, the participants grasped the approximate position by hearing the sound and determined the exact position by touching the sound-image icon [7].

As additional information, it was possible to localize the position of the icon only by tactile sensation. To clarify this, we also conducted an additional experiment for the tactile-only condition. The participants and procedures were the same as those described in the Experiment section, and the start cue was an extra monaural audio. In this case, the time to perform localization was 5.30 ± 3.39 s. Although the average of the required time was comparable to that for the "auditory+tactile" condition, the variance was significantly longer.

Before this additional experiment, the time cost was expected to be the shortest in the case of "auditory+tactile." However, this hypothesis was not observed in this experiment. Searching the entire space without prior information was not a time-consuming task, as it only took approximately 2 s for the participants to fumble around the area with their hands. Nevertheless, considering that the standard deviation for the "auditory+tactile" condition is the smallest, it was confirmed that the combination of sound and haptic feedback facilitated the search of the icon.

To avoid confusion, we reconfirm the purpose of the combination of auditory and tactile sensations as follows. The role of the sound is to notify the user of the existence and attributes of the icon around the user. Tactile sensation is necessary to determine the exact location and operate the icon. Therefore, even if the localization time for the tactile-only condition is short, it does not mean that the auditory cue is unnecessary.

5 Conclusion and Future Works

In this study, we proposed a sound-image icon and examined the basic feasibility of icon localization. The sound-image icon represents the virtual existence of sound and haptics without visual presentation. Through the research, we confirmed that the participants could search and estimate the location of a single icon in an efficient manner using their tactile and auditory senses. For future work, we will investigate the possibility of efficiently displaying multiple icons.

We used a headset as a sound display device, as this was a feasibility study to examine the effectiveness of the auditory–haptic integration. However, it is also possible to produce binaural sounds in a non-contact manner using airborne ultrasound [12]. Performing a detailed operation using the sound-image icon was beyond the scope of this paper, and it would be the next important challenge of the sound-image icon.

References

1. Blauert, J.: Spatial Hearing: The Psychophysics of Human Sound Localization. MIT Press, Cambridge (1997)
2. Carter, T., Seah, S.A., Long, B., Drinkwater, B., Subramanian, S.: UltraHaptics: multi-point mid-air haptic feedback for touch surfaces. In: Proceedings of the 26th Annual ACM Symposium on User Interface Software and Technology, pp. 505–514 (2013)
3. Hoshi, T., Takahashi, M., Iwamoto, T., Shinoda, H.: Noncontact tactile display based on radiation pressure of airborne ultrasound. IEEE Trans. Haptics **3**(3), 155–165 (2010)
4. Inoue, S., Makino, Y., Shinoda, H.: Scalable architecture for airborne ultrasound tactile display. In: Hasegawa, S., Konyo, M., Kyung, K.-U., Nojima, T., Kajimoto, H. (eds.) AsiaHaptics 2016. LNEE, vol. 432, pp. 99–103. Springer, Singapore (2018). https://doi.org/10.1007/978-981-10-4157-0_17
5. Iwamoto, T., Shinoda, H.: Two-dimensional scanning tactile display using ultrasound radiation pressure. In: 2006 14th Symposium on Haptic Interfaces for Virtual Environment and Teleoperator Systems, pp. 57–61. IEEE (2005)
6. Iwamoto, T., Tatezono, M., Shinoda, H.: Non-contact method for producing tactile sensation using airborne ultrasound. In: Ferre, M. (ed.) EuroHaptics 2008. LNCS, vol. 5024, pp. 504–513. Springer, Heidelberg (2008). https://doi.org/10.1007/978-3-540-69057-3_64

7. Kaul, O.B., Rohs, M.: Haptichead: a spherical vibrotactile grid around the head for 3D guidance in virtual and augmented reality. In: Proceedings of the 2017 CHI Conference on Human Factors in Computing Systems, pp. 3729–3740 (2017)
8. Korres, G., Eid, M.: Haptogram: ultrasonic point-cloud tactile stimulation. IEEE Access **4**, 7758–7769 (2016)
9. Makino, Y., Furuyama, Y., Inoue, S., Shinoda, H.: Haptoclone (haptic-optical clone) for mutual tele-environment by real-time 3D image transfer with midair force feedback. In: Proceedings of the 2016 CHI Conference on Human Factors in Computing Systems, pp. 1980–1990 (2016)
10. Møller, H.: Fundamentals of binaural technology. Appl. Acoust. **36**(3–4), 171–218 (1992)
11. Monnai, Y., Hasegawa, K., Fujiwara, M., Yoshino, K., Inoue, S., Shinoda, H.: HaptoMime: mid-air haptic interaction with a floating virtual screen. In: Proceedings of the 27th Annual ACM Symposium on User Interface Software and Technology, pp. 663–667 (2014)
12. Ochiai, Y., Hoshi, T., Suzuki, I.: Holographic whisper: rendering audible sound spots in three-dimensional space by focusing ultrasonic waves. In: Proceedings of the 2017 CHI Conference on Human Factors in Computing Systems, pp. 4314–4325 (2017)
13. Pulkki, V.: Virtual sound source positioning using vector base amplitude panning. J. Audio Eng. Soc. **45**(6), 456–466 (1997)
14. Storek, D., Rund, F., Suchan, R.: Virtual auditory space for the visually impaired-experimental background. In: 2011 International Conference on Applied Electronics, pp. 1–4. IEEE (2011)
15. Takahashi, R., Hasegawa, K., Shinoda, H.: Tactile stimulation by repetitive lateral movement of midair ultrasound focus. IEEE Trans. Haptics **13**, 334–342 (2019)

Stiffness Discrimination by Two Fingers with Stochastic Resonance

Komi Chamnongthai(✉), Takahiro Endo, Shohei Ikemura, and Fumitoshi Matsuno

Kyoto University, Kyoto 615-8540, Japan
chamnongthai.komi.46m@st.kyoto-u.ac.jp

Abstract. This paper focuses on Stochastic Resonance (SR) for stiffness discrimination by two fingers. In particular, we show that the subthreshold vibrotactile noise applied on a remote position can improve tactile sensations of both index finger and thumb for a task requiring multiple fingers. We evaluate the user performances in a virtual environment (VE) by Weber fraction for stiffness perception under one of three different vibration source positions: on the index finger, on the thumb, and between the index finger and thumb. The results show that the stiffness discrimination ability increase under all three vibration source positions with the best performance obtained for the source location between index finger and thumb. The finding indicates the potential of using a single vibration source to enhance sensation of multiple fingers by the effect of SR.

Keywords: Stiffness discrimination · Stochastic Resonance · Haptic interface · Multiple fingers

1 Introduction

In clinical field, palpation is one process to identify properties such as size, texture, location, etc., of an organ by a physical examination with multiple fingers. The ability to address the abnormality of the organ stiffness is one vital skill before operating other processes. However, to achieve the expert level performance to distinguish the difference accurately, the medical students must be trained strenuously with various models and difficulties. Furthermore, training with traditional methods consume a lot of time to obtain new skills. In addition, it also costs a large amount of money to prepare the instruction equipment for covering the various difficulties of training. To solve this problem, a haptic training system which combines a virtual environment (VE) and haptic interface is one interesting solution.

The common approach followed to communicate with haptic training system is to design a finger holder into which the user will insert his or her finger. However, the use of a finger holder decreases the force-detection ability at the finger [1] and therefore it is necessary to enhance the force-detection capability of the finger through other mechanisms in the presence of finger holder.

I. Nisky et al. (Eds.): EuroHaptics 2020, LNCS 12272, pp. 497–505, 2020.
https://doi.org/10.1007/978-3-030-58147-3_55

There are several techniques to enhance tactile sensation: transcutaneous electrical nerve stimulation [2], temporary deafferentation [3], and passive sensory stimulation [4]. One effective solution is the addition of sensory noise which provides vibrotactile noise through the human skin in order to boost the sensitivity in several body parts such as the feet [5] and fingers [6]. This phenomena, which boosts the weak signal to be detectable by adding white noise, is called Stochastic Resonance (SR).

There are numerous studies that show the increase of haptic sensitivity by the effect of SR. According to Kurita *et al.* [8], the effect of SR is applied directly to the tip of the index finger, which shows the improvement of the user performance in grasping task. Furthermore, the effect of SR does not only occur when the vibration source is close to the tip of the index finger, but also happen while the vibration is generated from remote positions, which is away from the tip of the finger [9]. The study shows that the haptic sensation of the user is enhanced when a subthreshold vibration with a remote position is applied to the stroke patients. The fingertip perception is also shown to improve using SR when the finger is enclosed in the finger holder [10]. Therefore, the effect of SR has a potential to raise the haptic sensation at the user fingertip. However, there is no study that investigates the effects of SR on the fingertip when multiple fingers are enclosed within finger holders while doing the motor task.

The goal of this study is to determine the effect of SR in stiffness discrimination task using two fingers in VE. For this, we propose a novel method which consists of haptic feedback generated by haptic devices and the effect of SR provided by a piezoelectric actuator, in a stiffness discrimination task which is manipulated by multi-fingers through VE. The user performances are evaluated with three different vibration source positions (on the index finger (Position 1), between the index finger and thumb (Position 2), on the thumb (Position 3)) in order to find the possibility of the enhancement of the sensitivities at the fingertips. Many papers address that the haptic sensation of one finger is increased by one vibration source, but in this paper, we newly reveal that even one vibration source can improve the sensations of the two fingers via the effect of SR.

2 Proposed Method

We propose a method which integrates haptic feedback and the effect of SR for a better haptic sensitivity in the stiffness discrimination task. To provide the effect of SR, the mechanical vibration is applied to one of the three positions shown in Fig. 1 with varying vibration intensities. We would like to investigate the possibility that one vibration source can enhance the sensation of both fingers via the effect of SR. Two haptic devices, both being Geomagic Touch haptic devices, are used in this study. The original end-effectors of the haptic devices are customized for the task to operate with a finger as shown in Fig. 2(b). The modified end-effector consists of the finger holder, made from polyoxymethylene, and a force sensor (Leptrino, CFS018CA101U). On the other hand, the VE is programmed by using the CHAI3D library [11].

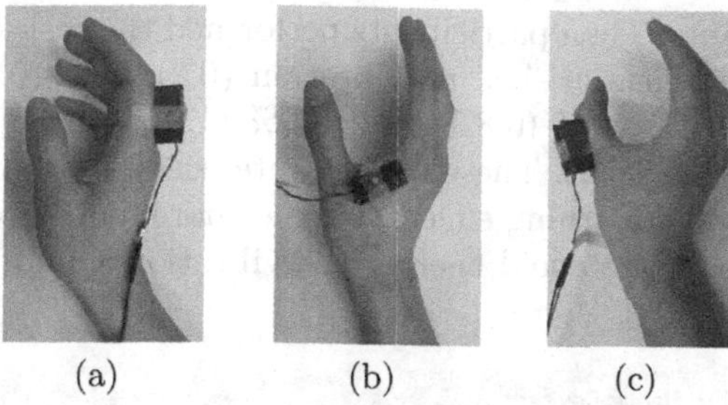

(a) (b) (c)

Fig. 1. The vibration source position in this study (a) on the index finger (Position 1), (b) between the index finger and the thumb (Position 2), (c) on the thumb (Position 3).

A piezoelectric actuator (Cedrat Technology Inc.: APA120S) is placed at the one of the three positions of the user hands to generate additive white Gaussian noise and gain the effect of SR. Additionally, the piezoelectric actuator can control the amplitude and the frequency of displacement freely, thus, the complicated signal is possible to generate. The generated vibration frequency is low-pass filtered at 400 Hz in order to activate all mechanoreceptors, as Pacinian corpuscles are active at frequencies between 0.5 and 400 Hz [12]. The Box–Muller method [13] is used to generate the white gaussian noise vibration $x(t)$ through the piezoelectric actuator:

$$x(t) = \sigma\sqrt{-2\ln\alpha(t)}\sin(2\pi\beta(t)), \tag{1}$$

where t is time, σ is the noise intensity, and α and β are independent random variables in the interval (0,1), and $x(t)$ corresponded to the voltage.

3 Evaluation Method

3.1 Subjects

Six healthy participants (mean age $\pm$ SD: 24.5 $\pm$ 1.88 years, all male) participated in the study. Before doing the experiment, all participants understood and consented to the experimental protocol approved by the Institutional Review Board of the Graduate School of Engineering, Kyoto University (No. 201707). The Weber Fraction (WF) value of stiffness perception was used to compare the user performance to detect the fingertip force. The participants inserted their index finger and thumb in the finger holders which are attached to the haptic devices, while VE in the computer monitor displayed two virtual objects to the participants.

3.2 Experimental Procedure

In this experiment, the sensory threshold (T) of each participant was measured using the *the staircase-method* [7] as the lowest vibration intensity that could be felt by the participants. Then participants performed the task with each of seven different vibration intensities; i.e., no vibration ($0T$), 40% ($0.4T$), 50% ($0.5T$), 60% ($0.6T$), 70% ($0.7T$), 80% ($0.8T$), and 100% ($1.0T$) of the sensory threshold with three different positions. The vibration intensities were provided randomly to participants to avoid learning effects. Furthermore, participants wore passive noise-cancelling headset to avoid hearing the vibration sound.

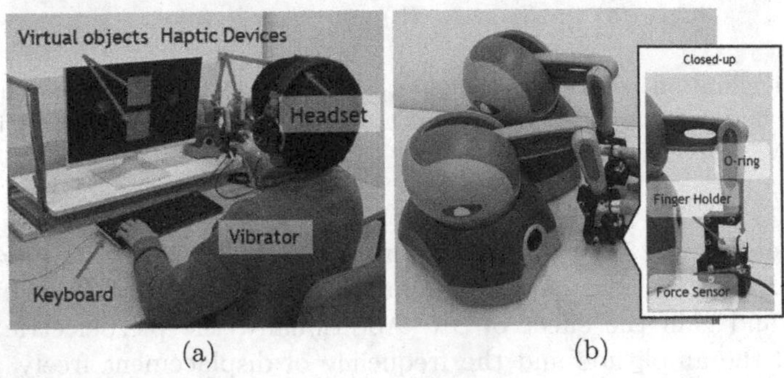

Fig. 2. Overview of experimental setup. (a) experimental setup when the participant was performing the task. The participants gets a force feedback through the haptic devices when he or she touches a virtual object through the red cursors which represent the finger positions on the screen, (b) the modified haptic devices were used in the study. The finger holder is attached to the device arm.

In VE, two virtual objects were presented as shown in Fig. 2(a). The objects were displayed as non-deformable in order to avoid the effect of visual feedback. The reference stiffness values were selected to be close to that of body fat of breast tissue (35 N/m [14]) as one of 25, 35 and 55 N/m. One virtual object had one of four reference stiffness value whereas the other object started with corresponding difference of 10 N/m and changed afterward as described later. The participants could touch the virtual objects for as long as they wanted. The force feedback was calculated using Hook's law; i.e., by multiplying the depth of the penetration of the finger into the virtual object and the stiffness of the touched object. The participants touched the two virtual objects and then chose the stiffer object by pressing a designated key on the keyboard, before starting the next trial. The task was completed by the white objects turning red. Average time spent on the experiment was around 2.5 h for one participant, including rest time.

For each reference value if the first object, Wald rule [15] was used to decide when to change the comparison stiffness for altering the stiffness of the second object. Furthermore, the change of the stiffness amount was decided by PEST rule [15], because the fixed step size takes longer time to complete the task, which lead the participant frustrated. A reversal point is defined as the turning point after which change in stiffness goes to the opposite direction of the previous direction. The average of the last four reversal values is used to determine the just noticeable difference (JND) of stiffness perception. Then we calculated WF of the subject in each condition as the average value between JND and reference stiffness:

$$WF = \frac{JND}{Reference\ Stiffness}. \tag{2}$$

4 Experimental Results and Discussion

The average WF values of each vibration intensity and stiffness difference are shown in three sub-figures of Fig. 3. The vertical axes show the WF value while the horizontal axes show the reference stiffness in each session. For example, a 35 N/m of reference stiffness with $0.6T$ is a point at 35 N/m. A lower WF value implies a higher haptic sensitivity of the user.

The results show that the WF values in all vibration-existed conditions tend to be less than the WF in no-vibration condition. As shown in Fig. 3, there are significant differences ($p < 0.05$) in the average WF values at $0T$ and $0.6T$ in all three sub-figures, confirmed by a two-tailed paired t-test. Moreover, the average of performance at $0.6T$ has the lowest WF values among other vibration levels.

Table 1. Results of Anova (Single Factor) in each vibration level

ANOVA (Single Factor)				
Vibration Intensity	Reference Stiffness (N/m)			
	25	35	45	55
$0T$	$F<F_{crit}$	$F<F_{crit}$	$F<F_{crit}$	$F<F_{crit}$
$0.4T$	$F<F_{crit}$	$F<F_{crit}$	$F<F_{crit}$	$F<F_{crit}$
$0.5T$	$F<F_{crit}$	$F>F_{crit}$	$F<F_{crit}$	$F<F_{crit}$
$0.6T$	$F>F_{crit}$	$F>F_{crit}$	$F>F_{crit}$	$F>F_{crit}$
$0.7T$	$F<F_{crit}$	$F<F_{crit}$	$F<F_{crit}$	$F<F_{crit}$
$0.8T$	$F<F_{crit}$	$F<F_{crit}$	$F<F_{crit}$	$F<F_{crit}$
$1.0T$	$F<F_{crit}$	$F<F_{crit}$	$F<F_{crit}$	$F<F_{crit}$

To compare the performances between the three positions, Analysis of Variance (ANOVA) was used to confirm a significant difference, shown in Table 1. In this table, $F-$value is calculated by the ANOVA test to examine whether the means between two conditions are significantly different or not, where the threshold is taken as $F_{crit(0.05,3,15)} = 3.28$. According to Table 1, $F > F_{crit}$ indicates a significant difference is observed for this comparison. Furthermore, Tukey's range test is used to find a pair of difference after the ANOVA test. $q-$value is calculated by the differences between means of two conditions to compare with $q_{crit(0.05,3,15)} = 3.67$. In Table 2, $q > q_{crit}$ shows the significant difference in each pair. Darker-color blocks in both tables show the significant difference in each comparison with $p < 0.05$. The ANOVA test reveals a significant difference from

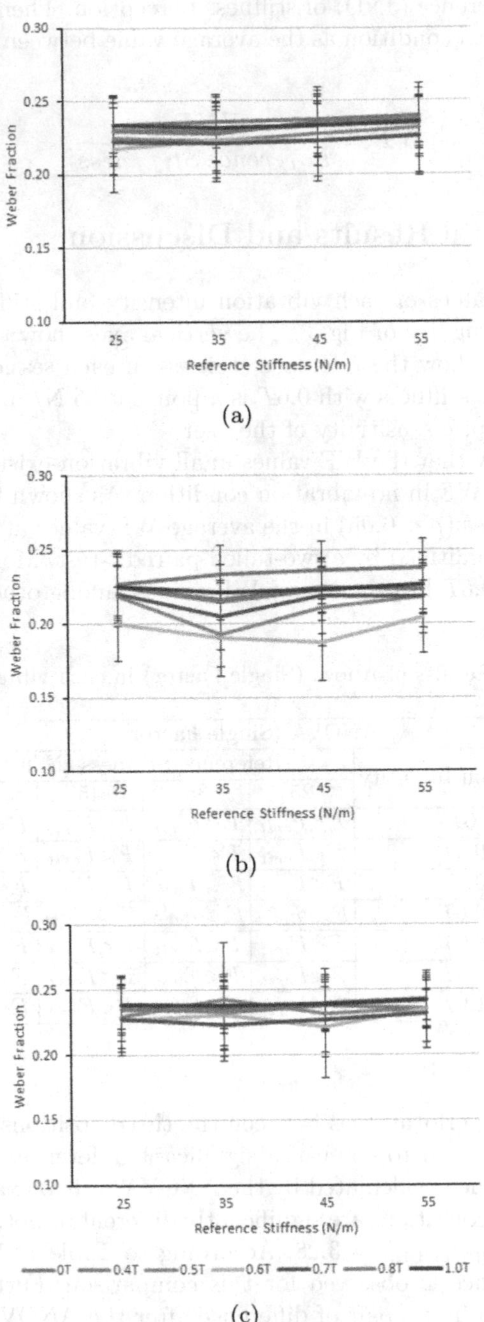

Fig. 3. User performance in each position of the vibration source (a) on the index finger, (b) between index finger and the thumb, (c) on the thumb.

Table 2. Results of Tukey's range test in the comparison of the position of vibration source

Tukey's Range Test				
Vibration Intensity	Reference Stiffness (N/m)	Pos.1 - Pos.2	Pos.1 - Pos.3	Pos.2 - Pos.3
$0.5T$	35	$q<q_{crit}$	$q<q_{crit}$	$q>q_{crit}$
$0.6T$	25	$q<q_{crit}$	$q<q_{crit}$	$q>q_{crit}$
	35	$q>q_{crit}$	$q<q_{crit}$	$q>q_{crit}$
	45	$q>q_{crit}$	$q<q_{crit}$	$q>q_{crit}$
	55	$q<q_{crit}$	$q<q_{crit}$	$q>q_{crit}$

Pos.1, Pos.2, and Pos.3 are Position 1, Position 2, and Position 3, respectively.

the position when the intensity is at $0.5T$ in 35 N/m and $0.6T$ in all reference stiffnesses. Then the results of the Tukey's test in Table 2 also show significant differences against Position 2 for Position 1 ($q > q_{crit}$) when the vibration is at $0.6T$ in all reference stiffnesses and $0.5T$ at 35 N/m. Furthermore, in comparison between Position 1 and Position 2, significant differences ($q > q_{crit}$) are also observed when the intensity is at $0.6T$ in both 35, and 45 N/m.

The present study showed the SR improves the fingertip sensation of both fingers even when the fingers are within the holders. It is hypothesized that when the vibration source is on Position 2, the vibration propagates to both fingers and not only a single finger as the other two conditions. Furthermore, the results indicated the potential of one vibration source at a remote position being able to improve the sensations of the two fingers. In addition, the limitation of this study is that the test subjects are still small in number and there is no variety of gender and age of the participants. Moreover, the subject in other conditions, such as stroke and other disability in haptic sensation, are not examined. Further investigations would be necessary, including the possibility of the improvement of the haptic performance through stochastic resonance.

5 Conclusion

We proposed a novel method for emphasizing the stiffness discrimination ability with multiple fingers in VE. The proposed method combined haptic feedback and the effect of SR in order to enhance the haptic performance while carrying out the task. The experimental results show the increase of the performance while applying a subthreshold vibration. Therefore, the proposed method is believed to enhance the sensation of two fingers in stiffness discrimination task with one vibration source when the fingers are inserted in the holders by the effect of SR. In future work, we will investigate the potential of this application and other possibilities in order to enhance haptic performances.

Acknowledgment. This work was supported in part by KAKENHI Grant No. 17K00270 and 20H04227.

References

1. Lederman, S.J., Klatzky, R.L.: Sensing and displaying spatially distributed fingertip forces in haptic interfaces for teleoperator and virtual environment systems. Presence Teleoperators Virtual Environ. **8**(1), 86–103 (1999)
2. Karol, S., Koh, K., Kwon, H.J., Park, Y.S., Kwon, Y.H., Shim, J.K.: The effect of Frequency of Transcutaneous Electrical Nerve Stimulation (TENS) on maximum multi-finger force production. Korean J. Sport Biomech. **26**(1), 93–99 (2016)
3. E, Sens., et al.: Effects of temporary functional deafferentation on the brain, sensation, and behavior of stroke patients. J. Neurosci. **32**(34), 11773–11779 (2012)
4. Smith, A.: Effects of caffeine in chewing gum on mood and attention. Hum. Psychopharmacol. Clin. Exp. **24**(3), 239–247 (2009)
5. Dettmer, M., Pourmoghaddam, A., Lee, B.C., Layne, C.S.: Effects of aging and tactile stochastic resonance on postural performance and postural control in a sensory conflict task. Somatosens. Motor Res. **32**(2), 128–135 (2015)
6. Collins, J.J., Imhoff, T.T., Grigg, P.: Noise-mediated enhancements and decrements in human tactile sensation. Phys. Rev. E **56**(1), 923 (1997)
7. Cornsweet, T.N.: The staircase-method in psychophysics. Am. J. Psychol. **75**(3), 485–491 (1962)
8. Kurita, Y., Shinohara, M., Ueda, J.: Wearable sensorimotor enhancer for fingertip based on stochastic resonance effect. IEEE Trans. Hum. Mach. Syst. **43**(3), 333–337 (2013)
9. Enders, L.R., Hur, P., Johnson, M.J., Seo, N.J.: Remote vibrotactile noise improves light touch sensation in stroke survivors' fingertips via stochastic resonance. J. Neuroengineering. Rehabil. **10**(1), 105 (2013)
10. Chamnongthai, K., Endo, T., Nisar, S., Matsuno, F., Fujimoto, K., Kosaka, M.: Fingertip force learning with enhanced haptic sensation using stochastic resonance. In: Proceedings of the IEEE World Haptics Conference, pp. 539–544 (2019)
11. Conti, F., et al.: The CHAI libraries. In: Proceedings of the Eurohaptics, pp. 496–500, (2003)
12. Johansson, R.S., Landstro, U., Lundstro, R.: Responses of mechanoreceptive afferent units in the glabrous skin of the human hand to sinusoidal skin displacements. Brain Res. **244**(1), 17–25 (1982)
13. Box, G.E.P.: A note on the generation of random normal deviates. Ann. Math. Stat. **29**, 610–611 (1958)
14. Samani, A., Zubovits, J., Plewes, D.: Elastic moduli of normal and pathological human breast tissues: an inversion-technique-based investigation of 169 samples. Phys. Med. Biol. **52**(6), 1565 (2007)
15. Taylor, M., Creelman, C.D.: PEST: efficient estimates on probability functions. J. Acoust. Soc. Am. **41**(4A), 782–787 (1967)

Adaptive Fuzzy Sliding Mode Controller Design for a New Hand Rehabilitation Robot

Alireza Abbasimoshaei[1], Majid Mohammadimoghaddam[2], and Thorsten A. Kern[1(✉)]

[1] University of Technology Hamburg Harburg, Eissendorferstr. 38, 21073 Hamburg, Germany
{al.abbasimoshaei,t.a.kern}@tuhh.de
[2] Tarbiatmodares University, Amirabad, 14115 Tehran, Iran
m.moghadam@modares.ac.ir
https://www.tuhh.de/imek

Abstract. Hand rehabilitation is one of the most important rehabilitation procedures. Due to the repetitive nature of rehabilitation training, a full robotic system could help the physiotherapists to gain time for creating new training schemes for a larger number of patients. Such a system can be based on live or recorded data and consists of the operator-device, patient-device, and control mechanism. This paper focuses on the design of the patient-device and its control-system in a decoupled training scenario. It presents a robot for hand rehabilitation training fingers and wrist independently based on only two actuators. These two actuators are configurable to allow consecutive training on the wrist and all joints of the fingers. To overcome uncertainties and disturbances, a sliding mode controller has been designed and an adaptive fuzzy sliding mode controller is used to reduce the chattering effects and compensate the varying forces of the patients. The experimental results show an approximate 80% improvement in tracking the desired trajectory by the adaptation.

Keywords: Adaptive fuzzy control · Rehabilitation robot · Haptics

1 Introduction

The need for rehabilitation of hand-fractures origins from two sources. One is hand-fractures, in general occurring among all ages including boys and girls, among which one third face fractures before the age of 17 [1,2]. The second source is rehabilitation after surgery or plastering to regain mobility. The traditional method usually requires the active involvement of a physiotherapist and requires a lot of time with repetitive training. Due to a lack of resources for therapy, new methods and equipment such as rehabilitation robots and actuated home-rehabilitation [3,4] are under strong development.

I. Nisky et al. (Eds.): EuroHaptics 2020, LNCS 12272, pp. 506–517, 2020.
https://doi.org/10.1007/978-3-030-58147-3_56

Acceptance of such active systems is usually good if the patient feels to be in charge due to understandable and expectable motions and the possibility for an emergency stop. An additional benefit is always the opportunity to record and collect data about the progress of therapy. Combining robotic therapy with other methods, such as motor learning, control or bio-signal processing, helps to develop the potentiality of rehabilitation [5]. Although some items such as device-accuracy in the medical robots need to be considered, a large number of clinical studies confirmed the efficiency of robotic rehabilitation robots [6].

Coming into technical details, a lot of different systems for hand and finger-rehabilitation were proposed by researchers and commercial vendors for therapeutic systems. A device-taxonomy can be given by the number of actuated DOFs, the physiological joints in therapeutic focus, whether they are grounded or wearable, mode of rehabilitation exercises and in general the complexity of the device according to Table 1.

The scope of this paper is about a combined wrist and finger rehabilitation robot with the capability to exercise each phalanx individually, with a maximum of multi DOFs combined in one device at an affordable price-point. Despite all systems from Table 1 have their benefits, nearest to the scope of this paper is [16], [12], and [19] from different points of view. [16] shows a reconfigurable system but it was not for fingers. [12] is for the rehabilitation of wrist and fingers and it differs from our system because it was not for each finger individually. [19] is for each finger but it was not for each phalanges.

Concerning the underlying control algorithm especially [20] shows an interesting approach by force control and [19] due to using impedance control, in this system an adaptive fuzzy sliding mode controller is used which will be described in the following.

2 Design and Prototyping

Figure 1a shows the schematic view of the designed wrist and finger rehabilitation robot, which moves finger joints and wrist with two motors. As can be seen in Fig. 1a, the hand is located in the upper section that includes the green finger part and two ball bearings. Because during the rotation, the joints center of motion changes, a flexible system is used for the finger part. In this system, the first motor (motor 1) moves the cable and rotates the finger. The wrist is rotated by the second motor (motor 2) while the ball bearings and a shaft transfer the rotation of the motor to the wrist [21,22].

A detailed view of the finger part is shown in Fig. 1b. There is a bar at the backside of the system to lock or unlock the joints. The configuration of the bar shown in Fig. 1b is for DIP training of the index finger. The finger part is adjusted by changing the engaged track (Fig. 1b) and the circular end of the bar is for making the movement of the bar easier. By changing the unlocked joint, the rehabilitation can be applied to each phalanx. As it is shown in Fig. 1c, DIP sits at the tip of the finger part and according to the size of the finger, the tracks will be fixed. The cable connected to the tip of the system moves the finger to the palm and the spring moves it in the reverse direction.

Table 1. Main features of some of the most common previous devices for wrist and finger rehabilitation

Name	DOF	Joints	Fixation	Mode of Rehabilitation	Source
Rutgers Master II	4	Four fingers (without little finger)	Wearable	Active	[7]
Wristbot	3	Wrist	Grounded	Active and Passive	[8]
Gloreha	5	Fingers	Wearable	Active	[9]
CR2-Haptic	1	Forearm and Wrist	Grounded	Active	[10]
Hand of Hope	5	Fingers	Wearable	Active	[11]
HWARD	3	Wrist and Fingers	Wearable	Active	[12]
Reha-Digit	4	Four Fingers (without thumb)	Grounded	Passive	[9]
CyberGrasp	5	Fingers	Wearable	Active	[13]
ARMin	2	Forearm and Wrist	Wearable	Additional hand module	[14]
GENTLE/G	3	Fingers	Wearable	Additional hand module	[15]
WReD	1	Forearm and Wrist	Grounded	Active	[16]
Amadeo	5	Fingers	Grounded	Passive	[17]
BiManu Track	1	Forearm and Wrist	Grounded	Active and Passive	[18]
Hand Robot Alpha-Prototype II	1	Fingers	Grounded	Additional hand module	[19]
HandCARE	5	Five fingers	Grounded	Active	[20]

In Fig. 2 the manufactured robot and rehabilitation procedure of the wrist and finger are shown. The system includes two actuators, rotating and fixed plates. Furthermore, an emergency key is designed to stop the system in an emergency condition. Due to the decoupled degrees of freedom, the actuation system moves the fingers and wrist separately. For rehabilitation, the patient's hand should be placed at the hand holder and according to the desired movement, finger or wrist will be trained.

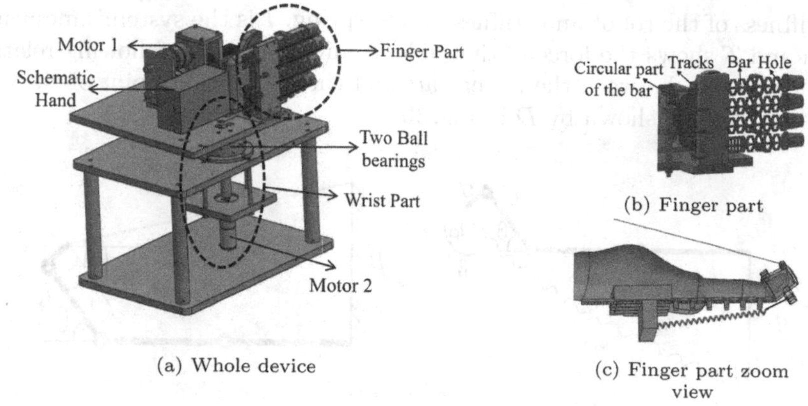

(a) Whole device (b) Finger part (c) Finger part zoom view

Fig. 1. Schematic design of the system.

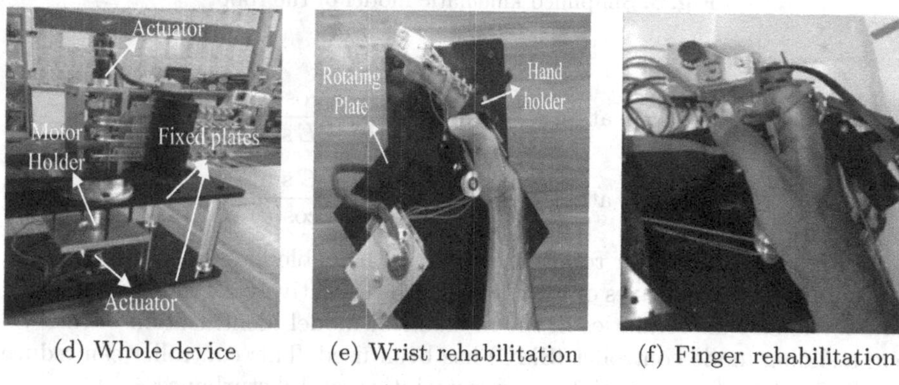

(d) Whole device (e) Wrist rehabilitation (f) Finger rehabilitation

Fig. 2. Prototype of the rehabilitation robot

3 Mathematical Model of the Device and SMC Design

The dynamic equation of the rehabilitation robot is the result of Newton's law applied to the fingertip. The overall equation of the system is obtained as follows:

$$\begin{aligned} I\ddot{\theta} = T &\times \sin(\alpha) \times l_3 + T \times \cos(\alpha) \times E - K \times ((\sqrt{A} - \sqrt{B}) \times \cos(\beta) \times l_3 \\ &+ (\sqrt{A} - \sqrt{B}) \times \sin(\beta) \times G) - C\dot{\theta} - K_1\theta \end{aligned} \tag{1}$$

$$A = (H + l_3 \sin(\theta))^2 + (l_1 + l_2 + l_3 \cos(\theta))^2 \tag{2}$$

$$B = H^2 + (l_1 + l_2 + l_3)^2 \tag{3}$$

$$I\ddot{\theta} = T \times R. \tag{4}$$

Figure 3a shows the simplified kinematic model of the robot for Fig. 1 and E, G, and H are the distances shown in the picture. In Eq. 1, l_1, l_2, and l_3 are the length of the phalanges, I is the inertia of the rotating part, R is the motor shaft, and θ is the rotation angle of the finger. C, K_1, and K represent the damping and stiffness of the robot and stiffness of the spring. I is the system's moment of inertia and T shows the force of the cable. α and β have the following relations and the distance between the finger part and the connection point of the cable with the system is shown by D in Fig. 3b.

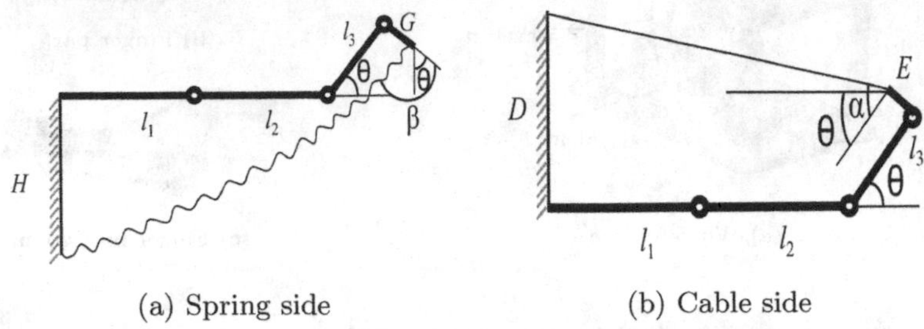

(a) Spring side (b) Cable side

Fig. 3. Simplified kinematic model of the robot

$$\alpha = \theta + \operatorname{atan}(\frac{D - l_3\sin(\theta) - E\cos(\theta)}{l_1 + l_2 + l_3\cos(\theta) - E\sin(\theta)}) \tag{5}$$

$$\beta = \theta + \operatorname{atan}(\frac{l_1 + l_2 + l_3\cos(\theta) + G\sin(\theta)}{H + l_3\sin(\theta) - G\cos(\theta)}) \tag{6}$$

In Eq. 1, θ is the finger rotation angle, l is the cable length, x and y are the horizontal and vertical axes of the cable length respectively. Because of unknown parameters and uncertainties in the mechanical model identification of the system, a sliding mode controller (SMC) has been used. This controller can reduce the effects of parameter variations, uncertainties, and disturbances.

For designing the SMC, it should be considered that the sliding mode controller could guarantee the stability of the system and it consists of two sub-controllers u_{eq} and u_{rb}. To make the system stable in the Lyapunov sense, $S\dot{S}$ should be less than zero. u_{eq} is the equivalent controller and u_{rb} is used to control the uncertainties and disturbances. In this system, u_{eq} and u_{rb} are considered as Eq. 8 and Eq. 9. The final design of the SMC found as follows [23].

$$u = u_{eq} + u_{rb} \tag{7}$$

$$u_{eq} = g^{-1}(\ddot{x}_d - f - k(\dot{x} - \dot{x}_d) - \eta s) \tag{8}$$

$$u_{rb} = -g^{-1}\rho.\text{sgn}(s) \tag{9}$$

In which, η is a positive constant. If we consider the general equation of the system as Eq. 10 and Eq. 11, g and f formula for this system would be obtained as Eq. 12 and Eq. 13.

$$\ddot{x} = f(x,t) + g(x,t)u + \lambda \tag{10}$$

$$y = x \tag{11}$$

$$g = (\frac{1}{I})(\sin(\alpha)l_3 + \cos(\alpha)E) \tag{12}$$

$$\begin{aligned} f = (\frac{1}{I})(-K \times ((\sqrt{A} - \sqrt{B}) \times \cos(\beta) \times l_3 \\ + (\sqrt{A} - \sqrt{B}) \times \sin(\beta) \times G) - C\dot{\theta} - K_1\theta) \end{aligned} \tag{13}$$

Where $g(x,t)$ and $f(x,t)$ are unknown functions of the system dynamic equation. Moreover, λ is unknown disturbances satisfying Eq. 14.

$$|\lambda| < \rho \tag{14}$$

4 Adaptive Fuzzy Sliding Mode Controller Design

The sliding mode controller reduces the error of the system. But it has a sign function which leads to an undesired chattering phenomenon. To overcome this error, fuzzy controller design is proposed. In the previous work [23], a fuzzy sliding mode controller (FSMC) is designed for the rehabilitation robot by integrating a fuzzy controller into an SMC. $S(t)$ and $\dot{S}(t)$ are the inputs of the fuzzy system and the output of the system is u_{fa} [23].

In this configuration, undesired chattering is overcome. However, during the experiments, it is shown that due to the different stiffness of the patients' hands, various interactive forces are entered into the hand and robot. To overcome this, the robot needs an adaptive controller according to Fig. 4.

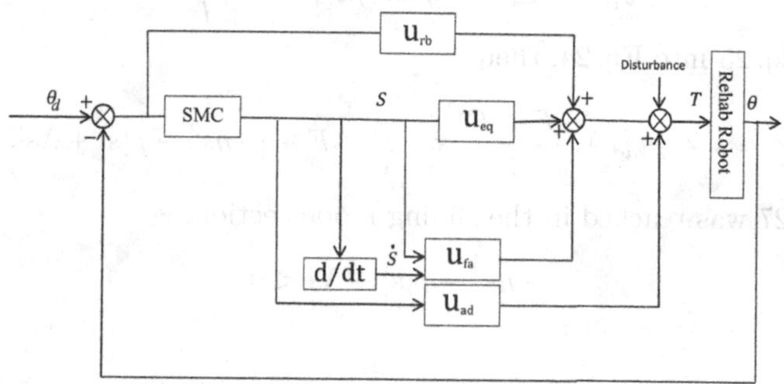

Fig. 4. The overall block diagram of the system containing adaptive controller

The adaptive controller tuning law is derived based on the Lyapunov theory to guaranty the system stability. This adaptive law is designed to approximate the indeterminacy and the interaction force and drives the trajectory tracking error to zero. After considering disturbances, unknown parameters, and patients' interaction force, the mathematical model of the system can be expressed as follows.

$$x_1 = \theta \tag{15}$$

$$\dot{x_1} = x_2 \tag{16}$$

$$\ddot{x} = (\frac{1}{I})(T \times \sin(\alpha) \times l_3 + T \times \cos(\alpha) \times E - K \times ((\sqrt{A} - \sqrt{B}) \times \cos(\beta) \times l_3 + (\sqrt{A} - \sqrt{B}) \times \sin(\beta) \times G) - C\dot{\theta} - K_1\theta + F_{int} + \lambda) \tag{17}$$

$$y = x_1 \tag{18}$$

Where x_1,x_2, and y are the state vectors and F_{int} is the interaction force of the robot and the hand. We also suppose that the interaction force changes slowly.

$$\dot{F}_{int} = 0 \tag{19}$$

The proposed Lyapunov function is defined as follow.

$$v = (\frac{1}{2})(s^2) + (\frac{1}{2})(\tilde{F}^2) \tag{20}$$

$$\tilde{F} = F_{int} - \hat{F} \tag{21}$$

In which $\hat{F}$ is the estimation of the interaction force(F_{int}), and $\tilde{F}$ is the error of this estimation. Thus,

$$\dot{\tilde{F}} = \dot{F}_{int} - \dot{\hat{F}} = -\dot{\hat{F}} \tag{22}$$

$$\dot{V} = S \times \dot{S} + \tilde{F} \times \dot{\tilde{F}} = S \times \dot{S} - \tilde{F} \times \dot{\hat{F}} \tag{23}$$

$$\dot{V} = S \times ((\ddot{x}_1 - \ddot{x}_d) + K(\dot{x}_1 - \dot{x}_d)) - \tilde{F} \times \dot{\hat{F}} \tag{24}$$

$$\ddot{x}_1 = f(x,t) + g(x,t)u + \lambda + \frac{F_{int}}{I}. \tag{25}$$

Taking Eq. 25 into Eq. 24, then

$$\dot{V} = s \times gu_{ad} + \frac{(\tilde{F} + \hat{F})}{I}s - \tilde{F} \times \dot{\hat{F}} + [-\eta s^2 - \rho|s| + \lambda s]. \tag{26}$$

The Eq. 27 was reached in the sliding mode section.

$$-\eta s^2 - \rho|s| + \lambda s < 0 \tag{27}$$

So,

$$\dot{V} < s \times g \times u_{ad} + \frac{(\tilde{F} + \hat{F})}{I}s - \tilde{F} \times \dot{\hat{F}} \tag{28}$$

$$u_{ad} = -\frac{\hat{F}}{I \times g} \tag{29}$$

$$\hat{F} = \int \frac{s}{I} \tag{30}$$

$$u_{ad} = -\frac{1}{l_3 \sin(\alpha) + E\cos(\alpha)} \int \frac{s}{I} \tag{31}$$

Thus, with this adaptive signal which is added to the other signals, $\dot{V}$ is always negative and the system stability is guaranteed.

5 Experiments and Results

Our robot design was developed with the focus on easy application in daily professional life and by advice from physiotherapists. Two trajectory tracking experiments were done to validate the controlling system and robot. The first one was for wrist and the second one was for fingers. In the first experiment, ten subjects (seven males and three females) did the tracking training three times and every training session takes time between 2 to 4 min. Each volunteer was encouraged to keep relaxed for training. In this experiment, a cosine wave function is defined as the desired trajectory for the wrist. Figure 5 shows the average error of the experiments and depicts that the error of the system for wrist trajectory tracking is reduced and the movement of the system became smooth and near to the desired. As shown in this figure, the fuzzy sliding mode controller reduced the error and the measured trajectory follows the desired trajectory better. But adaptive fuzzy SMC can reduce the errors resulting from the differences between patients. Therefore, this controller is used to reduce the effects of the different patients' interactions with the robot.

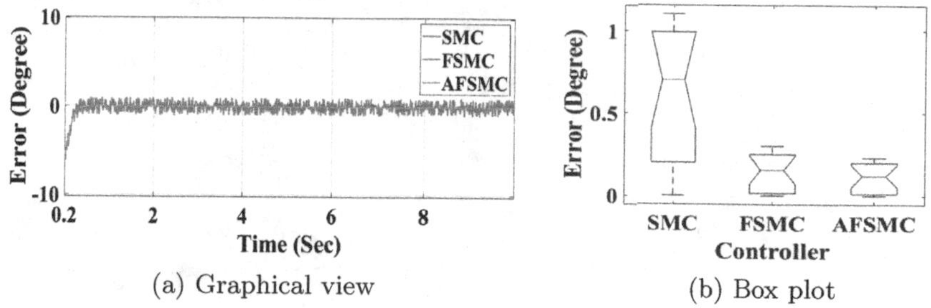

Fig. 5. Different controllers error

In the other experiment, the slow movement of each phalanx was explored. To find the desired trajectory of phalanges, the movements kinematic of all of them were analyzed during their tasks. Ten healthy subjects, seven males and three females, with different finger sizes, performed finger trials under the supervision of a physician [24]. They moved their phalanges (without robot) according to the physician instructions and an attached gyro sensor measured angle of rotation. The average of the collected data was found and fitted with a polynomial. Then, the experiments were done with the robot and different control algorithms. Figure 6 shows the results of the experiments with sliding mode controller and adaptive fuzzy SMC.

It can be understood from these experiments that the sliding mode controller improves the tracking performance, but there are some errors because of the chattering effects. The fuzzy controller reduces the chattering effects and makes the performance of the system better because it adjusts the output of the system according to the errors and disturbances. Adaptive fuzzy SMC decreases the error because this controller adapts the robot with different patients. According to the average data of the experiments, it is computed that using an adaptive fuzzy sliding mode controller (AFSMC) reduces the average errors in the wrist and phalanges about 80%.

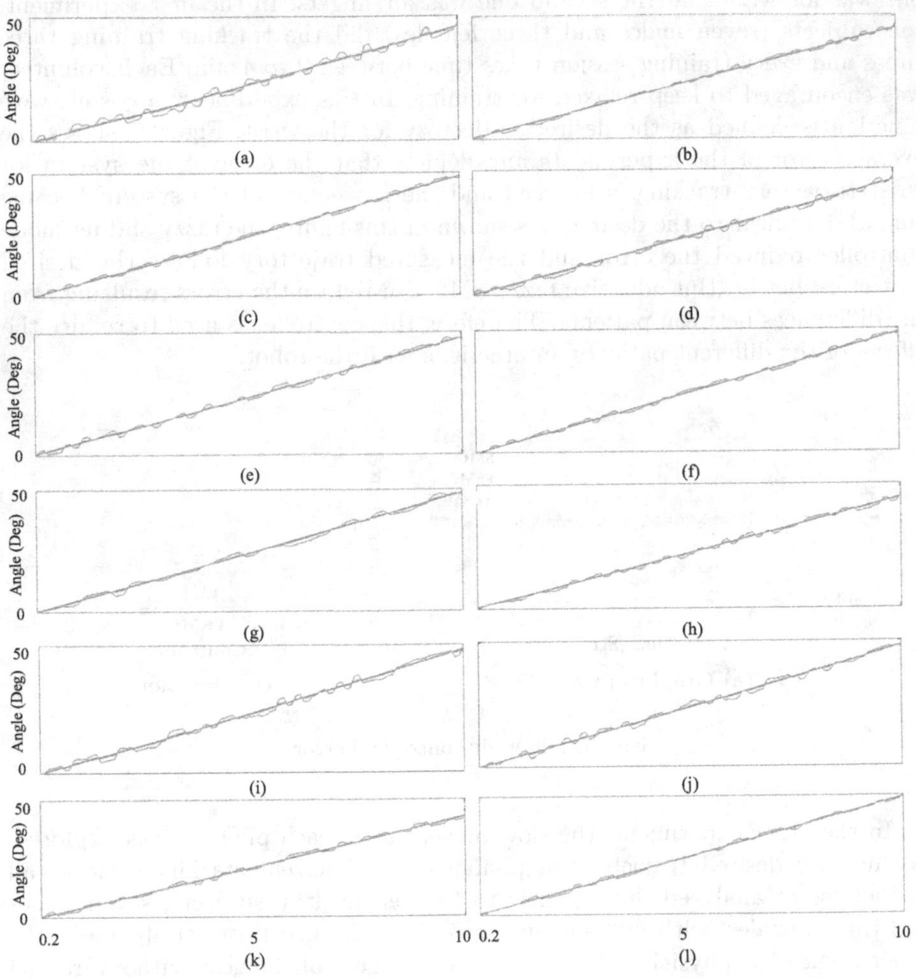

Fig. 6. Trajectory (blue line), output of the system with SMC (red line), and adaptive fuzzy SMC (green line) for each phalanx: a) DIP phalanx of index b) DIP phalanx of middle c) DIP phalanx of ring d) DIP phalanx of little e) PIP phalanx of index f) PIP phalanx of middle g) PIP phalanx of ring h) PIP phalanx of little I) MCP phalanx of index j) MCP phalanx of middle k) MCP phalanx of ring l) MCP phalanx of little finger (Color figure online)

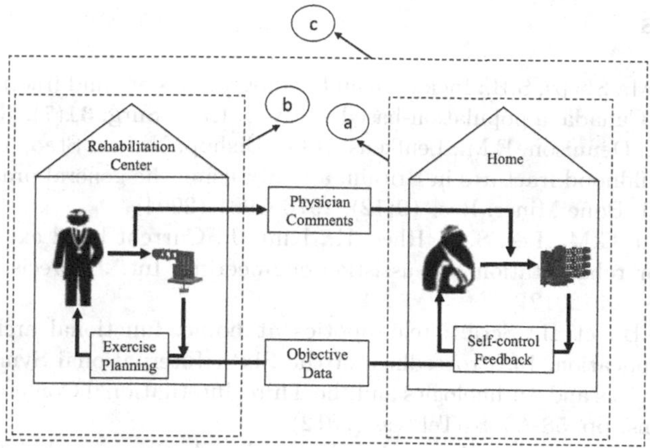

Fig. 7. Full telemanipulation system

6 Conclusion

In this paper, a novel mechanism for wrist and fingers is presented. In this system, the number of motors is reduced and the robot can rehabilitate all of the joints and wrist with only two motors. Still, each phalanx can be rehabilitated by this robot that makes rehabilitation very cost-efficient.

Furthermore, an AFSMC design method is proposed to control this robot. This controller can deal with unknown parameters and uncertainties and it enhances the system robustness. In this controller, the output of the fuzzy controller is calculated based on the error. Thus, the controller is more robust and independent of the system model. On the other hand, with different patients, there are different interaction forces between hand and robot and it is an important parameter that should be considered in the dynamic equation of the system. Therefore, there is a requirement for adapting the parameters, thus an adaptive controller beside the fuzzy SMC is designed to eliminate the effects of these forces. The effectiveness of the control system is examined with some trajectory tracking experiments. The experiments results show that the proposed AFSMC has much less error and as an average, the system performance improves by 80%.

For the future, following previous works like [4], a haptic system will be designed to make communication between the physician and the patient. This system can improve the efficiency of the rehabilitation procedure and has the potential not only to make the expert involvement more efficient but also to reduce the total duration of the rehabilitation due to only partially supervised offline-training capabilities (Fig. 7).

References

1. Feehan, L.M., Sheps, S.B.: Incidence and demographics of hand fractures in British Columbia, Canada: a population-based study. J. Hand Surg. **31**(7), 1068–e1 (2006)
2. Cooper, C., Dennison, E.M., Leufkens, H.G., Bishop, N., van Staa, T.P.: Epidemiology of childhood fractures in Britain: a study using the general practice research database. J. Bone Miner. Res. **19**(12), 1976–1981 (2004)
3. Heo, P., Gu, G.M., Lee, S.-J., Rhee, K., Kim, J.: Current hand exoskeleton technologies for rehabilitation and assistive engineering. Int. J. Precis. Eng. Manuf. **13**(5), 807–824 (2012)
4. Prange, G.B., et al.: Script: tele-robotics at home; functional architecture and clinical application. In: Proceedings of the Sixth International Symposium on e-Health Services and Technologies and the Third International Conference on Green IT Solutions, pp. 58–63. SciTePress (2012)
5. Iandolo, R., et al.: Perspectives and challenges in robotic neurorehabilitation. Appl. Sci. **9**(15), 3183 (2019)
6. Bogue, R.: Rehabilitation robots. Ind. Robot Int. J. **45**(3), 301–306 (2018)
7. Bouzit, M., Burdea, G., Popescu, G., Boian, R.: The Rutgers Master II-new design force-feedback glove. IEEE/ASME Trans. Mechatron. **7**(2), 256–263 (2002)
8. Masia, L., Casadio, M., Giannoni, P., Sandini, G., Morasso, P.: Performance adaptive training control strategy for recovering wrist movements in stroke patients: a preliminary, feasibility study. J. Neuroeng. Rehabil. **6**(1), 44 (2009)
9. Bos, R.A., et al.: A structured overview of trends and technologies used in dynamic hand orthoses. J. Neuroeng. Rehabil. **13**(1), 62 (2016)
10. Khor, K., Chin, P., Hisyam, A., Yeong, C., Narayanan, A., Su, E.: Development of CR2-haptic: a compact and portable rehabilitation robot for wrist and forearm training. In: 2014 IEEE Conference on Biomedical Engineering and Sciences (IECBES), pp. 424–429. IEEE (2014)
11. Balasubramanian, S., Klein, J., Burdet, E.: Robot-assisted rehabilitation of hand function. Curr. Opin. Neurol. **23**(6), 661–670 (2010)
12. Takahashi, C.D., Der-Yeghiaian, L., Le, V., Cramer, S.C.: A robotic device for hand motor therapy after stroke. In: 9th International Conference on Rehabilitation Robotics, ICORR 2005, pp. 17–20. IEEE (2005)
13. Nikolakis, G., Tzovaras, D., Moustakidis, S., Strintzis, M.G.: Cybergrasp and phantom integration: enhanced haptic access for visually impaired users. In: 9th Conference Speech and Computer (2004)
14. Nef, T., Mihelj, M., Colombo, G., Riener, R.: Armin-robot for rehabilitation of the upper extremities. In: Proceedings 2006 IEEE International Conference on Robotics and Automation, ICRA 2006, pp. 3152–3157. IEEE (2006)
15. Loureiro, R.C., Harwin, W.S.: Reach & grasp therapy: design and control of a 9-DOF robotic neuro-rehabilitation system. In: 2007 IEEE 10th International Conference on Rehabilitation Robotics, pp. 757–763. IEEE (2007)
16. Xu, D., et al.: Development of a reconfigurable wrist rehabilitation device with an adaptive forearm holder. In: 2018 IEEE/ASME International Conference on Advanced Intelligent Mechatronics (AIM), pp. 454–459. IEEE (2018)
17. Helbok, R., Schoenherr, G., Spiegel, M., Sojer, M., Brenneis, C.: Robot-assisted hand training (Amadeo) compared with conventional physiotherapy techniques in chronic ischemic stroke patients: a pilot study. DGNR Bremen, November 2010

18. Hesse, S., Schulte-Tigges, G., Konrad, M., Bardeleben, A., Werner, C.: Robot-assisted arm trainer for the passive and active practice of bilateral forearm and wrist movements in hemiparetic subjects. Arch. Phys. Med. Rehabil. **84**(6), 915–920 (2003)
19. Masia, L., Krebs, H.I., Cappa, P., Hogan, N.: Design and characterization of hand module for whole-arm rehabilitation following stroke. IEEE/ASME Trans. Mechatron. **12**(4), 399–407 (2007)
20. Dovat, L., et al.: Handcare: a cable-actuated rehabilitation system to train hand function after stroke. IEEE Trans. Neural Syst. Rehabil. Eng. **16**(6), 582–591 (2008)
21. Dehghan Neistanak, V., Moghaddam, M.M., Abbasi Moshaei, A.: Design of a hand tendon injury rehabilitation system using a DOF constrainer mechanism. Modares Mech. Eng. **20**(1), 1–12 (2019)
22. Niestanak, V.D., Moshaii, A.A., Moghaddam, M.M.: A new underactuated mechanism of hand tendon injury rehabilitation. In: 2017 5th RSI International Conference on Robotics and Mechatronics (ICRoM), pp. 400–405. IEEE (2017)
23. Abbasi Moshaii, A., Mohammadi Moghaddam, M., Dehghan Niestanak, V.: Fuzzy sliding mode control of a wearable rehabilitation robot for wrist and finger. Ind. Robot **46**(6), 839–850 (2019). https://www.emerald.com/insight/content/doi/10.1108/IR-05-2019-0110/full/html
24. Moshaii, A.A., Moghaddam, M.M., Niestanak, V.D.: Analytical model of hand phalanges desired trajectory for rehabilitation and design a sliding mode controller based on this model. Modares Mech. Eng. **20**(1), 129–137 (2020)

Interest Arousal by Haptic Feedback During a Storytelling for Kindergarten Children

Mina Shibasaki[1(✉)], Youichi Kamiyama[1], Elaine Czech[1], Koichi Obata[2], Yusuke Wakamoto[2], Keisuke Kishi[2], Takayuki Hasegawa[2], Shinkuro Tsuchiya[3], Soichiro Matsuda[4], and Kouta Minamizawa[1]

[1] Keio University Graduate School of Media Design, Yokohama, Japan
mina0415@kmd.kelo.ac.jp
[2] Toppan Printing Co., Ltd., Tokyo, Japan
[3] Froebel-Kan Co., Ltd., Tokyo, Japan
[4] University of Tsukuba, Tsukuba, Japan

Abstract. In this paper, we introduce "Kinder BURU BURU cushion" a vibrotactile cushion that is used during storytelling time to grab the attention of children to help them to focus on the story. We conducted field tests using our system, which confirmed that children were more interested in a story when using our system compared to normal storytelling. For this research, we collaborated with a printing company and a picture book company, designed our system based on user studies at a kid's space, and conducted empirical field experiments.

Keywords: Haptics · Embodied interaction · Storytelling · Edu-tech · Behavior analysis of children

1 Storytelling and Embodied Interaction

Vygotsky and Bandura [5] stated that the physicality of words and gestures should also be considered as types of tools for interaction for children. They believed communicating with other people was one method to help a child develop. A picture book is one example of a tool that helps children to gain verbal ability through communication with adults. It has been confirmed that children are able to learn languages from adults through picture books [1, 2]. Moreover, according to Bruner [6], the phenomenon of *joint attention* is how children display empathy for others. Bruner's infant study suggested that from an early age, children actively use their bodies to engage with others. Thus, collaborative interaction plays an essential role in the child's development for the child to understand themselves and others. Therefore, storytelling events at libraries and kids' spaces are beneficial to a child's development as they encourage collaborative interaction. However, the primary way that children explore the world using their bodies is through touch. As a result, educators, such as Frobel [3] and Montessori [4], attempted to demonstrate how interacting with "things" was essential for child development. Frobel [3], for example, founded a kindergarten that used the "Gifts," or educational play materials, to encourage children to learn while playing

I. Nisky et al. (Eds.): EuroHaptics 2020, LNCS 12272, pp. 518–526, 2020.
https://doi.org/10.1007/978-3-030-58147-3_57

freely with objects. Montessori [4], similarly, developed teaching tools that trained a child's senses. Reggio Emilia [7] took this notion further by introducing the importance of creativity and developed a preschool education method that stressed physical learning through art expression. Iverson et al. [9, 10] and Antle et al. [11] also showed that exploratory activity that allowed a child to use their body influenced the depth of their understanding. By utilizing human physicality, especially one's tactile sense, the brain is more actively stimulated, which can help comprehension. Therefore, many museums have incorporated hands-on [8] activities to encourage a deep understanding of exhibits, particularly exhibitions related to cultural properties. Already there are many studies in which tactile technology has been adapted to children's education. Several of which focus on storytelling and reading. Yannier [12] et al. and Zhao et al. [13] verified that children reading with a vibrotactile tablet could achieve a deeper understanding of the content they were reading. However, a limitation to these studies was that they focused on individual or one-on-one readers. Through this research, we hope to explore how to focus children's interest in picture books in a group reading setting. During storytelling time, at public spaces like the library or kid space, children of various backgrounds listen to a story together. However, there are usually differences in age, the ability of understanding, and the interest level of the children. Also, there is a field problem that it was difficult to grab children's attention all of them because children's attention was different when long storytelling. As a result, selecting a picture-book that is suitable for everyone is a difficult task. In this study, we aimed to design experience to grab children's attention to picture books when multiple children listen to storytelling together. In the next chapter, we did fieldwork to explore the issue of a typical kindergarten setting and to find what interaction needed to make the system. Based on this result we developed a cushion that multiple children can feel picture book world through their body by haptic feedback same time. And then we conducted experiment to verification the effective of this system, and we analyzed children behavior to use behavior psychology method. Finally, we discussed what value the embodied experience of picture book could provide to children.

2 Design

2.1 Fieldwork and Proposed Method

We interviewed stakeholders, such as storytelling staff, to explore the issues of a typical kindergarten setting (The Froebel-Kan staff who worked in kindergarten and public library staff). Also, we observed children playing in a kid's space operated by Frobel-Kan as well as the storytelling activity provided by them. As a result, we found that staff tended to continually choose the same books because finding books that will interest all the children was difficult. Also, the books selected were one's that sold many copies as this aligned with the books parents were choosing to read to their children. In other words, there was a lack of diversity in the range of picture book experiences being introduced to the children. Additionally, based on our observations of the kid's space, fewer children were interested in the picture books than those that played with the play equipment and toys. Furthermore, during the storytelling activity, a large number of children got bored immediately, as opposed to those who retained interest in the picture book. Thus, we proposed the following system to get children interested in the story of picture books as in Fig. 1.

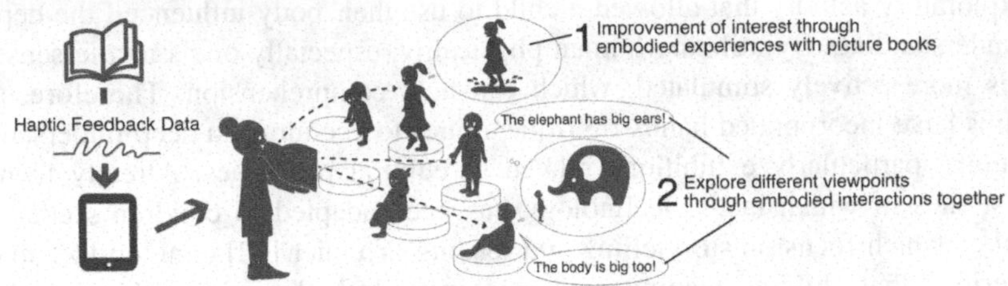

Fig. 1. Concept design of "Kinder BURU BURU Cushion"

1. *Improvement of Interest Through Embodied Experiences with Picture Books*
   As shown in previous studies, children use their bodies to explore objects and phenomena. Such learning methods have proven to be successful for expanding interest and deepening understanding. We thought it is possible to use such methods during storytelling in the kindergarten. To make children think about the story beyond the words on the page, we needed to help them embody the actions of the story.
2. *Explore Different Viewpoints Through Embodied Interactions Together*
   We thought if children can share their interests with each other, this would help expand their perspective of others. Allowing them to enjoy picture books more collectively. We decided to use a cushion, because if it were a wearable device, the operational costs would be too high. By simply sitting on the cushion it was possible for multiple users to experience the tactile sensations of the picture book all together.

2.2 Hardware Design and Haptic Design

We designed the configuration to satisfying our concept as in Fig. 2. We used a PC or tablet device uploaded with the tactile data for each picture book, and then transmitted the tactile data with a transmitter. The tactile data was received by multiple receivers and multiple Vibro-transducers (Acouve Co., Ltd./Vp6) operating all at same time. We created the transmit module and receive module. These modules could send and receive tactile information without cables by using wireless audio unit PLL synthesizer (Circuit Design, Inc./WA-TX-03, WA-RX-03). We also created the Haptic amp, and equipped it with a battery to eliminate cable. We then created corresponding haptic experiences for five picture books published by Froebel-Kan Co.,Ltd. In creating our haptic experiences, we did haptic design in the following process: First, we made haptic design storyboard to consider listener's perspective (subjectivity or objectivity). Second, we selected vibration from instrument sample data in Abltone (Ableton Co., Ltd.). For example, in the scene where a bear hit a tree, we used instruments such as drum kicks. We mainly used percussion instruments samples and rhythms to explain the body sensation of character. (Footsteps of characters, feeling of character knocked on the door etc.). Because percussion instruments could feel vibration clearly by our system. Finally, we adjusted the volume of all the contents to balance the vibrations, especially we strengthened data below 20 Hz. We aimed to put at least one sensation on each page to prevent the children from getting bored.

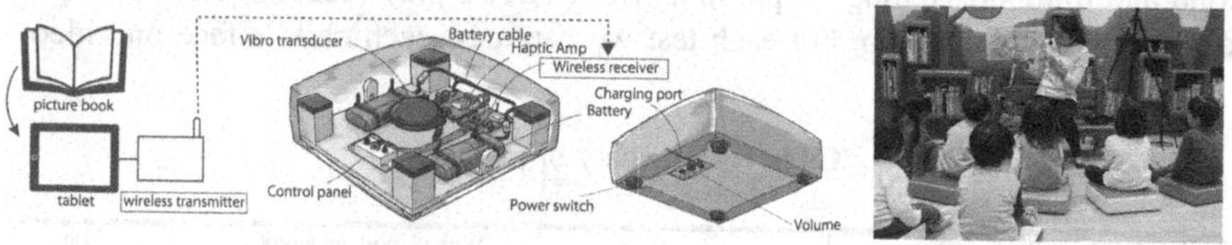

Fig. 2. System design and implementation

3 Validation

We conducted experiment to verification the effective of this system. Drew et al. [17] reported haptic feedback can draw children's attention and conversation about the sensation explain in picture book between parents and children. Therefore, we thought if children get information through the sensation, children's behavior like attention to the picture book and sitting on the cushion will change. For our study, we compared storytelling using our "BURU BURU" cushion and normal storytelling. And we analyzed children behavior to use behavior psychology method. The test was performed in four groups over two days (am/pm), and we analyzed what changes occurred in the children's behavior.

3.1 Experimental Method

Participants

Participants were recruited using flyers in a kid space operated by Froebel-Kan Co., Ltd. For the first test, we recruited a total of 10 children and their parents, 6 children in the morning (3 boys and 3 girls, aged 3 years) and 4 children in the afternoon (2 boys and 2 girls, aged 3 years) on July 23th 2019. In the second test, we recruited a total of 21 children and their parents, 10 pairs of children and parents in the morning (4 boys and 6 girls, aged 3–4 years) and 11 pairs of children and parents in the afternoon (6 boys and 5 girls, aged 3–4 years) on October 19th 2019. Participants were paid 3,000 JPY per hour for their participation. The total 31 children (average aged: 43.4 months) were used for the analysis, excluding those who were absence or attendees who joined in midway. Additionally, parents were asked to complete a questionnaire: "How many picture books do you read a week?" to help us gauge the participating children's existing level of familiarity with picture books. The majority, 21 people, answered that they read picture books every day. On the other hand, two participants stated that they rarely read picture books. The remaining eight people said they read picture books a few days a week. Overall, many of the children who participated were familiar with picture books.

Procedure

Testing was performed in a kid's space operated by Froebel-Kan Co., Ltd. We had children sit on the cushions as picture books were read aloud. We alternated between normal storytelling and storytelling with the system (see Table 1). We switched the order (with and without) between morning test and afternoon test to prevent preference bias due to the ordering. Also, we told parents to not stop a child should they decide to move off the cushion. Children were free to come and go as they wanted. Between the second and third storytelling we put in a brief exercise play break. In total, four picture books were read per group. For each test we captured each child's face on video.

Table 1. Summary of procedure

Order of storytelling and Title			With or Without vibrotactile representation (Time of storytelling)			
			7.23.am Group A	7.23.pm Group B	10.19.am Group C	10.19.pm Group D
First book	Train of Heteroconger hassi	Written by Kenta Ostuka Illustrate by Minako Kusaka	With (05:45)	Without (04:00)	With (04:30)	Without (04:00)
Second book	Hurry up!Train of Heteroconger hassi (2019.07.23)	Written by Kenta Ostuka Illustrate by Minako Kusaka	Without (04:30)	With (04:30)	Without (03:00)	With (03:15)
	Kinder book(2019.10.19)					
Third book	Nekozakana's egg	Yuichi Watanabe	With (06:30)	Without (05:00)	With (06:45)	Without (06:15)
Fourth book	"TON KOTO TON"	Written by Estuko Bushika Illustrate by Shigeki Suezaki	Without (04:30)	With (04:30)	Without (04:30)	With (05:15)

Video Coding and Depend Variables

Video Coding began when the reader opened the picture book and started to read then continued until the picture book completed. The video was divided into 15 s increments to elicit the rate of how often a set of variables occurred during that timeframe. We measured five different variables (*attention to picture books, looking at mother, pointing at picture books, utterances, sitting on the cushion.*): *Attention to picture books*: It was defined as the child looking at a picture book for 15 s. *Looking at mother*: It was scored when the child looks at the mother at least once every 15 s. Also, *pointing at picture books* and *utterances* was scored when these actions occurred at least once every 15 s. *Utterances:* it was scored that children talk about a picture book. *Sitting on the cushion*: It was defined as the child sitting on the cushion for 15 s. Coding was performed with the assistance of people not familiar with the contents of the experiment (hereinafter referred to naïve). The test on July 23, 2019, was analyzed by three people (author and two naïve) while the test on October 19, 2019, was analyzed by seven people (author and six naïve raters). And then we examined the agreement value between authors and naïve observers: *attention to picture books* of agreement value average is 88% (minimum/maximum values from 55% to 100%), *looking at the mother* is 96% (min/max values from 61% to 100%), *pointing at picture books* is 100% (min/max values from 91% to 100%), *utterances* is 96% (min/max values from 75% to 100%), and *sitting on the cushion* 95% (min/max values from 28% to 100%), all of the agreement value average maintained at least 85% or higher.

3.2 Result

Figure 3 shows average data of the percentage of 15 s intervals with *Attention to picture books*, *sitting on the cushion*, *Looking at mother*. We analyzed for haptic feedback factor (with vibration and without vibration) and order factor (the first half and the second half) using an ANOVA in each different variable. We found an interaction between haptic feedback factor and order factor in *attention to picture books* [F (1,30) = 9.272, p = .005]. There is no measurable effect of haptic feedback [F (1, 30) = 0.057, p = .813] and order [F (1,30) = 2.420, p = .130]. We noticed children's attention to picture books reduced in the second half when there was no haptic feedback. Also, there was significant interaction in *looking at mother* [F (1, 30) = 4.937, p = .034]. There is no main effect of haptic feedback [F (1, 30) = 0.080, p = .779] and order [F (1, 30) = 2.564, p = .120]. *Looking at mother* increased in the second half when there was no haptic feedback. Regarding *sitting on the cushion*, we noticed it decreased in the second half regardless of haptic feedback (first half vs second half), [F (1, 30) = 15.498, p < .001]. Also, *utterance* had the same result (6% first half, 3% second half with haptic feedback condition and 7% first half, 4% second half, without haptic feedback condition), [F (1,30) = 5.094, p = .031]. Additionally, most children did not *point to the picture books*. Only four children pointed to picture book during the vibrotactile sessions, and this behavior occurred only once during the normal storytelling. Finally, we got feedback from their parents by using a free-form questionnaire about differences in the child's behavior from the parent's perspective. Table 2 is categorized feedback from parents.

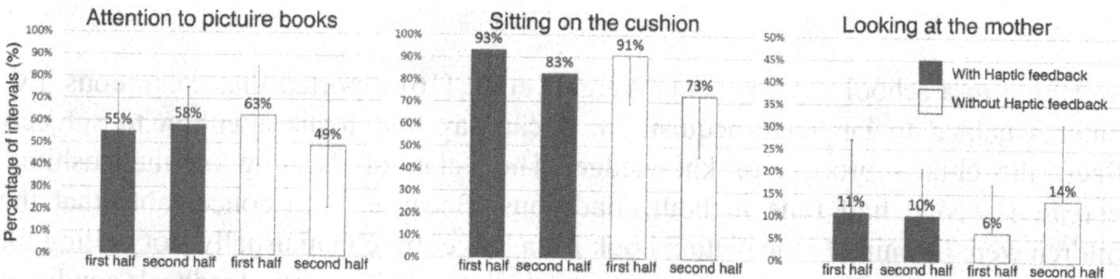

Fig. 3. Percentage of 15 s intervals with attention to picture books, sitting on the cushion, looking at the mother. Error bars is Standard Deviation

4 Discussion

We concluded the embodied experience with picture books using haptic feedback can reduced the declined children's attention to picture book when long storytelling with multiple children. Because, children's attention to picture books decreased in second half when normal storytelling condition but there was no change when with haptic feedback condition. Previous studies by Andreson et al. [14] that children watched TV-content a longer time when they were concentrating. Ruff et al. [15] defined this concentration "*Sustained attention*", it is an important ability when they learn

Table 2. Examples of feedback comments from parents

Categories	Example of feedback
Discovering the child's concentration	• *"she gets bored of long stories, but I confirmed that she was concentrating because she thought it was fun." (3 years 8 months old girls and mother)* • *"She can't listen to storytelling too long at kids' space, but today she listened to four books and I was surprised (though during the third one she seemed a little tired)." (3 years 8 months old girl and mother)* • *It was possible to return his attention to the picture book by the sound and vibration when he was distracted by toys. He won't come back from toys at home. (3 years 3 months old boy and mother)* • *He always pays attention to the details of the picture, but today he was concentrating on getting the whole story. (3 years 5 months boy and mother)*
Discovering children's positiveness through story-telling experiences	• *Normally he is shy and never leaves from me, but I was surprised because he sat alone on the cushion (3 years and 0 months boy/mother)* *He is such a baby, so I expected him get on my knees, but I was surprised when he returns to cushion himself. (3 years 7 months boy and mother)*
Discovery of children's sharing	• *I was impressed when she looked at me sitting behind her when the cushion vibrated. (4 years 7 months girl and mother)*

something in a school environment. Weighrt et al. [16] reported that continuous TV-contents helped to language acquisition, it can say that focus attention to subjects affects the child's process of knowledge. The value of "seating on the cushion" decreased second half time in both conditions. Because, it is conceivable that the children were listening to the picture book for a longer time than usually storytelling for the experiment. But Fig. 4 shows it recovered slightly with haptic feedback condition in the third book and fourth book. Also, we could observation that children who began to walk around kid space at third book normal storytelling, but they returned to their seat when fourth book by haptic feedback because children noticed that cushion was vibrating. In other words, it was possible to recover children's attention by haptic feedback. Additionally, this system might be more effective for children who normally have little ability to pay attention to picture books as it showed a significant shift in the child's behavior. Also, we could get these comments from parents that "she gets bored of long stories, but I confirmed that she was concentrating because she thought it was fun (3 years 8 months old girls and mothers)." And yet another, "She can't listen to storytelling too long at kids' spaces, but today she listened to four books and I was surprised (though during the third one she seemed a little tired) (3 years 8 months old girl and mother)." These comments support our results. We predicted the children's behavior that *looking at mother* occurred when children wanted to tell their mother about a picture book. It was occurred when first half and second half. However, it increased in the second half with no feedback. It was also observed that children then went to their mother after performing this behavior. Therefore, it seems that the behavior has two meaning: a sharing picture book interest and an expression of boredom. It is need to more detailed to measure these behaviors separately.

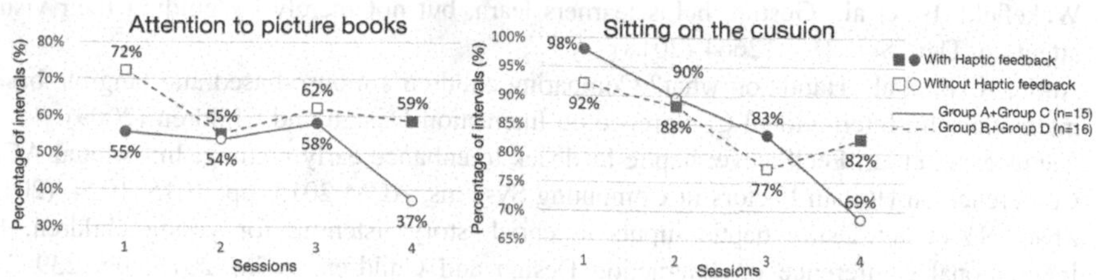

Fig. 4. Compared to order with or without haptic feedback

5 Conclusion

In this study, we aimed to design experience to grab children's attention to picture books when multiple children listen to storytelling together. Then we developed the Kinder "BURU BURU" cushion which multiple children can experience the world of picture books using their body. And we investigated the effects of our system using haptic feedback to use behavior analyzed method. As a result of experiment that we confirmed that our system can reduced the declined children's attention to picture book when long storytelling with multiple children. However, we found that the effect of the group experience was not enough to variables the children *looking at mother*. During the experiment, there were situations where children looked at each other who participated with friends. It is the possibility that more behavior will lead to *joint attention*. In the future, we would like to more clarify the effects of a group experience to storytelling, our system will be using continuous such as the environment where children know each other. Also, we believed that this method can helped other group work like elderly activity.

References

1. Ninio, A., et al.: The achievement and antecedents of labelling. J. Child Lang. **5**(1), 1–15 (1978)
2. Snow, C.E., et al.: Routines in mother-child interaction. In: The language of Children Reared in Poverty, pp. 53–72. (1982)
3. Wilson, S.: The "Gifts" of Friedrich Froebel. J. Soc. Archit. Hist. **26**(4), 238–241 (1967)
4. Montessori, M.: The Montessori Method. Transaction Publishers, Piscataway (2013)
5. Tudge, J.R., et al.: Vygotsky, Piaget, and Bandura: perspectives on the relations between the social world and cognitive development. Hum. Dev. **36**(2), 61–81 (1993)

6. Bruner, J.: From joint attention to the meeting of minds: an introduction. In: Joint attention: Its Origins and Role in Development, pp. 1–4 (1995)
7. Edwareds, C., et al.: The Hundred Languages of Children. Ablex Publishing Corporation, Norwood (1993)
8. Boston Children's Museum. http://www.bostonchildrensmuseum.org/
9. Iverson, J.M., et al.: What's communication got to do with it? Gesture in children blind from birth. Dev. Psychol. **33**(3), 453 (1997)
10. Wakefield, E., et al.: Gesture helps learners learn, but not merely by guiding their visual attention. Dev. Sci. **21**, e12664 (2018)
11. Antle, A.N., et al.: Hands on what? Comparing children's mouse-based and tangible-based interaction. In: International Conference on Interaction Design and Children (2009)
12. Yannier, N., et al.: FeelSleeve: haptic feedback to enhance early reading. In: Annual ACM Conference on Human Factors in Computing Systems, ACM 2015, pp. 1015–1024. (2015)
13. Zhao, S., et al.: Using haptic inputs to enrich story listening for young children. In: International Conference on Interaction Design and Children, ACM 2015, pp. 239–242 (2015)
14. Anderson, D.R., Levin, S.R.: Young children's attention to "Sesame Street". Child Dev. **47**, 806–811 (1976)
15. Ruff, H.A., Capozzoli, M., et al.: Age, individuality, and context as factors in sustained visual attention during the preschool years. Dev. Psychol. **34**(3), 454 (1998)
16. Wright, J.C., et al.: The relations of early television viewing to school readiness and vocabulary of children from low-income families: the early window project. Child Dev. **72** (5), 1347–1366 (2001)
17. Cingel, D., et al.: How parents engage children in tablet-based reading experiences: an exploration of haptic feedback. In: ACM Conference on Computer Supported Cooperative Work and Social Computing, pp. 505–510. (2017)

Investigating the Influence of Haptic Feedback in Rover Navigation with Communication Delay

Marek Sierotowicz(✉), Bernhard Weber, Rico Belder, Kristin Bussmann, Harsimran Singh, and Michael Panzirsch

Institute of Robotics and Mechatronics, German Aerospace Center, Muenchener Street 20, 82234 Wessling, Germany
marek.sierotowicz@dlr.de

Abstract. Safe navigation on rough terrain in the presence of unforeseen obstacles is an indispensable element of many robotic applications. In such conditions, autonomous navigation is often not a viable option within certain safety margins. Yet, a human-in-the-loop can also be arduous to include in the system, especially in scenarios where a communication delay is present. Haptic force feedback has been shown to provide benefits in rover navigation, also when confronted with higher communication delays. Therefore, in this paper we present the results of a user study comparing various performance metrics when controlling a rover with a car-like interface with and without fictitious force feedback, both with no communication delay and with a delay of 800 ms. The results indicate that with force feedback the navigation is slower, but task performance in the proximity of obstacles is improved.

Keywords: Teleoperation · Rover navigation · Wheeled mobile robot · Haptic feedback · TDPA

1 Introduction

With the renewed interest in manned exploration of celestial bodies, first and foremost the Moon, robot-assisted surface exploration is going to play a major role in the near future in many space programs. In various cases, however, full robot autonomy is not a viable option within certain margins of safety, and higher level commands designed for high communication delays may be overly time consuming and complex. Some form of teleoperation has been shown to be possible in scenarios where the operator has to control the robotic platform from orbit with communication delays in the order of magnitude of a second or more. Therefore, teleoperation is deemed preferable whenever possible in space related tasks involving remote functionalities [1,2]. Different space missions, such as ESA Haptics-1 [3], Kontur-2 [4] and Analog-1 [5], have investigated the feasibility of telemanipulation and telenavigation in microgravity conditions. In [6], a

I. Nisky et al. (Eds.): EuroHaptics 2020, LNCS 12272, pp. 527–535, 2020.
https://doi.org/10.1007/978-3-030-58147-3_58

fictitious force feedback principle for high communication delays was developed for the Kontur-2 mission. A space qualified DLR force feedback joystick with two degrees of freedom (DoF) along with a car-like curvature and longitudinal velocity interface was used to access the three planar DoFs of the omni-directional rover. A comparable 2-DoF interface without force feedback was used in the Analog-1 mission in 2019 [7], which involved telenavigation as well as sample picking and placing with an earth-based rover from the ISS at >800 ms round-trip delay. The rover was navigated through relatively obstacle-free environment, such that force-feedback is of minor importance during telenavigation. To name an example in the foreseeable future, a similar interface to the one proposed in [7] is planned to be used in the upcoming *Arches* experiment [8] on mount Etna, albeit with both the pilot and the rover being based on the ground, and with no force feedback. However, some form of corrective feedback is necessary for telenavigation tasks in non-deterministic environments in the presence of communication delays, which otherwise may lead to performance deterioration. Previous studies [9], show the benefits of haptic force feedback in rover navigation in such conditions in terms of collision avoidance. However, introducing a closed loop force feedback can lead to instability at high delays. Different control principles as Routh-Hurwitz [10], Llewellyn approach [11] and Time Domain Passivity Control (TDPA, [12]) have been proposed to guarantee stability in delayed telenavigation with haptic feedback. In [13] and [6], the TDPA was extended for telemanipulation and different types of force feedbacks.

In consideration of these previous studies, we propose that a force feedback setup with a fitting TDPA control can bring more sensible benefits in a complex to navigate physical environment. Therefore, in this paper, we present a user study involving the DLR Joystick [14] and the DLR Lightweight Rover Unit (LRU, [15]) with the goal to evaluate the advantages of telenavigation with fictitious force feedback against telenavigation without force feedback. The TDPA is applied in this work due to its robustness to varying delay, packet loss and jitter. The main goal of this study is to investigate the effects of our force feedback setup on rover navigation in close proximity to physical obstacles.

2 Materials and Methods

2.1 Sample

The user study was conducted with 16 subjects (3 females, 13 males) with an average age of $M = 25.2$ yrs. ($SD = 2.9$ yrs.; range of 21 to 32 yrs.). All of them signed an informed consent form prior to the experimental session.

2.2 Apparatus

Rover. Subjects controlled the LRU wheeled mobile robot (WMR) [15], a prototypical rover system specifically designed for rough terrain. The LRU has 12 DoF, with four wheel actuators, four steering actuators, two series-elastic joints and two joints in the pan-tilt unit (which is equipped with a black-and-white stereo and a colour camera). The total weight is 30 kg with a maximal payload of 5 kg. Two battery packs allow for more than 120 min operation time. The maximal speed is 1.11 m/s. The stereo camera images are processed by performing a dense stereo matching using an FPGA implementation of the so-called Semi-Global Matching (SGM) algorithm [16]. Additionally, the colour camera images are mapped onto the resulting depth image for object detection. The resulting estimates are used for generating a danger map indicating insurmountable obstacles.

Joystick. The DLR's Kontur-2 force feedback joystick [14] was used to drive the rover. The joystick has a 2-DoF, ±20° workspace, a maximum force of 15 N and an update rate of 1 kHz. A car-like mapping of the 2 DoFs was implemented, i.e. longitudinal velocity was commanded by moving the joystick forwards and backwards, whereas curvature (steering the LRU's front and back wheels) was commanded by lateral movements of the joystick. In order to navigate, the user has to press the dead-man button on the joystick. In order to switch the current movement direction of the LRU between forward and backward, the user has to press the lower side of the switch on top of the handle.

2.3 Force Feedback Controller

Figure 1 shows the signal flow diagram of a delayed bilateral teleoperator (BT), where T_1 and T_2 are the forward and backward delays, respectively. The coupling controller *Ctrl* ensures that the LRU (slave robot) follows the delayed master reference v_m^{del} and in turn generates a fictitious force F_f, which is felt by the operator through the master haptic-device (DLR's force feedback joystick) after a communication delay.

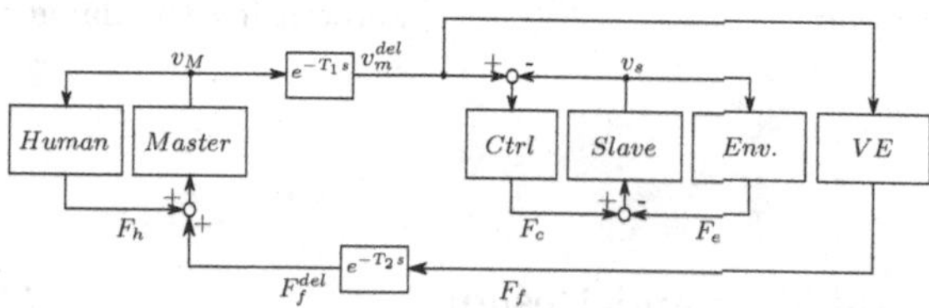

Fig. 1. Signal flow diagram of a bilateral teleoperator with fictitious force feedback

Persuasive force-feedback from the virtual environment VE is produced via the direction dependent curvature polygons P_L and P_R (Fig. 2) overlapping with the danger map (Fig. 3) which is generated using stereo vision. The danger map associates to any pixel (x, y) in the vicinity of the rover a binary danger index $D(x, y)$ (0 if there is no obstacle, 1 if an obstacle is present). The force components are computed according to:

$$F_d \propto \Sigma_{(x,y)\in P_d} D(x, y) \quad \text{for} \quad d = R, L \tag{1}$$

$$F_B \propto \Sigma_{(x,y)\in P_R \cup P_L} D(x,y) = F_L + F_R \tag{2}$$

An equally valid way of describing the calculation of the fictitious force feedback component is considering the obstacles and curvature polygons as sets containing the corresponding pixels on the danger map as elements. This way, the force feedback components can be computed using set operators. Figure 2 uses this notation to show a particular example with the calculation of force components for a specific obstacle and curvature polygon configuration. Notice that the curvature polygons are placed in a manner such that the subscripts of the variables F, P and O are consistent in the shown formulas, rather than respecting the actual right and left side in the rover coordinate frame. Such a system is prone to instability due to the energy generation by the delayed communication channel. In order to passivate the communication channel, Time-Domain-Passivity-Approach (TDPA) was implemented [6]. TDPA introduces adaptive virtual damping elements, both in a series and parallel fashion, at the master and slave side, to dissipate the exact amount of energy necessary to stabilize the overall system.

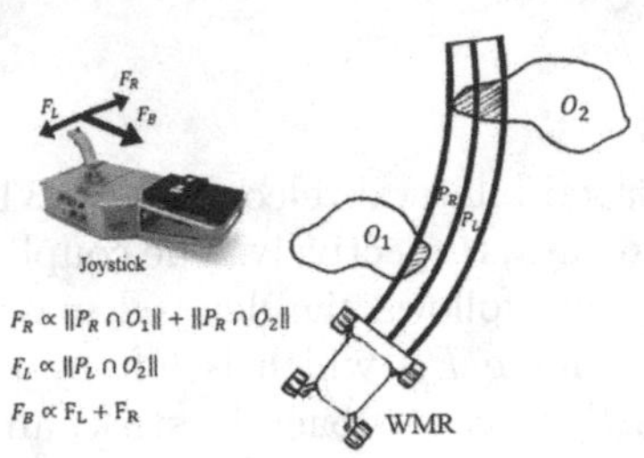

Fig. 2. An example of fictitious force component computation [6]

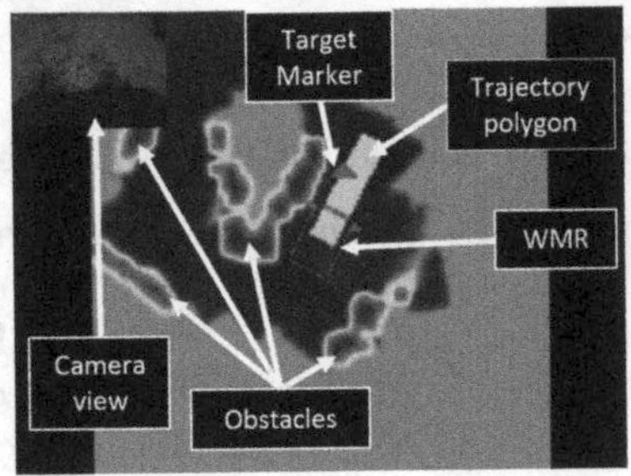

Fig. 3. Screenshot of the user interface showcasing the danger map

2.4 Experimental Task and Design

Participants had to drive the rover along or through different rock formations avoiding collisions (see Fig. 4). One complete experimental trial consisted of three subtasks: 1) navigating the rover through a narrow passage, 2) navigating along a curved rock formation as close as possible to the concave side, 3) navigating as close as possible along a convex boulder (see Fig. 5). Each trial was started from the same, predefined starting position and after having completed subtask 2, the rover was re-positioned by the experimenters to have an optimal starting position for the final subtask.

The two experimental factors Force Feedback (FF vs. noFF) and Delay (0 ms vs. 800 ms) resulted in four experimental conditions. For each condition, there

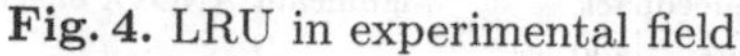

Fig. 4. LRU in experimental field

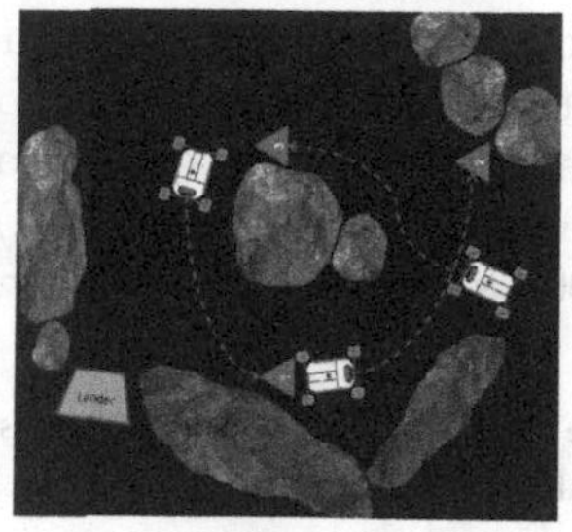

Fig. 5. Schematic subtask representation

were two subsequent experimental trials, for a total of eight trials per subject. While the order of the two force feedback conditions was counterbalanced across subjects, these always started with the no delay condition and then proceeded with the 800 ms delay condition, as pre-tests showed that navigation with 800 ms delay was too demanding for inexperienced users.

2.5 Procedure

First, subjects were given instructions on the experimental task and procedure. They conducted the experiment at a table, sitting on a chair. The joystick was positioned to the right of the subject. Subjects were asked to adjust chair height so that their right arm rested comfortably on the padded arm rest of the joystick module. The experimental GUI (see Fig. 3) was displayed on a 23" monitor in front of the subjects. Since the participants controlled the rover remotely (i.e. from a separated room), the rover's position and a danger map were displayed in the GUI together with a video stream shown in the upper left corner of the window. After having finished two training trials (with force feedback and without delay), the eight trials of the main experiment were started. In the room where the LRU was located, a technician and a supervisor managed the experiment parameters, and provided feedback about collisions and successful reaching of the intermediate milestones for each subtask.

2.6 Measures and Data Analysis

The metrics chosen to assess user performance are the number of actual collisions between rover and obstacles, task completion times, mean rover speed, mean and standard deviation of lateral forces on the master side of the control loop. Additionally, the centroid of all the odometry-measured path points followed by the LRU was used as a reference point in order to obtain the mean radius of curvature at which the subjects drove for subtask 2. For this subtask specifically, this metric can be used to estimate task fulfillment in terms of mean proximity of the followed path to the concave rock formation. For subtask 3, an analogous metric was computed based on the mean minimal distance for each path point from the line connecting the LRU's average starting position for subtask 3 to the centroid of all recorded path points. Data was analysed using a repeated measures ANOVA (rmANOVA) with Force Feedback (FF vs. noFF), Delay (0 ms vs. 800 ms) and Subtask as within factors. See Table 1 for a result overview.

Table 1. Result overview; means, standard deviations (in parentheses) for all experimental conditions and significant ANOVA main effects.

Measure [Unit]		No force feedback		Force feedback		Significant ANOVA effects
		No delay	800 ms	No delay	800 ms	
Collisions [#]		0.19 (0.54)	0.44 (0.81)	0.50 (0.89)	0.13 (0.34)	None
Compl. time [s]	All tasks	75.9 (16.6)	72.9 (34.1)	81.6 (28.6)	82.9 (30.2)	Force feedback:
	Task 1	70.5 (21.4)	82.3 (16.1)	84.5 (60.4)	83.7 (46.2)	$F(1,14) = 3.3$; $p<.10$
	Task 2	74.4 (32.3)	63.0 (34.3)	68.5 (26.8)	63.0 (18.6)	Task:
	Task 3	82.7 (16.1)	73.5 (14.4)	91.6 (25.0)	102 (36.1)	$F(2,28) = 3.7$; $p < .10$
Mean Speed [m/s]		0.13 (0.02)	0.14 (0.03)	0.12 (0.03)	0.12 (0.03)	Force feedback: $F(1,14) = 23.7$; $p < .001$
Mean Force [N]		3.29 (2.32)	3.16 (3.28)	2.67 (1.43)	2.08 (1.59)	Force feedback: $F(1,14) = 7.4$; $p < .05$
SD Force [N]		6.02 (3.13)	5.26 (3.21)	5.36 (2.12)	4.11 (2.47)	Force feedback: $F(1,14) = 8.0$; $p < .05$
Mean radius for subtask 2 [m]		4.62(0.16)	4.68 (0.12)	4.73 (0.15)	4.72 (0.10)	Force feedback: $F(1,14) = 7.1$; $p < .05$
Mean deviation for subtask 3 [m]		2.72 (0.34)	2.75 (0.25)	2.68 (0.33)	2.67 (0.36)	None

3 Results

While no significant effects were evident for the actual collisions, completion times tended to be longer when force feedback (FF) was activated compared to the noFF baseline, although the conventional level significance was not reached ($p < .10$). Similarly, there was a trend for the Subtask factor ($p < .10$). Specifically, completion times were longer for Subtask 1 (80.3 s) and Subtask 3 (87.4 s) compared to subtask 2 (67.2 s). A post-hoc contrast analysis, comparing the effect of FF on vs. off on completion times in each subtask revealed a (marginally) significant effect for subtask 3 only ($t(15) = 2.09$; $p = .05$), i.e. times were longer with FF compared to noFF. No significant differences were found for subtask 1 and 2. The mean speed of the rover was significantly reduced when FF was activated compared to the noFF condition ($p < .001$). The mean as well as the standard deviation (SD) of the lateral mean force was reduced when FF was provided at the joystick (both ps $< .05$). The task number has no observable effect on this metric, i.e. the positive effect of decreased mean and SD of force in the presence of FF was evident for all three subtasks. Finally, we checked the statistical power of the current study design since sample size was comparably small. Post-hoc statistical power analysis showed a power $1 - \beta = .99$ (well above .80 which is the desired probability) for the utilized design, sample and determined effect size.

4 Discussion

The fact that lateral force consistently decreases in the presence of FF for all subtasks indicates less overlap between the control polygon and obstacles on the danger map and therefore a safer navigation. This is to be expected, as the FF would exert forces on the input device inducing the subject to adjust the commanded trajectory in such a way that the overlap with obstacles is minimized. Despite the force feedback, the position drift of the TDPA brings a positive effect which was also found for measured force feedback in the telemanipulation setup of [17]. Since forces only appear in case of overlaps of the polygon with the obstacles, the position drift appears specifically in the presence thereof and prevents the overlap from increasing. In the absence of this feedback, the subjects are unhindered in commanding trajectories which intersect with obstacles.

The mean average velocity is consistently lower in the presence of FF. This indicates that the subjects tended to drive the rover more cautiously in this case. In fact, FF prevents the user to drive at higher velocities. The fact that the radius of curvature for subtask 2 tends to be higher in the presence of FF (ANOVA yields effect significance with $p < .05$), while the number of collision remains constant, shows that FF does provide the driver with information allowing for safe navigation even if closer to the outer rock formation. Interestingly, ANOVA shows no significant effect of the FF/noFF condition on completion time for subtask 1 and 2. When looking at this result, together with the reduced velocity in the presence of FF, this indicates that a more efficient path was followed using FF. Specifically, considering that the mean radius of curvature increases and mean velocity decreases in the presence of FF, the only way of explaining the absence of a significant difference in completion time is to infer that the subjects tended to change the commanded path curvature more often in the absence of FF, thus leading to a less efficient overall trajectory. In subtask 3, no significant effects of the FF condition were observed on the mean distance from the central rock formation. It would therefore seem that the FF/noFF condition has less bearing on navigation performance when confronted with convex structures, such as the one present in subtask 3. However, the lower incidence of lateral force in this subtask with FF, even though task completion was equivalent with or without FF, could be considered an index of safer navigation. These results show that some sensible benefits on rover navigation of using a force feedback setup with a fitting TDPA controller are measurable in a hard to navigate physical environment, even with sensible communication delays.

Acknowledgements. We want to thank Carina Schweiger, Karin Brüch and Margit Kanter for their support while conducting the experiments.

References

1. ISECG. The global exploration roadmap (2018). www.nasa.gov/sites/default/files/atoms/files/ger_2018_small_mobile.pdf. Accessed 20 May 2020

2. ISECG. Telerobotic control of systems with time delay gap assessment report (2018). https://www.globalspaceexploration.org/wordpress/docs/Telerobotic%20Control%20of%20Systems%20with%20Time%20Delay%20Gap%20Assessment%20Report.pdf. Accessed 20 May 2020
3. Schiele, A., et al.: Haptics-1: preliminary results from the first stiffness JND identification experiment in space. In: Bello, F., Kajimoto, H., Visell, Y. (eds.) EuroHaptics 2016. LNCS, vol. 9774, pp. 13–22. Springer, Cham (2016). https://doi.org/10.1007/978-3-319-42321-0_2
4. Artigas, J., et al.: Kontur-2: force-feedback teleoperation from the international space station. In: ICRA, pp. 1166–1173. IEEE (2016)
5. DLR. An astronaut controls a rover on earth (2019). https://www.dlr.de/content/en/articles/news/2019/04/20191125_astronaut-controls-rover-on-earth.html. Accessed 31 January 2020
6. Panzirsch, M., Singh, H., Stelzer, M., Schuster, M.J., Ott, C., Ferre, M.: Extended predictive model-mediated teleoperation of mobile robots through multilateral control. In: IEEE IV, pp. 1723–1730. IEEE (2018)
7. Krueger, T., et al.: How to design a rover cockpit for operation onboard the ISS. In: ASTRA, ESA (2019)
8. Wedler, A., et al.: Analogue research from robex etna campaign and prospects for arches project: advanced robotics for next lunar missions. In: EPSC-DPS Joint Meeting, EPSC (2019)
9. Ma, L., Xu, Z., Schilling, K.: Robust bilateral teleoperation of a car-like rover with communication delay. In: 2009 ECC, pp. 2337–2342. IEEE (2009)
10. Farkhatdinov, I., Ryu, J.-H.: Improving mobile robot bilateral teleoperation by introducing variable force feedback gain. In: International Conference on Intelligent Robots and Systems, pp. 5812–5817. IEEE (2010)
11. Li, W., Liu, Z., Gao, H., Zhang, X., Tavakoli, M.: Stable kinematic teleoperation of wheeled mobile robots with slippage using time-domain passivity control. Mechatronics **39**, 196–203 (2016)
12. Ryu, J.-H., Artigas, J., Preusche, C.: A passive bilateral control scheme for a teleoperator with time-varying communication delay. Mechatronics **20**(7), 812–823 (2010)
13. Van Quang, H., Farkhatdinov, I., Ryu, J.-H.: Passivity of delayed bilateral teleoperation of mobile robots with ambiguous causalities: time domain passivity approach. In: IROS, pp. 2635–2640. IEEE (2012)
14. Riecke, C., et al.: Kontur-2 mission: The DLR force feedback joystick for space telemanipulation from the ISS. In: i-SAIRAS, December 2016
15. Wedler, A., et al.: LRU-lightweight rover unit. In: 2009 ASTRA (2015)
16. Ernst, I., Hirschmüller, H.: Mutual information based semi-global stereo matching on the GPU. In: Bebis, G., et al. (eds.) ISVC 2008. LNCS, vol. 5358, pp. 228–239. Springer, Heidelberg (2008). https://doi.org/10.1007/978-3-540-89639-5_22
17. Panzirsch, M., Singh, H., Krüger, T., Ott, C., Albu-Schäffer, A.: Safe interactions and kinesthetic feedback in high performance earth-to-moon teleoperation. In: IEEE Aerospace Conference (2020)

Shared Haptic Perception for Human-Robot Collaboration

Kazuki Katayama[1(✉)], Maria Pozzi[2,3], Yoshihiro Tanaka[1], Kouta Minamizawa[4], and Domenico Prattichizzo[2,3]

[1] Nagoya Institute of Technology, Nagoya, Japan
k.katayama.806@nitech.jp, tanaka.yoshihiro@nitech.ac.jp
[2] University of Siena, Siena, Italy
[3] Istituto Italiano di Tecnologia, Genoa, Italy
[4] Keio University, Tokyo, Japan

Abstract. To obtain a fluent human-robot collaboration, reciprocal awareness is fundamental. In this paper, we propose to achieve it by creating a haptic connection between the human operator and the collaborative robot. Data coming from a wearable skin vibration sensor are used by the robot to recognize human actions, and vibrotactile signals are used to inform the human about the correct recognition of her/his actions. It is shown that the proposed communication paradigm, based on *shared haptic perception*, allows to improve cycle time performance in a complex human-robot collaborative task.

Keywords: Shared perception · Human-Robot Collaboration · Wearable haptics

1 Introduction

Human-Robot Collaboration (HRC) is expected to significantly advance manufacturing by introducing high flexibility in assembly cells [1], but also promises to enhance human capabilities in other fields, including domestic welfare and assistance to medical doctors [2,3]. To achieve a smooth collaboration between a human operator and a collaborative robot, *reciprocal awareness* is fundamental: the robot has to be aware of the human actions and the human has to know the robot state to fluently proceed with the collaborative task. This need was underlined by Drury *et al.*, in a review on awareness in Human-Robot Interaction [4].

This work was supported by Progetto Prin 2017 "TIGHT: Tactile InteGration for Humans and arTificial systems", prot. 2017SB48FP, and by the IEEE RAS Technical Committee on Haptics under the "Innovation in haptics" research programme.

Electronic supplementary material The online version of this chapter (https://doi.org/10.1007/978-3-030-58147-3_59) contains supplementary material, which is available to authorized users.

I. Nisky et al. (Eds.): EuroHaptics 2020, LNCS 12272, pp. 536–544, 2020.
https://doi.org/10.1007/978-3-030-58147-3_59

Recent advances in interfaces for improved human and robot perception in HRC were surveyed in [5]. On the one hand, human sensorimotor information can be used to monitor human behaviour and plan appropriate robot responses in different phases of a collaborative task [6–8]. Peternel *et al.*, for example, used a vision system and EMG electrodes to detect human motion and muscular activity [6], whereas Ishida *et al.*, used a wearable vibration sensor to discriminate human actions [7]. On the other hand, visual, auditory, and tactile feedback can be employed to improve human situation awareness in HRC [1,9]. In [9], for instance, human intention was inferred based on visual monitoring, and mutual understanding was achieved by alerting the human through haptic cues when the robot understood human intention with a certain level of confidence.

In this paper, we present a human-robot collaborative set-up where the human sensorimotor system is virtually connected to the system of sensors and actuators of the robot through wearable devices. Human actions are recognized thanks to a wearable skin vibration sensor, and successful recognition is communicated to the human through the activation of a vibrotactile ring. The proposed collaboration paradigm is sketched in Fig. 1. The idea is to integrate the benefits of shared human perception [7] with those of operator awareness [9], creating a bilateral haptic connection between humans and robots, that we call *shared haptic perception*. The effectiveness of the proposed communication paradigm was demonstrated through an experimental validation involving 8 trained volunteers performing a complex collaborative task with a robot arm.

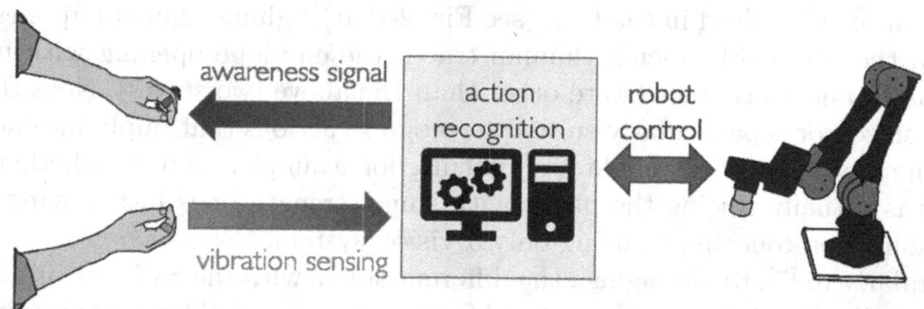

Fig. 1. Shared haptic perception between humans and robots: general idea. Human perception is shared because the same vibrations that are sensed by human touch receptors during the collaborative task are also detected by the wearable sensor and, thus, by the robot. Robot perception (enhanced by an action recognition algorithm) is shared with the human thanks to the tactile signal sent by the robot through the haptic device.

2 Methodology

The proposed collaboration paradigm is based on the use of two wearable devices, a vibration sensor and a vibrotactile ring, and on an action recognition algorithm.

Wearable Devices. In this study, the wearable skin vibration sensor developed by Tanaka *et al.* [10] is used for sending tactile information from the human to the robot. The sensor uses polyvinylidene difluoride (PVDF) film and detects vibrations propagating on the human skin surface. The acquired data are used to detect the current human action. The PVDF sensor does not hinder the natural movements of the human hand and allows to directly touch objects, because it is light (about 20 g) and can be worn by wrapping it around one of the fingers, as a ring. In [7], authors showed the advantages of putting the sensor on the human finger, with respect to applying it on the manipulated object. Not only the sensor output "directly represents operator's perception" [7], but instrumenting the human makes it possible to apply the proposed framework in different situations, without having to modify the environment around the user.

To send tactile information from the robot to the human, a wearable vibrotactile ring embedding a HAPTIC^{TM} Reactor (ALPS ALPINE CO., LTD.), is used. Two vibration bursts separated by an interval of 20 ms were sent to the participant to alert her/him that her/his action was recognized. We chose a frequency of 200 Hz for the vibration, as in [9] this kind of feedback was found to be easily recognizable and helpful to proceed smoothly with a HRC task.

Action Recognition. A paradigmatic task in which the human closes an envelope and the robot applies a stamp over it was chosen to show the effectiveness of the proposed tactile communication strategy. A Support Vector Machine (SVM) was used to recognize, based on the PVDF sensor output, the three different human actions involved in the task (see Fig. 2-(left)): gluing (human applies the glue on the envelope), tracing (human traces the envelope opening with index fingernail), and no contact (state other than the above two states). Note that a vibration sensor is particularly suited to recognize actions that imply interaction with the environment. It might be difficult, for example, to infer whether the human is actually tracing the paper with some strength or is just moving over it without even touching it, using only a vision system.

Similarly to [7], to distinguish the different states with the SVM, we used two features: vibration intensity (i_{RMS}) and frequency ratio (r). They were computed based on the power spectral density (PSD) of the sensor output calculated in the range between $f_1 = 100$ Hz and $f_2 = 1000$ Hz[1]: $i_{RMS} = log\sqrt{\int_{f_1}^{f_2} PSD(f)df}$, $r = \frac{A}{B}$. The value i_{RMS} indicates the root mean square (RMS) of the PSD of each sample, A is the log(RMS) of the PSD in the range [850–1000 Hz] and B is the log(RMS) of the PSD in the range $[F_{peak} \pm 75$ Hz$]$. F_{peak} is the frequency at which the PSD reaches its maximum value. Before each experiment, participants were asked to wear the vibration sensor and perform the three different actions, five times each. The collected data were used to create a linear SVM model based on the values of the two indices defined above. Figure 2-(right) presents

[1] Data below 100 Hz were not considered as they could easily be affected by minimal body motions and heart beat. For frequencies above 1000 Hz, the sensor output hardly changes based on user body motions [7].

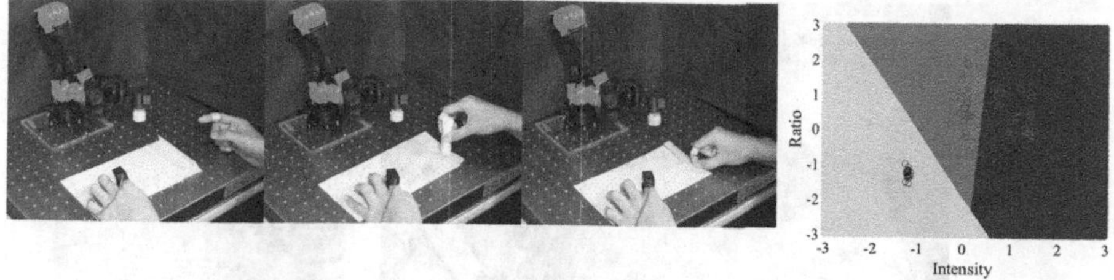

Fig. 2. (Left) Human states recognized by the SVM: no contact, gluing, tracing. (Right) Example of SVM model where data are well separated considering the intensity and ratio parameters (black: no contact, blue: gluing, red: tracing). (Color figure online)

an example of obtained SVM model. The top panel of the graphs in Fig. 3 show examples of complete acquisitions from the sensor for the gluing and the tracing state. From a total of 2 s, a central interval lasting 1 s was selected and divided into 5 samples of 0.2 s each (middle panel). For each sample, i_{RMS} and r were computed from the PSD (lower panel). The gluing action generates vibrations with a lower amplitude than those related to tracing.

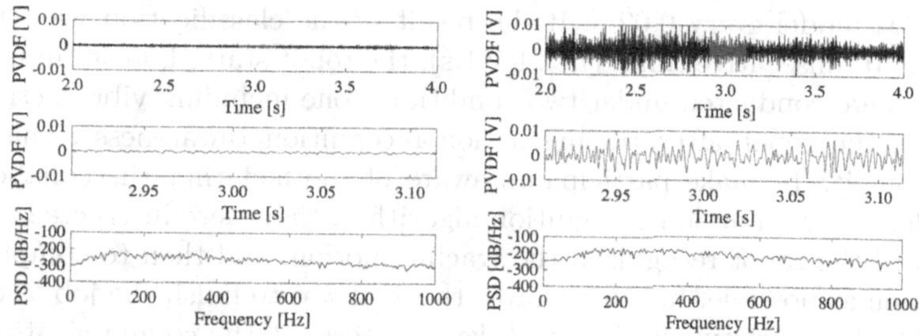

Fig. 3. Examples of PVDF sensor output and power spectral density (PSD) for gluing state (left) and tracing state (right). Top panel: complete acquisitions, middle and lower panels: sensor output and PSD for a sample of 0.2 s.

3 Experiments

3.1 Experimental Procedure

To design our experimental set-up (Fig. 4) we took inspiration from the previously described prototypical task of closing and stamping an envelope (Fig. 2), and made it more complex, so to better study the effectiveness of the proposed communication paradigm. In particular, we wanted to investigate how awareness vibrotactile signals affect the performance of well trained participants. This is an advancement with respect to [9], where participants only underwent a brief training, but were not expert in the performed task.

Fig. 4. Experimental set-up for the chosen HRC task.

Participants wore the haptic ring on the left hand and the PVDF sensor on the right, and listened to white noise while conducting the experiments. They sat in front of a collaborative robot arm, the open source manipulator Mikata arm (ROBOTIS Co., Ltd.), having four actuators and a stamp attached at the end-effector. In each experimental trial, the human operator had to trace with the right index finger a long piece of paper (size: 60×1000 mm), and the robot had to put a stamp in a predefined position upon recognition of the tracing state. In particular, the current PVDF sensor output was classified according to the found SVM model every 0.02 s. If the result of the classification was "tracing state" for 50 consecutive times (*i.e.*, for 1 s), the robot started its stamping task.

Trials were conducted under two conditions, one including vibrotactile feedback from the robot after tracing action recognition (awareness signal), and one without it. To make participants aware of the fact that the collaboration was mediated by an action recognition algorithm, they were instructed to trace slowly until the robot recognized the tracing motion, and then to complete the task as soon as possible. In other words, the goal was to finish the job as quickly as possible, but participants had to take into account the communication with the robot to be sure to get the paper stamped and thus successfully accomplish the task. Coordination between human and robot was important for two main reasons: *i*) the robot could put a stamp only after the human traced the part of the paper where the stamp had to be applied, and *ii*) the human had to be sure that the robot recognized the tracing action before it was actually completed. Note that when haptic feedback was not active, the user could infer robot state only by looking at it and waiting to see it moving towards the stamping position.

A within-subjects experimental design with complete counterbalancing was adopted. Each participant tested both conditions, with randomly assigned orderings. In particular, half of the participants initially conducted the experiment with feedback and then without, the other half did the opposite. Eight volunteers (6 males, 2 females, average age 26.5) participated in the study. They all had previous experience with wearable haptics. Informed consent was obtained from all of them and the experimental evaluation protocol followed the Declaration of Helsinki. Participants did not perceive any payment and were able to leave the experiment at any moment. Firstly, they were asked to record data to create the

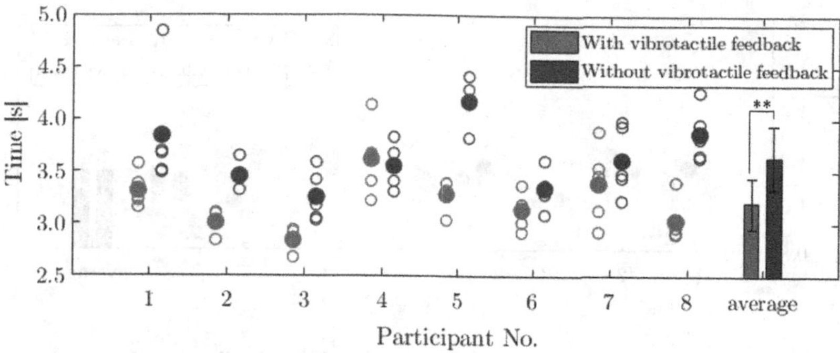

Fig. 5. Task execution time in two conditions (with/without feedback): single trial (empty circle) and average (filled circle) for each participant, and bar plot of mean and std of the averages. ** indicates $p < 0.01$ with the paired t-test.

SVM model, as described in Sect. 2. Then, each of them performed 15 trials per condition as *training phase* and, lastly, 5 trials for each condition as *test phase*. In the test phase, users' performance in terms of execution time was recorded. At the end of each trial, users had to press a button on the keyboard of a laptop placed on their right and then wait for a fixed amount of time (showed through a countdown on the screen), before starting the new trial.

In the first part of the training phase (10 trials), we were more interested in making the users learn the task, and thus we kept the robot stationary until the recognition of the tracing state. However, in real applications, the robot is never left idle and usually executes other actions while waiting for human operations. This is why, in the second part of the training phase (5 trials) and in the test phase, the arm was programmed to randomly reach four different poses, emulating other possible tasks, while waiting for the action recognition.

3.2 Experimental Results

The execution time of the 5 trials of the test phase, in the two conditions, is displayed for all participants in Fig. 5. The empty circles show the execution time of each trial, and the filled ones indicate the average execution time for each participant over five trials. The mean and standard deviation (3.20 ± 0.25 s with vibrotactile feedback, and 3.64 ± 0.31 s without vibrotactile feedback) of these average data are used to plot the bar plots on the right labeled as "average". Regarding these data, the Shapiro-Wilk test showed normal distribution and the paired t-test for each condition showed that there was a significant difference between the average execution times for the two conditions ($t_7 = 3.8$, $p = 7.0 \times 10^{-3}$). In other words, when haptic feedback was active, participants took significantly less time for completing their task, than when there was no haptic feedback.

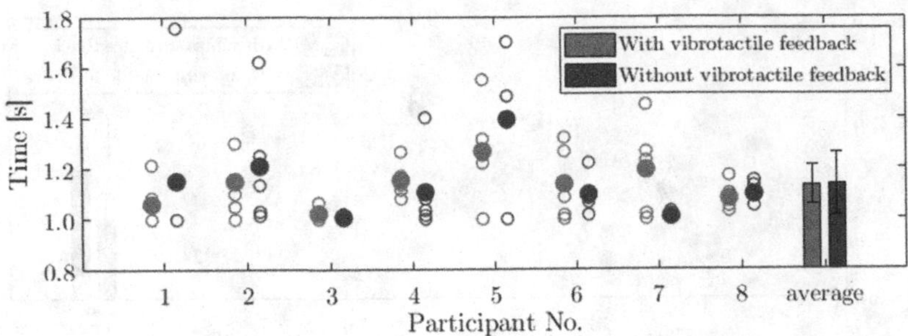

Fig. 6. Recognition time in two conditions (with/without feedback): single trial (empty circle) and average (filled circle) for each participant, and bar plot of mean and std of the averages.

In both conditions, the PVDF sensor worn by participants was active and was used to recognize user actions. The recognition was successful in all the trials. To ensure the validity of this result, we analysed the *recognition time* of the robot, *i.e.*, the time that it took to recognize that the human was tracing, in the two conditions. Figure 6 shows the recognition time for each participant for each trial (empty circles) and on average (filled circles). As before, the bar plots are built by considering mean and standard deviation (1.13 ± 0.08 s with vibrotactile feedback, and 1.13 ± 0.12 s without vibrotactile feedback) of the average values for all participants. In this case, no significant difference was found between the two conditions at a significance level of 5% for all participants. Thus, the recognition time did not significantly vary between the two conditions.

4 Discussion

Results presented in Sect. 3.2 show that not only the proposed communication paradigm offers a viable solution for implementing human-robot collaborative tasks, but also, and more importantly, that the vibrotactile feedback significantly improves human performance. The vibrotactile awareness signal allows operators to understand whether their action was successfully recognized, without having to wait to see the robot moving towards the stamping position. Besides, the fact that the robot performs other actions before the recognition, makes it even more difficult for users to understand robot next movements just from sight.

The advantages of enhancing operator awareness were initially observed in [9], and in this paper we show that awareness is important also in a completely different scenario, where human actions are not predicted but recognized, using skin vibration sensing and not visual monitoring, and, above all, where participants are not novice, but are well trained to perform the task.

5 Conclusions

This work presents a new human-robot communication paradigm based on the concept of *shared haptic perception*: the user sends to the robot haptic cues that allow the robot to recognize human actions, and the robot informs the human through symbolic vibrotactile signals (awareness signals) about the successful interpretation of the received data. This bilateral communication, achieved through the use of unobtrusive wearable sensing and actuation devices, allows to reach reciprocal awareness and mutual understanding between the two partners.

An experimental validation with 8 participants was conducted and showed that awareness signals allow well trained users to complete their task in significantly less time than without haptic feedback. Future work will focus on investigating other tactile feedback modalities (*e.g.*, continuous exchange of tactile information), on finding other collaborative tasks that can benefit from the proposed communication strategy, and on studying whether shared haptic perception can improve also the learning process of a task for untrained operators.

References

1. Valeria, V., Fabio, P., Francesco, L., Cristian, S.: Survey on human-robot collaboration in industrial settings: safety, intuitive interfaces and applications. Mechatronics **55**, 248–266 (2018)
2. Canal, G., Alenyà, G., Torras, C.: Adapting robot task planning to user preferences: an assistive shoe dressing example. Auton. Robots **43**(6), 1343–1356 (2018). https://doi.org/10.1007/s10514-018-9737-2
3. Grischke, J., Johannsmeier, L., Eich, L., Haddadin, S.: Dentronics: Review, first concepts and pilot study of a new application domain for collaborative robots in dental assistance. In: 2019 International Conference on Robotics and Automation (ICRA), pp. 6525–6532 (2019)
4. Drury, J.L., Scholtz, J., Yanco, H.A.: Awareness in human-robot interactions. In: 2003 IEEE International Conference on Systems, Man and Cybernetics. Conference Theme-System Security and Assurance, vol. 1, pp. 912–918. IEEE (2003)
5. Ajoudani, A., Zanchettin, A.M., Ivaldi, S., Albu-Schäffer, A., Kosuge, K., Khatib, O.: Progress and prospects of the human-robot collaboration. Auton. Robots **42**(5), 957–975 (2018)
6. Peternel, L., Tsagarakis, N., Ajoudani, A.: A human-robot co-manipulation approach based on human sensorimotor information. IEEE Trans. Neural Syst. Rehabil. Eng. **25**(7), 811–822 (2017)
7. Ishida, R., Meli, L., Tanaka, Y., Minamizawa, K., Prattichizzo, D.: Sensorymotor augmentation of the robot with shared human perception. In: Proceedings of IEEE/RSJ International Conference Intelligent Robots and Systems, Madrid, Spain, pp. 2596–2603 (2018)
8. DelPreto, J., Rus, D.: Sharing the load: human-robot team lifting using muscle activity. In: 2019 International Conference on Robotics and Automation (ICRA), pp. 7906–7912 (2019)

9. Casalino, A., Messeri, C., Pozzi, M., Zanchettin, A.M., Rocco, P., Prattichizzo, D.: Operator awareness in human-robot collaboration through wearable vibrotactile feedback. IEEE Robot. Autom. Lett. **3**(4), 4289–4296 (2018)
10. Tanaka, Y., Nguyen, D.P., Fukuda, T., Sano, A.: Wearable skin vibration sensor using a PVDF film. In: 2015 IEEE World Haptics Conference (WHC), pp. 146–151 (2015)

Two-Point Haptic Pattern Recognition with the Inverse Filter Method

Lucie Pantera(✉), Charles Hudin, and Sabrina Panëels

CEA, LIST, Sensory and Ambient Interfaces Laboratory, Palaiseau, France
{lucie.pantera,charles.hudin,sabrina.paneels}@cea.fr

Abstract. Touchscreens are widely used nowadays, yet still crucially lack haptic feedback for a rich interaction. Haptic feedback presents several benefits for touch interactions but can be difficult to achieve on a surface, due to issues of vibration propagation. The Inverse Filter Method enables to achieve localised multitouch haptic feedback on a glass surface by controlling the vibrations field over the entire surface. This recent method could enable a wide range of novel interactions. Yet, it has not been tested with users. This paper presents an initial study evaluating 2-point based pattern recognition using IFM with two fingers from each hand and with different timing difference in presentation, varying from 0 ms to 300 ms. The results are promising as participants could discriminate rather well the different patterns with averaged rates of 83% for simultaneous stimuli and up to 92% for stimuli separated by 300 ms.

Keywords: Surface haptics · Localised feedback · Pattern recognition

1 Introduction

Touchscreens have become a new standard for mobile devices as they enable a natural user interaction by directly touching and interacting with the items of interest, rather than using an external peripheral to map the gestures to the display and possible actions (e.g. a mouse). Unfortunately, current touchscreens are still devoid of rich haptic feedback, such as being able to render different textures and localised multitouch haptic feedback. Yet, rich haptic feedback on a surface presents several benefits. For instance, it is helpful for typing on a virtual keyboard as it increases performance and reduces typing errors [4, 11]. It can also enrich the interaction by providing different sensations to different actions to help differentiate them (e.g. long vs short click, etc.) or to enrich the overall experience (e.g. in a game) [9, 1]. Non-visual interactions could even be envisioned in order to help visually impaired access digital contents [15, 13].

Consequently, rich localised haptic feedback on a surface can open up many new interaction possibilities. However, delivering localised haptic feedback on a surface can be very challenging technically. The work conducted on user interactions often either directly equips the fingers of the users [13] or relies on a

I. Nisky et al. (Eds.): EuroHaptics 2020, LNCS 12272, pp. 545–553, 2020.
https://doi.org/10.1007/978-3-030-58147-3_60

single or two actuators at most [15, 4, 11], thus avoiding the issues of vibration propagation when multiple actuators are required, but then with limited feedback possibilities. Few methods are currently available to render localised haptic feedback on a surface [6, 5, 2, 8]. The time reversal approach focuses on bending waves to produce localised impulsive displacements on transparent glass plates [6]. Another method relies on a phased actuator array in a surface, which can focus ultrasounds to create localised mid-air sensations [5]. These two methods work in ultrasonic frequencies, far beyond the tactile sensitivity range. The perception of such stimulation then relies on nonlinear demodulation phenomena that are not easily controlled. Another method uses an array of electromagnets and a magnetorheological fluid [8] but this technique is not very accurate and thus, it cannot isolate the effect to areas as small as a single fingertip and thus is not suitable for multitouch. Another approach investigated the use of the vibration modes of a surface to localise haptic feedback [2] in the tactile sensitivity range. It combines the different vibration modes of the surface to vibrate chosen locations while canceling others. However, this method does not permit to choose the desired frequency of the haptic feedback. The Inverse Filter Method has been proposed [7] that provides such flexibility with a similar idea. This method uses a glass surface equipped with piezoelectric actuators glued on the bottom of the glass and produces a localised haptic feedback at the locations of these actuators while canceling the signals at the other actuator positions. A user study was conducted demonstrating that the stimulus on a finger was better perceived than without the method. However, no studies were conducted evaluating the recognition of multiple simultaneous stimuli. Recently, this method has been improved [14] by permitting localised haptic feedback at any point of a surface without the necessity to put the fingers on top of the actuators. Furthermore, this method provides a good resolution of about 1.5 cm, far below the wavelength $\lambda_{250Hz} = 19$ cm. However, this novel method has not yet been tested with users, which is the contribution presented in this paper.

Therefore, this paper presents a user study of two-point based pattern discrimination using the Inverse Filter Method (IFM) to investigate the effect of varying the temporal difference between stimuli on pattern recognition and subjective evaluation. The setup and results are presented in the following sections.

2 Experimental Setup

2.1 Setup Description

The user study carried out in this paper was conducted on the setup depicted on the left of Fig. 1. Eleven piezoelectric actuators (Murata 7BB-35-3, 35 mm diameter, 0.51 mm thickness) were glued on the bottom of a $96 \times 162 \times 1$ mm touchscreen (7" pingbo PB70DR8272-R1), which can detect up to five fingers. An Arduino receives the finger localisation data through an I2C protocol. After decoding the data with Python, the number of control points can be obtained as well as their X, Y coordinates. Actuators were driven individually with piezo haptic drivers (DRV8662-Texas Instrument) delivering up to 200 Vpp. Surface displacements were measured by a laser vibrometer (Polytec OFV-5000/MLV-100) mounted on a motorised three axis platform. The acquisition device (NI-9264, NI-9205 and cDAQ-9174) allowed for a synchronous emission and acquisition of actuators and vibrometer signals. All signals were sampled at $F_s = 10\,\text{kHz}$.

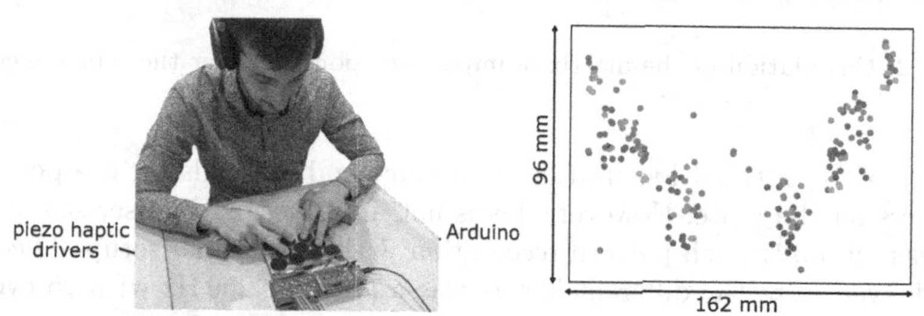

Fig. 1. Left: experimental setup. Right: finger positions of the participants.

2.2 System Calibration

In order to apply the IFM, a matrix of impulse response is needed (more details in [14]). It captures the mechanical transduction of the actuators as well as the propagation, reverberation and attenuation of waves into the touch surface at calibration points spaced at 2 mm. These entries of the matrix of impulse response are calculated as the ratio between the displacement above a calibration point and the driving signal sent to an actuator, in the frequency domain for each actuator on the surface. We chose to acquire experimentally the matrix with the multiple sweep method [12]. Each entry was measured by sending an exponential sweep sine signal sequence of duration $T = 2\,\text{s}$, with frequency varying linearly from 0 to $F_s/2 = 5\,\text{kHz}$ to all the actuators. All actuators were driven simultaneously with a time shifted version of the same exponential sweep while measuring synchronously the resulting displacement at the calibration point. Then, the matrix is the ratio of the displacement by the sweep signal in the frequency domain. The same procedure was repeated for all the actuators and all the calibration points. When the user places his fingers on the screen, the matrix of impulse response at the current positions of the fingers is computed. Using the initial matrix of impulse response and an interpolation function, we can calculate the matrix of impulse response of the current different fingers (Fig. 2).

When the output signal sent to the different actuators is calculated thanks to the IFM, a rescaling is applied to this signal in order to avoid saturation at the output of the amplifiers, which in turn causes inaccurate localisation rendering. However, this rescaling induces an amplitude variation between 1.5 and 5 μm under the different fingers. As a consequence, this amplitude is not controlled,

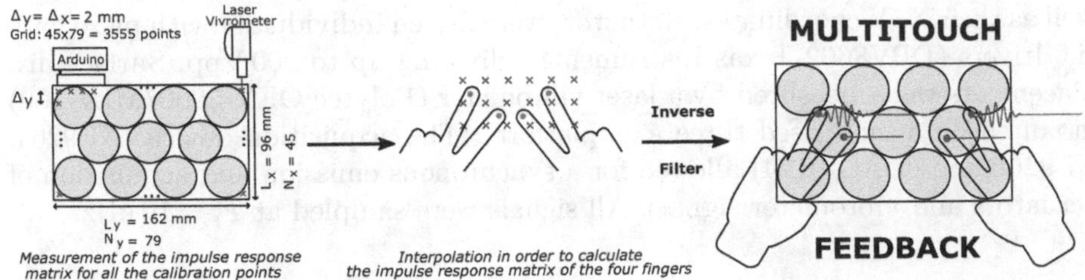

Fig. 2. Calculation of the matrix of impulse responses under the four fingers.

in favour of an optimal localisation rendering, and depends on the position of the fingers on the plate. However, this is not an issue for the user study as the focus was on multitouch pattern recognition with an optimal setup. The haptic signal delivering to the different fingers was a burst at 250 Hz with 25 cycles.

3 User Study

The purpose of this user study was to conduct preliminary investigations of multitouch pattern recognition on a tactile surface using the IFM. Specifically, this study investigated whether users could discriminate vibrotactile stimuli on two different fingers and in particular simultaneous stimuli. To that effect, four different timing differences were chosen, i.e. 0 ms, 100 ms, 200 ms and 300 ms.

3.1 Methodology

Participants. The study was conducted with 12 participants (4f–8m), aged between 14 and 45 ($M = 28.25$, $SD = 7.5$). 11 participants were aged between 25 and 45 and only one minor participated. All but two participants were right-handed. Half of the participants were very familiar with haptic technologies (researchers recruited within the laboratory), while the rest had limited or no knowledge about haptics (a college student, a security coordinator, a project manager in construction project management and three researchers from the vision laboratory). None reported any issues with their fingers or sensitivity.

Technical Settings. The device was placed on a table, in front of the participant (see Fig. 1). The participants were instructed to place their fingers onto the tactile screen for the trials, at the positions that were comfortable to them, as displayed on Fig. 1, with their wrists on resting supports to minimize the fatigue. They wore noise-canceling headphones during the trials with pink noise to cancel any biais due to the noise generated by the setup. The experimenter used a standard Windows laptop both for running the Python application controlling the feedback and for logging the verbal answers.

Procedure. The experiment was a within-subject repeated measures design with four conditions (0, 100, 200, 300 ms), tested in different sessions. The order of the sessions was counterbalanced between participants as well as the direction of the stimuli within a session (i.e. either playing from left-to-right, or from right-to-left). There were 10 trials per stimuli/finger combination (depicted on Fig. 3) with half from each direction, accounting for 60 trials per session. In total, each participant performed 240 trials. The experiment lasted one hour on average.

In each trial, the task was to identify the positions/finger combination that received the haptic stimuli with the fingers numbered from 1 to 4 (e.g. stimuli on the index fingers of each hand corresponded to '2–3', see Fig. 3). The participants were instructed to provide the answer verbally as soon as they recognised the stimuli, which was provided only once, whilst their hands remained on the device. The experimenter logged the answer by first, pressing a button to measure the response time and then, typing the given answer. This was a forced-choice experiment: if participants had doubts, they were asked to answer the most likely option. To accustom participants and reduce the impact of learning effects, prior to each condition, participants were presented with each of the stimuli twice per direction and performed a blind test.

After each session, the participants were asked whether they perceived a difference with the previous condition and to describe it. They were also asked to rate the difficulty of discrimination on a continuous numeric scale from 0 to 10 (10: very difficult). At the end of the last session, the participants were asked which timing difference they preferred and general comments about the perception. As for quantitative measures, the interface collected the responses and the response times. The response time was collected to provide trends, as the experimenter logging the responses induced a bias, in particular in terms of longer hesitations to answer for a condition.

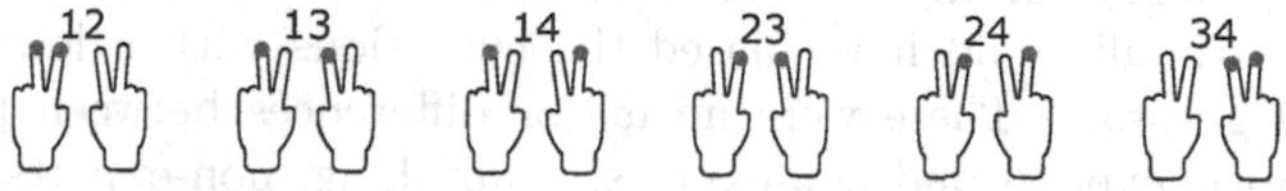

Fig. 3. Stimulation number corresponding to the stimulated fingers.

3.2 Results

Recognition Rates. The average recognition rates are displayed left of Fig. 4. The distribution was normal for all the conditions except for 0 ms. Therefore, a Friedman's ANOVA was conducted and revealed that the recognition rates were significantly different between the timing conditions, $\chi^2(3) = 13.07$, $p < .05$ ($M_0 = 49.58$ or 82.64%, $SE_0 = 6.1$, $M_{100} = 51.83$ or 86.39 %, $SE_{100} = 4.78$, $M_{200} = 54.75$ or 91.25%, $SE_{200} = 3.52$, $M_{300} = 55$ or 91.67%, $SE_{300} = 4.59$). Post hoc tests were conducted based on the following inequality [3]: $|\overline{R_u} - \overline{R_v}| \geq$

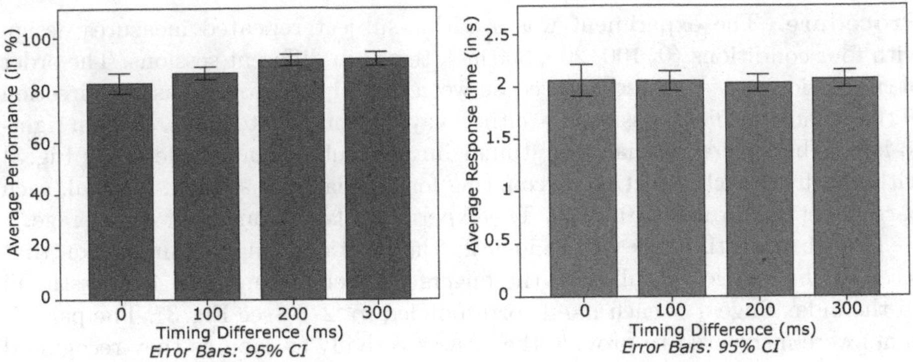

Fig. 4. Left: average performance (in %). Right: average response time (in s).

$z_{\alpha/k(k-1)}\sqrt{k(k+1)/6N}$ (Eq. 1), with $R_{u,v}$ the mean rank of a group, z the statistic from the table of the standard normal distribution, k the number of conditions and N the total sample size. We computed the critical difference (right side of Eq. 1) as being equal to 1.39 with a z value of 2.64. We then calculated the differences between the mean ranks of the groups with the most likely significant differences, i.e. 0–200, 0–300, 100–200 and 100–300. The inequality of Eq. 1 indicates that if the differences between mean ranks is greater than or equal to the critical difference, then that difference is significant. In this case, the pairs 0–200 ($|1.67 - 3.13|$) and 0–300 ($|1.67 - 3.17|$) have values of 1.45 and 1.5, greater than 1.39, thus their difference is significant. Overall, the results indicate that participants recognised well patterns made of two distinct stimuli on the surface using the IFM, even simultaneous ones (82.64% recognition, chance level at 17%). However, to further confirm this hypothesis, studies involving mixed stimuli on one to several fingers need to be conducted as the knowledge of having only two stimuli could have biased the conditions with a lower temporal difference by guessing. There were no major differences between participants judged as haptic experts and non-experts, if anything, non-experts had higher scores than experts. As expected, participants performed better at longer timing differences with a significant difference between 0 and 200 ms onwards.

For further analysis of the participants' performance, we computed confusion matrices for each of the conditions (see Fig. 5). They show that the pairs '12' and '34' are nearly always recognised, no matter the timing difference between stimuli. Most of the confusion happened when involving pairs of fingers from different hands. In particular, for 0 and 100 ms, the pairs with the lowest scores were '13', '14' and '24' and the confusion most often happened with one adjacent finger. Further analysis of the direction of the stimuli could inform us whether it had any effect on the confusions. Preliminary analysis of the results of amplitude differences between patterns did not indicate any notable impact on the perception. For 0 ms, the amplitude varied on average between 2.95 (patterns '13' and '24') and 4.16 µm (pattern '23'), for 100 ms between 2.75 (pattern '24') and 3.72 (pattern '23'), for 200 ms between 2.69 (patterns '13' and '14') and 3.75 (pattern '23') and for 300 ms between 2.78 (patterns '12' and '13') and 3.8 (pattern '23'). There was no clear correlation between this difference in amplitude and the recognition rates, as for instance '23' always had the highest amplitudes, but not the highest recognition rates. For 100 ms the lowest recognition was for '13' and yet had an amplitude of 3.06 μm. Further analysis will be conducted to assess the impact of the different amplitudes.

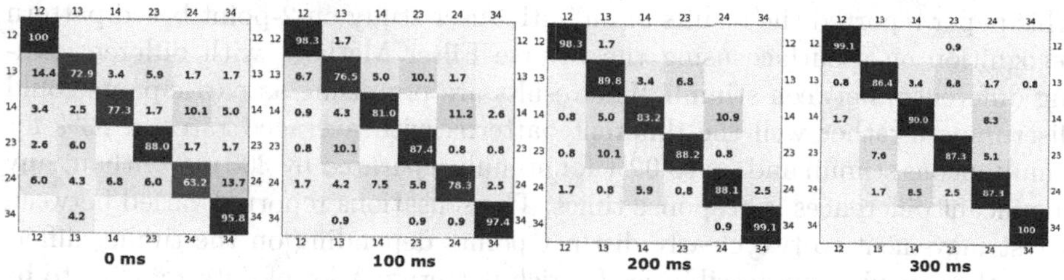

Fig. 5. Confusion matrices. From left to right: 0 ms, 100 ms, 200 ms and 300 ms

Response Times. The average response times are displayed right of Fig. 4. The distribution was normal for all the conditions except for the 300 ms condition. Therefore, a Friedman's ANOVA was conducted and revealed that the response times did not significantly change between the different timing conditions, $\chi^2(3) = 3.6$, $p > .05$ ($M_0 = 2.06$, $SE_0 = 0.57$, $M_{100} = 2.05$, $SE_{100} = 0.37$, $M_{200} = 2.03$, $SE_{200} = 0.41$, $M_{300} = 2.07$, $SE_{300} = 0.37$). This shows that participants had no particular difficulty according to the timing difference, even with simultaneous stimuli, which is confirmed by the relatively high recognition rates.

Qualitative Feedback. During the study, after each session, participants were asked to rate the difficulty of discrimination for each condition. 0 ms obtained an average score of 6.42 out of 10, 100 ms a score of 3.63, 200 ms a score of 3.36 and 300 ms an average score of 2.33. This shows that despite good recognition rates for the 0 ms condition, participants felt less confident, some participants reported feeling a movement rather than two distinct points. This echoes the work on 'out of the body' phantom sensations on a surface [10] and warrants further exploration. On the contrary, the 300 ms condition was deemed the less difficult as participants could clearly feel and distinguish the two stimuli. From a pattern recognition point of view, this is an interesting result as the 0 ms could produce patterns or textures that need to be perceived as continuous movements, whereas timing delays above 200 ms could be used to ensure the discrimination of several points. In concordance with the perceived difficulty ratings, 7 participants preferred the 300 ms condition as they were more confident about the perception, with stimuli well separated, whereas 5 participants preferred the 100 ms where the stimuli were still well perceived and the rhythm was faster. Some participants reported perceiving different intensities on their fingers in a trial, though not consistently. This could be explained by the uncontrolled amplitude of the setup, though the difference in amplitude of the stimuli were constant for a given pattern in a trial, which contradicts user perceptions.

4 Conclusion

This paper reported the results of an initial user study on 2-point based pattern recognition on a surface using the Inverse Filter Method, with different timing differences between stimuli. The results are promising as participants could discriminate rather well the different patterns with averaged rates of 83% for simultaneous stimuli and up to 92% for stimuli separated by 300 ms, without any significant differences in response times. The sensations reported varied between a fast movement to two clearly distinct points depending on the timing difference, thus opening up possibilities for rich patterns. A lot of data remains to be analysed to assess the impact of stimuli direction and uncontrolled amplitudes on the observed confusions. This initial study also opens up many future leads for experiments by evaluating 3 to more actuated fingers at a time, perceptual illusions and for future design of patterns using this method.

References

1. Chen, H.Y., Park, J., Dai, S., Tan, H.Z.: Design and evaluation of identifiable key-click signals for mobile devices. IEEE TOH **4**(4), 229–241 (2011)
2. Emgin, S.E., Aghakhani, A., Sezgin, T.M., Basdogan, C.: Haptable: an interactive tabletop providing online haptic feedback for touch gestures. IEEE TVCG **25**(9), 2749–2762 (2018)
3. Field, A.: Discovering Statistics Using IBM SPSS Statistics, 3rd edn. Sage, Thousand Oaks (2009)
4. Hoggan, E., Brewster, S.A., Johnston, J.: Investigating the effectiveness of tactile feedback for mobile touchscreens. In: CHI 2008, pp. 1573–1582 (2008)
5. Hoshi, T., Takahashi, M., Iwamoto, T., Shinoda, H.: Noncontact tactile display based on radiation pressure of airborne ultrasound. IEEE TOH **3**(3), 155–165 (2010)
6. Hudin, C., Lozada, J., Hayward, V.: Localized tactile feedback on a transparent surface through time-reversal wave focusing. IEEE TOH **8**(2), 188–198 (2015)
7. Hudin, C., Panëels, S.: Localisation of vibrotactile stimuli with spatio-temporal inverse filtering. In: Prattichizzo, D., Shinoda, H., Tan, H.Z., Ruffaldi, E., Frisoli, A. (eds.) EuroHaptics 2018. LNCS, vol. 10894, pp. 338–350. Springer, Cham (2018)
8. Jansen, Y., Karrer, T., Borchers, J.: MudPad: localized tactile feedback on touch surfaces. In: Adjunct proceedings of UIST 2010, pp. 385–386 (2010)
9. Kim, S., Lee, G.: Haptic feedback design for a virtual button along force-displacement curves. In: UIST 2013, pp. 91–96. ACM (2013)
10. Kim, Y., Lee, J., Kim, G.J.: Extending "Out of the Body" tactile phantom sensations to 2D and applying it to mobile interaction. Pers. Ubiquit. Comput. **19**(8), 1295–1311 (2015)

11. Ma, Z., Edge, D., Findlater, L., Tan, H.Z.: Haptic keyclick feedback improves typing speed and reduces typing errors on a flat keyboard. In: WHC 2015, pp. 220–227. IEEE (2015)
12. Majdak, P., Balazs, P., Laback, B.: Multiple exponential sweep method for fast measurement of head-related transfer functions. J. Audio Eng. Soc. **55**(7/8), 623–637 (2007)
13. Nicolau, H., Guerreiro, J., Guerreiro, T., Carriço, L.: Ubibraille: designing and evaluating a vibrotactile braille-reading device. In: ACM SIGACCESS Conference on Computers and Accessibility, pp. 1–8 (2013)
14. Pantera, L., Hudin, C.: Sparse actuator array combined with inverse filter for multitouch vibrotactile stimulation. In: WHC 2019, pp. 19–24. IEEE (2019)
15. Rantala, J., et al.: Methods for presenting braille characters on a mobile device with a touchscreen and tactile feedback. IEEE TOH **2**(1), 28–39 (2009)

Author Index

Printed in the United States
By Bookmasters

Printed in the United States
By Bookmasters